Environmental Science
&
Climate Change

About the Book

Environmental science is a multidisciplinary academic field that integrates physical, biological and information sciences to the study of the environment, and the solution of environmental problems. Environmental science provides an integrated, quantitative, and interdisciplinary approach to the study of environmental systems. Related areas of study include environmental studies and environmental engineering. Environmental studies incorporate more of the social sciences for understanding human relationships, perceptions and policies towards the environment.

About the Author

Saurabh Bhatia is the Professor and Head of Nagaland Institute of Disaster. He did his studies in the field of Ecology and Environment, Disaster Mitigation, Sustainable Development and Geoinformatics. He has experience of teaching from last 23 years in this field. He has organized many Congresses and Conventions like Eco-Revival Summit 2001, Peace, Mercy and Tolerance Summit 2003 and World Disaster Education Congress 2006. He has been participating in different Congresses and conventions besides presenting his papers related to Ecology, Pollution, Biotechnology and Biochemistry. He has successfully designed a master plan and a new paradigm for developing the resources and infrastructure in different states in general and in the State of Orissa in particular.

Environmental Science
&
Climate Change

Sourabh Bhatia

AMIGA PRESS INC.

(DELHI (INDIA

AMIGA PRESS INC.
H.O.: 28-A, Indra Park,
Chander Nagar, Delhi-110051 (India)
Tel.: 011-22504028, +91-98736-57000
E-mail: amigapress@gmail.com
amigaeditorial@gmail.com
amigapub@gmail.com
info@amigapress.co.in
Website : www.amigapress.co.in

This edition is published by Amiga Press Inc.

ISBN: 978-93-87295-51-3

First Published 2020

Printed and bound in India by Replika Press Pvt. Ltd.

Preface

Environmental science is a multidisciplinary academic field that integrates physical, biological and information sciences to the study of the environment, and the solution of environmental problems. Environmental science provides an integrated, quantitative, and interdisciplinary approach to the study of environmental systems. Related areas of study include environmental studies and environmental engineering. Environmental studies incorporates more of the social sciences for understanding human relationships, perceptions and policies towards the environment.

Environmental engineering focuses on design and technology for improving environmental quality in every aspect. Environmental scientists work on subjects like the understanding of earth processes, evaluating alternative energy systems, pollution control and mitigation, natural resource management, and the effects of global climate change. Environmental issues almost always include an interaction of physical, chemical, and biological processes.

Environmental scientists bring a systems approach to the analysis of environmental problems. Key elements of an effective environmental scientist include the ability to relate space, and time relationships as well as quantitative analysis.

Environmental science came alive as a substantive, active field of scientific investigation in the 1960s and 1970s driven by the need for a multi-disciplinary approach to analyze complex environmental problems, the arrival of substantive environmental laws requiring specific environmental protocols of investigation and the growing public awareness of a need for action in addressing environmental problems.

Events that spurred this development included the publication of Rachael Carson's landmark environmental book Silent Spring along with major environmental issues becoming very public, such as the 1969 Santa Barbara oil spill, and the Cuyahoga River of Cleveland, Ohio, "catching fire" and helped increase the visibility of environmental issues and create this new field of study. This encyclopedia will prove to be useful and informative for students, research scientists, engineers and teachers in the field of environment science and technology.

–Author

Contents

❒

1

Environment Policy and Management System

Harmony between man and environment is the essence of healthy life and growth. Therefore, maintenance of ecological balance and a pristine environment has been of utmost importance to the Union Ministry of Power (MoP). NTPC being the leading organization under the ministry in the areas of power generation, has been taking various measures discussed below for mitigation of environment pollution due to power generation.

Driven by its commitment for sustainable growth of power, NTPC has evolved a well defined environment management policy and sound environment practices for minimising environmental impact arising out of setting up of power plants and preserving the natural ecology.

NATIONAL ENVIRONMENT POLICY

At the national level, the Ministry of Environment and Forests had prepared a draft Environment Policy (NEP) and the Ministry of Power along with NTPC actively participated in the deliberations of the draft NEP. The NEP 2006 has since been approved by the Union Cabinet in May 2006.

NTPC ENVIRONMENT POLICY

As early as in November 1995, NTPC brought out a comprehensive document entitled "NTPC Environment Policy and Environment Management System". Amongst the guiding principles adopted in the document are company's proactive approach to environment, optimum utilisation of equipment, adoption of latest technologies and continual environment improvement. The policy also envisages efficient utilisation of resources, thereby minimising waste, maximising ash utilisation and providing green belt all around the plant for maintaining ecological balance.

ENVIRONMENT MANAGEMENT, OCCUPATIONAL HEALTH AND SAFETY SYSTEMS

NTPC has actively gone for adoption of best international practices on environment, occupational health and safety areas. The organization has pursued the Environmental Management System (EMS) ISO 14001 and the Occupational Health and Safety Assessment System OHSAS 18001 at its different establishments.

As a result of pursuing these practices, all NTPC power stations have been certified for ISO 14001 and OHSAS 18001 by reputed national and international Certifying Agencies.

POLLUTION CONTROL SYSTEMS

While deciding the appropriate technology for its projects, NTPC integrates many environmental provisions into the plant design. In order to ensure that NTPC comply with all the stipulated environment norms, various state-of-the-art pollution control systems/devices as discussed below have been installed to control air and water pollution.

Electrostatic Precipitators

The ash left behind after combustion of coal is arrested in high efficiency Electrostatic Precipitators (ESPs) and particulate emission is controlled well within the stipulated norms. The ash collected in the ESPs is disposed to Ash Ponds in slurry form.

Flue Gas Stacks

Tall Flue Gas Stacks have been provided for wide dispersion of the gaseous emissions (SOX, NOX etc) into the atmosphere.

Low-NOX Burners

In gas based NTPC power stations, NOx emissions are controlled by provision of Low-NOx Burners (dry or wet type) and in coal fired stations, by adopting best combustion practices.

Neutralisation Pits

Neutralisation pits have been provided in the Water Treatment Plant (WTP) for pH correction of the effluents before discharge into Effluent Treatment Plant (ETP) for further treatment and use.

Coal Settling Pits/Oil Settling Pits

In these Pits, coal dust and oil are removed from the effluents emanating from the Coal Handling Plant (CHP), coal yard and Fuel Oil Handling areas before discharge into ETP.

DE & DS Systems

Dust Extraction (DE) and Dust Suppression (DS) systems have been installed in all coal fired power stations in NTPC to contain and extract the fugitive dust released in the Coal Handling Plant (CHP).

Cooling Towers

Cooling Towers have been provided for cooling the hot Condenser cooling water in closed cycle Condenser Cooling Water (CCW) Systems. This helps in reduction in thermal pollution and conservation of fresh water.

ASH DYKES AND ASH DISPOSAL SYSTEMS

Ash ponds have been provided at all coal based stations except Dadri where Dry Ash Disposal System has been provided. Ash Ponds have been divided into lagoons and provided with garlanding arrangements for change over of the ash slurry feed points for even filling of the pond and for effective settlement of the ash particles. Ash in slurry form is discharged into the lagoons where ash particles get settled from the slurry and clear effluent water is discharged from the ash pond. The discharged effluents conform to standards specified by CPCB and the same is regularly monitored. At its Dadri Power Station, NTPC has set up a unique system for dry ash collection and disposal facility with Ash Mound formation. This has been envisaged for the first time in Asia which has resulted in progressive development of green belt besides far less requirement of land and less water requirement as compared to the wet ash disposal system.

ASH WATER RECYCLING SYSTEM

Further, in a number of NTPC stations, as a proactive measure, Ash Water Recycling System (AWRS) has been provided. In the AWRS, the effluent from ash pond is circulated back to the station for further ash sluicing to the ash pond. This helps in savings of fresh water requirements for transportation of ash from the plant.

The ash water recycling system has already been installed and is in operation at Ramagundam, Simhadri, Rihand, Talcher Kaniha, Talcher Thermal, Kahalgaon, Korba and Vindhyachal. The scheme has helped stations to save huge quantity of fresh water required as make-up water for disposal of ash.

Dry Ash Extraction System (DAES)

Dry ash has much higher utilization potential in ash based products (such as bricks, aerated autoclaved concrete blocks, concrete, Portland pozzolana cement, etc.). DAES has been installed at Unchahar, Dadri, Simhadri, Ramagundam, Singrauli, Kahalgaon, Farakka, Talcher Thermal, Korba, Vindhyachal, Talcher Kaniha and BTPS.

Liquid Waste Treatment Plants and Management System

The objective of industrial liquid effluent treatment plant (ETP) is to discharge lesser and cleaner effluent from the power plants to meet environmental regulations. After primary treatment at the source of their generation, the effluents are sent to the ETP for further treatment. The composite liquid effluent treatment plant has been designed to treat all liquid effluents which originate within the power station e.g., Water Treatment Plant (WTP), Condensate Polishing Unit (CPU) effluent, Coal Handling Plant (CHP) effluent, floor washings, service water drains etc.

The scheme involves collection of various effluents and their appropriate treatment centrally and re-circulation of the treated effluent for various plant uses. NTPC has implemented such systems in a number of its power stations such as Ramagundam, Simhadri, Kayamkulam, Singrauli, Rihand, Vindhyachal, Korba, Jhanor Gandhar, Faridabad, Farakka, Kahalgaon and Talcher Kaniha. These plants have helped to control quality and quantity of the effluents discharged from the stations.

Sewage Treatment Plants and Facilities

Sewage Treatment Plants (STPs) sewage treatment facilities have been provided at all NTPC stations to take care of Sewage Effluent from Plant and township areas. In a number of NTPC projects modern type STPs with Clarifloculators, Mechanical Agitators, sludge drying beds, Gas Collection Chambers etc have been provided to improve the effluent quality.

The effluent quality is monitored regularly and treated effluent conforming to the prescribed limit is discharged from the station. At several stations, treated effluents of STPs are being used for horticulture purpose.

ENVIRONMENTAL INSTITUTIONAL SET-UP

Realizing the importance of protection of the environment with speedy development of the power sector, the company has constituted different groups at project, regional and Corporate Centre level to carry out specific environment related functions.

The Environment Management Group, Ash Utilisation Group and Centre for Power Efficiency and Environment Protection (CENPEEP) function from the Corporate Centre and initiate measures to mitigate the impact of power project implementation on the environment and preserve ecology in the vicinity of the projects. Environment Management and Ash Utilisation Groups established at each station, look after various environmental issues of the individual station.

Environment Management During Operation Phase

NTPC's environment friendly approach to power has already begun to show results in conservation of natural resources such as water and fuel (coal, oil and gas) as well as control of environmental pollution. As already mentioned earlier, NTPC has chalked out a set of well defined activities that are envisaged right from the project conceptualisation stage so that during the entire life cycle of the power plant, NTPC is fully compliant with various environment regulations and a pristine environment and ecological balance is maintained in and around its power station and townships.Following is brief description of some of the measures taken during the operation phase of the stations. Performance enhancement and up-gradation measures are undertaken by the organisation during the post operational stage of the stations. These activities have greatly helped to minimise the impact on environment and preserve the ecology in and around its power projects. These measures have been enumerated as follows.

Monitoring of Environmental Parameters

A broad based Environment Monitoring Programmes has been formulated and implemented in NTPC. All pollutants discharged from the power plant such as stack emission, ash pond effluent, main plant effluent, domestic effluent and Condenser Cooling Water (CCW) effluent are monitored at the stipulated frequency at the source itself and at the points of discharge. In addition to the above, ambient air, surface water and ground water quality in and around NTPC plants are regularly monitored to assess any adverse impacts as a result of operation of the power plant.

On-Line Data Base Management

In order to have better control on pollution and to achieve effective environment management in and around NTPC stations, it is imperative to have an on-line, reliable and efficient environment information system on the plant operational and environmental performance parameters at all three levels i. e generating Stations, Regional Headquarters and Corporate Centre. In consideration of the above, a computerized programmes, namely "Paryavaran Monitoring System"-PMS, which could provide reliable storage, prompt and accurate flow of information on environmental performance of Stations was developed and installed in NTPC. This software facilitates direct transfer of environment reports and other environment related information from stations to the Regional Headquarters and Corporate Centre. The PMS has already been implemented at Corporate Centre, the Regional Headquarters and most of the Stations. This system has helped in achieving continuous improvement in NTPCs environment performance through improved monitoring and reporting system by using the trend analysis and advanced data management techniques.

ENVIRONMENT REVIEWS

To maintain constant vigil on environmental compliance, Environmental Reviews are carried out at all operating stations and remedial measures have been taken wherever necessary. As a feedback and follow-up of these Environmental Reviews, a number of retrofit and up-gradation measures have been undertaken at different stations. Such periodic Environmental Reviews and extensive monitoring of the facilities carried out at all stations have helped in compliance with the environmental norms and timely renewal of the Air and Water Consents.

Upgradation and Retrofitting of Pollution Control Systems

In order to keep pace with the changing norms and ensure compliance with statutory requirements in the field of pollution control, NTPC keeps an open mind for Renovation and Modernisation (R & M) and Retrofitting and Upgradation of pollution monitoring and control facilities in its existing stations. It is important to mention that such modifications/retrofit programs not only helped in betterment of environment but also in resource conservation. High efficiency Electro-Static Precipitators (ESPs) of the order of 99.5% and above have been provided at NTPC stations for control of stack particulate emissions. However, the ESPs of a number of stations were built prior to the promulgation of the Environment (Protection) Act, 1986 and notification of emission control standards under this Act. Remedial measures have already been taken up and implemented to improve the efficiency of the existing ESPs at various NTPC stations. ESP performance enhancement programmes by adopting advanced microprocessor based Electrostatic Precipitator Management System (EPMS) was installed at its power stations at Singrauli, Ramagundam, Korba, Farakka, Rihand, Vindhyachal and Unchahar.

Additional ESPs were retrofitted in the older power stations, namely at Badarpur and Talcher Thermal. As a result of the above retrofits, the emission of Suspended Particulate Matter (SPM) has been brought down appreciably at the above stations and is maintained within the present statutory limit of 150 mg/Nm3. In new projects, the ESPs have been designed for a maximum permissible outlet dust emission of 50 mg/Nm3 to meet the likely stringent emission norms in the near future.

RESOURCES CONSERVATION

With better awareness and appreciation towards ecology and environment, the organization is continually looking for innovative and cost effective solutions to conserve natural resources and reduce wastes.

Some of the measures include:

- Reduction in land requirements for main plant and ash disposal areas in newer units.
- Capacity addition in old plants, within existing land.

- Reduction in water requirement for main plant and ash disposal areas through recycle and reuse of water.
- Efficient use of Fuel (Coal, Natural gas and Fuel oil)
- Reduction in fuel requirement through more efficient combustion and adoption of state-of-the-art technologies such as super critical boilers

WASTE MANAGEMENT

Various types of wastes such as Municilal or domestic wastes, hazardous wastes, Bio-Medical wastes get generated in power plant areas, plant hospital and the townships of projects. The wastes generated are a number of solid and hazardous wastes like used oils and waste oils, grease, lead acid batteries, other lead bearing wastes (such as garkets etc.), oil and clarifier sludge, used resin, used photochemicals, asbestos packing, e-waste, metal scrap, C&I wastes, electricial scrap, empty cylinders (refillable), paper, rubber products, canteen (bio-degradable) wastes, buidling material wastes, silica gel, glass wool, fused lamps and tubes, fire resistant fluids etc. These wastes fall either under hazardous wastes category or non-hazardous wastes category as per classification given in Government of Indias notification on Hazardous Wastes (Management and Handling) Rules 1989 (as amended on 06.01.2000 & 20.05.2003). Handling and manegement of these wastes in NTPC stations have been discussed below.

Municipal Waste Management

Domestic or municipal waste is generated in households at townships. This waste is seggregated into bio-degradable and nonbiodegradable wastes at source itself in different coloured containers and thereafter the two types are disposed separately. Bio-degradable waste is spread uniformly in identified low lying areas and thereafter it is covered with soil for use later as manure after composting. The seggregated non bio-degradable waste is disposed off separately in other identified low lying areas and is spread out uniformly.

Hazardous Waste Management

NTPC being a proactive organization, the handling and disposal of hazardous wastes are done as per the Hazardous Wastes (Management and Handling) Rules 1989 (as amended in 2003) guidelines issued by Government of India for the treatment, storage and disposal of hazardous wastes. Scientific study on management and handling of hazardous wastes was carried out at a few NTPC stations to adopt the best practices so that there is a complete compliance with statutory requirements. In NTPC sataione, the Hazardous Wastes (Recyclable) are sold/auctioned to registered recyclers/refiners.

The other hazardous wastes such as the activated carbon resins, used drums (hazardous) chromium (Cr-III electrolytes, used petro-chemicals, asbestos packings, used torch batteries, ribbon, toners/cartridges, mixed wastes (waste oil, water and cotton) filters, earth contaminated with synthetic oil (FQF) glass used and sodium silicate, lamps and tubes etc. fall under the category of Hazardous Wastes (Non-Recyalable). These wastes are small in quantity and are stored in properly identified locations As per the notification, hazardous wastes (non-recyalable) are to be sent to State Pollution Control Board (SPCB) approved common treatment storage and disposal facility (TSDF).

Bio-Medical Waste Management

Hospital (or Bio-medical) wastes get generated from hospitals and they include urine bags, human anatomical wastes, plaster of aris waste, empty plastic bottles of water and glucose, blood and chemical mixed cotton, blood and urines tubes etc. these wastes are segregated and are placed in buckets of different colours as per the notification for Bio-Medical Waste (Management and Handling) Rules.

The seggregated bio-medical wastes are either disposed through the SPCB approved agency or they are treated in autoclaves before disposal into bio- medical waste disposal pits. The treated bio-medical waste is spread uniformly and covered with 10 cm thick soil in bio-medical waste disposal pits.

LAND USE/BIO-DIVERSITY

As a policy, NTPC lays special emphasis on land use and Bio-diversity by way of development of green belts, energy plantations, reclmation of abandoned Ash Ponds and EIA and ecological monitoring in the project areas and its surroundings.

Reclamation of Abandoned Ash Ponds

The reclamation of abandoned ash pond sites is a challenging task. NTPC has reclaimed temporary ash disposal areas at some of its projects namely Ramagundam, Talcher Thermal, Rihand, Singrauli and Unchahar through plantation and converted these sites into lush green environments. Extensive plantations have also been undertaken on dry ash mound at NTPC-Dadri. It is planned to reclaim all the abandoned ash disposal areas by plantation.

Green Belts, Afforestation and Energy Plantations

Whats more, in a concerted bid to counter the growing ecological threat, NTPC is undertaking afforestation programmes covering vast areas of land in and around its projects. Appropriate afforestation programmes for plant, township and green belt areas of the project have been implemented at all projects. In order to enhance green cover in the areas around our projects, as a responsible corporate citizen, NTPC till date has planted more than 18 million trees at its projects throughout the country. The afforestation has not only contributed to the aesthetics but also has been serving as a 'sink' for the pollutants released from the station and thereby protecting the quality of ecology and environment in and around the projects. Thrust has also been given to bio-diesel plantation and around 4.8 lakh energy plants including Pongamia and Jatropha have already been planted. A pilot project for extraction of seeds from these bio-diesel plants has also been set up.

USE OF WASTE PRODUCTS AND SERVICES-ASH UTILIZATION

Ash is the main solid waste which is put into use for various products and services. NTPC has adopted user friendly policy guidelines on ash utilisation. In order to motivate entrepreneurs to come forward with ash utilisation schemes, NTPC offers several facilities and incentives. These include free issue of all types of ash viz.

Dry Fly Ash/Pond Ash/Bottom Ash and infrastructure facilities, wherever feasible. Necessary help and assistance is also offered to facilitate procurement of land, supply of electricity etc from Government Authorities. Necessary techno-managerial assistance is given wherever considered necessary.

Besides, NTPC uses only ash based bricks and Fly Ash portland pozzolana cement (FAPPC) in most of its construction activities. Demonstration projects are taken up in areas of Agriculture, Building materials, Mine filling etc. The utilisation of ash and ash based products is progressively increasing as a result of the concrete efforts of these groups.

Advanced/Eco-Friendly Technologies

NTPC has gained expertise in operation and management of 200 MW and 500 MW Units installed at different Stations all over the country and is looking ahead for higher capacity Unit sizes with super critical steam parameters for higher efficiencies and for associated environmental gains.

At Sipat, higher capacity Units of size of 660 MW and advanced Steam Generators employing super critical steam parameters have already been implemented as a green field project. Higher efficiency Combined Cycle Gas Power Plants are already under operation at all gas-based power projects in NTPC. Advanced clean coal technologies such as Integrated Gasification Combined Cycle (IGCC) have higher efficiencies of the order of 45% as compared to about 38% for conventional plants. NTPC has initiated a techno-economic study under USDOE/USAID for setting up a commercial scaledemonstration power plant by using IGCC technology.

These plants can use low grade coals and have higher efficiency as compared to conventional plants. With the massive expansion of power generation, there is also growing awareness among all concerned to keep the pollution under control and preserve the health and quality of the natural environment in the vicinity of the power stations. NTPC is committed to provide affordable and sustainable power in increasingly larger quantity.

NTPC is conscious of its role in the national endeavour of mitigating energy poverty, heralding economic prosperity and thereby contributing towards India s emergence as a major global economy.

ENVIRONMENT DESTRUCTION

It looks as if environment is an idea whose time has come. Newspapers give prominent display to environmental horror stories. Editorials demand better management of natural resources. Government statements on the need to preserve the environment are commonplace. Government programmes, too, are quite numerous and increasing in number day by day. There are massive schemes for afforestation, for instance. In the last four years, some 1,000 crore seedlings are said to have been distributed or planted. There are new laws for control of air and water pollution and for the conservation of forests.

India has been praised all over the world for what it has done to preserve tigers. Nearly three per cent of India's giant land mass is now protected national parks and wildlife sanctuaries, and there are demands to strengthen their protection and increase their area. Plan documents and party manifestoes take care to mention the importance of environment. But there is a major problem with this entire range of activities and concerns: it does not appear to be based on a holistic understanding of the relationship between environment and the development process in this country.

The programmes are ad hoc, without clear priorities, and there is too much of a policeman's attitude. They seem to be based on the belief that concern for the environment essentially means protecting and conserving it, partly from development programmes but mainly from the people themselves. There is little effort to modify the development process itself in a manner that will bring it into greater harmony with the needs of the people and with the need to maintain ecological balance, while increasing the productivity of our land, water and forest resources. The post-independence political debate in India has centred on two major issues: equity and growth. The environmental concern has added a third dimension: sustainability. India's biggest challenge today is to identify and implement a development process that will lead to greater equity, growth and sustainability. The environment is not just pretty trees and tigers, threatened plants and ecosystems. It is literally the entity on which we all subsist, and on which entire agricultural and industrial development depends. Development can take place at the cost of the environment only upto a point.

Beyond that point, it will be like the foolish person who was trying to cut the very branch on which he was sitting. Development without concern for the environment can only be short-term development. In the long term, it can only be anti-development and can go on only at the cost of enormous human suffering, increased poverty and oppression. India may be rapidly approaching that point. Amongst the hundreds of voluntary groups working at the micro level in India, there has been a remarkably rapid growth of interest in environmental problems.

So rapid has been this growth that sometimes it is loosely described as the beginnings of an environmental movement in the country. Hundreds of field-level groups today take a keen interest in environmental issues and their experiences and interests are extremely diverse. Some are interested in preventing deforestation.

Others are interested only in afforestation. Many want to prevent the construction of one dam or another. There are others who want to prevent water pollution. There is the famous Chipko movement in the UP Himalaya, probably the oldest and most famous of all the groups, which has played a major role in bringing the issue of deforestation to the forefront of public opinion.

There is its counterpart in the south, the Appiko movement in the Western Ghats of Karnataka. Dams like the Silent Valley and Bedthi have already been stopped because of strong people's protests, and the well-known social worker Baba Amte is leading a major campaign against the proposed Bhopalpatnam and Inchampalli dams on the borders of MP, Andhra Pradesh and Maharashtra.

The Kerala Sastra Sahitya Parishad has had a long acrimonious battle over the pollution of the Chiliyar river in Kerala by a rayon mill. The India Development Service finds itself embroiled in another case of river pollution by a rayon mill in Karnataka.

Meanwhile, the Shahdol Group has worked against the pollution of a river in Shahdol district by a paper mill. There is, also, the Mitti Bachao Abhiyan to organise farmers against water-logging caused by faulty irrigation systems.

While all these are relatively well known groups and have attracted varying degrees of media attention, there are many others which are doing excellent work in mobilising people, both to prevent further ecological destruction, often in the face of determined government policies, and to bring about ecological regeneration. One thing, however, that binds most of these groups is their concern to put the environment at the service and the control of the people, the people usually being defined as the local communities who live within that environment.

Environmental protection per se is of least concern to most of these groups, including the well known Chipko movement, for example. Their main concern is about the use of the environment: how should the environment be used and who should use it and benefit from it. It is this growing understanding of the relationship between the people and their environment, born out of a concern for a more equitable and sustainable use of the environment, that is probably the most fascinating development in India today.

ENVIRONMENTAL DESTRUCTION BY THE RICH

To understand the nature of the environmental problems in India, it may be useful to compare and contrast certain environmental trends and concerns in India with those in the West, especially since the environment crusade began in the West and since many groups in India, including political parties, have for long dismissed it as a petty Western concept. The argument has always been that too much concern for the environment can only retard economic and industrial development. The UN Conference on the Human Environment held in Stockholm in 1972 was the landmark conference that created worldwide consciousness about environment.

No UN conference has ever been able to collect so many luminaries at one place. Many delegations from developing countries had argued that the solution to environmental problems lay in economic development. “Smoke is a sign of progress,” the Brazilian delegation had thundered, then representing a country witnessing an economic boom. India’s Prime Minister Indira Gandhi, who made a major impression on the conference, is still remembered for her oft-quoted statement: “Poverty is the biggest polluter.” In all those who came from the Third World, both leftists and rightists, there was a sneaking suspicion that the Western countries were up to some trick.

The West may simply be pushing the environmental concern onto an unsuspecting Third World to retard its technological modernisation and industrial development. It was even argued that having got their riches and their affluent lifestyles, Westerners were now simply asking for more affluence: clean air, clean water and large tracts of nature for enjoyment and recreation, many of which were going to be preserved in the tropical forests and savannas of Asia, Africa and South America. But exactly 10 years later when the UN organised a meeting to commemorate the Stockholm conference, few non- governmental groups from the Third World were prepared to argue in favour of the development process as it is.

The Third World today faces both an environment crisis and a development crisis, and both these crises seem to be intensifying and interacting to reinforce each other. On one hand, there does not seem to be any end to the problems of inequality, poverty and unemployment, the crucial problems that the development process is meant to solve. On the other, environmental destruction has grown further apace. But what is interesting is that while many environmental problems, expecially those related to air and water pollution have tended to become less severe in many parts of the industrialised world, because of the introduction of highly capital-intensive pollution control technologies, these problems have continued to grow and become critical in many parts of the developing world.

In other words, while the economic development process in the world is only worsening our environmental problems, it is tending to solve them in the West. Michael Heseltine, then minister of environment in the anti-environment British government of Margaret Thatcher, even went so far in a UN meeting in 1982 as to say that all environmental problems in the West have been solved and they now remain mainly in the Third World! Heseltine, indeed, had a point. London, for instance, has not seen for years any of those smogs it saw regularly in the 1940s and’50s, which led to thousands of deaths, and the Thames now even boasts of salmon. But Michael Heseltine, like all representatives of vested interests, was only hiding something of deep significance: the role of the Western world in destroying the Third World’s environment.

Very simply speaking, the major environmental problems in the West are those arising out of waste disposal—problems of air and water pollution and of disposal of highly toxic industrial and nuclear wastes. Though problems of acid rain have definitely increased and there does not yet seem to be any solution to the problem of toxic wastes, it is true that some cities and rivers do look cleaner. In the Third World, however, as its own industrialisation proceeds, these waste disposal problems are getting worse day by day but they are still not the major or the only environmental problem.

In the Third World, the major environmental problems are those which arise out of the misuse of the natural resource base: the soils, forests and water resources. These problems are created to a great extent because of the pressure to produce raw materials for modern industry. The Third World's environment not only provides raw materials for its own industries but also for the industries of the West. For instance, the Japanese and Western timber industries have been the biggest source of forest destruction in Southeast Asia. Having turned countries like Thailand from net exporters into net importers of wood, Japanese companies are now turning to the last great wooded frontier of the world: the Amazon basin of South America.

FEEDING THE RICH

The food needs of the Western world have played equal havoc with the lands of the Third World. No statistics on this are available, but if someone did collect them, we will definitely find that despite the worldwide process of decolonisation, there is today many times more land being used in the developing world to meet the food and other biomass needs of the Western countries than in the 1940s, before the process of decolonisation began.

More than a quarter of all central American forests have been destroyed since 1960 for cattle ranching: 85 per cent to 95 per cent of the beef produced as a result has gone to the US while domestic consumption of beef in central America has fallen dramatically. In the US, this beef has been mainly used to make tinned and pet foods and cheap hamburgers because central American beef is half the price of the grass-fed beef produced in the US. The price of central American beef does not represent its correct ecological cost. Cattle ranching has proved to be the worst form of land use for the fragile soils on which these tropical moist forests existed.

Within five to seven years, their productivity has dropped dramatically and cattle ranchers have had to move on. The Sahelian drought of 1968-74 in Africa which grabbed world headlines and claimed the lives of approximately 100,00 nomadic people, was caused by the French colonial policy to drive the region's countries into peanut farming to secure its own source of vegetable oils. Through heavy taxation policies, the French colonial authorities forced west African peasants to grow groundnuts at the expense of subsistence cr ps. Groundnut cultivation rapidly depleted the soil. It soon spread to traditionally fallow and forest zones and encroached on land previously used for grazing, upsetting the delicate balance between the farmers and nomadic herders.

The expansion of groundnuts was encouraged by artificially high prices but when US soya production began to hit the European market and vegetable oil prices began to fall, the newly independent west African countries had no alternative but to increase area under groundnut to keep up their foreign exchange reserves. As this area increased by leaps and bounds under the pressure of government policies, the nomads were slowly pushed further and further north into the desert, something for which they were not prepared, their traditional relationships with the settled farmers having been totally disturbed. When the long period of drought set in and thousands of animals and human beings began to die, the nomads and their overgrazing was blamed.

Nobody blamed the French or the Sahelian elite which worked hand-in-glove with the French. The UN Environment Programmes (UNEP) in a recent report points to the impact of the heavy debt burden of the Third World and high interest rates in the West on the environment of the Third World. The debt burden and regressive terms of trade have forced many developing countries to put enormous pressure on their natural resources, sometimes even to the point of overexploitation.

In 1981, for instance, it took one Latin American country 9.8 times as much beef to buy a barrel of oil as it did in 1981. At the end of the 1970s, profits from the export of one tonne of bananas were enough to purchase only half the steel they would have bought at the end of the 1960s. When interest rates are high, there is a tendency to discount long-term issues like environment for short-term gains. A one per cent increase in interest rates adds approximately US $5 billion to the current debt burden of developing countries. To have increased its export earnings (not profits) by $1 billion in 1981, South America as a whole would have had to increase its banana exports three-fold, Ecuador three-fold and Colombia ninefold, while leading cotton exporters like Egypt and Turkey would have had to double and triple their cotton exports respectively.

This would have meant bringing millions of additional hectares into production to grow these export crops. And, it can be added, this would have pushed millions of marginal peasants into marginal lands like desert fringes and steep hill slopes for their survival, leading to accelerated desertification and soil erosion. In our own country, the first major attack on the forests of the northeast came with the establishment of tea plantations. Destruction of forests goes on for coffee and other export crop plantations even today. The current overfishing on India's coasts, as on the coasts of almost all Southeast Asian countries, is taking place because of the heavy demand for prawns in Western and Japanese markets.

This overfishing is leading to considerable tensions between traditional fisherfolk and trawler owners, and violent encounters between the two are regularly reported. Recently, Indonesia completely banned the operation of trawlers in its coastal waters and several countries, including India, have decreed regulations to prevent trawler operators from fishing in the first few kilometres from the coast.

This zone is reserved for traditional fisherfolk. But policing trawlers over such an extensive coastline is an expensive proposition and regulations are, therefore, seldom observed or enforced. The export of frog's legs to cater to the palates of

Western consumers and its impact on the agricultural pest populations in affected areas is now an old story. Fortunately, India not being a major export-oriented country, there are not many such examples. But the pattern of environmental exploitation by industry that we see on the global scale simply reproduces itself on the national scale in India.

What Western industry does to the Third World environment, Indian industry does to the Indian environment. Just to get an idea of how heavily dependent modern industry is on the natural environment, it may be useful to point out that nearly half the industrial output in India is accounted for by industries which can be called biomass-based: that is, industries like cotton textiles, rayon, paper, plywood, rubber, soap, sugar, tobacco, jute, chocolate, food processing and packaging, and so on. Each of these industries exerts an enormous pressure on the country's cultivated and forest lands. They need crop lands, they need forests, and they need energy and irrigation.

The Indian paper industry has ruthlessly destroyed the forests of India. Paper companies in Karnataka, having destroyed all the bamboo forests, are now getting their raw materials from the last major forested frontier of India: the northeast. The government's own public sector paper companies are coming up in the northeast itself. The Andhra Pradesh government has meanwhile set its sights on the forests of Andaman and Nicobar Islands for the paper mill that it wants to build in Kakinada. A leading soap manufacturer has proposed that the Great Nicobar Islands be denuded to plant palm oil. The shortage of raw materials for wood pulp has already forced the government to liberalise import of pulp for the country's paper industry, thus adding to the pressure on the forests of other Third World countries. One lesson is, therefore, clear: the main cause of environmental destruction in the world is the demand for natural resources generated by the consumption of the rich (whether they are rich nations or rich individuals and groups within nations) and because of their gargantuan appetite, it is their wastes mainly that contribute to the global pollution load.

THE ROLE OF SCIENCE AND TECHNOLOGY

If we want to see this process of 'resource colonisation' in a historical light, we can trace it back to the beginnings of the industrial revolution. The cultural diversity that existed in the world at that time was no historical accident. It was born out of the world's biological diversity and people's ability to shape a society with the use of that biological diversity. People came to adopt a lifestyle that was in harmony with the dictates of the immediate environment. Those who lived in the deserts practised nomadism and those who lived in the hills practised shifting cultivation.

Houses and cities built with local materials and skills were built differently to satisfy the conditions set by the local climate. But as modern science and technology grew, it gave people enormous power. Unfortunately the new nowledge created by scientific and technological development fell on an highly unequal world, where access to this knowledge was and still is, unequal, and the infrastructure to use it very uneven. The major result of scientific and technological development, embedded as it is within an unequal global society, is that it has given few people more and more power to exploit resources from further and further away.

First, distant lands (distant ecosystems) were turned into colonies for resource exploitation; when decolonisation took place, multinational corporations took over the role of global resource exploitation and management within a framework of liberal international trade; and now it is the turn of such hitherto unfathomable global commons as the deep sea bed, Antarctica and space to be exploited.

The UN even has a committee to discuss resource exploitation on the moon. But at least on the terrestrial level, with hardly any uninhabited frontiers left, this relentless search for resources from others' ecosystems can only mean extraordinary conflict. In a more and more densely populated world, that which is a resource in a distant ecosystem for one human being is simply a resource within someone else's immediate ecosystem. If that someone's very survival is dependent on the resources of the immediate ecosystem, then there will be conflict, and resource colonisation by the powerful will end in the dispossession and impoverishment of the less powerful, in their own homeland, so to say.

These trends raise serious doubts about the sustainability of the Western industrialisation model based on global management of resources for the consumption of a few. The growth of science and technology has indeed been humankind's most magnificent achievement but definitely not the ends to which this knowledge has been used. The most unfortunate part of this process is that we have now got a large number of consumers in the world whose consumption is highly destructive of the environment but who cannot perceive this destruction psychologically. These are consumers of distant resources: those who have access to all the fruits of modern science and technology. Just imagine a resident of Delhi who uses shirts made of cotton which has been produced in a field in Maharashtra heavily sprayed with pesticides leading to multiple resistance in mosquitoes; electricity from a dam in the Himalaya that has destroyed forests and blocked migration of fish; paper produced in Madhya Pradesh by a factory that has polluted the local river and logged forests in an ever-widening circle, disrupting the life of tribals; cereals from Punjab where food is produced using a technology that drains soil fertility, and so on. It is humanly impossible for such distant consumers to appreciate what their purchasing power is doing to distant areas.

Just imagine, then, the situation of the Western consumer whose beef and fruits come from Latin America and Africa, peanuts from west Africa, coffee from east Africa, tea from India, prawns from Asia and timber from all over the world. But what is probably worse is that such consumers, no longer dependent in any crucial sense on the immediate environment for their consumption needs, slowly become oblivious of the destruction even of the immediate environment. Thus, we get a world economic system in which individuals with power have almost no interest in the fate of their environment, distant or immediate–and only a global environmental crisis, situated amidst a sea of economic poverty, has been able to evoke some concern, and we still talk of the need for environmental education in a big way.

Modern science and technology is not always used as a brutish tool by the rich and powerful to dispossess the poor. It has also been used subtly to undermine the confidence of the poor in their own resources and resource- use patterns. Particularly, if a natural resource used traditionally by people is so widely available that it just cannot be monopolised, every effort is made in the name of scientific progress to discredit that natural resource as archaic and useless. This drive to discredit a resource is undertaken to the extent that even the poor who often have no alternative except to use that resource, look forward to the day when they can do without it. The classic example is mud.

Over half the world's population lives in mud structures. Prices of cement and bricks required for modern buildings continue to rise, being extremely energy intensive, faster than the purchasing power of the world's poor. The chances of all mud dwellers ever moving from their buildings is pretty low, at least not in the next 50 years. The majority of the urban population in the Third World cannot afford even the cheapest modern housing (about US $500-$3,000). In most cities of the developing world, including India, the portion of the population living in slums and squatter settlements has been rising rapidly.

Unable to meet the housing needs of the relatively small urban populations, the governments of the Third World have not even dared to launch housing programmes for rural areas, where the majority of the population lives. The importance of traditional building materials to meet the needs of the world's population is ignored by housing authorities everywhere, and their use is never taught to architects or civil engineers. Mud homes are synonymous with extreme poverty, so much so that they are despised even by the poor and virtually every effort to introduce improved mud housing has failed.

While architects and civil engineers only neglect mud, medical professionals adopt an extremely hypocritical attitude towards herbal medicine. The majority of drugs available on the US market today, for instance, are plant-derived (that is, the drugs were either made directly from a plant-based raw material or they were first discovered in a plant before a chemical company began to market their chemical analogues). Drug companies take great care to screen herbs for pharmacological properties. The moment an useful herb is discovered, the company tries to identify the active ingredient and develop its chemical analogue. This is produced, packaged and marketed and prescribed by the medical profession.

Yet doctors dismiss herbal medicine as outright quackery and make no effort to study medical systems based on herbal medicine. Drug companies make no effort to promote herbs directly. Colonial authorities in Africa had even outlawed the practice of herbal medicine. These administrations had such an effect on the colonised cultures that even after independence, many African countries continued to ban the practice of folk medicine, still the only real option available to a large part of the world's people.

Even more dramatic is the process by which commercial and scientific propaganda has been used to discredit a readily available resource like breast milk, and promote commercial substitutes. Now scientific studies show the superiority of breast milk, but it took a long campaign to formulate appropriate public policies to protect breast milk and even today, vested interests oppose these policies.

THE POOR AND THEIR ENVIRONMENT

The above discussion makes it clear that it is the poor that are affected the most by environmental destruction. The field experience of voluntary groups confirms that eradication of poverty in a country like India is simply not possible without the rational managemet of our environment and that conversely, environmental destruction will only intensify poverty. Environmental destruction goes hand-in-hand with social injustice. A major reason for this is seldom recognised.

The vast majority of the rural households meet their daily household needs through biomass or biomass-related products, which are mostly collected free from the immediate environment. In short, they live within nothing other than a biomass-based, subsistence economy. Food, fuel (firewood, cowdung, crop wastes), fodder, fertilizers (organic manure, forest litter, leaf mulch), building materials (poles, thatch), herbs and clothing are all biomass products. Water is another crucial product for survival.

Water is not biomass itself, but its availability is closely related to the level of biomass available in the surrounding environment. Once the forest disappears, the local pond silts up, the village well dries up, and the perennial stream gets reduced to a seasonal one. The water balance gets totally upset with the destruction of vegetation; in a monsoon climate like ours with highly uneven rainfall over the year, this means greatly increased run-off and floods during the peak water season and greatly increased drought and water scarcity in the lean dry season. There is reason to believe that the number of 'problem villages' from the point of view of availability of drinking water may be increasing. The magnitude of India's dependence on biomass for meeting crucial household needs can be appreciated by looking at the energy situation. We as Indians love to point out that India is the world's 10th largest industrial power, or something like that. But even then, over 50 per cent of the fuel consumption in India is for such a fundamental survival activity as cooking.

In developed countries, cooking accounts for less than 10 per cent of total national fuel consumption. Equally important for India is the fact that 90 per cent of the cooking fuel in India is biomass: that is, firewood, cowdung and crop wastes. Even urban households are heavily dependent on firewood as fuel. Few people, energy planners and government officials alike, had any idea until recently of the dimensions of the rural-urban fuelwood trade.

Annual urban purchases of fuelwood are well over `500 crore in India. Over the last 10 years, Madhya Pradesh, the only state left in the heart of India with any reasonable degree of forest cover, has emerged as major supplier of fuelwood to the cities of northern and western India. But as the state gets more and more deforested every year, fuelwood prices have been rising within Madhya Pradesh itself. Madhya Pradesh has now become India's first state and probably the world's first state too, to ban the export of firewood.

As a result, Delhi, which uses about `15 crore worth of firewood on the retail market, is now getting less firewood, and an increasingly higher proportion from as far away as Assam, the same place from which paper mills are now getting their bamboo. Firewood prices in Delhi are today higher than ever before and according to FAO figures, are among the highest in the world. This firewood is used by the poorest of the poor in Delhi, by construction workers and others. When the effective heat provided by firewood is compared to kerosene, because of the low efficiency chullhas in which firewood is used, firewood becomes four to five times more expensive than kerosene. LPG, incidentally, a fuel used by the richest sections of the society and supplied entirely by government sources, is the most convenient and cheapest fuel available. Which state will become the next to ban firewood exports is anyone's guess. What happens when Assam also says no?

Biomass resources not only meet crucial household needs but also provide a range of raw materials for traditional occupations and crafts and are, hence, a major source of employment: firewood and cowdung are important sources of fuel for potters; bullock carts and catamarans are made from wood; bamboo is a vital raw material for basket weavers, and so on. Traditional craft are not just being threatened by the introduction of modern products but also by the acute shortage of biomass-based raw materials.

A study from the Indian Institute of Science–the first in India on the changing market for bullock carts–reports that people in Ungra village in Karnataka can now longer afford to buy new bullock carts with the traditional wooden wheel because wood has become extremely expensive. A recent report from the Murugappa Chettiar Research Centre from Madras says that traditional fisherfolk now find it very difficult to make new catamarans because the special wood they use is extremely scarce and expensive. Several reports from all over the country–from Madhya Pradesh, from Maharashtra, from Tamil Nadu–portray the extreme difficulty of hundreds of thousands of basket weavers in eking out an existence because of the acute shortage of bamboo.

In the Bhandara and Chandrapur districts of Maharashtra, nearly 70,000 mat and basket weavers have been protesting against the discriminatory prices and small quota of bamboos given to them whereas big paper mills have been leased large bamboo forests. In Karnataka, Madhav Gadgil undertook a study of the use of the state's bamboo forests by paper mills, after a series of protests by basket weavers. Gadgil found that whereas bamboo was available to paper mills at `15 a tonne, it was available to basket weavers and other small bamboo users in the market at `1,200 a tonne. Social activists in Saharanpur have pointed out the travails of the baan makers who have now been deprived of their earlier sources of bhabhar grass. The Uttar Pradesh Forest Development Corporation discriminates in favour of paper mills and this policy has turned thousands of these baan workers into destitutes, landless labourers and urban migrants. Wood is now difficult to get for making even agricultural implements like the plough, especially wood that has been traditionally used for these implements.

Few people know that one of the things that led to the Chipko movement was the anger of the local people over the forest department's refusal to provide ash wood, wood that has been traditionally used for making ploughs, whereas the forest departments happily allocated the same wood to sports goods manufacturers.

Even biomass resources like thatch have become so scarce that maintenance and repair cycles of mud-and-thatch huts have increased considerably. A government report from Bastar, of all places, as it is still one of the heavily forested districts in the country, points to a village where no new hut has been built over the last two decades because the entire area around the village has been deforested. Traditional mud roofs have almost disappeared from many parts of the country because of the large quantities of timber needed by them.

They are being replaced by tiled roofs, but baking tiles still requires large quantities of firewood. Fodder is another vital resource that is in acute shortage. With only 2.45 per cent of the world's land mass, India supports 15 per cent of its cattle, 52 per cent of its buffaloes and 15 per cent of its goats, and these animals play an extremely important role in the integrated system of agriculture and animal husbandry that Indian farmers practise.

Shortage of fodder, especially from public lands, means, as a study from the tribal areas of Gujarat shows, poor landless households and marginal farmers do not benefit much from the milk cooperatives and animal improvement schemes in the region. In such a situation where millions of people are heavily dependent on biomass sources for their daily existence, the destruction of the environment or any policy that reduces access to biomass resources like the proposed Forest Bill and Forest Policy, will have an extremely adverse impact on the daily lives of the people.

ENVIRONMENT AND WOMEN

The destruction of the environment clearly poses the biggest threat to marginal cultures and occupations like that of tribals, nomads and fisherfolk which have always been heavily dependent on their immediate environment for their survival. But the maximum impact of the destruction of biomass sources is on women. Women in all rural cultures are affected, especially women from poor landless, marginal and small farm families. Seen from the point of view of these women, it can even be argued that all development is ignorant of women's needs at best, and anti-women at worst, literally designed to increase their work burden.

Given the culturally accepted division of labour within the family, the collection of household needs like fuel, fodder and water is left to women. As the environment degrades, and this collection work becomes increasingly difficult, women have to spend an extraordinary amount of time foraging for fuel, fodder and water in addition to household work, agricultural work and caring for animals. There is almost no data which shows how the time spent by women on their daily household activities is increasing and how this increase differs across different eco-climatic zones of India. But the data that is already available on the existing work burden is downright shocking. In many parts of the country, women may have literally reached the end of their 'carrying capacities'. The worst situation is in the arid and semi-arid parts of the country and in the hill and mountain villages. In all these areas, trees and forests have been steadily destroyed. Because of a number of factors–soil and climatic conditions, very small land holdings, lack of irrigation, etc–the Green Revolution has also not reached these areas unlike Punjab and Haryana, where trees are few but where the Revolution has meant an enormous increase in biomass from croplands.

As a result, there is now an acute biomass famine in these areas. In all such areas women can spend as much as five to six hours every day–in some households as much as 10 hours every day–just collecting fuel and fodder. On the contrary, in a state like Kerala, where tropical eco-climatic conditions permit a rich green cover, the work burden of women is much smaller–probably the least in the country. Even the minimal land reforms in which landless families have been given a 10th of an acre, has meant access to a few dozen coconut trees, which helps to provide at least half the fuel requirements.

But in the rest of the country, with deforestation, the increasing work burden on women is affecting everything else in their lives. Women do not even have time to seek health care when they are ill or take rest for few days after a tubectomy operation. A study by the Operations Research Group in western Uttar Pradesh found that even pregnant women work for 14 hours a day. They do this almost till a few hours before the delivery and begin normal work three to four days after the birth.

Just as deforestation affects women most, laws that prevent access to biomass, like the proposed Forest Bill which provides inordinate powers to forest officials, will mean maximum oppression of women: increased demands for bribes and harassment. The penetration of the cash economy is also affecting the relationship between men and women in a peculiar way and is creating a real dichotomy in their respective relationships with nature. Men have become more involved with the cash economy than women.

Women continue to deal with the non-monetised, biomass-based subsistence economy of the household. Even within the same household, we can find cases of men happy to destroy nature to earn cash even though it would create greater hardships for the women in collecting the daily fuel and fodder needs.

The Chipko movement has given us numerous examples of this dichotomy in male-female interests. Even the Chipko experience with afforestation confirms this dichotomy and stresses the role of women in ecological regeneration. Even though many crucial household needs could be met by rehabilitating the local village ecosystem–by planting fuel and fodder trees, Environment Policy and for instance–the men do not show much interest in doing so. It is the women who are doing most of the afforestation work organised by the Chipko movement. The new culture created by the penetration of the cash economy has slowly but steadily alienated the men psychologically from their ecosystem. Employment for them means work which can bring in cash. Since employment can be found mainly in the city, there is mass male migration. But even when the men are in a village, a job is still something that earns cash. If such jobs are not available they prefer to remain unemployed and do little to enrich thelocal ecosystem.

The modern economic system is teaching them that household needs can be met only by earning cash, unlike women, who find that in their daily reality, many vital needs are still collected as a free good. There are few of those caste and class barriers in the Himalayan villages which prevent people from working together as a community. But still women continue to walk miles and miles to fetch fuel, fodder and water every day, while men do little to plant trees in the denuded areas around them. This psychological and economic marginalisation of men greatly helps to expand the availability of cheap labour to the cash economy.

It is not surprising that the eucalyptus-based social forestry, trotted out to be such a great success by the World Bank and the forest departments, is all in the hands of men, all planting trees with the cash motive. Other than employing women as cheap labour in nurseries, these agencies have nothing to show in terms of involvement of women–the very people who deal with fuel and fodder–and the government still calls it social forestry. But maybe this should not be surprising.

Making a fast buck, even at the expense of society and ecology, is probably the most social thing that the World Bank can teach us in a cash economy.

Male migration–another major phenomenon in modern society–also seems to be increasing the work burden of women, who then have to take care not only of household needed to collect fuel and fodder grows, agriculture must get neglected. As time for fuel and fodder collection grows and firewood becomes scarcer, the traditional practice of manuring fields will also be given up. Cowdung will be used as fuel, as in the plains, but with the lack of manuring, these fragile hill soils will be exhausted very soon.

It will be a disastrous situation both for the local people and the environment. Already in Ratangiri district, a major source of male workers for the city of

Bombay, an enormous amount of cultivable land lies fallow. In this way, women from migrant households suffer a double exploitation: that of their own and of their husbands, who are not given enough of a wage to keep a household together. While the husband sweats for a pittance in the town, the wife keeps the household going in the village by collecting free goods from nature and by working harder on the family farm or as a farm labourer. Because of the increasing incidence and intensity of floods, there has been considerable talk in recent years about integrated watershed management in the Himalaya and in the Ghats. The Himalaya is being described as one of the most threatened ecosystems in the world, which in turn determines the fate of several hundred million people in the Indo-Gangetic plains. But if any action for ecological reconstruction has to be taken in the hills, it cannot be done without the involvement of women. The census data of 1981 shows that all the districts in the country which have high rates of female work participation are situated in the Himalaya or in the Ghats.

In the Himalaya a very high proportion of women workers are also classified as cultivators. Therefore, any programmes which aims at ecological rehabilitation in these areas will have to involve heavily overworked women unless, of course, labour is brought into these areas from outside, which will create tensions of another kind. Fortunately, the experience of the Chipko movement shows that women in these parts, despite their 14 to 16 hour back-breaking work schedule, are extremely keen to participate in such work, especially in tree planting.

Once the women are organised and mobilised, the evidence is that they work with great keenness and fight any obstacles that may be created by men, and we get as a result some of the highest tree survival rates found in afforestation efforts. The biggest ally in the demand for an ecologically and socially sound nature is, therefore, womankind. As exactly similar experiences have been noted in east and west Africa, in Kenya and in the Sahelian countries, there is every reason to believe that this differential interest in nature between men and women is cross-cultural in character.

Male trees and female trees are now common jargon amongst those interested in involving communities in afforestation. All this should not be taken to mean that poor rural households do not have any need for cash. The unfortunate thing is that much of the cash generated by the male does not get spent on household needs. A reasonable proportion of this cash gets spent on products like alcohol and tobacco and artefacts of modernisation like transistorradios.

The situation is creating a new demand on the time of poor women and in some cases, is resulting in a new militancy against nature. Women now have to go out and also earn some cash. Millions of rural women today sell firewood in the towns and cities. If we look at the quantum of firewood consumed in the cities, we can say that at least 2.3 million people must be engaged in headloading–bringing wood on their heads to sell in the towns– making the firewood trade the largest employer in the commercial energy sector of the country. Surveys show that selling firewood is a profession that has grown rapidly in the last 10 to 15 years with the growth in landlessness and joblessness. Most of these headloaders are women, and mostly tribal women. Selling firewood is back-breaking work and brings in extremely little money.

But the women do this because they cannot rely on their men to bring any cash back home. The big advantage with headloading is that this work is generally available round the year. So when no other work is available, headloading at least provides some income. Every headloading woman knows that the forests will be soon destroyed and even this horrible occupation will come to an end.

But they are afraid that if they do not take advantage of the forest now, the foresters will soon sell it off to a contractor. The forest departments are extremely keen to get headloading banned. They have made no study of the phenomenon of headloading. They have not made any long-term plans to meet the firewood needs of the cities. They have obviously not made any effort to connect the two issues.

Foresters have become environmentalists: they say that these women destroy young trees and lop trees excessively: therefore, ban them. No wonder forests are a major issue in most tribal agitations. If a head count is taken at this stage, we will find that the destruction of the environment and its transformation is already affecting on an immediate and a daily basis at least the following groups: artisans, nomads, tribals, fisherfolk and women from almost all landless, marginal and small farm households.

These groups add up to no less than half to threequarters of the country's entire rural population. And unlike the situation in the West, the question of environmental destruction is not an issue related to quality of life but a matter of survival.

TOWARDS HOLISTIC MANAGEMENT

If these are the problems, then what do we do about them? First of all, there must be much more holistic thinking regarding the management of our land and water resources. And this will not be easy unless a determined effort is made. For all the talk about the need for a scientific temper, it must be recognised that the current methodology of scientific analysis carries within itself an extremely unscientific practice: reductionism. It is this reductionist approach that has today produced both natural and social scientists who know more and more about less and less, who know how to cure a disease but create another disease in the process.

Ecology is the first scientific discipline that has actually forced people to integrate and not reduce. This can be illustrated by describing what is appening to the three major components of our lands: forest lands, croplands and grazing lands. The destruction of forests has a major impact on the productivity of our croplands. This happens in two ways. Soil erosion increases manifold and the soil literally gets washed away, leading to an accentuated cycle of floods and droughts. But equally important is the impact of the shortage of firewood on the productivity of croplands. When firewood becomes scarce, people begin to burn cowdung and crop wastes.

In many places, cowdung and crop wastes are now the major sources of cooking energy. Thus, slowly every part of the plant gets used and nothing goes back to the soil. Over a period of time, this nutrient drain affects crop productivity. Add to this the technology of the Green Revolution: the technology of growing high yielding varieties on a limited diet of chemical fertilizers like nitrogen, phosphates and potash. The total biomass production goes up and so does the drain of the nutrients from the soil.

Plants need some 25-odd elements to grow, not just N, P and K. They keep taking these nutrients from the soil and the more intensive the agriculture, without any manure and crop wastes going back to the soil, the faster is the nutrient drain. Today, the district of Ludhiana has the highest yields of many cereals but it also has the highest deficiencies of many micronutrients in its soil. In Punjab, many farmers have routinely begun to use zinc as a fertilizer. Soon it will be sulphur, manganese and iron deficiencies.

The micronutrient fertilizer industry, already a `10 crore industry, is set to become the boom industry of the future. But the net result is that Indian agriculture, because of this and many other factors, most of them related to ecological factors, is showing a consistently declining trend in output when compared to the rapidly rising levels of inputs.

If existing croplands and irrigation water resources are not used well, then faced with a rising population, the demand for colonisation of marginal lands for agriculture will grow. As large parts of the country have excellent soils and enormlus sunlight and the only shortage is of water, government programmes have also promoted the cultivation of marginal lands, especially through the spread of irrigation. Fortunately, the rate of expansion of the cropped area has now come down as compared to the'50s and'60s but enormous ecological damage has already been done.

Even more than forest lands, croplands have expanded on to grazing lands. The result is that graziers have been pushed on to lesser and lesser grazing lands. This has, in turn, led to the overstocking of grazing lands, the destruction of their productivity and the impoverishment of the graziers in the process. The graziers have taken recourse to two strategies in this situation. As the environment becomes more and more hostile, they get rid of the more vulnerable cattle and start keeping goats.

The number of goats in Rajasthan has expanded dramatically–much faster than any other livestock. Environmentalists may howl that the goat is highly destructive of the environment but it is a far better suited animal to the hostile environment that human beings are creating in Rajasthan, Gujarat and Maharashtra. It makes economic sense for the grazier to reduce risk during a period of drought, which is common in these areas. The herders try to solve their problem in yet another way:

they begin to use forests as grazing lands. This infuriates foresters who see goats and cattle as the worst evil ever devised.

It is true that India's forests are among the most heavily grazed in the world. But it is not true that forests disappear because of goats and sheep: they disappear only when they are chopped down and this is done usually under the aegis of the forest departments. Animals then move in to stop the regeneration of the logged forest. Animals cannot destroy a closed forest by overgrazing. As forests are disappearing in Rajasthan and Gujarat, nomads in large numbers from these states now enter Madhya Pradesh, still a heavily forested state.

The state's foresters, thereatened by this invasion, banned these 'foreign' invaders by law but the Supreme Court struck down the ban saying that Indians cannot be restricted from going from one state to another. So the inter-state movement goes on but clashes are accelerating. Madhya Pradesh's foresters killed 5,000 sheep in one extremely brutal incident on the Madhya Pradesh-Rajasthan border last year and local politicians rushed to rouse passions against the nomads, making every effort to break down the highly synergistic relationship that once prevailed between nomads and settled farmers.

The nomadic animals would earlier bring manure to the fields of settled farmers. Like a walking vacuum cleaner, these animals would gather nutrients from all over and dump them on to the farmers' fields where they would rest and the nomad would even be paid for this service. But today there is open hostility. The Madhya Pradesh forest secretary recently wrote an article in the Times of India saying: "For the love of Mike, please keep these animals out of our forests." If only the animals could be kept out, the forests would regenerate themselves.

But such despair is no answer. Animals cannot obviously be kept out without creating extraordinary conflict. Meanwhile, however, the foresters have found an ingenious solution: plant trees like the eucalyptus which cannot be browsed by animals. Eucalyptus is loved by foresters exactly for this reason. There are many who doubt the ability of eucalyptus to produce more biomass than many indigenous species. But it is indeed ironic that when the country faces an acute fodder crisis, the forester can only plant eucalyptus and produce non-browsable biomass, doing exactly the opposite of what the people need. In fact, eucalyptus is the true weed from the point of view of the landless. It is non-browsable like all fast-spreading weeds and does not benefit the poor unless they own land. But in this manner the cycle of destruction is complete.

The forest departments have destroyed forests by selling off timber to the industrial and urban interests. The firewood shortage and the resulting soil erosion is keeping the productivity of Indian agricultural lands low. Croplands have expanded on to marginal lands and have reduced grazing lands. Animals have moved into forests and are preventing regeneration. Meanwhile, as landlessness and joblessness grow, even groups like the tribals who from times immemorial have lived in total harmony with forests, are turning against forests and want to sell them off as fast as they can.

All the chickens are coming home to roost. Experts sit in grand isolation. Foresters have no interest in fuelwood or in croplands. Agricultural experts have no interest in animals or in grazing lands. Animal husbandry people never tell foresters that they must produce fodder banks. In the life of the proverbial last man–or as I would prefer, the last person of Gandhi–all this isolated thinking brings hovoc.

Life for the poor becomes terrible. Finally, even nature turns against itself. Once denuded, the high solar energy and temperature–the very factors that an bring high productivity and prosperity–begin to bring high desiccation, erosion, destruction and social and ecological poverty. FAO figures show that the effect of unchecked soil erosion on soil productivity is one of the highest in the world, given the ecoclimatic conditions of south and Southeast Asia.

IMPROVING THE GROSS NATURE PRODUCT

Nothing could be more important for planners and politicians today than to rebuild nature. But this can only be done if we re-establish a healthy relationship between the people and their environment. Then only a nature that is useful to the millions, not for making millions, can be re-established.

Regardless of what happens in the West in the next two decades, for allits electronics revolution, its efforts to mine the oceans and its efforts to build solar cells and windmills, and how dramatically this changes the face of the world technostructure, and regardless of how much we may want to catch up with the West in the name of modernisation or out of the sheer compulsions of geopolitics, rebuilding nature and rebuilding its relationship with the people will remain the only way to solve the problem of poverty and possibly even unemployment.

With some 100 millon to 150 million ha of waste and near-wastelands and with a crying need to produce biomass, this country can never get a better opportunity to harness the power of its people to the productivity of its land, to strike at the roots of landlessness, poverty and unemployment, all at the same time. If enough biomass was available, poverty, that is, lack of cash, as defined by economists and by the modern civilization, will not disappear. But the rigours of poverty, the increasing susceptibility to natural emergencies like floods and droughts will be arrested by creating more biomass.

In fact, conventional measurements of poverty based on income data or on food calories are clearly inadequate in a situation where the rest of the biomass needs are becoming increasingly difficult to meet and collecting them on a daily basis constitutes the worst–and growing–drudgery humankind, especially womankind, has ever known. These calculations are not only just inadequate but they also reflect a strong gender bias because they deal mainly with those aspects of poverty–lack of cash–that the male is generally concerned with, but not with those aspects of poverty that the woman deals with–lack of fuel, fodder, water etc.

If we were to construct a concept like gross nature product, we would find that for the poor it is this indicator, which is many times more important than the conventional gross national product (GNP). In fact, we can even say that those who do not get much from the conventional GNP–the poor–are the ones who are most critically dependent on the gross nature product. The GNP cannot be allowed to destroy or transform the gross nature product. Just as economists get very worried about the structure of the GNP, it is equally important, if they have the poor in mind, that they get worried about the structure of the gross nature product.

It is not just the quantity of biomass that is important for meeting basic household needs but also its diversity; sources of biomass within any village ecosystem must be diverse enough to meet the various household needs of fuel, fodder, building materials and herbs, and of raw materials for artisans.

The diversity in nature has also acted as insurance during periods of emergency by reducing societal vulnerability. During periods of drought and resulting crop failures, which are recurring phenomena in many parts of India, fruits from trees and roots, leaves and wild animals in the forests used to become an important, alternative source of nutrition. In 1983, the tribals of Chotanagpur survived a drought not because of government assistance but despite government callousness. It is the forests which gave them their nutrition. Surviving on the forest during a drought is common in Bastar.

Wild animals caught in the jungles become a significant source of food during periods of crop failure. A study from Africa found that in times of drought, traditional societies had nearly 150 responses. They even fed the thatch of their huts to the goats. But in a modern village there were only two responses: pray to God (which even the Tamil Nadu government recommended during the Madras water crisis) or migrate to the towns. The combination of trees, grasses, crops, animals and ponds, which we found in almost every village was an extraordinarily interactive and resilient system for emergencies.

Instead of destroying this complex and interrelated system, science must be used to build on it. In other words, it is not enough to preserve biological diversity in just those areas of our country where the flora and fauna are genetically rich and diverse by setting up biosphere reserves and national parks, but that biological diversity must be preserved and/or recreated in every village ecosystem. Concentrating on the production of a few commodities (cereals, for instance) is totally inadequate in a society which is only partly monetarised and where the vast majority still has to depend on access to free biomass resources from the immediate environment.

Every village has to become a biosphere reserve. If suff-icient biomass could be generated, there would also be none of those tensions we see around nature protection areas today. At a time when biomass is in acute shortage, it is obviously foolhardy to attempt such protection without massive schemes for largescale biomass generation. Understanding of the gross nature product and how it is changing within the national ecological space (or shall we say, national economy) is extremely limited, despite its crucial importance for the poor. We know nothing, for instance, about the importance of so-called weeds for the poor. The reason we know so little about the gross nature product is because the 'growing stock' of biomass does not get reflected in GNP calculations anywhere.

Therefore, whether the 'growing stock' lives or dies, exists or disappears, does not make any difference to economists. Most economists, not surprisingly, have no clue of what hapens to a subsistence economy when its biomass resources are affected. The increasing hardships and tensions only get reflected in studies of 'political economy which record growing conflicts over land, forests, fodder, grazing lands, water sources, etc, or in totally unexpected indicators like the sex ratio of the country. Just why 22 million women failed to show up in India's 1981 census is anybody's guess.

Surely their extraordinary work burden, in a situation of malnutrition, must be an important factor. It is for this reason that an indicator like gross nature product, and changes in that indicator, would probably reflect far better the changing reality in the subsistence sector of a country like India. Unfortunately, we do not know as yet how to construct such an indicator. But if we did, we will definitely find that while the conventional GNP has gone up, the gross nature product has steadily gone down, the former acting as a parasite on the latter.

The answer to India's immediate problem of poverty lies in increasing the biomass available in nature in a manner that access to it is ensured on an equitable basis. But giving a relevant 'green cover' to the country—the real Green Revolution—would probably require the most holistic thinking that planners economists and scientists have ever known.

The conflicts and complementarities in the existing land use patterns have to be extremely well understood. Otherwise, landuse patterns will remain as chaotic as today. Landless and poorer peasants will continue to oppose planting trees on comunity lands under so-called social forestry programmes because they are afraid this will take away their grazing lands, while forest departments and richer peasants will only plant those trees which animals cannot touch (like eucalyptus) even though there is a stark fodder crisis all around.

Nothing could take us closer to Gandhiji's concept of gram swarajya than striving to create village ecosystems which are biologically diverse and self-reliant in their local biomass needs to the maximum extent possible. This will clearly demand an extremely intensive use of our natural resources like land and water to create huge and diverse growing stock of biomass. Any science which teaches how to do this will truly have the right to be called a people's science—and indeed it will have to begin with the traditional knowledge of the people.

Even more so, planning for the enhancement of village ecosystems will call for village-level planning with the involvement of the people–a level of decentralisation that has never been attempted either in resource planning or resource management. The biggest challenge, therefore, is before social workers and politicians who have to play a crucial role in ensuring that people can participate in this biomass-based development process. No biomass-based strategy can succeed without the involvement of the people, especially women.

If useful nature is to grow, it will have to be protected and nurtured. The walls built to protect nature will be respected by the people only if they are people's walls and if the people know that what grows beyond, belongs to them. Easy availability of biomass leading to a reduction in women's work burden could create the appropriate conditions for many desirable social changes. Kerala, for instance, is often cited as an interesting case of an economically poor state moving ahead with its demographic transition. But is Kerala poor in environmental terms also?

The availability of biomass, in fact, appears to be the maximum in India and women's work burden low. This may sound an exaggerated proposition but would it be right to surmise that, among many other things, it was also the easy availability of biomass and relatively lower work burden that created the appropriate conditions for literacy programmes for women to succeed? If women in Kerala had to spend as much time working as women do in large parts of Rajasthan, Uttar Pradesh and Bihar, wouldn't there have been immense male pressure to keep women working in crucial household activities and away from schools?

Immediately, at least, the country must realise that a clear biomass policy is desperately needed, which recognises the competing uses for biomass in society, especially between biomass-based industry and poor households and sets clear priorities on the use of biomass in a situation of scarcity. The needs of the poor must be specified as a priority use of biomass in the existing situation of environmental degradation. A beginning definitely needs to be made with the proposed Forest Policy and with rural energy planning in general.

If India fails to recreate nature on a massive scale in a manner that generates employment and equity, not only its villages but also its cities will become unlivable. Many people prefer to call the urban migrants economic refugees from the countryside. But to my mind many of them are really ecological refugees, displaced by dams, by mines, by deforestation, by destruction of grazing lands, by floods, by droughts, by urban expansion, and what not. We have today the world's fourth largest urban population. Before the end of the century we will be the largest. Managing this huge urban population will call for extraordinary political and managerial sagacity and altogether new approaches, something we cannot learn from the rest of the world. But one thing is definite–if the process of urbanisation continues to create the same demands on our rural environment, it will only accelerate the destruction of that environment and in turn make the urban environment impossible to manage.

India cannot survive without a low-energy, low-resource input urbanisation. In its absence, no law or laws, like the so-called Delhi Bills, which try to turn the incoming ecological and economic refugees into our cities into criminals, will work. Only a holistic approach to our problems and dedicated political will to solve them will. ❐

2

Pollution

Pollution is the introduction by man into the environ-ment of contaminants that cause harm or discomfort to humans or other living organisms, or damage the environment Pollution can be in the form of chemical substances, or energy such as noise, heat, or light. Pollutants can be naturally occurring substances or energies, but are considered contaminants when in excess of natural levels. Pollution is often categorized into point source and no point source pollution. An important aspect of the notion of pollution is that ecological change must actually be demonstrated. If some potentially polluting substance is present at a concentration or intensity that is less than the threshold required to cause a demonstrable ecological change, then the situation would be referred to as contamination, rather than pollution.

Some other elements can be present in very large concentrations, aluminum and iron, which are important constituents of rock and soil. Aluminum constitutes 8-10% of the earth's crust and iron 3-4%. However, almost all of the aluminum and iron present in minerals are insoluble in water and are therefore not readily assimilated by biotic community and cannot cause toxicity. In acidic environments, however, ionic forms of aluminum are solubilized, and these can cause toxicity in concentrations of less than one part per million. Therefore, the bio-availability of a chemical is an important determinant of whether its presence in some concentration will cuse pollution. Most instances of pollution result from the activities of humans can be caused by:

- The emission of sulfur dioxide and metals from a smelter, causing toxicity to vegetation and acidifying surface waters and soil,

- The emission of waste heat from an electricity generating station into a river or lake, causing community change through thermal stress, or
- The discharge of nutrient-containing sewage wastes into a water body, causing eutrophication. Most instances of anthropogenic pollution have natural analogues, that is, cases where pollution is not the result of human activities.

Pollution can be caused by the emission of sulfur dioxide from volcanoes, by the presence of toxic elements in certain types of soil, by thermal springs or vents, and by other natural phenomena. In many cases, natural pollution can cause an intensity of ecological damage that is as severe as anything caused by anthropogenic pollution. An interesting case of natural air pollution is the Smoking Hills, located in a remote and pristine wilderness, virtually uninfluenced by humans.

However, at a number of places along the 18.63 miles (30 km) of seacoast, bituminous shales in sea cliffs have spontaneously ignited, causing a fumigation of the tundra with sulfur dioxide and other pollutants. The largest concentrations of sulfur dioxide (more than two parts per million) occur closest to the combustions. Further away from the sea cliffs the concentrations of sulfur dioxide decrease rapidly.

The most-important chemical effects of the air pollution are acidification of soil and fresh water, which in turn causes a solubilization of toxic metals. Surface soils and pond waters commonly have pHs less than 3, compared with about Ph 7 at non-fumigated places. The only reports of similarly acidic water are for volcanic lakes in Japan, in which natural pHs as acidic as 1 occur, and pH less than 2 in waters affected by drainage from coal mines. At the Smoking Hills, toxicity by sulfur dioxide, acidity, and water-soluble metals has caused great damage to ecological communities.

The most-intensively fumigated terrestrial sites have no vegetation, but further away a few pollution-tolerant species are present. About one kilometre away the toxic stresses are low enough that reference tundra is present. There are a few pollution-tolerant algae in the acidic ponds, with a depauperate community of six species occurring in the most-acidic pond in the area. Other cases of natural pollution concern places where certain elements are present in toxic amounts. Surface mineralizations can have toxic metals present in large concentrations, copper at 10% in peat at a copper-rich spring in New Brunswick, or surface soil with 3% lead plus zinc on Baffin Island. Soils influenced by nickel-rich serpentine minerals have been well-studied by ecologists.

The stress-adapted plants of serpentine habitats form distinct communities, and some plants can have nickel concentrations larger than 10% in their tissues.

Similarly, natural soils with large concentrations of seleium support plants that can hyperaccumulate this element to concentrations greater than 1%. These plants are poisonous to livestock, causing a toxic syndrome known as blind staggers. There are many well-known cases where pollution is caused by anthropogenic emissions of chemicals.

Some examples include:

- Emissions of sulfur dioxide and metals from smelters can cause damage to surrounding terrestrial and aquatic ecosystems. The sulfur dioxide and metals are directly toxic. In addition, the deposition of sulfur dioxide can cause an extreme acidification of soil and water, which causes metals to be more bio-available, resulting in important, secondary toxicity. Because smelters are point sources of emission, the spatial pattern of chemical pollution andecological damage displays an exponentially decreasing intensity with increasing distance from the source.
- The use of pesticides in agriculture, forestry, and around homes can result in a non-target exposure of birds and other wildlife to these chemicals. If the non-target biota is vulnerable to the pesticide, then ecological damage will result. During the 1960s urban elm trees in the eastern United States were sprayed with large quantities of the insecticide DDT, in order to kill beetles that were responsible for the transmission of Dutch elm disease, an important pathogen. Because of the very large spray rates, many birds were killed, leading to reduced populations in some areas. Birds and other non-target biota have also been killed by modern insecticide-spray progammes in agriculture and in forestry.
- The deposition of acidifying substances from the atmosphere, mostlyas acidic precipitation and the dry deposition of sulfur dioxide can cause an acidification of surface waters. The acidity solubilizes metals, most notably aluminum, making them bio-available. The acidity in combination with the metals causes toxicity to the biota, resulting in large changes in ecological communities and processes. Fish are highly intolerant of acidic waters.
- Oil spills from tankers and pipelines can cause great ecological damage. When oil spilled at sea washes up onto coastlines, it destroys seaweeds, invertebrates, and fish, and their communities are changed for many years. Seabirds are very intolerant of oil and can die of hypothermia if even a small area of their feathers is coated by petroleum.
- Most of the lead shot fired by hunters and skeet shooters miss their target and are dispersed into the environment. Waterfowl and other avian wildlife actively ingest lead shot because it is similar in size and hardness to the grit that they ingest to aid in the mechanical abrasion of hard seeds in their gizzard.

Humans can also cause pollution by excessively fertilizing natural ecosystems with nutrients.Freshwaters can be made eutrophic by fertilization with phosphorus in the form of phosphate. The most conspicuous symptoms of eutrophication are changes in species composition of the phytoplankton community and, especially, a large increase in algal biomass, known as a bloom. In shallow water bodies there may also be a vigourous growth of vascular plants. These primary responses are usually accompanied by secondary changes at higher trophic levels, including arthropods, fish, and waterfowl, in response to greater food availability and other habitat changes. However, in the extreme cases of very eutrophic waters, the blooms of algae and other microorganisms can be noxious, producing toxic chemicals and causing periods of oxygen depletion that kill fish and other biota. Extremely eutrophic water bodies are polluted because they often cannot support a fishery, cannot be used for drinking water, and have few recreational opportunities and poor esthetics.

CURRENT PROTOCOLS

Protocols exist for avoiding pollution of other worlds. Payloads are scrupulously sterilized. Rocket stages that loft probes towards these targets are diverted to avoid trajectories that would follow the probe. Pollution is a harmful change in the natural environment caused by human activities: This may be the release of substances which are toxic to either animals or plants, or it may be the release of energy (heat, light, radiation, or sound) which interferes with the development of animals or plants. Pollution is usually harmful to one or more species of animal or plant. However, releasing nutrients into rivers may be beneficial to some organisms while it is harmful to others.
Here are some general examples:

- Automobile exhaust emissions
- Oil spills
- The dumping of industrial wastes in the water supply
- The overuse of pesticides
- The overuse of chemical fertilizers
- Improper disposal of solid wastes Releasing raw sewage into a river causes pollution. Sewage contains lots of organic substances which are a source of food for bacteria and fungi. The result is that the bacteria and fungi thrive.

Since they use up all the oxygen in the water, most of the animals will die. Pollution may occur naturally, as when an erupting volcano emits sulphur dioxide, but the term usually refers to the negative effect of human activities.

POLLUTION PREVENTION

Pollution prevention is reducing or eliminating waste at the source by modifying production processes, promoting the use of non-toxic or less-toxic substances, implementing conservation techniques, and re-using materials rather than putting them into the waste stream. Pollution prevention means "source reduction," as defined under the Pollution Prevention Act, and other practices that reduce or eliminate the creation of pollutants through:

- Increased efficiency in the use of raw materials, energy, water, or ther resources, or
- Protection of natural resources by conservation. The Pollution Prevention Act defines "source reduction" to mean any practice which
- Reduces the amount of any hazardous substance, pollutant, or contaminant entering any waste stream or otherwise released into the environment prior to recycling, treatment, or disposal; and
- Reduces the hazards to public health and the environment associated with the release of such substances, pollutants, or contaminants. The term includes: equipment or technology modifications, process or procedure modifications, reformulation or redesign or products, substitution of raw materials, and improvements in housekeeping, maintenance, training, or inventory control.

Specific Pollution Prevention Approaches

Pollution prevention approaches can be applied to all pollution-generating activities, including those found in the energy, agriculture, Federal, consumer, as well as industrial sectors. The impairment of wetlands, ground water sources, and other critical resources constitutes pollution, and prevention practices may be essential for preserving these resources.

These practices may include conservation techniques and changes in management practices to prevent harm to sensitive ecosystems. Pollution prevention does not include practices that create new risks of concern. In the agricultural sector, pollution prevention approaches include:

- Reducing the use of water and chemical inputs;
- Adoption of less environmentally harmful pesticides or cultivation of crop strains with natural resistance to pests; and
- Protection of sensitive areas. In the energy sector, pollution prevention can reduce environmental damages from extraction, processing, transport, and combustion of fuels. Pollution prevention approaches include:
- Increasing efficiency in energy use;
- Substituting environmentally benign fuel sources; and
- Design changes that reduce the demand for energy. Some Types of
- Pollution are: Air,Water and thermal, Radioactive.

AIR POLLUTION

One of the formal definitions of air pollution is as follows— 'The presence in the atmosphere of one or more contaminants in such quality and for such duration as is injurious, or tends to be injurious, to human health or welfare, animal or plant life.' It is the contamination of air by the discharge of harmful substances. Air pollution can cause health problems and it can also damage the environment and property. It has caused thinning of the protective ozone layer of the atmosphere, which is leading to climate change. Modernisation and progress have led to air getting more and more polluted over the years. Industries, vehicles, increase in the population, and urbanization are some of the major factors responsible for air pollution.

The following industries are among those that emit a great deal of pollutants into the air: thermal power plants, cement, steel, refineries, petro chemicals, and mines. Air pollution results from a variety of causes, not all of which are within human control. Dust storms in desert areas and smoke from forest fires and grass fires contribute to chemical and particulate pollution of the air. The source of pollution may be in one country but the impact of pollution may be felt elsewhere. The discovery of pesticides in Antarctica, where they have never been used, suggests the extent to which aerial transport can carry pollutants from one place to another.

Sources of Air Pollution

The combustion of gasoline and other hydrocarbon fuels in automobiles, trucks, and jet airplanes produces several primary pollutants: nitrogen oxides, gaseous hydrocarbons, and carbon monoxide, as well as large quantities of particulates, chiefly lead. In the presence of sunlight, nitrogen oxides combine with hydrocarbons to form a secondary class of pollutants, the photochemical oxidants, among them ozone and the eye-stinging peroxyacetylnitrate (PAN). Nitrogen oxides also react with oxygen in the air to form nitrogen dioxide, a foul-smelling brown gas.

In urban areas like Los Angeles where transportation is the main cause of air pollution, nitrogen dioxide tints the air, blending with other contaminants and the atmospheric water vapour to produce brown smog. Although the use of catalytic converters has reduced smog-producing compounds in motor vehicle exhaust emissions, recent studies have shown that in so doing the converters produce nitrous oxide, which contributes substantially to global warming. Air may be severely polluted not only by transportation but also by the burning of fossil fuels (oil and coal) in generating stations, factories, office buildings, and homes and by the incineration of garbage.

The massive combustion produces tons of ash, soot, and other particulates responsible for the gray smog of cities like New York and Chicago, along with enormous quantities of sulfur oxides (which also may be result from burning coal and oil). These oxides rust iron, damage building stone, decompose nylon, tarnish silver, and kill plants. Air pollution from cities also affects rural areas for many miles downwind. Every industrial process exhibits its own pattern of air pollution. Petroleum refineries are responsible for extensive hydrocarbon and particulate pollution. Iron and steel mills, metal smelters, pulp and paper mills, chemical plants, cement and asphalt plants—all discharge vast amounts of various particulates. Uninsulated high-voltage power lines ionize the adjacent air, forming ozone and other hazardous pollutants. Airborne pollutants from other sources include insecticides, herbicides, radioactive fallout, and dust from fertilizers, mining operations, and livestock feedlots.

Sources And Major Air Pollutants

Carbon monoxide (CO) is a colourless, odourless gas that is produced by the incomplete burning of carbon-based fuels including petrol, diesel, and wood. It is also produced from the combustion of natural and synthetic products such as cigarettes. It lowers the amount of oxygen that enters our blood. It can slow our reflexes and make us confused and sleepy. Carbon dioxide (CO_2) is the principle greenhouse gas emitted as a result of human activities such as the burning of coal, oil, and natural gases. Chloroflorocarbons (CFC) are gases that are released mainly from air-conditioning systems and refrigeration. When released into the air, CFCs rise to the stratosphere, where they come in contact with few other gases, which leads to a reduction of the ozone layer that protects the earth from the harmful ultraviolet rays of the sun. Lead is present in petrol, diesel, lead batteries, paints, hair dye products, etc. Lead affects children in particular.

It can cause nervous system damage and digestive problems and, in some cases, cause cancer. Ozone occur naturally in the upper layers of the atmosphere. This important gas shields the earth from the harmful ultraviolet rays of the sun. However, at the ground level, it is a pollutant with highly toxic effects. Vehicles and industries are the major source of ground-level ozone emissions. Ozone makes our eyes itch, burn, and water. It lowers our resistance to colds and pneumonia. Nitrogen oxide (Nox) causes smog and acid rain. It is produced from burning fuels including petrol, diesel, and coal. Nitrogen oxides can make children susceptible to respiratory diseases in winters.

Suspended particulate matter (SPM) consists of solids in the air in the form of smoke, dust, and vapour that can remain suspended for extended periods and is also the main source of haze which reduces visibility. The finer of these particles, when breathed in can lodge in our lungs and cause lung damage and respiratory problems. Sulphur dioxide (SO_2) is a gas produced from burning coal, mainly in thermal power plants. Some industrial processes, such as production of paper and smelting of metals, produce sulphur dioxide. It is a major contributor to smog and acid rain. Sulfur dioxide can lead to lung diseases.

Chemical Pollution

In some parts of the world, the bodies of whales and dolphins washing ashore are so highly contaminated that they qualify as toxic waste and have to be specially disposed of.

There are many different sources of chemical pollution, including:

- Domestic sewage
- Industrial discharges

- Seepage from waste sites
- Atmospheric fallout
- Domestic run-off
- Accidents and spills at sea
- Operational discharges from oil rigs
- Mining discharges and
- Agricultural run-off.
- Health Effects of Air Pollution

The human health effects of poor air quality are far reaching, but principally affect the body's respiratory system and the cardiovascular system. Individual reactions to air pollutants depend on the type of pollutant a person is exposed to, the degree of exposure, the individual's health status and genetics. People who exercise outdoors on hot smoggy days increase their exposure to pollutants in the air.The health effects caused by air pollutants may range from subtle biochemical and physiological changes to difficulty breathing, wheezing, coughing and aggravation of existing respiratory and cardiac conditions. These effects can result in increased medication use, increased doctor or emergency room visits, more hospital admissions and even premature death.

- Human Respiratory System
- Human Cardiovascular System
- Heart and Lung Diseases
- Pyramid of Health Effects
- Populations at Risk
- Leading Causes of Hospitalization
- Leading Causes of Death
- Estimating Health Benefits

HUMAN RESPIRATORY SYSTEM

The health of our lungs and entire respiratory system is affected by the quality of the air we breathe. In addition to oxygen, this air contains other substances such as pollutants, which can be harmful. Exposure to chemicals by inhalation can negatively affect our lungs and other organs in the body. The respiratory system is particularly sensitive to air pollutants because much of it is made up of exposed membrane. Lungs are anatomically structured to bring large quantities of air (on average, 400 million litres in a lifetime) into intimate contact with the blood system, to facilitate the delivery of oxygen. Lung tissue cells can be injured directly by air pollutants such as ozone, metals and free radicals.

Ozone can damage the alveoli — the individual air sacs in the lung where oxygen and carbon dioxide are exchanged. More specifically, airway tissues which are rich in bioactivation enzymes can transform organic pollutants into reactive metabolites and cause secondary lung injury. Lung tissue has an abundant blood supply that can carry toxic substances and their metabolites to distant organs. In response to toxic insult, lung cells also release a variety of potent chemical mediators that may critically affect the function of other organs such as those of the cardiovascular system. This response may also cause lung inflammation and impair lung function.

Structure and Function

The human respiratory system is dominated by our lungs, which bring fresh oxygen (O_2) into our bodies while expelling carbon dioxide (CO_2). The oxygen travels from the lungs through the bloodstream to the cells in all parts of the body. The cells use the oxygen as fuel and give off carbon dioxide as a waste gas. The waste gas is carried by the bloodstream back to the lungs to be exhaled.The lungs accomplish this vital process - called gas exchange - using an automatic and quickly adjusting control system. This gas exchange process occur in conjunction with the central nervous system (CNS), the circulatory system, and the musculature of the diaphragm and the chest. The human respiratory system can be divided into the upper respiratory tract and the lower respiratory tract.

The upper respiratory tract includes the following rigid structures:

- Nasal cavities: Filter the air we breathe and provide a sense of smell.
- Pharynx: Acts in the respiratory and the digestive system.
- Larynx: Link between the pharynx and the trachea. Generates the voice with the presence of vocal folds. Trachea: The trachea is the bond with the lower respiratory tract.

This is a flexible structure allowing the air to go down to the lungs. The lungs and the other parts of the respiratory system have important jobs to do related to breathing.

These include:

- Bringing all air to the proper body temperature.
- Moisturizing the inhaled air for necessary humidity.
- Protecting the body from harmful substances by coughing, sneezing, filtering or swallowing them, or by alerting the body through the sense of smell. Defending the lungs with cilia (tiny hair-like structure), mucus and macrophages.

Causes of Air Pollution

Humans probably first experienced harm from air pollution when they built fires in poorly ventilated caves. Since then we have gone on to pollute more of the earth's surface. Until recently, environmental pollution problems have been local and minor because of the Earth's own ability to absorb and purify minor quantities of pollutants. The industrialization of society, the introduction of motorized vehicles, and the explosion of the population, are factors contributing towards the growing air pollution problem. At this time it is urgent that we find methods to clean up the air. The primary air pollutants found in most urban areas are carbon monoxide, nitrogen oxides, sulfur oxides, hydrocarbons, and particulate matter (both solid and liquid). These pollutants are dispersed throughout the world's atmosphere in concentrations high enough to gradually cause serious health problems. Serious health problems can occur quickly when air pollutants are concentrated, such as when massive injections of sulfur dioxide and suspended particulate matter are emitted by a large volcanic eruption.

Air Pollution in the Home

You cannot escape air pollution, not even in your own home. "The Environmental Protection Agency (EPA) reported that toxic chemicals found in the air of almost every American home are three times more likely to cause some type of cancer than outdoor air pollutants". The health problems in these buildings are called "sick building syndrome". The EPA has found that the air in some office buildings is 100 times more polluted than the air outside.

Poor ventilation causes about half of the indoor air pollution problems. The rest come from specific sources such as copying machines, electrical and telephone cables, mold and microbe-harboring air conditioning systems and ducts, cleaning fluids, cigarette smoke, carpet, latex caulk and paint, vinyl molding, linoleum tile, and building materials and furniture that emit air pollutants such as formaldehyde. A major indoor air pollutant is radon-222, a colourless, odourless, tasteless, naturally occurring radioactive gas produced by the radioactive decay of uranium-238. "The EPA and the National Research Council, exposure to radon is second only to smoking as a cause of lung cancer".

WATER

Water is life ! It is a precondition for human, animal and plant life as well as an indispensable resource for the economy. Water also plays a fundamental role in the climate regulation cycle.Protection of water resources, of fresh and salt water ecosystems and of the water we drink and bathe in is therefore one of the cornerstones of environmental protection in Europe. The stakes are high and the issues transcend national boundaries and concerted action at the level of the EU is necessary to ensure an effective protection.

WASTE AND WATER POLLUTION

When toxic substances enter lakes, streams, rivers, oceans, and other water bodies, they get dissolved or lie suspended in water or get deposited on the bed. This results in the pollution of water whereby the quality of the water deteriorates, affecting aquatic ecosystems. Pollutants can also seep down and affect the groundwater deposits. The effects of water pollution are not only devastating to people but also to animals, fish, and birds. Polluted water is unsuitable for drinking, recreation, agriculture, and industry. It diminishes the aesthetic quality of lakes and rivers. More seriously, contaminated water destroys aquatic life and reduces its reproductive ability. Eventually, it is a hazard to human health. Nobody can escape the effects of water pollution.

Importance of Water

Water is our lifeline that bathes us and feeds us. In ancient cultures waterrepresented the very essence of life. The Romans were the first to pipe water into their growing cities, especially with their aqueducts. They also realised that sewage water could cause damage to their people, and needed to be removed from large areas of people. Water has played a role not only in the history of countries, but in religion, mythology, and art.

Water in many religions cleanses the soul through holy water. The water at Lourdes, France is thought by many religions to be sacred water with healing powers. In Egyptian mythology, the Nu was the beginning of everything and represented water. It brought life to their people, but in drought, produced chaos. The water or hydrologic cycle explains interactions between the atmosphere, hydrosphere, and lithosphere. The water or hydrologic cycle is a major driving force on our planet. Water is in constant motion, evapourating into the atmosphere from oceans, lakes, rivers and streams.

When the atmosphere can no longer support the moisture within the clouds, we experience rain, snow, hail, or sleet. Some water is locked in the form of ice at the polar caps and in glaciers. Water melts in the spring, producing runoff, that percolates through the Earth as groundwater (subsurface) or makes its way back to the sea (surface). The oceans contain most of the water, but it is salt water which is unusable by most organisms. Only pure H_2O (water) can interact with organisms. The movement of the oceans also has a direct effect on the atmosphere.

The atmosphere is that envelope of gas that keeps organisms living on this planet. Oceans and atmosphere interact to give us weather. Water provides the Earth with the capacity of supporting life. An organism doesn't have to be told how important water is to their existence. An amphibian knows to lay their eggs in water or else there will be no new born. Even flies know to lay their eggs in fresh water.The only organism that doesn't understand the importance of water is humans, especially in industrialized countries.

Children in those societies turn on the water in a sink and never think about the trouble someone has gone for that "miracle" to occur. Dams, reservoirs, filtering plants, and pipes all bring clean water when the facet is turned on. Sewage water is only mixed with recycled water supplies after the water goes through rigourous cleaning methods. Water borne diseases are any illnesses caused by drinking contaminated water. Diseases can include infection from bacteria (Salmonella), viruses, or by small parasites (Cryptosporida, Giardia, and Toxoplasma). These organisms and viruses cause diseases like cholera, typhoid fever, malaria, botulism, polio, dysentery, giardia, and hepatitis. A one of the first symptoms of these diseases is diarrhea, which cause about three million deaths throughout the world.Sewage is sometimes discharged into rivers, where children downstream might be taking a bath or using the water to drink. The simplest treatment method is boiling. Just bring the water to a boil for at least one minute, then allow it to cool. But this is not always effective in heavily chemical polluted water supplies. Without water, organisms could not exist.

Water is a resource that should not be taken for granted. It needs to be conserved, just as we save other valuable resources. Water is one of the weirdest compounds known to humans. The difference between the boiling point and freezing point of water is one of the largest ranges of any compound. It is this span of temperature that mirrors the range of where life can exist, from bacteria to humans. Water also has a very high specific heat, which means that it can absorb or lose much heat before its temperature changes. This is important in maintaining body heat in mammals. It also takes a lot of energy before vapourization can occur. For this reason, water evapourates slowly from ponds and lakes, where many life forms are dependent on a stable, warm environment. Water is less dense in its solid state than in its liquid state, so that ice floats instead of sinking. This property permits life to develop in polar and subpolar regions where ice floats and allows life to continue living below the surface. If ice were heavier than water, it would sink, and more ice would form on top of it. As a result, all life in the waters would be trapped in the ice in the many areas of the world where it gets cold enough to freeze water.

Resistance of Water to a Disturbance

Water also exhibits viscosity. One can observe the effects of viscosity alongside a stream or river with uniform banks. The water along the banks is nearly still, while the current in the centre may be swift. This resistance between the layers is called viscosity. This property allows smaller fish to live near the shore, while larger fish are able to swim efficiently in strong currents. Viscosity is also responsible for the formation of eddies, creating turbulence that leads to good mixing of air in the water and more uniform distribution of microscopic organisms. Domestic sewage refers to waste water that is discarded from households. Also referred to as sanitary sewage, such water contains a wide variety of dissolved and suspended impurities. It amounts to a very small fraction of the sewage by weight. But it is large by volume and contains impurities such as organic materials and plant nutrients that tend to rot. The main organic materials are food and vegetable waste, plant nutrient come from chemical soaps, washing powders, etc. Domestic sewage is also very likely to contain disease-causing microbes. Thus, disposal of domestic waste water is a significant technical problem.

Industrial Effluents

Waste water from manufacturing or chemical processes in industries contributes to water pollution. Industrial waste water usually contains specific and readily identifiable chcmical compounds. During the last fifty years, the number of industries in India has grown rapidly. But water pollution is concentrated within a few subsectors, mainly in the form of toxic wastes and organic pollutants. Out of this a large portion can be traced to the processing of industrial chemicals and to the food products industry. In fact, a amount of large- and medium-sized industries in the region covered by the Ganga Action Plan do not have adequate effluent treatment facilities. Most of these defaulting industries are sugar mills, distilleries, leather processing industries, and thermal power stations. Most major industries have treatment facilities for industrial effluents. But this is not the case with small-scale industries, which cannot afford enormous investments in pollution control equipment as their profit margin is very slender.

Effects of Run-off Pollution

Rain picks up dirt and silt and carries it into the water. If the dirt and silt settle in the water body, then these sediments prevent sunlight from reaching aquatic plants. If the Sun can't reach the plants, these perish. These sediments also clog fish gills and smother organisms that live on the bottom of the body of the water.

Effects of Oil Pollution and Antifreeze

If oil is spilled on the water, the effects on the ecosystem and the components are harmful. Many animals can be annihilated in case they ingest oil. Oil contaminated prey may be a reason of death for many. If the oil coats the feathers of birds, these may die. Oil and antifreeze makes the water have a foul odour and there is a sticky film on the surface of water that kills animals. Oil is the most harmful pollutant in the water.

Contaminated Ground Water Effects

If contaminated water enters the ground, there may be serious effects. People may become very sick and there is a probability of developing liver or kidney problems and cancer or other illnesses. Fertilizers and Other Chemicals Nitrates in drinking water leads to diseases of infants that may lead to their death. Cadmium is a metal in sludge-derived fertilizer. This can be absorbed by crops. When people ingest this, they may cause diarrheal disorders, liver and kidney damage. The inorganic substances like mercury, arsenic and lead are the causes of pollution. Other chemicals can also lead to problems concerning the taste, smell and colour of water. Pesticides, PCBs and PCPs are all poisonous to all sorts of life. Pesticides are used in farming, homes and forestry. PCBs are found as insulators in old electrical transformer. PCPs are found in products like wood preservatives.

Effects of Agricultural Water Pollution

Rain and irrigation water drains off cultivated land that has been fertilized and treated with pesticides, the excess nitrogen and poisons are mixed with it into the water supply. These pesticides are toxic and pollute the water in a different mode. Aquatic plants growth cause deoxyge-nation of water and annihilate flora and fauna in a stream, lake and river. Fertilizers enhance the growth of bacteria that are in water and increase the concentration of bacteria to hazardous levels.

Effects of Thermal Water Pollution

Machinery in the industries are cooled with water from lakes and rivers. This water reaches the river in a heated state. This water decreases the ability of the aquatic system to hold oxygen and raises the growth of warm water species.

Effects of Heavy Metal Water Pollution

Heavy metals like lead, mercury, iron, cadmium, aluminum and magnesium are present in water sources. If these metals are present in the sediment, these reach the food chain through plants and aquatic animals. This causes heavy metal poisoning in case the level in the water is very high.

THERMAL POLLUTION

Thermal pollution describes the introduction of waste heat into water bodies such as rivers, estuaries and coastal waters. Although other industries abstract water and then discharge it, the coastal power plants produce the greatest amount of water and thus waste heat. Coastal power stations of up to 1000MW capacity discharge up to 50cumecs (m3 s-1) at elevated temperatures. The effects of that discharge depend on the time and place of discharge and the characteristics of the receiving area. In particular, the effects depend on the ability to disperse and absorb the additional energy.

Power stations with a ready supply of cold water, such as those on the coast and in estuaries, use this to condense the steam used to turn the turbines. This flow of direct cooling, as opposed to the use of cooling towers, returns water to an adjacent area up to 10 degrees centigrade above ambient. The fossil and nuclear fuel power plants require this cooling. The high cost-effective value of direct cooling explains the large amount of directly cooled estuarine and coastal power plants worldwide.

Fig. 2.1: Thermal Power Plant

ECOSYSTEMS CAUSED BY HEAT (THERMAL POLLUTION)

There are various effects on the biology of the ecosystems when heated effluents reach the receiving waters. The species that are intolerant to warm conditions may disappear, while others, rare in unheated water, may thrive so that the structure of the community changes. Thermal pollution can have a great influence on the aquatic ecosystem. Species that are restricted to heated waters, can build up large populations in the receiving waters.

Respiration and growth rates may be changed and these may alter the feeding rates of organisms. The reproduction period may be brought forward and development may be speeded up. Parasites and diseases may also be affected. An increase of temperature also means a decrease in oxygen solubility. Any reduction in the oxygen concentration of the water, particularly when organic pollution is also present, may result in the loss of sensitive species.

Major Sources

The major sources of thermal pollution are electric power plants and industrial factories. In most electric power plants, heat is produced when coal, oil, or natural gas is burned or nuclear fuels undergo fission to release huge amounts of energy. This heat turns water to steam, which in turn spins turbines to produce electricity. After doing its work, the spent steam must be cooled and condensed back into water. To condense the steam, cool water is brought into the plant and circulated next to the hot steam. In this process, the water used for cooling warms 5 to 10 Celsius degrees (9 to 18 Fahrenheit degrees), after which it may be dumped back into the lake, river, or ocean from which it came. Similarly, factories contribute to thermal pollution when they dump water used to cool their machinery.

The second type of thermal pollution is much more widespread. Streams and small lakes are naturally kept cool by trees and other tall plants that block sunlight. People often remove this shading vegetation in order to harvest the wood in the trees, to make room for crops, or to construct buildings, roads, and other structures. Left unshaded, the water warms by as much as 10 Celsius degrees (18 Fahrenheit degrees). In a similar manner, grazing sheep and cattle can strip streamsides of low vegetation, including young trees.

Even the removal of vegetation far away from a stream or lake can contribute to thermal pollution by speeding up the erosion of soil into the water, making it muddy. Muddy water absorbs more energy from the sun than clear water does, resulting in further heating. Finally, water running off of artificial surfaces, such as streets, parking lots, and roofs, is warmer than water running off vegetated land and, thus, contributes to thermal pollution.

Impact

All plant and animal species that live in water are adapted to temperatures within a certain range. When water in an area warms more than they can tolerate, species that cannot move, such as rooted plants and shellfish, will die. Species that can move, such as fish, will leave the area in search of cooler conditions, and they will die if they can not find them. Typically, other species, often less desirable, will move into the area to fill the vacancy. In general, cold waters are better habitat for plants and animals than warm ones because cold waters contain more dissolved oxygen. Many freshwater fish species that are valued for sport and food, especially trout and salmon, do poorly in warm water. Some organisms do thrive in warm water, often with undesirable effects.

Algae and other plants grow more rapidly in warm water than in cold, but they also die more rapidly; the bacteria that decompose their dead tissue use up oxygen, further reducing the amount available for animals.

The dead and decaying algae make the water look, taste, and smell unpleasant. Effects of a thermal pollution Heat and hot water result from many industrial processes. They are in particular by-products of the activity of the power stations and nuclear. The water rejected into the marine mediums has harmful effects, primarily on the marine animal-life.The pollution thermal has effects difficult to define with precision. One observes significant interferences of the reduction in the salinity of water and variations of this one in time and space. Several actions of such a pollution were however determined.

It was in particular established that the hotter the temperature of rejected water is and more harmful is its effect on the benthic organizations.On the algae, the action of a thermal pollution is very variable. One noted opposite effects is strong proliferations or on the contrary a significant mortality. The metabolism of the algae, the phytoplanktons, the benthic macrophages is very faded by heated water.One will thus be satisfied to point out some facts of observation which seem to have characters of general data:

ECOLOGICAL EFFECTS

Warm Water

Warm water typically decreases the level of dissolved oxygen in the water. The decrease in levels of dissolved oxygen can harm aquatic animals such as fish, amphibians and copepods. Thermal pollution may also increase the metabolic rate of aquatic animals, as enzyme activity, resulting in these organisms consuming more food in a shorter time than if their environment were not changed. An increased metabolic rate may result in food source shortages, causing a sharp decrease in a population. Changes in the environment may also result in a migration of organisms to another, more suitable environment, and to in-migration of organisms that normally only live in warmer waters elsewhere.

This leads to competition for fewer resources; the more adapted organisms moving in may have an advantage over organisms that are not used to the warmer temperatures. As a result one has the problem of compromising food chains of the old and new environments. Biodiversity can be decreased as a result.It is known that temperature changes of even one to two degrees Celsius can cause significant changes in organism metabolism and other adverse cellular biology effects. Principal adverse changes can include rendering cell walls less permeable to necessary osmosis, coagulation of cell proteins, and alteration of enzyme metabolism. These cellular level effects can adversely affect mortality and reproduction. Primary producers are affected by warm water because higher water temperature increases plant growth rates, resulting in a shorter lifespan and species overpopulation.

This can cause an algae bloom which reduces the oxygen levels in the water. The higher plant density leads to an increased plant respiration rate because the reduced light intensity decreases photosynthesis. This is similar to the eutrophication that occurs when watercourses are polluted with leached agricultural inorganic fertilizers. A large increase in temperature can lead to the denaturing of life-supporting enzymes by breaking down hydrogen- and disulphide bonds within the quaternary structure of the enzymes. Decreased enzyme activity in aquatic organisms can cause problems such as the inability to break down lipids, which leads to malnutrition.In limited cases, warm water has little deleterious effect and may even lead to improved function of the receiving aquatic ecosystem.

This phenomenon is seen especially in seasonal waters and is known as thermal enrichment. An extreme case is derived from the aggregational habits of the manatee, which often uses power plant discharge sites during winter.

Projections suggest that manatee populations would decline upon the removal of these discharges. The added heat lowers the dissolved oxygen content and may cause serious problems for the plants and animals living there. In extreme cases, major fish kills can result. Warm water may also increase the metabolic rate of aquatic animals, as enzyme activity, meaning that these organisms will consume more food in a shorter time than if their environment was not changed.

Cold Water

Releases of unnaturally cold water from reservoirs can dramatically change the fish and macroinvertebrate fauna of rivers, and reduce river productivity. In Australia, where many rivers have warmer temperature regimes, native fish species have been eliminated, and macroinvertebrate faunas drastically altered and impoverished, from large stretches of rivers due to cold water thermal pollution.

A solution for Thermal Pollution

Using nearby water for cooling in industrial applications and then returning that water to the source is a traditional maneuver, but it can also damage local aquatic life and violate EPA regulations. Now more than ever, cooling towers are being used to precool that water before discharge, diverting rejected heat into the atmosphere instead. Learn the costs and benefits of a few configurations, along with tips for dealing with blowdown and effluent within the facility itself. Warm water can be found in a amount of different areas in nature- hot springs or water warmed by volcanic activity.

Warm water can become a problem, however, when it is created by man and introduced into nature. An example of this is when the water used to cool power plants or other industrial applications is discharged into streams, rivers, and lakes. This is known as thermal pollution or thermal discharge, and it is the introduction of waste heat into bodies of water that support aquatic life. The addition of heat reduces the water's ability to hold dissolved gases, including the oxygen required for aquatic life. If the water temperature is greater than 95[degrees]F, the dissolved oxygen content may be too low to support some species.

If the differential temperature is too large, the difference can also stress some species. As a result, thermal pollution can wreak havoc on native fish species, such as trout, that require cold water with high levels of dissolved oxygen. When the water becomes warmer, other non-native fish that thrive in the warmth can take over habitats from native fish. In addition, warmer water allows bacterial populations to increase and thrive, and algae "blooms" may occur.Regulators and lawmakers in the United States long ago recognized that thermal pollution is a problem and addressed the issue of the EPA Clean Water Act. States and other regulatory agencies use those guidelines to require power plants and industries to limit warm water discharges back into surface waters, sometimes by way of cooling towers.

GROWING NEED FOR TOWERS

Geological Survey about 48% of all freshwater and saline-water withdrawals for 2000 were used for thermoelectric power. Most of this water was derived from surface water and used for once through cooling at power plants. About 52% of fresh surface water withdrawals and about 96% of saline water withdrawals were for thermoelectric power use.This large amount of water is needed by power plants due to the fact that over the years, there has been an ever-increasing need for electricity. This means power plants are expected to run at near maximum output for a large part of the year. The cheapest and easiest method for power plants to operate has always been to withdraw water from a nearby body of water, pass it through the plant, and return the heated water to the same body of water. These once-through cooling systems now require very strict environmental permits, issued in accordance with the National Pollution Discharge Elimination System.

The permits vary from state to state and location to location and may have different requirement .Some permits require that the plant must discharge the water within a temperature differential limit over the temperature of the intake water.

Other permits have an ultimate limit; in other words, they can't ever exceed a specific temperature. Still other permits have both differential and ultimate limits. Depending on the permit, restrictions are often magnified during low river or lake levels, or drought conditions. That's because most utilities see their peak loads in the summer months, when air conditioning loads are high. In addition, water temperatures are at their highest in the summer, which can make it difficult for power plants to comply with permit requirements.

RADIOACTIVE POLLUTION

Radioactive pollution can be defined as the release of radioactive substances or high-energy particles into the air, water, or earth as a result of human activity, either by accident or by design. The sources of such waste include:
Nuclear weapon testing or detonation;

- The nuclear fuel cycle, including the mining, separation, and
- Production of nuclear materials for use in nuclear power plants or nuclear bombs;
- Accidental release of radioactive material from nuclear power plants.

Sometimes natural sources of radioactivity, such as radon gas emitted from beneath the ground, are considered pollutants when they become a threat to human health.

Since even a small amount of radiation exposure can have serious (and cumulative) biological consequences, and since many radioactive wastes remain toxic for centuries, radioactive pollution is a serious environmental concern even though natural sources of radioactivity far exceed artificial ones at present.

The problem of radioactive pollution is compounded by the difficulty in assessing its effects. Radioactive waste may spread over a broad area quite rapidly and irregularly from an abandoned dump into an aquifer and may not fully show its effects upon humans and organisms for decades in the form of cancer or other chronic diseases. Surface waters are a powerful factor that causes migration of radionuclides across the territory of Belarus. For this reason it is essential to take into due account the transit role of rivers in the transportation of radionuclides, including transboun-dary transfer. In watercourses and flowing water bodies concentration of radionuclides are reducing every year, but they tend to accumulate in sttatic water bodies (lakes, ponds, reservoirs, especially in bottom sediments).

Due to the accident at the Chernobyl nuclear power station (CNPS) one quarter of the country's territory was contaminated by caesium-137 (23%), strontium-90 (10%) and plutonium (about 2 %). Concentrations of caesium-137 early 90's presented. The monitoring data the radiation situation in the Dnieper-Sozh and Pripyat basins is stable. The annual average concentration of caesium-137 has decreased significantly in large and small rivers for the observed early 1990's period. Exceeding of the National permissible levels for caesium-137 and strontium-90 were not observed. Caesium-137 concentration in surface waters are closely connected with the annual volume of river flow, as can be seen from the increase in concentrations of caesium-137 in some rivers, where water supply was lower than the perennial average values.

The data analysis of concentration of caesium-137 during the spring flood in the Pripyat basin shows that concentration of caesium-137 in a dissolved form in the Pripyat basin remains at the level of the average indices for the prior period. Yet concentration of this radionuclide has considerably increased in dredges. This means that caesium-137 is washed out and transported by flood flow with sediments. Since the radiation situation has stabilized, transboundary transport of radioactive elements through river flow has significantly decreased. Mainly it is the Pripyat that transports radionuclides, in particular strontium-90, as they are washed out from the 30-kilometre Chernobyl zone. Due attention is devoted to studies of the radiation state of small rivers, which are tributaries of the Pripyat and the Sozh in the most contaminated areas of the Gomel and Mogilev regions. Over the years radioactivity in water tends to decrease.

The exception is the dissolved strontium -90, which is a specific feature for the CNPS zone. Annual data on the content of caesium-137 and strontium-90 (in soluble and suspended forms) in the bottom sediments and water biota suggest that bottom sediments and water biota are significant contributors to the total radioactivity of surface water systems. A tendency towards a reduction in radioactivity of bottom sediments and water biota is minor. During spring high water and summer-autumn flood, migration of radionuclides into open water systems occurs both in soluble and in absorbed forms on organic and mineral carrieres. The ratio between concentrations of caesium-137 and strontium-90 at the end checkpoints of the Braginka and Senna Rivers suggests that starting from1992-1993, the concentration of strontium-90 has begun to exceed that of caesium-137. That phenomenon is characteristic for surface watercourses close to the CNPS zone and is explained by an increase in migration ability of strontium-90 due to its release from active particles.

RADIOACTIVE MATERIALS REGULATIONS

Many facilities use radioactive material (RAM) in diverse ways or have radioactive wastes.Exit signs may contain tritium or radioactive hydrogen-3 gas, and smoke detectors may contain radium. Other examples of industrial uses of RAM include devices to measure the density of concrete or blacktop, determine the thickness of paper and rolled steel as it is made, find cracks in pipes or airplane surfaces, test the amount of lead in paint, or monitor the flow of sludge through pipes at a sewage treatment plant.

Research facilities and academic institutions use RAM during the development of new pharmaceuticals to "tag" certain molecules to follow their progress through chemical or biological processes and in other research activities. Medical facilities inject patients with RAM to diagnose medical conditions and for therapeutic treatments. Medical facilities also use large radiation sources for cancer treatment.

Radium paint was once used on aircraft instruments, naval compasses,ilitary vehicle instruments, and on clocks and watches to make the amounts and lines glow in the dark. Naturally occurring radioactive material is found as uranium in clay and bricks, granite, shale, or other rocks. It is also found as radium in soils or as radium sulfate scales on some pipes and fittings from the oil and gas industry and as the naturally radioactive constituent of potassium, potassium-40.

Environmental Monitoring and Radon Gas

Environmental Monitoring Unit (EMU) of the RPMWS is responsible for DEQ's Radon Progamme activities. Environmental Protection Agency's State Indoor Radon Grant Progamme. This progamme provides education on the radiological risks posed by public exposure to radon gas and works closely with local health departments throughout the state for outreach at the local level.

The EMU operates an environmental monitoring network around each of Michigan's nuclear power plant sites. The progamme collects and performs radioanalyses of several types of samples, including direct radiation, air, surface water, precipitation, and milk from the environs of the nuclear plants. Unit labouratory analyses also include samples collected by other section staff during investigations of potentially contaminated sites, during emergency response activities, and from routine staff compliance investigations.

NUCLEAR FACILITIES

The Nuclear Facilities Unit of the RPMWS develops and implements the DEQ's Nuclear Facilities Emergency Response Procedures and the nuclear accident aspects of the Michigan Emergency Management Plan as they relate to the DEQ's responsibilities to respond to accidents or emergencies at any of Michigan's commercial nuclear power plants or to large-scale radiological ncidents. These efforts are conducted in cooperation with other state agencies and under the overall emergency response coordination of the Michigan State Police. Unit staff also interact with nuclear plant utility staff and staff of the NRC concerning the day-to-day operations of nuclear power reactors to assure radiological protection of the public and the environment. Radioactive Materials Regulations Waste

Industrial Smoke Detectors

Remove any batteries from the detector and handle the battery as a universal waste or under the applicable hazardous waste regulations for that company's hazardous waste generator status The specific requirements a company would havc to follow would depend on if the smoke detectors were subject to the federal nuclear regulations or if it was a hazardous waste. There are two types of materials commonly found in used smoke detectors.

Radioactive Materials Regulations

The older models may contain non-exempt radium-226 sources that are regulated under the state of Michigan. These detectors should not go to a solid waste landfill but returned to the manufacturer or disposed as radioactive waste.

The newer models contain a small americium source. The combined smoke detector and americium source have a specific exemption in the federal regulations. Large quantities such as resulting from a major construction renovation project should not be disposed without first checking with officials of the NRC or the WHMD radioactive materials staff. If the smoke detectors are not recycled for etal, some smoke detectors could be subject to the hazardous waste regulations because the amount of metal in the detectors may fail the Toxicity Characteristic Leaching Procedure.

Small quantity generators and large quantity generators can not put hazardous waste smoke detectors in the trash. Conditionally exempt small quantity generators may dispose smoke detectors in licensed solid waste landfills if the landfill will accept them. If these smoke detectors are not classified as a hazardous waste, then they may be sent to a licensed landfill. However, companies should contact the landfill if disposing of large amounts because the waste load may set off the landfill's radiation detectors. Contact the DEQ Waste and Hazardous Materials Division regarding potential safety concerns when numerous smoke detectors are disposed of at the same time or regarding nuclear regulations.

The Inspection and Mappig of Radioactive Sources

During the ten years since the Norwegian-Russian cooperation was established in form of The Group of Experts, a magnificent effort in mapping the present environmental situation has been carried out. That includes sources of radioactivity in the environment On these missions water samples were collected, along with sediments and living organisms which have given new knowledge about the present environmental status of the area. At two occasions closer investigations were done at dumping places for radioactive materials, and due to this work future risk of pollution has been evaluated. At the moment effort is being made to establish a co-ordinated progamme for surveying the northern sea areas, a system for exchanging data, and an extension of a network for automatic metering in the county of Murmansk.

Experiences from the cooperation in the Expert Group gave a solid foundation for joint inspections and risk evaluation In addition to the work done by the Expert group, inspections of different kinds of pollution in the Arctic are carried out by the organizations called AMAP (Arctic Monitoring and Assessment Progammeme). AMAP has established an international data centre, localized at the Norwegian Radiation Protection Authority (NRPA) for the purpose of monitoring sources and levels of radioactive pollution. The work of AMAP is close to the subjects investigated by the Expert group, resulting in close contact between the two organisations.

RISK EVALUATION

The joint projects between Norway and Russia include risk analysis of potential scenes of accidents at several nuclear plants in Russia. Substantial amounts of radioactive waste was released from nuclear plants and directed cout in rivers and ocean, causing severe pollution of the environment. Today, storing of radioactive waste at plants with poor securing is a big problem. There is great concern about the consequences if accidents take place, what the effects might be on the local environment, and even for other countries. Calculations and risk evaluations have been made in case of potential accidents occur at the nuclear plant in Majak in Ural. In management of the Expert group, fieldwork has been carried out twice, and potential accidents have been considered.

Two of the most important potential scenarios at Majak are an explosion in a container for high-level waste, and destruction of the lower dam at the plant. This scenario has potential for large releases of radioactive substances with scattering by the river Ob to the Kara Sea, but the level of pollution will be relatively limited A similar evaluation is being done at the nuclear plant of Krasnojarsk close to the river Jenisej. Norwegian authorities have given special attention to the nuclear plant at Kola. This plant is placed 200 km from the Norwegian border. Analysis has been made on long-term effects on the environment and population in the Nort-West of Russia from possible accidents at this plant. The worst thinkable scenarios have been used in the modelling; meaning large fallouts, and bad weather conditions during an accident. The results indicate that the negative effects will be long lasting.

Radioactive Mine Waste Polluting Colourado River

Water tests reveal that "uranium mill waste" leaching into the Colourado River has made the water radioactive at "one-third the level considered dangerous," the San Diego Union Tribune 1/10. The mine's owner, Atlas Corporation, has declared bankruptcy, leaving the bulk of the enormous clean up costs to taxpayers.

The huge pile of mine waste "sits 750 feet from the river," and is leaking "an estimated 28,800 gallons of radioactive pollution and toxic chemicals" into the river each day.

Types of Radiation

Radiation is classified as being ionizing or nonionizing. Both types can be harmful to humans and other organisms. Radioactive Pollution-Nonionizing Radiation Radioactive Pollution - Ionizing Radiation Radioactive Pollution - Lifestyle and Radiation Dose Radioactive Pollution - Nuclear Weapons Testing Radioactive Pollution - Nuclear Power Plants Radioactive Pollution - Biological Effects of Radioactivity

NONIONIZING RADIATION

Nonionizing radiation is relatively long-wavelength electro-magnetic radiation, such as radio waves, microwaves, visible radiation, ultraviolet radiation, and very low-energy electromagnetic fields. Nonionizing radiation is generally considered less dangerous than ionizing radiation.

However, some forms of nonionizing radiation, such as ultraviolet, can damage biological molecules and cause health problems. Scientists do not yet fully understand the longer-term health effects of some forms of nonionizing radiation, such as that from very low-level electromagnetic fields (e.g., high-voltage power lines), although the evidence to date suggests that the risks are extremely small.

Ionizing Radiation

Ionizing radiation is the short wavelength radiation or particulate radiation emitted by certain unstable isotopes during radioactive decay. There are about 70 radioactive isotopes, all of which emit some form of ionizing radiation as they decay from one isotope to another. A radioactive isotope typically decays through a series of other isotopes until it reaches a stable one. As indicated by its name, ionizing radiation can ionize the atoms or molecules with which it interacts.

In other words, ionizing radiation can cause other atoms to release their electrons. These free electrons can damage many biochemicals, such as proteins, lipids, and nucleic acids (including DNA). In intense, this damage can cause severe human health problems, including cancers, and even death. Ionizing radiation can be either short-wavelength electromagnetic radiation or particulate radiation. Gamma radiation and X-radiation are short-wavelength electromagnetic radiation.

Alpha particles, beta particles, neutrons, and protons are particulate radiation. Alpha particles, beta particles, and gamma rays are the most commonly encountered forms of radioactive pollution. Alpha particles are simply ionized helium nuclei, and consist of two protons and two neutrons. Beta particles are electrons, which have a negative charge. Gamma radiation is high-energy electromagnetic radiation. Scientists have devised various units for measuring radioactivity.

A Curie (Ci) represents the rate of radioactive decay. One Curie is $3.7 \times 10_{10}$ radioactive disintegrations per second. A rad is a unit representing the absorbed dose of radioactivity. One rad is equal to an absorbed energy dose of 100 ergs per gram of radiated medium. One rad = 0.01 Grays. A rem is a unit that measures the effectiveness of radioactivity in causing biological damage.

One rem is equal to one rad times a biological weighting factor. The weighting factor is 1.0 for gamma radiation and beta particles, and it is 20 for alpha particles. One rem = 1000 millirem = 0.01 Sieverts. The radioactive half-life is a measure of the persistence of radioactive material. The half-life is the time required for one-half of an initial quantity of atoms of a radioactive isotope to decay to a different isotope.

Sources of Radioactive Pollution

In the United States, people are typically exposed to about 350 millirems of ionizing radiation per year. On average, 82% of this radiation comes from natural sources and 18% from anthropogenic sources (i.e., those associated with human activities). The major natural source of radiation is radon gas, which accounts for about 55% of the total radiation dose. The principal anthropogenic sources of radioactivity are medical X-rays and nuclear medicine.

Radioactivity from the fallout of nuclear weapons testing and from nuclear power plants make up less than 0.5% of the total radiation dose, i.e., less than 2 millirems. Although the contribution to the total human radiation dose is extremely small, radioactive isotopes released during previous atmospheric testing of nuclear weapons will remain in the atmosphere for the next 100 years.

Lifestyle and Radiation Dose

People who live in certain regions are exposed to higher doses of radiation. Residents of the Rocky Mountains of Colourado receive about 30 millirems more cosmic radiation than people living at sea level. This is because the atmosphere is thinner at higher elevations, and therefore less effective at shielding the surface from cosmic radiation. Exposure to cosmic radiation is also high while people are flying in an airplane, so pilots and flight attendants have an enhanced, occupational exposure. In addition, residents of certain regions receive higher doses of radiation from radon-222, due to local geological anomalies. Radon-222 is a colourless and odourless gas that results from the decay of naturally occurring, radioactive isotopes of uranium.

Radon-222 typically enters buildings from their basement, or from certain mineral-containing construction materials. Ironically, the trend towards improved home insulation has increased the amount of radon-222 which remains trapped inside houses. Personal lifestyle also influences the amount of radioactivity to which people are exposed. Miners, who spend a lot of time underground, are exposed to relatively high doses of radon-222 and consequently have relatively high rates of lung cancer. Cigarette smokers expose their lungs to high levels of radiation, since tobacco plants contain trace quantities of polonium-210, lead-210, and radon-222. These radioactive isotopes come from the small amount of uranium present in fertilizers used to promote tobacco growth. Consequently, the lungs of a cigarette smoker are exposed to thousands of additional millirems of radioactivity, although any associated hazards are much less than those of tar and nicotine.

NUCLEAR WEAPONS TESTING

Nuclear weapons release enormous amounts of radioactive materials when they are exploded. Most of the radioactive pollution from nuclear weapons testing is from iodine-131, cesium-137, and strontium-90. Iodine-131 is the least dangerous of these isotopes, although it has a relatively half-life of about eight days. Iodine-131 accumulates in the thyroid gland, and large doses can cause thyroid cancer. Cesium-137 has a half-life of about 30 years. It is chemically similar to potassium, and is distributed throughout the human body. Based on the total amount of cesium already in the atmosphere, all humans will receive about 27 millirems of radiation from cesium-137 over their lifetime. Strontium-90 has a half-life of 38 years. It is chemically similar to calcium and is deposited in bones. Strontium-90 is expelled from the body very slowly, and the uptake of significant amounts increases the risks of developing bone cancer or leukemia.

NUCLEAR POWER PLANTS

Many environmentalists are critical of nuclear power generation. They claim that there is an unacceptable risk of a catastrophic accident, and that nuclear power plants generate large amounts of unmanageable nuclear waste. Nuclear Regulatory Commission has strict requirements regarding the amount of radioactivity that can be released from a nuclear power reactor. In particular, a nuclear reactor can expose an individual who lives on the fence line of the power plant to no more than 10 millirems of radiation per year.

Actual measurements at U.S. nuclear power plants have shown that a person who lived at the fence line would actually be exposed to much less that 10 millirems. Thus, for a typical person who is exposed to about 350 millirems of radiation per year from all other sources, much of which is natural background, the proportion of radiation from nuclear power plants is extremely small. In fact, coal- and oil-fired power plants, which release small amounts of radioactivity contained in their fuels. Although a nuclear power plant cannot explode like an atomic bomb, accidents can result in serious radioactive pollution. During the past 45 years, there have been a amount of not-fully controlled or uncontrolled fission reactions at nuclear power plants in the United States and elsewhere, which have killed or injured power plant orkers. These accidents occurred in Los Alamos, New Mexico; Oak Ridge, Tennessee; Richland, Washington; and Wood River Junction, Rhode Island.

The most famous case was the 1979 accident at the Three Mile Island nuclear reactor in Pennsylvania, which received a great deal of attention in the press. However, nuclear scientists have estimated that people living within 50 mi (80 km) of this reactor were exposed to less than two millirems of radiation, most of it as iodine-131, a short-lived isotope. This exposure constituted less than 1% of the total annual radiation dose of an average person. However, these data do not mean that the accident at Three Mile Island was not a serious one; fortunately, technicians were able to reattain control of the reactor before more devastating damage occurred, and the reactor system was well contained so that only a relatively small amount of radioactivity escaped to the ambient environment. By far, the worst nuclear reactor accident occurred in 1986 in Chernobyl, Ukraine. An uncontrolled build-up of heat resulted in a meltdown of the reactor core and combustion of graphite moderator material in one of the several generating units at

Chernobyl, releasing more than 50 million Curies of radioactivity to the ambient environment. The disaster killed 31 workers, and resulted in the hospitalization of more than 500 other people from radiation sickness. The Ukrainian authorities, during the decade following the Chernobyl disaster an estimated 10,000 people in Belarus, Russia, and Ukraine died from cancers and other radiation-related diseases caused by the accident.

In addition to these relatively local effects, the atmosphere transported radiation from Chernobyl into Europe and throughout the Northern Hemisphere. More than 500,000 people in the vicinity of Chernobyl were exposed to dangerously high doses of radiation, and more than 300,000 people were permanently evacuated from the vicinity. Since radiation-related health problems may appear decades after exposure, scientists expect that many thousands of additional people will eventually suffer higher rates of thyroid cancer, bone cancer, leukemia, and other radiation-related diseases.

Unfortunately, a cover-up of the explosion by responsible authorities, including those in government, endangered even more people. Many local residents did not known that they should flee the area as soon as possible, or were not provided with the medical attention they needed. The large amount of radioactive waste generated by nuclear power plants is another an important problem.

This waste will remain radioactive for many thousands of years, so technologists must design systems for extremely long-term storage. One obvious problem is that the long-term reliability of the storage systems cannot be fully assured, because they cannot be directly tested for the length of time they will be used (i.e., for thousands of years). Another problem with nuclear waste is that it will remain extremely dangerous for much longer than the expected lifetimes of existing governments and social institutions.

Thus, we are making the societies of the following millennia, however they may be structured, responsible for the safe storage of nuclear waste that is being generated in such large quantities today.

Biological Effects of Radioactivity

The amount of injury caused by a radioactive isotope depends on its physical half-life, and on how quickly it is absorbed and then excreted by an organism. Most studies of the harmful effects of radiation have been performed on single-celled organisms. Obviously, the situation is more complex in humans and other multicellular organisms, because a single cell damaged by radiation may indirectly affect other cells in the individual. The most sensitive regions of the human body appear to be those which have many actively dividing cells, such as the skin, gonads, intestine, and tissues that grow blood cells (spleen, bone marrow, lymph organs). Radioactivity is toxic because it forms ions when it reacts with biological molecules.

These ions can form free radicals, which damage proteins, membranes, and nucleic acids. Radioactivity can damage DNA (deoxyribonucleic acid) by destroying individual bases (particularly thymine), by breaking single strands, by breaking double strands, by cross-linking different DNA strands, and by cross-linking DNA and proteins. Damage to DNA can lead to cancers, birth defects, and even death. However, cells have biochemical repair systems which can reverse some of the damaging biological effects of low-level exposures to radioactivity.

This allows the body to better tolerate radiation that is delivered at a low dose rate, such as over a longer period of time. In fact, all humans are exposed to radiation in extremely small doses throughout their life. The biological effects of such small doses over such a long time are almost impossible to measure, and are essentially unknown at present. There is, however, a theoretical possibility that the small amount of radioactivity released into the environment by normally operating nuclear power plants, and by previous atmospheric testing of nuclear weapons, has slightly increased the incidence of certain cancers in human populations. However, scientists have not been able to conclusively show that such an effect has actually occurred. Currently, there is disagreement among scientists about whether there is a threshold dose for radiation damage to organisms.

In other words, is there a dose of radiation below which there are no harmful biological effects? Some scientists maintain that there is no such threshold, and that radiation at any dose carries a finite risk of causing some biological damage. Furthermore, the damage caused by very low doses of radiation may be cumulative, or additive to the damage caused by other harmful agents to which humans are exposed. Other scientists maintain that there is a threshold dose for radiation damage.

They believe that biological repair systems, which are presumably present in all cells, can fix the biological damage caused by extremely low doses of radiation. Thus, these scientists claim that the extremely low doses of radiation to which humans are commonly exposed are not harmful. One of the most informative studies of the harmful effects of radiation is a long-term investigation of the survivors of the 1945 atomic blasts at Hiroshima and Nagasaki by James Neel and his colleagues.

The survivors of these explosions had abnormally high rates of cancer, leukemia, and other diseases. However, there seemed to be no detectable effect on the occurrence of genetic defects in children of the survivors. The radiation dose needed to cause heritable defects in humans is higher than biologists originally expected. Radioactive pollution is an important environmental problem. It could become much worse if extreme vigilance is not utilized in the handling and use of radioactive materials, and in the design and operation of nuclear power plants.

RADIATION EXPOSURE

Radiation exposure occurs any time that energy in the form of electromagnetic rays or particles interacts with biological tissue. Ionizing radiation is particularly energetic; examples include: X rays, gamma radiation, and subatomic particles. Biological damages caused by exposure to ionizing range from mild tissue burns to cancer, genetic damage, and ultimately, death.

However, there are potential benefits of controlled exposures to certain kinds of radiation, which can be used for the detection, diagnosis, and treatment of certain diseases. Exposure to many types of radiation is routinely monitored using sensitive devices, such as film badges and dosimeters. Effects of Radiation Exposure

With many forms of ionizing radiation, energy is transferred to electrons that surround atomic nuclei. Atoms affected by x rays usually absorb enough energy to lose some of their electrons, and so become ionized. (An atom is ionized when it gains or loses electrons and acquires a net electric charge.) Ultraviolet radiation causes electrons to absorb energy and jump to a higher energy orbit around the atomic nucleus.

The sun and sunlamps emit enough ultraviolet radiation to cause sunburn, premature aging of the skin, and skin cancers. Exposure of humans and animals to ultraviolet radiation also results in the production of vitamin D, a biochemical necessary for good health. Radiation that consists of charged particles can knock electrons out of their orbit around atoms. This also creates ions. Such radiation can also cause atoms to enter an exited state, if the electrons are bumped into higher-energy orbits.

These changes result in atoms and molecules (including biochemicals) that are chemically reactive. Seeking to become stable, they interact with unaffected atoms and molecules, which may be damaged (i.e., changed) in the process, and are then unable to perform their usual metabolic functions. Nuclear material such as DNA molecules may be damaged to the degree that they can no longer be accurately copied. This may lead to impaired cell function, cell death, or genetic abnormalities.

RADIOACTIVE FALLOUT

Radioactive fallout is material produced by a nuclear explosion or a nuclear reactor accident that enters the atmosphere and eventually falls to Earth. This fallout consists of minute, radioactive particles of dust, soil, and other debris. While some fallout results from natural sources, the term is usually used in reference to radioactive particles that were released into the atmosphere by a nuclear explosion or reactor accident. Fallout refers to material that has fallen to Earth, and also to material that is still suspended in the atmosphere.

Types of Fallout

Particles that make up radioactive fallout can be as small as the invisible droplets produced by an aerosol spray can, or as large as ash that falls close to wood fire. The type of radioactivity in fallout depends on the nature of the nuclear reaction that emitted the particles into the atmosphere. More than 60 different types of radioactive substances may be initially present in fallout. Some of these decay into non-radioactive products in seconds, while others take centuries or longer to become non-radioactive. It takes 28 years, for a sample of strontium-90 to lose one-half of its initial radioactivity. Strontium-90 is one of the most dangerous elements in fallout because it is treated by the metabolism of humans in the same manner as calcium, an important component of bone. If animals or humans eat food contaminated with strontium-90, it will accumulate in their bodies. Other particularly harmful products in fallout include cesium-134, cesium-137, and iodine-131.Radiation damages and kills cells in the body. Large doses of radiation can result in burns, vomiting, and damage to the nervous system, digestive system, and bone marrow. Smaller doses can cause genetic mutations and cancer years after exposure.

MARINE POLLUTION

Marine pollution occurs when harmful effects, or potentially harmful effects, can result from the entry into the ocean of chemicals, particles, industrial, agricultural and residential waste, or the spread of invasive organisms. Most sources of marine pollution are land based. The pollution often comes from nonpoint sources such as agricultural runoff and wind blown debris. Many potentially toxic chemicals adhere to tiny particles which are then taken up by plankton and benthos animals, most of which are either deposit or filter feeders. In this way, the toxins are concentrated upward within ocean food chains. Many particles combine chemically in a manner highly depletive of oxygen, causing estuaries to become anoxic.

When pesticides are incorporated into the marine ecosystem, they quickly become absorbed into marine food webs. Once in the food webs, these pesticides can cause mutations, as well as diseases, which can be harmful to humans as well as the entire food web. Toxic metals can also be introduced into marine food webs. These can cause a change to tissue matter, biochemistry, behaviour, reproduction, and suppress growth in marine life. Also, many animal feeds have a high fish meal or fish hydrolysate content. In this way, marine toxins can be transferred to land animals, and appear later in meat and dairy products.

Pathways of Pollution

Bjorn Jenssen, "Anthropogenic pollution may reduce biodiversity and productivity of marine ecosystems, resulting in reduction and depletion of human marine food resources" . There are many different ways to categorize, and examine the inputs of pollution into our marine ecosystems. Patin (n.d.) notes that generally there are three main types of inputs of pollution into the ocean: direct discharge of waste into the oceans, runoff into the waters due to rain, and pollutants that are released from the atmosphere. One common path of entry by contaminants to the sea are rivers. The Hudson in New York State and the Raritan in New Jersey, which empty at the northern and southern ends of Staten Island, are a source of mercury contamination of zooplankton (copepods) in the open ocean.

The highest concentration in the filter-feeding copepods is not at the mouths of these rivers but 70 miles south, nearer Atlantic City, because water flows close to the coast. It takes a few days before toxins are taken up by the plankton. Pollution is often classed as point source or nonpoint source pollution. Point source pollution occurs when there is a single, identifiable, and localized source of the pollution. An example is directly discharging sewage and industrial waste into the ocean. Pollution such as this occurs particularly in developing nations. Nonpoint source pollution occurs when the pollution comes from ill-defined and diffuse sources. These can be difficult to regulate. Agricultural runoff and wind blown debris are prime examples.

Pollution from Ships

Ships can pollute waterways and oceans in many ways. Oil spills can have devastating effects. While being toxic to marine life, polycyclic aromatic hydrocarbons (PAHs), the components in crude oil, are very difficult to clean up, and last for years in the sediment and marine environment. Discharge of cargo residues from bulk carriers can pollute ports, waterways and oceans. In many instances vessels intentionally discharge illegal wastes despite foreign and domestic regulation prohibiting such actions. Ships create noise pollution that disturbs natural wildlife, and water from ballast tanks can spread harmful algae and other invasive species.

Meinesz believes that one of the worst cases of a single invasive species causing harm to an ecosystem can be attributed to a seemingly harmless jellyfish. Mnemiopsis leidyi, a species of comb jellyfish that spread so it now inhabits estuaries in many parts of the world. It was first introduced in 1982, and thought to have been transported to the Black Sea in a ship's ballast water.

The population of the jellyfish shot up exponentially and, by 1988, it waswreaking havoc upon the local fishing industry. "The anchovy catch fell from 204,000 tonnes in 1984 to 200 tons in 1993; sprat from 24,600 tonnes in 1984 to 12,000 tonnes in 1993; horse mackerel from 4,000 tonnes in 1984 to zero in 1993." Now that the jellyfish have exhausted the zooplankton, including fish larvae, their numbers have fallen dramatically, yet they continue to maintain a stranglehold on the ecosystem. Invasive species can take over once occupied areas, facilitate the spread of new diseases, introduce new genetic material, alter underwater seascapes and jeopardize the ability of native species to obtain food. Invasive species are responsible for about $138 billion annually in lost revenue and management costs in the US alone.

PLASTIC DEBRIS

Marine debris is mainly discarded human rubbish which floats on, or is suspended in the ocean. Eighty per cent of marine debris is plastic - a component that has been rapidly accumulating since the end of World War II.

The mass of plastic in the oceans may be as high as one hundred million metric tons. Discarded plastic bags, six pack rings and other forms of plastic waste which finish up in the ocean present dangers to wildlife and fisheries.

Aquatic life can be threatened through entanglement, suffocation, and ingestion. Fishing nets, usually made of plastic, can be left or lost in the ocean by fishermen. Known as ghost nets, these entangle fish, dolphins, sea turtles, sharks, dugongs, crocodiles, seabirds, crabs, and other creatures, restricting movement, causing starvation, laceration and infection, and, in those that need to return to the surface to breathe, suffocation.

Many animals that live on or in the sea consume flotsam by mistake, as it often looks similar to their natural prey. Plastic debris, when bulky or tangled, is difficult to pass, and may become permanently lodged in the digestive tracts of these animals, blocking the passage of food and causing death through starvation or infection. Plastics accumulate because they don't biodegrade in the way many other substances do. They will photodegrade on exposure to the sun, but they do so properly only under dry conditions, and water inhibits this process.

In marine environments, photodegraded plastic disintegrates into ever smaller pieces while remaining polymers, even down to the molecular level. When floating plastic particles photodegrade down to zooplankton sizes, jellyfish attempt to consume them, and in this way the plastic enters the ocean food chain. Many of these long-lasting pieces end up in the stomachs of marine birds and animals, including sea turtles, and black-footed albatross.

Plastic debris tends to accumulate at the centre of ocean gyres. In particular, the Great Pacific Garbage Patch has a very high level of plastic particulate suspended in the upper water column. In samples taken in 1999, the mass of plastic exceeded that of zooplankton (the dominant animal life in the area) by a factor of six. Toxic additives used in the manufacture of plastic materials can leach out into their surroundings when exposed to water.

Waterborne hydrophobic pollutants collect and magnify on the surface of plastic debris, thus making plastic far more deadly in the ocean than it would be on land. Hydrophobic contaminants are also known to bioaccumulate in fatty tissues, biomagnifying up the food chain and putting pressure on apex predators.

Some plastic additives are known to disrupt the endocrine system when consumed, others can suppress the immune system or decrease reproductive rates. Floating debris can also absorb persistent organic pollutants from seawater, including PCBs, DDT and PAHs. Aside from toxic effects, when ingested some of these are mistaken by the animal brain for estradiol, causing hormone disruption in the affected wildlife.

TOXINS

Apart from plastics, there are particular problems with other toxins that do not disintegrate rapidly in the marine environment. Examples of persistent toxins are PCBs, DDT, pesticides, furans, dioxins and phenols. Heavy metals are metallic chemical elements that have a relatively high density and are toxic or poisonous at low concentrations. Examples are mercury, lead, nickel, arsenic and cadmium.

Such toxins can accumulate in the tissues of many pecies of aquatic life in a process called bioaccumulation. They are also known to accumulate in benthic environments, such as estuaries and bay muds: a geological record of human activities of the last century.

Eutrophication

Eutrophication is an increase in chemical nutrients, typically compounds containing nitrogen or phosphorus, in an ecosystem. It can result in an increase in the ecosystem's primary productivity (excessive plant growth and decay), and further effects including lack of oxygen and severe reductions in water quality, fish, and other animal populations. The biggest culprit are rivers that empty into the ocean, and with it the many chemicals used as fertilizers in agriculture as well as waste from livestock and humans.

An excess of oxygen depleting chemicals in the water can lead to hypoxia and the creation of a dead zone. Estuaries tend to be naturally eutrophic because land-derived nutrients are concentrated where runoff enters the marine environment in a confined channel. The World Resources Institute has identified 375 hypoxic coastal zones around the world, concentrated in coastal areas in Western Europe, the Eastern and Southern coasts of the US, and East Asia, particularly in Japan. In the ocean, there are frequent red tide algae blooms that kill fish and marine mammals and cause respiratory problems in humans and some domestic animals when the blooms reach close to shore.

In addition to land runoff, atmospheric anthropogenic fixed nitrogen can enter the open ocean. A study in 2008 found that this could account for around one third of the ocean's external (non-recycled) nitrogen supply and up to three per cent of the annual new marine biological production. It has been been suggested that accumulating reactive nitrogen in the environment may have consequences as serious as putting carbon dioxide in the atmosphere.

Acidification

The oceans are normally a natural carbon sink, absorbing carbon dioxide from the atmosphere. Because the levels of atmospheric carbon dioxide are increasing, the oceans are becoming more acidic. The potential consequences of ocean acidification are not fully understood, but there are concerns that structures made of calcium carbonate may become vulnerable to dissolution, affecting corals and the ability of shellfish to form shells.

A report from NOAA scientists published in the journal Science in May 2008 found that large amounts of relatively acidified water are upwelling to within four miles of the Pacific continental shelf area of North America. This area is a critical zone where most local marine life lives or is born. While the paper dealt only with areas from Vancouver to northern California, other continental shelf areas may be experiencing similar effects.

Specific Examples

- Chinese and Russian industrial pollution such as phenols and heavy metals in the Amur River have devastated fish stocks and damaged its estuary soil.
- Wabamun Lake in Alberta, Canada, once the best whitefish lake in the area, now has unacceptable levels of heavy metals in its sediment and fish.
- Acute and chronic pollution events have been shown to impact southern California kelp forests, though the intensity of the impact seems to depend on both the nature of the contaminants and duration of exposure.
- Ballast water taken up at sea and released in port is a major source of unwanted exotic marine life. The invasive freshwater zebra mussels, native to the Black, Caspian and Azov seas, were probably transported to the Great Lakes via ballast water from a transoceanic vessel.
- Due to their high position in the food chain and the subsequent accumulation of heavy metals from their diet, mercury levels can be high in larger species such as bluefin and albacore. As a result, in March 2004 the United States FDA issued guidelines recommending that pregnant women, nursing mothers and children limit their intake of tuna and other types of predatory fish.
- Some shellfish and crabs can survive polluted environments, accumulating heavy metals or toxins in their tissues. For example, mitten crabs have a remarkable ability to survive in highly modified aquatic habitats, including polluted waters. The farming and harvesting of such species needs careful management if they are to be used as a food.
- Mining has a poor environmental track record. For example, United States Environmental Protection Agency, mining has contaminated portions of the headwaters of over 40% of watersheds in the western continental US. Much of this pollution finishes up in the sea.
- Heavy metals enter the environment through oil spills - such as the Prestige oil spill on the Galician coast - or from other natural or anthropogenic sources.

SOLUTIONS

Marine pollution is part of the problem of too much pollution by humans in general. There are only two ways to remedy this: either the human population is reduced, or the ecological footprint left behind by the average human is reduced. If we do not follow the second way, then the first way may be imposed upon us, as world ecosystems falter and cease to support us. The second way is for us, individually, to consume and pollute less than we do currently. For this there must be social and political will, together with a shift in awareness, so more people respect their environment and are less disposed to abuse it.

At an operational level, regulations, and international government participation is needed. It is often very difficult to regulate marine pollution because pollution spreads over international barriers, thus making regulations hard to create as well as enforce.

Perhaps the most important strategy for reducing marine pollution is education. Most are unaware of the sources, and harmful effects of marine pollution, and therefore little is done to address the situation. In order to inform the population of all the facts, in depth research must be done to provide the full scale of the situation.

Then this information must be made public. As expressed in Daoji and Dag's research, one of the reasons why environmental concern is lacking among the Chinese is because the public awareness is low and therefore should be targeted. Likewise, regulation, based upon such in-depth research should be employed.

In California, such regulations have already been put in place to protect Californian coastal waters from agricultural runoff. This includes the California Water Code, as well as several voluntary programs. Similarly, in India, several tactics have been employed that help reduce marine pollution, however, they do not significantly target the problem. In Chennai city, India, sewage has been dumped further into open waters. Due to the mass of waste being deposited, open-ocean is best for diluting, and dispersing pollutants, thus making them less harmful to marine ecosystems.

SOIL CONTAMINATION

Soil contamination is caused by the presence of man-made chemicals or other alteration in the natural soil environment. This type of contamination typically arises from the rupture of underground storage tanks, application of pesticides, percolation of contaminated surface water to subsurface strata, oil and fuel dumping, leaching of wastes from landfills or direct discharge of industrial wastes to the soil. The most common chemicals involved are petroleum hydrocarbons, solvents, pesticides, lead and other heavy metals. This occurrence of this phenomenon is correlated with the degree ofcindustrializations and intensities of chemical usage.

The concern over soil contamination stems primarily from health risks, both of direct contact and from secondary contamination of water supplies. Mapping of contaminated soil sites and the resulting cleanup are time consuming and expensive tasks, requiring extensive amounts of geology, hydrology, chemistry and computer modeling skills.

It is in North America and Western Europe that the extent of contaminated land is most well known, with many of countries in these areas having a legal framework to identify and deal with this environmental problem; this however may well be just the tip of the iceberg with developing countries very likely to be the next generation of new soil contamination cases.

The immense and sustained growth of the People's Republic of China since the 1970s has exacted a price from the land in increased soil pollution. The State Environmental Protection Administration believes it to be a threat to the environment, to food safety and to sustainable agriculture. A scientific sampling, 150 million mi (100,000 square kilometres) of China's cultivated land have been polluted, with contaminated water being used to irrigate a further 32.5 million mi (21,670 square kilometres) and another 2 million mi(1,300 square kilometres) covered or destroyed by solid waste. In total, the area accounts for one-tenth of China's cultivatable land, and is mostly in economically developed areas. An estimated 12 million tonnes of grain are contaminated by heavy metals every year, causing direct losses of 20 billion yuan . .

The United States, while having some of the most widespread soil ontamination, has actually been a leader in defining and implementing standards for cleanup. Other industrialized countries have a large amount of contaminated sites, but lag the U.S. in executing remediation. Developing countries may be leading in the next generation of new soil contamination cases. Each year in the U.S., thousands of sites complete soil contamination cleanup, some by using microbes that "eat up" toxic chemicals in soil, many others by simple excavation and others by more expensive high-tech soil vapour extraction or air stripping. Efforts proceed worldwide to identify new sites of soil contamination.

ECOSYSTEM EFFECTS

Not unexpectedly, soil contaminants can have significant deleterious consequences for ecosystems. There are radical soil chemistry changes which can arise from the presence of many hazardous chemicals even at low concentration of the contaminant species. These changes can manifest in the alteration of metabolism of endemic microorganisms and arthropods resident in a given soil environment. The result can be virtual eradication of some of the primary food chain, which in turn have major consequences for predator or consumer species. Even if the chemical effect on lower life forms is small, the lower pyramid levels of the food chain may ingest alien chemicals, which normally become more concentrated for each consuming rung of the food chain. Many of these effects are now well known, such as the concentration of persistent DDT materials for avian consumers, leading to weakening of egg shells, increased chick mortality and potentially species extinction.

Effects occur to agricultural lands which have certain types of soil contamination. Contaminants typically alter plant metabolism, most commonly to reduce crop yields. This has a secondary effect upon soil conservation, since the languishing crops cannot shield the earth's soil mantle from erosion phenomena. Some of these chemical contaminants have long half-lives and in other cases derivative chemicals are formed from decay of primary soil contaminants.

REGULATORY FRAMEWORK

United States of America

Until about 1970 there was little widespread awareness of the worldwide scope of soil contamination or its health risks. In fact, areas of concern were often viewed as unusual or isolated incidents. Since then, the U.S. has established guidelines for handling hazardous waste and the cleanup of soil pollution. In 1980 the U.S.Superfund/CERCLA established strict rules on legal liability for soil contamination. Not only did CERCLA stimulate identification and cleanup of thousands of sites, but it raised awareness of property buyers and sellers to make soil pollution a focal issue of land use and management practices.

While estimates of remaining soil cleanup in the U.S. may exceed 200,000 sites, in other industrialized countries there is a lag of identification and cleanup functions. Even though their use of chemicals is lower than industrialized countries, often their controls and regulatory framework is quite weak. For example, some persistent pesticides that have been banned in the U.S. are in widespread uncontrolled use in developing countries. It is worth noting that the cost of cleaning up a soil contaminated site can range from as little as about $10,000 for a small spill, which can be simply excavated, to millions of dollars for a widespread event, especially for a chemical that is very mobile such as perchloroethylene.

United Kingdom

Generic guidance commonly used in the UK are the Soil Guideline Values published by DEFRA and the Environment Agency. These are screening values that demonstrate the minimal acceptable level of a substance. Above this there can be no assurances in terms of significant risk of harm to human health.

These have been derived using the Contaminated Land Exposure Asseeement Model (CLEA UK). Certain input parameters such as Health Criteria Values, age and land use are fed into CLEA UK to obtain a probablistic output. Guidance by the Inter Departmental Committee for the Redevelopment of Contaminated Land (ICRCL) has been formally withdrawn by the Department for Environment, Food and Rural Affairs (DEFRA), for use as a prescriptive document to determine the potential need for remediation or further assessment. Therefore, no further reference is made to these former guideline values. Other generic guidance that exists (to put the concentration of a particular contaminant in context), includes the United States EPA Region 9 Preliminary Remediation Goals (US PRGs), the US EPA Region 3 Risk Based Concentrations (US EPA RBCs) and National Environment Protection Council of Australia Guideline on Investigation Levels in Soil and Groundwater. However international guidance should only be used in the UK with clear justification. This is because foreign standards are usually particular to that country due to drivers such as political policy, geology, flood regime and epidemiology. It is generally accepted by UK regulators that only robust scientific methods that relate to the UK should be used.

The CLEA model published by DEFRA and the Environment Agency (EA) in March 2002 sets a framework for the appropriate assessment of risks to human health from contaminated land, as required by Part IIA of the Environmental Protection Act 1990. As part of this framework, generic Soil Guideline Values (SGVs) have currently been derived for ten contaminants to be used as "intervention values". These values should not be considered as remedial targets but values above which further detailed assessment should be considered.

Three sets of CLEA SGVs have been produced for three different land uses, namely:

- Residential (with and without plant uptake).
- Allotments.
- Commercial/industrial.

It is intended that the SGVs replace the former ICRCL values. It should be noted that the CLEA SGVs relate to assessing chronic (long term) risks to human health and do not apply to the protection of ground workers during construction, or other potential receptors such as groundwater, buildings, plants or other ecosystems. The CLEA SGVs are not directly applicable to a site completely covered in hardstanding, as there is no direct exposure route to contaminated soils.

To date, the first ten of fifty-five contaminant SGVs have been published, for the following: arsenic, cadmium, chromium, lead, inorganic mercury, nickel, selenium ethyl benzene, phenol and toluene.

Draft SGVs for benzene, naphthalene and xylene have been produced but their publication is on hold. Toxicological data (Tox) has been published for each of these contaminants as well as for benzo[a]pyrene, benzene, dioxins, furans and dioxin-like PCBs, naphthalene, vinyl chloride, 1,1,2,2 tetrachloroethane and 1,1,1,2 tetrachloroethane, 1,1,1 trichloroethane, tetrachloroethene, carbon tetrachloride, 1,2-dichloroethane, trichloroethene and xylene.

The SGVs for ethyl benzene, phenol and toluene are dependent on the soil organic matter (SOM) content (which can be calculated from the total organic carbon (TOC) content). As an initial screen the SGVs for 1% SOM are considered to be appropriate.

CLEANUP OPTIONS

Cleanup or remediation is analysed by environmental scientists who utilize field measurement of soil chemicals and also apply computer models for analyzing transport and fate of soil chemicals. Thousands of soil contamination cases are currently in active cleanup across the U.S. as of 2006.

There are several principal strategies for remediation:

- Excavate soil and take it to a disposal site away from ready pathways for human or sensitive ecosystem contact. This technique also applies to dredging of bay muds containing toxins.
- Aeration of soils at the contaminated site (with attendant risk of creating air pollution).
- Thermal remediation by introduction of heat to raise subsurface temperatures sufficiently high to volatize chemical contaminants out of the soil for vapour extraction. Technologies include ISTD, electrical resistance heating (ERH), and ET-DSPtm.
- Bioremediation, involving microbial digestion of certain organic chemicals. Techniques used in bioremediation include landfarming, biostimulation and bioaugmentating soil biota with commercially available microflora.
- Extraction of groundwater or soil vapour with an active electromechanical system, with subsequent stripping of the contaminants from the extract.
- Containment of the soil contaminants (such as by capping or paving over in place).

LAND POLLUTION

Land Pollution is the degradation of Earth's land surfaces often caused by human activities and their misuse of land resources. It occurs when waste is not disposed properly.Haphazard disposal of urban and industrial wastes, exploitation of minerals, and improper use of soil by inadequate agricultural practices are a few factors. Urbanization and industrialisation are major causes of land pollution. The Industrial Revolution set a series of events into motion which destroyed natural habitats and polluted the environment, causing diseases in both humans and animals.

CAUSES

The increase in the concentration of population in cities, along with the internal combustion engine, led to the increased amount of roads and all the infra structure that goes with them. Roads cause visual, noise, light, air and water pollution, in addition to land pollution. The visual and noise areas are obvious, however light pollution is becoming more widely recognised as a problem. From outer space, large cities can be picked out at night by the glow of their lighting, so city dwellers seldom experience total darkness.

The contribution of vehicular traffic to air pollution is dealt with in another substance, but, suffice to that sulfur dioxide, nitrogen oxide and carbon monoxide are the main culprits. Water pollution is caused by the run off from roads of oil, salt and rubber residue, which enter the water courses and may make conditions unsuitable for certain organisms to live.

As the demand for food has grown very high, there is an increase in field size and mechanization. The increase in field size makes it economically viable for the farmer but results in loss of habitat and shelter for wildlife, as hedgerows and copses disappear. When crops are harvested, the naked soil is left open to wind after the heavy machinery has compacted it. Another consequence of more intensive agriculture is the move to monoculture. This is unnatural, it depletes the soil of nutrients, allows diseases and pests to spread and, in short, brings into play the use of chemical substances foreign to the environment.

Pesticides

A pesticide is a substance or mixture of substances used to kill a pest. A pesticide may be a chemical substance, biological agent (such as a virus or bacteria), antimicrobial, disinfectant or device used against any pest. Pests include insects, plant pathogens, weeds, molluscs, birds, mammals, fish, nematodes (roundworms) and microbes that compete with humans for food, destroy property, spread or are a vector for disease or cause a nuisance. Although there are benefits to the use of pesticides, there are also drawbacks, such as potential toxicity to humans and other animals. Pesticides are used to control organisms which are considered harmful. For example, they are used to kill mosquitoes that can transmit potentially deadly diseases like west nile virus, yellow fever, and malaria. They can also kill bees, wasps or ants that can cause allergic reactions. Insecticides can protect animals from illnesses that can be caused by parasites such as fleas. Pesticides can prevent sickness in humans that could be caused by mouldy food or diseased produce.

Herbicides

Herbicides are used to kill weeds, especially on pavements and railways. They are similar to auxins and most are biodegradable by soil bacteria.

However one group derived from trinitrophenol (2:4 D and 2:4:5 T) have the impurity dioxin, which is very toxic and causes fatality even in low concentrations. It also causes spontaneous abortions, haemorrhaging and cancer. Agent Orange (50% 2:4:5 T) was used as a defoliant in Vietnam. Eleven million gallons were used and children born since then to American soldiers who served in this conflict, have shown increased physical and mental disabilities compared to the rest of the population. It affects the head of the sperm and the chromosomes inside it.

Another herbicide, much loved by murder story writers, is Paraquat. It is highly toxic but it rapidly degrades in soil due to the action of bacteria and does not kill soil fauna.

Insecticides

Insecticides are used to rid farms of pests which damage crops. The insects damage not only standing crops but also stored ones and in the tropics it is reckoned that one third of the total production is lost during food storage. As with fungicides, the first insecticides used in the nineteenth century were inorganic e.g. Paris Green and other compounds of arsenic. Nicotine has also been used since the late eighteenth century. There are now two main groups of synthetic insecticides.

Organochlorines

Organochlorines include DDT, Aldrin, Dieldrin and BHC. They are cheap to produce, potent and persistent. DDT was used on a massive scale from the 1930s, with a peak of 72,000 tonnes used 1970. Then usage fell as the harmful environmental effects were realized. It was found worldwide in fish and birds and was even discovered in the snow in the Antarctic. It is only slightly soluble in water but is very soluble in the bloodstream. It affects the nervous and endocrine systems and causes the eggshells of birds to lack calcium causing them to be easily breakable. It is thought to be responsible for the decline of the numbers of birds of prey like ospreys and peregrine falcons in the 1950s - they are now recovering. As well as increased concentration via the food chain, it is known to enter via permeable membranes, so fish get it through their gills.

As it has low water solubility, it tends to stay at the water surface, so organisms that live there are most affected. DDT found in fish that formed part of the human food chain caused concern, but the levels found in the liver, kidney and brain tissues was less than 1ppm and in fat was 10 ppm which was below the level likely to cause harm. However, DDT was banned in Britain and America to stop the further build up of it in the food chain.

The USA exploited this ban and sold DDT to developing countries, who could not afford the expensive replacement chemicals and who did not have such stringent regulations governing the use of pesticides. Some insects have developed a resistance to insecticides - e.g. the which carries malaria.

Organophosphates

Organophosphates, e.g. parathion, methyl parathion and about 40 other insecticides are available nationally. Parathion is highly toxic, methyl-parathion is less so and Malathion is generally considered safe as it has low toxicity and is rapidly broken down in the mammalian liver. This group works by preventing normal nerve transmission as cholinesterase is prevented from breaking down the transmitter substance acetylcholine, resulting in uncontrolled muscle movements. Entry of a variety of pesticides into our water supplies causes concern to environmental groups, as in many cases the long term effects of these specific chemicals is not known.

Restrictions came into force in July 1985 and were so frequently broken that in 1987, formal proceedings were taken against the British government. Britain is still the only European state to use Aldrin and organochlorines, although it was supposed to stop in 1993. East Anglia has the worst record for pesticide contamination of drinking water. Of the 350 pesticides used in Britain, only 50 can be analysed, which is worrying for the global community.

MINING

Modern mining projects leave behind distrupted communities, damages landscapes, and polluted water. Mining also affects ground and surface waters, the aquatic life, vegetation, soils, animals, and the human health. Acid mine drainage can cause damage to streams which in return can kill aquatic life. The vast variety of toxic chemicals released by mining activities can harm animals and aquatic life as well as their habitat. The average mine disturbs over a thousand acres of land.

BURIAL

Burial is the technique used by Jews, Muslims, Christians and other religions with Abrahamic influence, to dispose off the corpse of dead humans and animals. This process leads to regular soil erosion due to loosening of soil. Also, the decomposing fluids act as poisonous herbicides, pesticides and may even lead to epidemics in surrounding areas. It leads to soil pollution, soil erosion and even water pollution.

CONSTRUCTION

Construction often puts sediments in rivers and bodies of water. By doing this, natural water filters are damaged. Natural water filters help break down many pollutants before they reach other water bodies. Some harmful chemicals that may run off with water and sediments from construction sites are oils, debris, and paint. This can cause damage to soil, aquatic life, and promote hazordous chemicals to get into drinking water. California Integrated Waste Management Board provides more "green building basics" that educates readers about healthy construction.

Increased Waste Disposal

In Scotland in 1993, 14 million tons of waste was produced. 100,000 tons was special waste and 260,000 tons was controlled waste from other parts of Britain and abroad. 45% of the special waste was in liquid form and 18% was asbestos - radioactive waste was not included. Of the controlled waste, 48% came from the demolition of buildings, 22% from industry, 17% from households and 13% from business - only 3% were recycled. 90% of controlled waste was buried in landfill sites and produced 2 million tons of methane gas. 1.5% was burned in incinerators and 1.5% were exported to be disposed of or recycled. There are 748 disposal sites in Scotland. Landfills produce leachate, which has to be recycled to keep favourable conditions for microbial activity, methane gas and some carbon dioxide. There are very few vacant or derelict land sites in the north east of Scotland, as there are few traditional heavy industries or coal/mineral extraction sites. However some areas are contaminated by aromatic hydrocarbons (500 cubic meters). The Urban Waste

Water Treatment Directive allows sewage sludge to be sprayed onto land and the volume is expected to double to 185,000 tons of dry solids in 2005. This has good agricultural properties due to the high nitrogen and phosphate content. In 1990/1991, 13% wet weight was sprayed onto 0.13% of the land, however this is expected to rise 15 fold by 2005. There is a need to control this so that pathogenic microorganisms do not get into water courses and to ensure that there is no accumulation of heavy metals in the top soil.

Impacts

A large percentage of material that is dumped into waters not only destroys habitats for animals but for humans as well. A list of some impacts of land pollution. This list was retrieved from the Victorian Litter Action Alliancewebsite.

- Smothered habitats, where aquatic animals feed and live, thus driving them out of a home.
- Reduced oxygen levels from decomposition of organic material dumped from drains.
- Reduced light penetration in the water column, from sediments and excessive algal growth, leading to oxygen depleted water.
- Massively changed flow regimes, large fast flows can scour existing habitat and wash species downstream.
- Increased human health risks, from syringes, broken glass and cans washing onto creek banks and beaches.
- Reduced aesthetic appeal of areas from litter strewn on banks and beaches, thus reducing recreation and tourist appeal.
- Acid rain damages trees and other plants.

Reduce Land Pollution

There are many ways to help reduce land pollution:

- Use reusable materials.
- Do not litter.
- Recycle.
- Purchase products with little packaging to throw away (buy in bulk if possible).
- Use safer alternative pesticides.
- Buy organic grown fruits.
- Buy biogradeable products.

ACID RAIN

Acid rain is rain or any other form of precipitation that is unusually acidic. It has harmful effects on plants, aquatic animals, and infrastructure. Acid rain is mostly caused by human emissions of sulfur and nitrogen compounds which react in the atmosphere to produce acids. In recent years, many governments have introduced laws to reduce these emissions.

"Acid rain" is a popular term referring to the deposition of wet (rain, snow, sleet, fog and cloudwater, dew) and dry (acidifying particles and gases) acidic components. A more accurate term is "acid deposition".

Distilled water, which contains no carbon dioxide, has a neutral pH of 7. Liquids with a pH less than 7 are acidic, and those with a pH greater than 7 are bases. "Clean" or unpolluted rain has a slightly acidic pH of about 5.2, because carbon dioxide and water in the air react together to form carbonic acid, a weak acid but unpolluted rain also contains other chemicals.

$$H_2O\ (l) + CO_2\ (g) \rightarrow H_2CO_3\ (aq)$$

Carbonic acid then can ionize in water forming low concentrations of hydronium and carbonate ions:

$$2\ H_2O\ (l) + H_2CO_3\ (aq) \rightarrow CO_{32}-\ (aq) + 2\ H_3O+\ (aq)$$

EMISSIONS OF CHEMICALS LEADING TO ACIDIFICATION

The most important gas which leads to acidification is sulfur dioxide. Emissions of nitrogen oxides which are oxidized to form nitric acid are of increasing importance due to stricter controls on emissions of sulfur containing compounds. 70 Tg(S) per year in the form of SO_2 comes from fossil fuel combustion and industry, 2.8 Tg(S) from wildfires and 7-8 Tg(S) per year from volcanoes.

Natural Phenomena

The principal natural phenomena that contribute acid-producing gases to the atmosphere are emissions from volcanoes and those from biological processes that occur on the land, in wetlands, and in the oceans. The major biological source of sulfur containing compounds is dimethyl sulfide.

Acidic deposits have been detected in glacial ice thousands of years old in remote parts of the globe.

Human Activity

The principal cause of acid rain is sulfur and nitrogen compounds from human sources, such as electricity generation, factories, and motor vehicles. Coal power plants are one of the most polluting. The gases can be carried hundreds of kilometres in the atmosphere before they are converted to acids and deposited. In the past, factories had short funnels to let out smoke, but this caused many problems locally; thus, factories now have taller smoke funnels. However, dispersal from these taller stacks causes pollutants to be carried farther, causing widespread ecological damage.

CHEMICAL PROCESSES

Gas Phase Chemistry

In the gas phase sulfur dioxide is oxidized by reaction with the hydroxyl radical via an intermolecular reaction:

$$SO_2 + OH \rightarrow HOSO_2$$

which is followed by:

$$HOSO_2 + O_2 \rightarrow HO_2 + SO_3$$

In the presence of water, sulfur trioxide (SO_3) is converted rapidly to sulfuric acid:

$$SO_3\ (g) + H_2O\ (l) \rightarrow H_2SO_4\ (l)$$

Nitric acid is formed by the reaction of OH with nitrogen dioxide:

$$NO_2 + OH \rightarrow HNO_3$$

Chemistry in Cloud Droplets

When clouds are present, the loss rate of SO2 is faster than can be explained by gas phase chemistry alone. This is due to reactions in the liquid water droplets.

Hydrolysis: Sulfur dioxide dissolves in water and then, like carbon dioxide, hydrolyses in a series of equilibrium reactions.

$$SO_2\ (g) + H_2O \rightarrow SO_2H_2O$$

$$SO_2\text{-}H_2O \rightarrow H_+ + HSO_{3-}$$

$$HSO_{3-} \rightarrow H_+ + SO_{32-}$$

Oxidation: There are a large amount of aqueous reactions that oxidize sulfur from S(IV) to S(VI), leading to the formation of sulfuric acid. The most important oxidation reactions are with ozone, hydrogen peroxide and oxygen.

ACID DEPOSITION

Processes involved in acid deposition (note that only SO2 and NOx play a significant role in acid rain).

Wet Deposition

Wet deposition of acids occurs when any form of precipitation (rain, snow, etc.) removes acids from the atmosphere and delivers it to the Earth's surface. This can result from the deposition of acids produced in the raindrops or by the precipitation removing the acids either in clouds or below clouds. Wet removal of both gases and aerosols are both of importance for wet deposition.

Dry Deposition

Acid deposition also occurs via dry deposition in the absence of precipitation. This can be responsible for as much as 20 to 60% of total acid deposition. This occurs when particles and gases stick to the ground, plants or other surfaces.

ADVERSE EFFECTS

Acid rain has been shown to have adverse impacts on forests, freshwaters and soils, killing insect and aquatic life-forms as well as causing damage to buildings and having impacts on human health.

Surface Waters and Aquatic Animals

Both the lower pH and higher aluminum concentrations in surface water that occur as a result of acid rain can cause damage to fish and other aquatic animals. At pHs lower than 5 most fish eggs will not hatch and lower pHs can kill adult fish. As lakes and rivers become more acidic biodiversity is reduced.

Acid rain has eliminated insect life and some fish species, including the brook trout in some lakes, streams, and creeks in geographically sensitive areas, such as the Adirondack Mountains of the United States. However, the extent to which acid rain contributes directly or indirectly via runoff from the catchment to lake and river acidity (i.e., depending on characteristics of the surrounding watershed) is variable. The United States Environmental Protection Agencyý's (EPA) web site states: "Of the lakes and streams surveyed, acid rain caused acidity in 75 per cent of the acidic lakes and about 50 per cent of the acidic streams".

Soils

Soil biology and chemistry can be seriously damaged by acid rain. Some microbes are unable to tolerate changes to low pHs and are killed. The enzymes of these microbes are denatured (changed in shape so they no longer function) by the acid. The hydronium ions of acid rain also mobilize toxins such as aluminium, and leach away essential nutrients and minerals such as magnesium.

$$2\ H_{+}\ (aq) + Mg2+\ (clay) \rightarrow 2\ H_{+}\ (clay) + Mg2+\ (aq)$$

Soil chemistry can be dramatically changed when base cations, such as calcium and magnesium, are leached by acid rain thereby affecting sensitive species, such as sugar maple (Acer saccharum).

Forests and Other Vegetation

Adverse effects may be indirectly related to acid rain, like the acid's effects on soil or high concentration of gaseous precursors to acid rain. High altitude forests are especially vulnerable as they are often surrounded by clouds and fog which are more acidic than rain. Other plants can also be damaged by acid rain but the effect on food crops is minimized by the application of lime and fertilizers to replace lost nutrients. In cultivated areas, limestone may also be added to increase the ability of the soil to keep the pH stable, but this tactic is largely unusable in the case of wilderness lands. When calcium is leached from the needles of red spruce, these trees become less cold tolerant and exhibit winter injury and even death.

Human Health

Scientists have suggested direct links to human health. Fine particles, a large fraction of which are formed from the same gases as acid rain (sulfur dioxide and nitrogen dioxide), have been shown to cause illness and premature deaths such as cancer and other diseases.

Other Adverse Effects

Acid rain can also cause damage to certain building materials and historical monuments. This results when the sulfuric acid in the rain chemically reacts with the calcium compounds in the stones (limestone, sandstone, marble and granite) to create gypsum, which then flakes off.

$$CaCO_3\ (s) + H_2SO_4\ (aq) \rightarrow CaSO_4\ (aq) + CO_2\ (g) + H_2O\ (l)$$

This result is also commonly seen on old gravestones where the acid rain can cause the inscription to become completely illegible. Acid rain also causes an increased rate of oxidation for iron. Visibility is also reduced by sulfate and nitrate aerosols and particles in the atmosphere.

AFFECTED AREAS

Particularly badly affected places around the globe include most of Europe (particularly Scandinavia with many lakes with acidic water containing no life and many trees dead) many parts of the United States (states like New York are very badly affected) and South Western Canada. Other affected areas include the South Eastern coast of China and Taiwan.

Potential Problem Areas in the Future

Places like much of South Asia (Indonesia, Malaysia and Thailand), Western South Africa (the country), Southern India and Sri Lanka and even West Africa (countries like Ghana, Togo and Nigeria) could all be prone to acidic rainfall in the future.

PREVENTION METHODS

Technical Solutions

In the United States, many coal-burning power plants use Flue gas desulfurization (FGD) to remove sulfur-containing gases from their stack gases.

An example of FGD is the wet scrubber which is commonly used in the U.S. and many other countries. A wet scrubber is basically a reaction tower equipped with a fan that extracts hot smoke stack gases from a power plant into the tower. Lime or limestone in slurry form is also injected into the tower to mix with the stack gases and combine with the sulfur dioxide present.

The calcium carbonate of the limestone produces pH-neutral calcium sulfate that is physically removed from the scrubber. That is, the scrubber turns sulfur pollution into industrial sulfates. In some areas the sulfates are sold to chemical companies as gypsum when the purity of calcium sulfate is high.

In others, they are placed in landfill. However, the effects of acid rain can last for generations, as the effects of pH level change can stimulate the continued leaching of undesirable chemicals into otherwise pristine water sources, killing off vulnerable insect and fish species and blocking efforts to restore native life. Automobile emissions control reduces emissions of nitrogen oxides from motor vehicles.

International Treaties

A amount of international treaties on the long range transport of atmospheric pollutants have been agreed e.g. Sulphur Emissions Reduction Protocol under the Convention on Long-Range Transboundary Air Pollution.

Emissions Trading

Every current polluting facility is given or may purchase on an open market an emissions allowance for each unit of a designated pollutant it emits. Operators can then install pollution control equipment, and sell portions of their emissions allowances they no longer need for their own operations, thereby recovering some of the capital cost of their investment in such equipment. The intention is to give operators economic incentives to install pollution controls. The first emissions trading market was established in the United States by enactment of the Clean Air Act Amendments of 1990. The overall goal of the Acid Rain Programmes established by the Act is to achieve significant environmental and public health benefits through reductions in emissions of sulfur dioxide (SO_2) and nitrogen oxides (NOx), the primary causes of acid rain. To achieve this goal at the lowest cost to society, the programmes employs both regulatory and market based approaches for controlling air pollution.

❐

3

Biodegradation of Waste

BIODEGRADATION OF WASTE

Biodegradation is nature's way of recycling wastes, or breaking down organic matter into nutrients that can be used by other organisms.

"Degradation" means decay, and the "bio-" prefix means that the decay is carried out by a huge assortment of bacteria, fungi, insects, worms, and other organisms that eat dead material and recycle it into new forms.

In nature, there is no waste because everything gets recycled. The waste products from one organism become the food for others, providing nutrients and energy while breaking down the waste organic matter. Some organic materials will break down much faster than others, but all will eventually decay. By harnessing these natural forces of biodegradation, people can castes and clean up some types of environmental contaminants. Through composting, we accelerate natural biodegradation and convert organic wastes to a valuable resource.

Wastewater treatment also accelerates natural forces of biodegradation. In this case the purpose is to break down organic matter so that it will not cause pollution problems when the water is released into the environment. Through bioremediation, microorganisms are used to clean up oil spills and other types of organic pollution. Composting and bioremediation provide many possibililites for student research.

Wastewater Treatment

Wastewater treatment uses microbes to decompose organic matter in sewage. If too much untreated sewage or other organic matter is added to a lake or stream, dissolved oxygen levels will drop too low to support sensitive species of fish and other aquatic life. Wastewater treatment systems are designed to digest much of the organic matter before the wastewater is released so that this will not occur. Treatment systems use physical, chemical, and biological processes:

Primary treatment physically removes large solids using grates, screens, and settling tanks.

- Secondary treatment promotes growth of bacteria and other microbes that break down the organic wastes. These biodegradation processes also take place in streams, lakes, and oceans, but the purification systems in nature can easily be overloaded with input of too much organic waste. Secondary treatment prevents this type of pollution by degrading most of the organic matter before the water is released into the environment.
- Tertiary treatment is used only where it is needed to protect the receiving waters from excess nutrients. In tertiary treatment, the concentrations of phosphorus or nitrogen are reduced through biological or chemical processes.
- Disinfection kills disease-causing organisms, most commonly through chlorination.

Sludge, the collection of solids that are removed during wastewater treatment, requires processing to reduce odour and water content. Depending on the disposal method, the sludge also may undergo treatment to decompose organic matter or kill disease-causing organisms.

Wastewater Treatment Principles

Sewage is the wastewater released by residences, businesses and industries in a community. It is 99.94 per cent water, with only 0.06 per cent of the wastewater dissolved and suspended solid material. The cloudiness of sewage is caused by suspended particles which in untreated sewage ranges from 100 to 350 mg/l. A measure of the strength of the wastewater is biochemical oxygen demand, or BOD5. The BOD5 measures the amount of oxygen microorganisms require in five days to break down sewage. Untreated sewage has a BOD5 ranging from 100 mg/l to 300 mg/l. Pathogens or disease-causing organisms are present in sewage. Coliform bacteria are used as an indicator of disease-causing organisms. Sewage also contains nutrients (such as ammonia and phosphorus), minerals, and metals.

Ammonia can range from 12 to 50 mg/l and phosphorus can range from 6 to 20 mg/l in untreated sewage. Sewage treatment is a multi-stage process to renovate wastewater before it reenters a body of water, is applied to the land or is reused. The goal is to reduce or remove organic matter, solids, nutrients, disease-causing organisms and other pollutants from wastewater.

Each receiving body of water has limits to the amount of pollutants it can receive without degradation. Therefore, each sewage treatment plant must old a permit listing the allowable levels of BOD5, suspended solids, coliform bacteria and other pollutants. The discharge permits are called NPDES permits which stands for the National Pollutant Discharge Elimination System.

Preliminary Treatment

Preliminary treatment to screen out, grind up, or separate debris is the first step in wastewater treatment. Sticks, rags, large food particles, sand, gravel, toys, etc., are removed at this stage to protect the pumping and other equipment in the treatment plant. Treatment equipment such as bar screens, comminutors (a large version of a garbage disposal), and grit chambers are used as the wastewater first enters a treatment plant. The collected debris is usually disposed of in a landfill.

Primary Treatment

Primary treatment is the second step in treatment and separates suspended solids and greases from wastewater. Waste-water is held in a quiet tank for several hours allowing the particles to settle to the bottom and the greases to float to the top. The solids drawn off the bottom and skimmed off the top receive further treatment as sludge. The clarified wastewater flows on to the next stage of wastewater treatment. Clarifiers and septic tanks are usually used to provide primary treatment.

Secondary Treatment

Secondary treatment is a biological treatment process to remove dissolved organic matter from wastewater. Sewage microorganisms are cultivated and added to the wastewater. The microorganisms absorb organic matter from sewage as their food supply. Three approaches are used to accomplish secondary treatment; fixed film, suspended film and lagoon systems.

Fixed Film Systems

Fixed film systems grow microorganisms on substrates such as rocks, sand or plastic. The wastewater is spread over the substrate, allowing the wastewater to flow past the film of microorganisms fixed to the substrate. As organic matter and nutrients are absorbed from the wastewater, the film of microorganisms grows and thickens. Trickling filters, rotating biological contactors, and sand filters are examples of fixed film systems.

Suspended Film Systems

Suspended film systems stir and suspend microorganisms in wastewater. As the microorganisms absorb organic matter and nutrients from the wastewater they grow in size and number. After the microorganisms have been suspended in the wastewater for several hours, they are settled out as a sludge.

Some of the sludge is pumped back into the incoming wastewater to provide "seed" microorganisms. The remainder is wasted and sent on to a sludge treatment process. Activated sludge, extended aeration, oxidation ditch, and sequential batch reactor systems are all examples of suspended film systems.

Lagoon Systems

Lagoon systems are shallow basins which hold the waste-water for several months to allow for the natural degradation of sewage. These systems take advantage of natural aeration and microorganisms in the wastewater to renovate sewage.

Final Treatment

Final treatment focuses on removal of disease-causing organisms from wastewater. Treated wastewater can be disinfected by adding chlorine or by using ultraviolet light. High levels of chlorine may be harmful to aquatic life in receiving streams. Treatment systems often add a chlorine-neutralizing chemical to the treated wastewater before stream discharge.

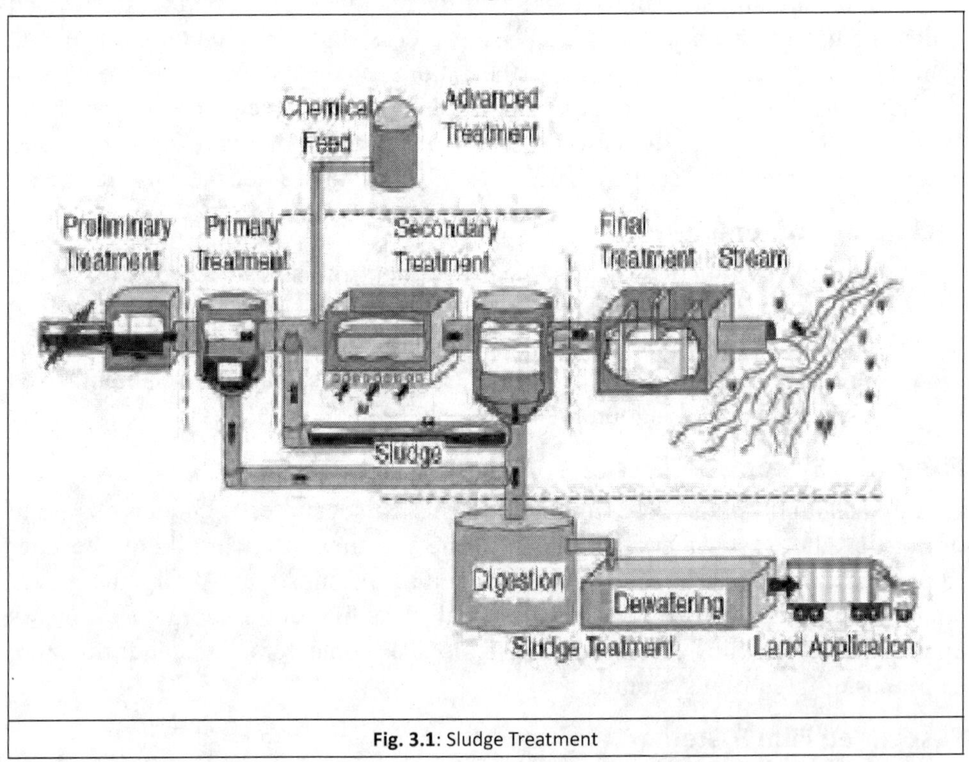

Fig. 3.1: Sludge Treatment

Advanced Treatment

Advanced treatment is necessary in some treatment systems to remove nutrients from wastewater. Chemicals are sometimes added during the treatment process to help settle out or strip out phosphorus or nitrogen. Some examples of nutrient removal systems include coagulant addition for phosphorus removal and air stripping for ammonia removal.

Sludges

Sludges are generated through the sewage treatment process. Primary sludges, material that settles out during primary treatment, often have a strong odour and require treatment prior to disposal. Secondary sludges are the extra microorganisms from the biological treatment processes. The goals of sludge treatment are to stabilize the sludge and reduce odours, remove some of the water and reduce volume, decompose some of the organic matter and reduce volume, kill disease causing organisms and disinfect the sludge. Untreated sludges are about 97 per cent water.

Settling the sludge and decanting off the separated liquid removes some of the water and reduces the sludge volume. Settling can result in a sludge with about 96 to 92 per cent water. More water can be removed from sludge by using sand drying beds, vacuum filters, filter presses, and centrifuges resulting in sludges with between 80 to 50 per cent water.

This dried sludge is called a sludge cake. Aerobic and anaerobic digestion are used to decompose organic matter to reduce volume. Digestion also stabilizes the sludge to reduce odours.

Caustic chemicals can be added to sludge or it may be heat treated to kill disease-causing organisms. Following treatment, liquid and cake sludges are usually spread on fields, returning organic matter and nutrients to the soil. Wastewater treatment processes require careful management to ensure the protection of the water body that receives the discharge. Trained and certified treatment plant operators measure and monitor the incoming sewage, the treatment process and the final effluent.

Wastewater Treatment Regulations

Clean water has been a concern nationwide Congress adopted the Clean Water Act to protect the waters of the nation. Through this act. Environmental Protection Agency and corresponding state agencies were given the responsibility to regulate activities that threaten the quality of the nation's water resources.

In the Federal Clean Water Act, Congress adopted a comprehensive water policy for the nation and set as a national goal the elimination of pollutant discharges to the navigable waters. An interim goal was set to insure that all navigable waters would be fishable and swimmable

To reach these goals:

- No one has the right to pollute the navigable waters of the United States. Dischargers are required to obtain permits.
- Permits shall set limits on the concentration of the pollutants being discharged. A violation of the limits carries a penalty of fines or imprisonment.
- The best technology available shall be used to control the discharge of pollutants.

Each state, through a designated regulatory agency, issues discharge permits and enforces the discharge limits.

The Ohio Water Quality Standards specify that all surface waters of the state must be free from the following pollutants as a result of human activity: suspended solids, floating debris, colour, odour, toxic substances and nutrients that create nuisance growths of weeds and algae.

Permits for wastewater treatment systems serving single family homes, two- or three-family dwellings are issued by local health departments. Permits for all wastewater treatment systems serving more than 2 homes or buildings larger than a 3-family dwelling are issued by the Ohio Environmental Protection Agency.

Two different permits are issued by the Ohio Environmental Protection Agency. All systems must first obtain a Permit to Install (PTI) before a treatment plant can be constructed. The Permit to Install is issued after the plans for construction, operation and management are examined to insure that the receiving stream will be protected.

If the discharge from the treatment plant enters a stream, a NPDES permit is required. The NPDES permit specifies the maximum allowable level of total suspended solids, biochemical oxygen demand, nutrients and bacteria that can be discharged to a stream as well as the minimum level of dissolved oxygen that must be present in the discharge. The levels specified in the NPDES permit are determined by the condition of the receiving stream. Therefore, NPDES permits are subject to change every 5 years as water quality concerns change throughout Ohio. Discharging raw sewage from a dwelling is considered a public nuisance. For systems serving a 1-, 2- or 3-family dwelling, the local health department is the responsible regulatory agency. The penalty for creating a public nuisance is a third degree misdemeanor which is punishable by not more than 60 days in jail and/or a $500 fine. In addition, the violator may be required to remove the sewage from public or private property or the waters of the state.

Discharges of pollutants from a multiple dwelling is under the jurisdiction of the Ohio Environmental Protection Agency. Discharges may be liable for civil penalties of up to $10,000 for each day of violation. In addition, criminal penalties can be assessed or up to one year of imprisonment or both. A "Connection Ban" can also be issued by the Ohio Environmental Protection Agency which prohibits the construction or installation of home sewage disposal systems when it is shown that pollutants from new homes will be discharged into the waters of the state.

The Ohio Environmental Protection Agency issued 25 administrative rders totaling $80,000 in fines to Ohio communities for wastewater treatment violations.

This publication may contain pesticide recommendations that are subject to change at any time. These recommendations are provided only as a guide. It is always the pesticide applicator's responsibility, by law, to read and follow all current label directions for the specific pesticide being used. Due to constantly changing labels and product registrations, some of the recommendations given in this writing may no longer be legal by the time you read them. If any information in these recommendations disagrees with the label, the recommendation must be disregarded.

Composting

Composting refers to the process of combining organic materials such as weeds, grass clippings, and food scraps under conditions that enhance the rate at which they decompose. In the United States, organic wastes make up 20% to 40% of what we throw away.

Composting provides an opportunity to keep these organic materials out of landfills and incinerators, using them instead to create a useful end product. Of course, composting isn't as easy as simply throwing organic waste into big piles and waiting for it to decompose. If moist food scraps or grass clippings are placed in a container and left to sit for a week or two, the result is likely to be a stinky mess that attracts flies.

Given the proper conditions, these same organic materials can be composted to produce a material that looks and smells like rich soil and is prized by gardeners. Clearly the second option is preferable, and the key question is how to provide the proper conditions for problem-free composting. The answer lies in paying attention to moisture content, airflow, and the mixture between materials that are high in carbon and those that are high in nitrogen..

Composting takes place at a wide range of scales, from huge commercial or municipal operations to simple backyard bins or heaps. It can also be carried out indoors in containers such as garbage can bioreactors or worm bins. For research purposes, composting can even take place in soda bottle bioreactors, which are small and inexpensive enough to enable students to build their own individualized systems. Once made, compost can be used for gardening projects or for plant growth experiments ranging from nutrient analysis of compost-enriched soils to use of composts to suppress plant diseases.

Compost Ingredients

Garden supply stores and catalogs often sell compost "starters," which supposedly speed up the composting process. Develop a recipe for a compost starter and design a research project to test its effect on the compost temperature profile.

How well do human nutrition concepts apply to compost microorganisms? The microbes get a "sugar high," demonstrated by a quick, high temperature peak when fed sugary foods, compared with a longer but lower peak for more complex carbohydrates.

Measure the pH of a number of different compost mixes. How does the pH of initial ingredients affect the pH of finished compost?

Some instructions call for adding lime to increase the pH when compost ingredients are mixed. Other instructions caution to avoid this because it causes a loss of nitrogen. How does adding various amounts of lime to the initial ingredients affect the pH of finished compost.

Microorganisms

Composting recipes sometimes call for inoculating the pile by mixing in a few handfuls of finished compost. Is there any observable difference in appearance of microbes between systems that have and have not been inoculated.

The pH of the initial compost ingredients affect the populations of microorganisms during composting.

Worm Composting

Do organic wastes in compost break down more readily in the presence of worms than through composting that depends solely on microbial decomposition.

In some experiments, plants have not show increased growth when planted in fresh worm castings. Does aging or "curing" worm castings increase their ability to enhance plant growth? Are there chemical differences between fresh and older worm castings? Should worm compost be mixed with soil before being used to grow plants.

Different food sources affect reproductive and growth rates of red worms (Eisenia fetida). Red worms grow best in wastes with pH between 5.0 and 8.0.

Effects of Compost on Plant Growth

Some leaves, such as those of black walnut or eucalyptus trees, contain chemicals that inhibit growth of other plants. Are these compounds broken down by composting. Finished compost is near neutral pH. Water in which compost has been soaked (often called compost tea) is said to be beneficial to plants. Can you design experiments to test whether different types, concentrations, and amounts of compost tea enhance plant growth?

In China, farmers dig parallel trenches and fill them with organic wastes mixed with cocoons of Eisenia fetida. Soybeans planted in rows between the trenches are highly productive.

Building a Two-Can Bioreactor

Purpose

Two-can bioreactors are designed to be used as small-scall indoor composting units for families, and for composting as an educational tool in the classroom.

- Materials
- 32-gallon plastic garbage can
- 20-gallon plastic garbage can
- Drill
- Brick
- Spigot (optional)

- Duct tape (optional)
- Insulation (optional)

Construction

- Using a drill, make 15 to 20 holes (0.5" to 1" diameter) through the bottom of the 20-gallon can. Next drill three rows of holes through the sides of this can, six to eight inches apart with four to five inches between rows, ending about two inches below where the can expands at the top.
- Place a brick or some other object in the bottom of the 32-gallon can. This is to separate the leachate from the compost and allow for its measurement and addition back into the compost pile. The leachate, often referred to as "compost tea," is rich in nutrients which may be in a form readily usable by plants. If not used right way on growing plants, pour the leachate back into the compost.
- Excessive leachate can be responsible for foul smells. If your sytem produces enough leachage to cause odour problems, your initialscompost mixture was probably too wet.

  Variations on the design:
  - — Add insulation to the barrels (inner and outer) with duct tape.
  - — Include a spigot to draw off the leachate.
  - — Add a layer of old compost, wood chips, or soil inside the outer barrel. This will allow the leachate to be absorbed and may cause fewer leachate/odour issues.

A system of 10-gallon plastic garbage cans that can fit inside 20-gallon cans can be substituted if space is a problem. The smaller system may operate at lower temperatures. This should not affect the final product; it will just take longer before the product can be used.

Bioremediation

"Remediate" means to solve a problem, and "bio-remediate" means to use biological organisms to solve an environmental problem such as contaminated soil or groundwater.

In a non-polluted environment, bacteria, fungi, protists, and other microorganisms are constantly at work breaking down organic matter. What would occur if an organic pollutant such as oil contaminated this environment?

Some of the microorganisms would die, while others capable of eating the organic pollution would survive.

Bioremediation works by providing these pollution-eating organisms with fertilizer, oxygen, and other conditions that encourage their rapid growth.

These organisms would then be able to break down the organic pollutant at a correspondingly faster rate. In fact, bioremediation is often used to help clean up oil spills.

Bioremediation of a contaminated site typically works in one of two ways. In the case described above, ways are found to enhance the growth of whatever pollution-eating microbes might already be living at the contaminated site. In the second, less common case, specialized microbes are added to degrade the contaminants. Bioremediation provides a good cleanup strategy for some types of pollution, but as you might expect, it will not work for all. Bioremediation may not provide a feasible strategy at sites with high concentrations of chemicals that are toxic to most microorganisms. These chemicals include metals such as cadmium or lead, and salts such as sodium chloride.

Nonetheless, bioremediation provides a technique for cleaning up pollution by enhancing the same biodegradation processes that occur in nature.

Depending on the site and its contaminants, bioremediation may be safer and less expensive than alternative solutions such as incineration or landfilling of the contaminated materials. It also has the advantage of treating the contamination in place so that large quantities of soil, sediment or water do not have to be dug up or pumped out of the ground for treatment.

Bioremediation is a process that can destroy pollutants by microbial action. Certain bacteria can use nutrients and oxygen to degrade volatile organic compounds and toxic compounds in soils. In the 1980s, bans on disposal of hazardous waste at landfills led to the increased primary use of disposal alternatives such as bioremediation.

However, little information has been available on the airborne emissions associated with these processes. This project was designed to help determine whether airborne emissions from bioremediation operations could pose an air pollution problem.

Decay and Renewal

Developed by the Cornell Environmental Inquiry Progammeme, Decay and Renewal consists of a student edition and teacher's guide designed to enable high school students to carry out authentic research. By using scientific protocols to investigate the processes in wastewater treatment, composting, landfilling, and bioremediation of contaminated sites, Decay and Renewal teaches important science concepts within the context of environmental issues.

Decay and Renewal includes:

- An overview of basic principles of biodegradation, the assortment of biological processes that cause organic matter to decay,
- Simple but authentic research protocols that measure biodegration in compost, landfills, wastewater treatment facilities, bioremediation sites, and streams,
- A progression of worksheets that guides students through each step research, providing structure but flexibility in designing and conducting meaningful experiments, and

- Guidelines for integrating peer review and other collabourative knowledge-building into classroom science.

Bacterial Biodegradation of High Molecular Weight Polycyclic Aromatic Hydrocarbons

Polycyclic aromatic hydrocarbons (PAH) are among the most common pollutants found at contaminated industrial sites, and in fact around the world.

Although PAH are naturally occurring compounds and essentially ubiquitous at low concentrations in the terrestrial environment, high concentrations of PAH are found in contaminated soils at wood treating facilities, at sites formerly used to produce manufactured gas, and at petroleum processing operations. There are a wide range of PAH, from the simple two-ring compound naphthalene to large multi-ring compounds. We arbitrarily define those PAH with two or three rings as low molecular weight (LMW) species, and those with four or more rings as high molecular weight (HMW) species. One of the characteristic features of PAH contamination is that there is always a complex mixture of PAH, often in conjunction with a variety of other hazardous chemicals.

Environmental Protection Agency currently regulates 16 PAH compounds as priority pollutants in water, and generally considers these same compounds as "total PAH" (tPAH) in contaminated soils. The 16 regulated PAH comprise both low and high molecular weight species, and seven of them are designated as known human carcinogens. All of the carcinogenic PAH (cPAH) are high molecular weight compounds.

In addition to their toxicity, PAH as a class are extremely hydrophobic chemicals. Naphthalene is among the most water-soluble of the PAH, and its solubility in water is only about 30 mg/L. The situation becomes much worse with increasing molecular weight: chrysene (a four-ring PAH) and benzo[a]pyrene (a five-ring compound) are soluble in water in the low part per billion (mg/L) range. By comparison benzene, normally considered to be a water-immiscible chemical, has an aqueous solubility of about 2,000 mg/L. Despite the very low solubility of PAH in water, PAH-contaminated materials have been shown to be toxic in a variety of bioassays.

Microbiologists have known for a long time that microorganisms can act on most of the PAH of concern. For this reason, PAH-contaminated soils and sediments are generally considered to be candidates for bioremediation. In any of the studies in which PAH degradation in contaminated soil or sediment has been studied, however, the higher molecular weight compounds have not been removed completely. We generally do not know whether this apparent recalcitrance is due to inherently greater difficulty in metabolism of these compounds by microorganisms, a lack of bioavailability caused by their low solubility in water, or both. It is also not usually known whether microbial activity leads to complete mineralization (conversion to carbon dioxide and water) of individual PAH or only partial transformation to other organic products. Some fungi, will metabolize PAH only partially in pathways that seem to be analogous to mammalian detoxification mechanisms. While fungi do not appear to use PAH as primary carbon sources for growth, bacteria have been known for many years to be able to use several of the low molecular weight PAH as growth substrates; in fact, some of the groundbreaking work on aerobic bacterial metabolism of aromatic hydrocarbons included studies on naphthalene metabolism.

Ironically the depth of study on bacterial metabolism of naphthalene and, to a lesser extent, phenanthrene has been accompanied by only sporadic study of the metabolism of other PAH. Until the late 1980s, it was generally believed that PAH with four or more rings could not serve as growth substrates for bacteria, and could only be degraded at the expense of other growth substrates.

Since then a handful of reports have appeared in the literature describing bacterial growth on the four-ring compounds fluoranthene, pyrene, and chrysene, but there is still no evidence of bacterial growth on PAH with more than four rings. There are also several studies in which the high molecular weight PAH were observed to be mineralized in the absence of growth, either by pure cultures or in complex environmental samples such as soil or sediment. We still know relatively little, however, about the mechanisms by which the high molecular PAH are degraded by bacteria.

Our work on bacterial degradation of the high molecular weight PAH began by isolating bacteria from PAH-contaminated soils. In all, we obtained ten isolates from eight different sites and one more PAH-degrading strain from another investigator. We have used these bacteria to study various aspects of PAH biodegradation, which focus on two key issues: whether the high molecular weight PAH are degraded by the same metabolic pathways a given bacterium uses to degrade the low molecular weight PAH; and how the low aqueous solubility of these compounds influences their availability to bacteria, which is prerequisite to degradation.

Mineralization of High Molecular Weight PAH

All of the bacteria we isolated from PAH-contaminated materials were obtained by enrichment culture techniques using phenanthrene as the sole arbon source. This clearly biases our observations to only those bacteria which can grow on phenanthrene, but we had hypothesized that such bacteria might be able to degrade a range of other PAH. Our hypothesis was based on the knowledge that these organisms are always exposed to a mixture of PAH, all of which are, to varying extents, structural analogues.

Over evolutionary time scales such conditions would be expected to select for microorganisms with versatile metabolic capabilities. We also knew that organisms able to degrade but not grow on the higher molecular weight compounds needed to grow on something, so the low molecular weight PAH seemed like the best choice with which to begin.

We have studied each of the 11 bacterial isolates for their ability to grow on or degrade 13 PAH, ranging from two- to five-ring compounds. Most of the bacteria have a broad range of PAH substrates they can metabolize. Of most interest in terms of maximizing the biodegradation of these compounds, however, is mineralization. So far we have studied the ability of nine of the strains to mineralize pyrene, chrysene, benzanthracene and benzopyrene (the latter three compounds are all cPAH).

First, none of the organisms was able to mineralize pyrene under the conditions used in this experiment. Second, every organism either mineralized all three of the remaining high molecular weight substrates or did not mineralize any of them. These findings suggest that pyrene metabolism may not be linked directly to the metabolism of the other PAH, and that there may be a relationship among the other three PAH.

Induction of High Molecular Weight PAH Degradation

Many heterotrophic bacteria have a broad range of organic compounds they can grow on or degrade. To conserve energy for vital growth-related processes, the enzymes needed by these bacteria to degrade a particular compound are not normally synthesized unless the compound is present in the medium. Thus, biodegradation pathways in bacteria are normally induced, or "turned on," by the presence of either the parent compound or one of the pathway intermediates. If we are to learn how to control the biodegradation of compounds such as high molecular weight PAH, we clearly must understand how the relevant enzymes are turned on. This is particularly important for those compounds that do not serve as growth substrates for bacteria, since they are not likely to be able to induce their own degradation.

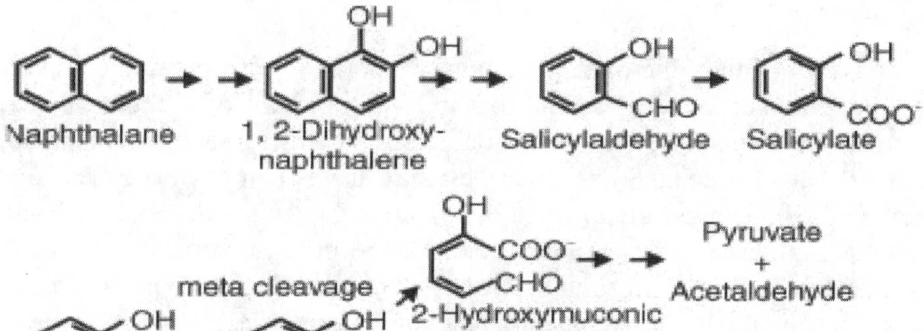

Fig. 3.2: The Principal Pathway for Acrobic Metabolism of Napthalene by Pseudomonads.

The only well studied example of the induction of PAH degradation in bacteria is that of naphthalene degradation. The complete pathway for conversion of naphthalene to simple, common metabolic end products is inducible by the intermediate salicylate in the bacteria for which it has been studied. In other words, the enzymes required for naphthalene metabolism are elevated to high levels in naphthalene-degrading bacteria when salicylate is introduced into the medium. (The astute reader might ask how salicylate is formed when naphthalene is the only carbon source in the medium; the answer is that in such cases the organism produces very low, or constitutive, levels of the required enzymes at all times, which then metabolize the naphthalene to form salicylate).

We have focused our attention on the inducibility of HMW PAH degradation in one of the organisms in our collection, Pseudomonas saccharophila P15. This organism was isolated several years ago by Will Stringfellow, who also conducted in-depth studies on its physiology. Both phenanthrene and salicylate induce the mineralization of benzanthracene, chrysene, and benzopyrene by P. saccharophila.

None of the three compounds are used as growth substrates by the organism, so these results suggest that the metabolism of high molecular weight PAH in P. saccharophila P15 is linked to the metabolism of a low molecular weight compound. We do not yet know, however, if the different PAH are metabolized by a common pathway or if different pathways are regulated through a common mechanism.

We hope to understand eventually if the metabolism of high molecular weight PAH by the other organisms in our collection is inducible by salicylate or other inducers of low molecular weight PAH metabolism. If we find such inducibility to be a general phenomenon, then we may have a tool with which to aid the removal of HMW PAH in bioremediation projects.

Bioavailability

Those who study the biodegradation of poorly soluble organic compounds in soil or sediment usually assume that only the amount of a chemical that is present in the aqueous phase is available to the microorganisms capable of degrading that chemical. Some microbes may have their own mechanisms of improving the bioavailability of these compounds, however. Many of the bacteria that degrade aliphatic hydrocarbons are known to produce surfactants (so-called "biosurfactants") which can solubilize or emulsify the hydrocarbons, presumably making them available for degradation at a much faster rate.

Nobody knows whether there is a similar involvement of biosurfactants in the natural biodegradation of PAH. We screened our collection of bacteria for their ability to reduce the surface tension in the culture medium (an indicator of surfactant production) during growth on phenanthrene or naphthalene, but only one organism was able to do so significantly. Thus, we did not see evidence that biosurfactant production is a common feature of PAH metabolism by these bacteria. We were, however, recently able to demonstrate how synthetic surfactants can improve the rate of phenanthrene biodegradation in a well-defined experimental system

Another mechanism by which bacteria might improve their access to poorly soluble chemicals in soil or sediment is through their motility. Motility is a very common feature of soil bacteria, yet its role in biodegradation of pollutants remains virtually unexplored.

Chemotaxis is a special form of motility in which an organism swims towards chemoattractants or away from chemorepellents, both cases requiring the existence of concentration gradients of the attractant or repellent for a chemotactic response to occur. The reliance on concentration gradients seems to make chemotaxis an ideal mechanism by which bacteria might be able to locate sources of "food," which in the case of hydrophobic chemicals is likely to be distributed heterogeneously in the porous medium.

We have become interested in the potential effects of motility and chemotaxis on PAH biodegradation. The lack of previous work on environmentally relevant situations has left a wide scientific gap which needs to be filled, to understand both the importance of such mechanisms in in situ biodegradation reactions and whether such mechanisms can eventually be manipulated to our advantage. If, motility and/or chemotaxis are found to be important aspects of PAH (or other pollutant) biodegradation by soil bacteria, then we need to begin incorporating such phenomena into models that purport to simulate biodegradation processes in the subsurface.

Most of the bacteria identified are motile, and we screened most of them for chemotaxis to naphthalene, phenanthrene and known metabolites of these low molecular weight PAH. Only one of them, Pseudomonas stutzeri P16, seemed to have a chemotactic response to the chemicals of interest. Interestingly, strain appears to be chemotactic to 1-hydroxy-2-naphthoate and to salicylate, both metabolites that have been found extracellularly in media containing active PAH degraders.

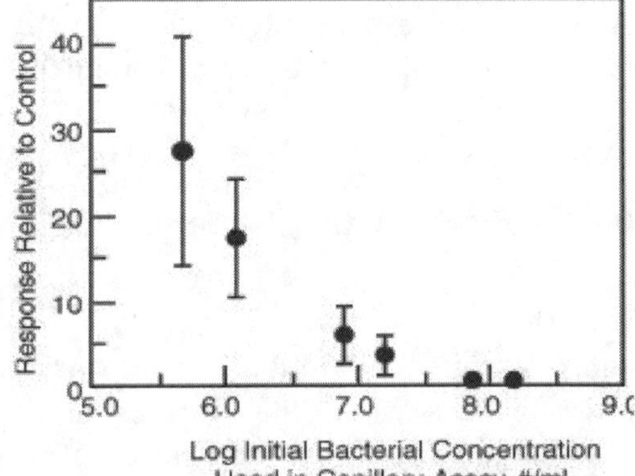

Fig.3.3: Effect of initial Bacterial Concentration

However, we are learning that some of the methods traditionally used to study chemotaxis are not well suited to the study of chemotaxis towards compounds of low aqueous solubility.

To test our methods with a known chemotactic organism, we recently obtained the strain Pseudomonas putida G7 from Professor Caroline. We learned that the chemotactic response of strain G7 in a standard quantitative assay becomes increasingly detectable as the starting concentration of bacterial cells is decreased.

The optimum response we observed was at a cell concentration nearly two orders of magnitude lower than the concentration typically used in the assay, which we attribute to the rapid loss of the naphthalene concentration gradient at high cell concentrations and the necessarily low starting concentration of naphthalene. Overall the response of strain G7 to naphthalene was variable yet strong. Consequently, we will focus on this organism in our subsequent efforts to understand the relevance of chemotaxis to PAH biodegradation. There is much to be learned about the bacterial biodegradation of polycyclic aromatic hydrocarbons, and we of course can only begin to scratch the surface. In another project, we are studying whether the concepts we have explored relative to the inducibility of PAH metabolism and the effects of surfactants on PAH bioavailability can be applied to the bioremediation of PAH-contaminated soils. It is always difficult to bridge the gap between basic and applied research, but we ultimately hope that each will provide insights that will shape the future direction of our own work and that of others, both in the research community and in practice.

Pollution Prevention in Resource-related Setting

Preventing pollution in a sensitive resource-related setting means thinking through all of the activities and services associated with the facility and planning them in a way that generates less waste. Waste prevention leads to thinking about materials in terms of reduce, reuse, recycle. The best way to prevent pollution is not to use materials that become waste problems. When such materials must be used, they should be reused onsite. Materials that cannot be directly reused should be recycled. Everyone associated with the facility must change their habits and adopt a more responsible attitude toward waste. This includes the ownership and management of the facility-the architects, engineers, designers, and contractors; the employees; and the visitors. Each of these groups needs to consider the issues so that no waste will be generated that adversely affects the environment.

Visible, Participatory Systems

The "out of sight, out of mind" mentality regarding waste is perpetuated because the systems that deal with waste problems are all behind the scenes and off-limits. Prevention can only be accomplished by paying attention to the issue up front, rather than waiting until after the problem has been created. An environmentally sound tourism facility would ensure the visibility of systems to prevent the generation of waste. Such systems require conscious participation by visitors, users, and operators. They should not dominate the experience of a visitor at the facility. If each visitor does his or her share, the facility can be operated in a more environmentally sound manner. An additional benefit is that this can lead to long-term changes in behaviour benefiting the participant and the earth.

Training and Maintenance

Waste prevention requires training the operators, educating all users of the system, and performing diligent maintenance. Most waste problems have been created because attention has not been paid. Conventional disposal systems are designed for ease of use with the user having little idea of how the systems operate or where the waste goes. Because waste prevention represents a change in the way activities are carried out, it requires an extra effort to ensure that these practices are maintained until they become routine. In situations with high turnover of both employees and visitors, continuous training and education would be essential.

Host and Visitor Attitudes

Preventing waste means taking a measure of responsibility for activities. This can conflict with the attitude that going on vacation means escaping responsibilities. Without an attempt to change this attitude, the necessary participation may be difficult to obtain.

Fostering a sense of responsibility in visitors is an important element of environmentally sound development in many other ways as well. One ay to encourage this attitude is to convey to visitors that they are privileged visitors, and as a result, they may naturally feel obliged to follow certain guidelines established by their hosts. Facility operators should behave as hosts that care about making waste prevention work. Employees of the facility must care about it, and that caring must be visible to the visitors.

Garbage/Solid Waste Prevention Strategies

In planning for facilities, a comprehensive design strategy is needed for preventing generation of solid waste. A good garbage prevention strategy would require that everything brought into a facility be recycled for reuse or recycled back into the environment through biodegradation. This would mean a greater reliance on natural materials or products that are compatible with the environment.

Any resource-related development is going to have two basic sources of solid waste-materials purchased and used by the facility and those brought into the facility by visitors. The following waste prevention strategies apply to both, although different approaches will be needed for implementation:

- Use products that minimize waste and are nontoxic
- Compost or anaerobically digest biodegradable wastes
- Reuse materials onsite or collect suitable materials for offsite recycling

Use of Products that Minimize Waste

Much of the growing volume of garbage is from the use of disposable consumer products and excess packaging. Consideration must be given to materials or products that minimize waste disposal needs. Purchase items with minimal packaging; buy in bulk; replace disposable products with durable, reusable items.

Durable, reusable products can be substituted for disposable ones, such as the use of returnable glass bottles instead of aluminum cans or the use of rechargeable batteries. Use of plastics for packaging isincreasing, thereby replacing recyclable products and materials. Plastics, which account for about 20% of solid waste by volume, are nonbiodegradable, difficult or impossible to recycle, have a high volume to weight ratio, and are toxic when burned.

When selecting materials and goods, nothing should be purchased that will ultimately become toxic. Nontoxic materials can often be substituted for products that cause contamination problems during disposal.

Materials should be purchased locally whenever possible. Locally produced goods needing less transport and less storage should have less waste packaging. No matter how diligent people are in purchasing materials, certain items will need to be handled. There are two methods to consider-biodegradation or recycling. Technology and economics are continually changing and the system selected must have the flexibility to adjust to market conditions. The scope of composting or anaerobic digestion of organic materials and the availability of markets for recycled materials should be considered in planning for a facility. These decisions are site-specific and must be made at the outset before final design is completed. Factors such as land availability, local markets, and isolation all contribute to that decision.

Biodegradation

In the process of biodegradation, microorganisms break down the products of other living things and incorporate them back into the ecosystem. Biodegradable or bioconvertible material includes anything that is organic. Plastics are not considered includable, despite industry contention that they are. Most of the organic components of garbage, such as paper and food wastes, can be eliminated through composting. Between 60% and 75% of the solid waste is bioconvertible.

In a tourism development, the greatest fraction of waste would be generated by the support and maintenance services. Wastes that arebioconvertible include newspapers, magazines, and wet wastepaper in thekitchen and restaurants. Also included are food waste, cardboard materials, paper office supplies, and leaves, grass, and tree trimmings. The solid fractionof the sewage waste is also considered available for the bioconversion process,and is often the most costly to dispose of otherwise, usually in a special landfill.

The biodegradable or bioconvertible percentage of the waste stream is large enough to consider at least two options for the conversion process. Two obvious options for conversion are composting and anaerobic digestion.

Composting

Composting is a familiar concept, and is used for handling yard waste and even sewage sludge. Both of these organic wastes require mixing of other materials to achieve a nutrient balance. Large chunks of relatively inert material (most commonly wood chips) add bulk and aeration to make the process work.

This is typically done in open windrows or piles, with mixing done daily to provide aeration and homogeneity. This takes land space on a drainable surface, and a collection of any runoff for redispersal of the liquid to the process.

It produces a quality soil amendment and reduces the bulk of the original material by approximately 40-50%. Composting does off-gas ammonia and carbon dioxide and produces offensive orders.

It typically takes 50-60 days to process. Screening of the final product is necessary to remove the bulking material, and provide granulation before use in the soil bed. The use of this product as a soil amendment is valuable, particularly in the tropics, because the soil is essentially sterile, with only about 2% organic content. Affected by humidity, rainfall, temperature, and normal soil activity, the organic material placed on the soil will typically last only 30- 40 days in the tropics. In a temperate climate, that same material may last as long as six months.

Most composting operations of the size typical for communities could be suitable for tourism development use, considering the opportunity to also process construction debris and landscape waste. This will typically be the smallest scale. The construction debris process equipment can be transferred directly to the operations management and be capable of handling the full volume of waste.

Anaerobic Digestion

Anaerobic digestion is used extensively worldwide for the processing of food waste, animal waste, the solids of human waste systems, and for the total array of solid waste such as waste paper, green waste, and landscape waste. This wet fermentation process converts the waste stream into three usable by-products:

- Biogas-an energy-rich gas stream, comparable to natural gas, that can be used to offset the cost of energy utilities of the development
- A high quality organic fertilizer solid that may be useful in landscaping efforts or even crop production;
- A dilute liquid organic fertilizer that may be used in drip irrigation as an additive to any planting progammeme, for foliar feeding of ornamentals, or in landscape plots for replenishing native or endangered plant species of plants.

The anaerobic digestion process does need management supervision in startup. Design of the system should be simple and modular as well as easy to operate. The modular design allows seasonal loading rather than using one large vessel. This also provides for stages of digestion that help the natural biology of the fermentation reach its optimum capacity. While this requires more capital investment, the savings to be made in maintaining operations through seasonal fluctuation will prove to be cost effective. The advantages of this system are in the smaller space it would require and in the energy it would produce, which may make the facility more self-sufficient. The value of the by-products cannot be understated, particularly in the case of remote development.

Recycling

A material doesn't become waste until it is thrown in the garbage can. If a material can be reused it is a resource, not waste. Reuse is the best form of recycling. There are markets in the United States, Europe, Asia and other parts of the world for many recyclable materials including aluminum, paper, glass, steel, and some types of plastics. These may be available to a facility depending on location and the quantity of the materials generated. Recycling can be maximized through the purchase of products for which there is a ready market as recycled materials. The feasibility of recycling a given material depends on the price offered by a buyer for that material and shipping costs, both of which may vary over time. However, if no other reuse or recycling option exists, materials used at the facility should be recycled even at a net loss rather than sent for disposal. In circumstances where there is no available market for a given material, often a beneficial end use can be developed locally. Every effort should be made to work with the local community to determine if any of the materials generated by the facility can be used.-e.g., glass beverage containers can be ground up and incorporated into materials for construction and road building. Efficient recycling requires sorting of materials; convenient bins should be provided at the facility for the materials being recycled.

Offsite Disposal

If a garbage prevention strategy has been fully executed, actual remaining waste should be minimal. Remaining residuals mean that the facility is not entirely environmentally sustainable. All residuals must be collected separately and disposed of offsite. Although disposal of residual waste anywhere would have an adverse impact on the environment, in an environmentally sound development special care must be taken that this will not be borne locally. In most cases residuals must be returned to their place of origin. Toxic material residuals must be segregated and disposed of separately.

Materials Brought in and/or Purchased Onsite

Visitors and guests normally bring various materials that are consumed onsite, often leaving behind disposal problems. Products purchased onsite and in the surrounding community can exacerbate the problem. Such materials may include toxic and difficult to dispose of items (e.g., batteries, flash bulbs, instant film refuse, and repellents. Education, including what is incorporated in the written materials about the facility and what is available, is important in minimizing these types of waste materials.

DOMESTIC WASTE MANAGEMENT STRATEGIES

General Considerations

Domestic wastes in park and tourism developments come from toilets and urinals, showers, bathroom sinks, kitchen sinks, laundry facilities, and most floor drains. Varying amounts of water are used to carry these wastes through pipes to a treatment point. Different processing techniques are used to convert the wastes for reuse or disposal. True sustainable development would not permit direct disposal of either liquids or solids before reuse.

Responsible waste management recognizes the value of reducing water needs. Properly treated wastewater can be used for toilet flushing (after approved disinfection) and irrigation (landscape needs or agricultural plantings).

The solids separated from domestic waste may be incorporated into the solid wastes (garbage) and composted into an excellent soil amendment product or anaerobically digested to produce a gaseous energy source and a residue that can also be used as a soil amendment.

If wastewater recycling for toilet flushing is used without concurrent irrigation of vegetation, there will be a volume of excess liquid that must be sensitively disposed of.

This will be the amount of liquid wastes coming in each day from all sources, less the recirculated amount used for toilet flushing. Best disposal practice would use a suitable area for subsurface movement of the liquid through soils that provide good filtering and additional treatment before reaching any body of water. Direct discharge to a waterway should be the lowest priority of alternatives investigated.

Using treated, recycled wastewater for toilet flushing, rather than an equal volume of drinking-quality water, would save a large amount of water. There are also strong economic and environmental reasons for using one of more than 20 different low-volume flush toilets currently on the market, rather than the more conventional water-wasting models. In seawater development sites, toilets may be flushed with seawater and processed with septic tank and subsurface treatment and disposal. This provides a reliable system if the corrosive nature of seawater is addressed in selection of materials used in the piping, plumbing, pumping, and treatment systems.

Similar water conservation can be realised through the use of flow-restricted, fine-spray shower heads, and sink faucet aerators. Spring-loaded handles should be installed on faucets. Laundry waste volume can be reduced through the use of machines with suds-saver cycles.

Park and tourism developments that make use of water conservation devices should make an effort to educate their visitors on the amount of water or energy) they have conserved during their stay, compared with the conventional development. It is important to monitor the conservation devices in order to evaluate their performance.

As with all pollution, sewage is best treated by not creating it in the first place. Human waste only becomes waste when there is no use for it.

Specific Waste Treatment Technologies

Conditions at sustainable developments favour simple, reliable, passive techniques for waste treatment. Systems that have few moving parts, controls, and monitoring requirements should be used. Continuity of operation and maintenance knowledge is essential. Waste treatment systems must also be able to withstand seasonal fluctuations in use. Preparation for an idle period should be simple, and no energy should be required when the system is not being used. If an innovative technology is to be employed for waste treatment, it should have a conventional backup system, or be sufficiently redundant that multiple failures are unlikely.

Dry Toilets

A composting toilet consists of a large tank located directly below the toilet room. Wastes enter the tank through a large diameter chute connecting to the toilet, and decompose in an oxygen rich environment. No water is used for the toilet, but a bulking agent (such as wood shavings) is added to improve liquid drainage and aeration, and to provide fuel. A small fan draws air through the tank and up the vent pipe to ensure adequate oxygen for decomposition and odourless operation. Internal components (such as ducts, baffles, and rotating tines) enhance the composting process. The finished compost can be removed from the lower end of the tank about once each year. It can be used as a fertilized for soil.

Composting toilets need a mild temperature, moisture, fuel, and air to function. Liquid may have to be added to the tank to keep the compost pile moist during periods of little use or a bulking agent added periodically to improve the compost texture.

Composting toilets have several advantages over other systems-e.g., no water is used and only a small amount of energy is needed for an exhaust fan; valuable nutrients are used to benefit soils; proper maintenance requires little time.

There are some disadvantages to this system. The user is close to the sewage treatment systems and this bothers some visitors. Without proper maintenance, the tank can become anaerobic and unpleasant odours arise. Undesirable pests can take up residence in the composting tank. The availability of a bulking agent in some areas can also be a problem.

Anaerobic Waste Treatment

Anaerobic waste treatment (septic systems) is accomplished through microorganisms living in the wastewater. Anaerobic microorganisms work in an environment where there is no free oxygen. Complex reactions and interactions take place with the resulting generation of some offensive gaseous by-products. These unpleasant odours are actually an indication of an efficient progression of the anaerobic process to remove the pollutants from the waste stream.

This type of system is reasonable to consider for smaller developments, where the by-product gases can be separated from occupancy areas, dissipated with good air movement, or neutralized with soil or carbon filters. Since slow treatment means longer holding periods (shallow depth tanks), large, isolated treatment and disposal areas are needed. Treated wastewater (effluent) is usually disposed of in an underground system that passes the effluent through carefully selected undisturbed soil profiles. These soils must further filter and remove nutrients as the effluent makes its way to the groundwater or other bodies of water.

One variation of this type of treatment uses part of its treated effluent for toilet flushing. The stored recycled effluent should receive some aeration to ensure odour-free recycle water in the toilets.

This type of waste treatment system has several advantages. It is relatively inexpensive to install, and is not complex to operate and maintain. It provides a consistently good quality effluent. Most components can be installed aboveground, or they can be buried or partly buried. Good quality flushing water can be provided to reduce water supply needs and final disposal volumes.

The disadvantages are that septic tank odours must be dealt with properly. Several pumps and blowers can be involved, which creates maintenance and power costs, and can cause noise in some developments. Septic tank solids require proper disposal. An aerobic digestion tank would reduce volumes for disposal. A sand drying bed can be used for digested sludge drying, and the dried sludge should be buried in a landfill. A fairly large area is required for installation. There is no known manufacturer of a complete package system.

Prior to committing time and resources, it is prudent to contact actual users of similar systems-talk with a designer, system operator, an owner, and possibly a regulatory agency inspector who has observed performance.

Aerobic Waste Treatment

Aerobic waste treatment is also accomplished with microorganisms. However in this system, air is bubbled into the treatment process to ensure plenty of free oxygen. Aerobic organisms work about 20 times faster than anaerobic organisms. No offensive gas odours are released in this treatment process. Since they process is so much faster, much less holding time is required and less treatment area is needed.

This type of treatment plant is reasonable to consider for all sizes of park and tourism development, and because offensive odours are not normally produced, this type of system can be located close to occupied areas. The normal effluent quality of a properly designed and operated system meets most all secondary treatment standards With a reasonably small filter, high quality effluent can be produced for irrigation and recycled toilet flushing water. With carefully planned effluent recycling and irrigation reuse, little or no effluent disposal would be required. If disposal is required, several options are possible.

There are several manufacturers of small aerobic treatment plants. The best plants for harsh environments and isolated areas are fabricated from durable, low maintenance materials, simple to operate and maintain, designed to use the fewest moving parts, and consistent in effluent quality. They are also quick to install and put in service, proven in similar installations, and completely assembled and warranted by one manufacturer. A good manufacturer also provides, at a reasonable cost, technical and monitoring services on the new plant for at least one year after initial startup.

Advantages of such a system are that all materials in contact with liquids are noncorrosive (plastic, fibreglass, rubber, stainless steel) and can be buried or installed aboveground. The blower is the only moving part and requires only simple, routine maintenance. Effluent quality is consistently good. Excellent recycle and reuse options are available. The process easily handles surge loadings, as well as underloaded periods. Land area required is about 0.1 acre per 10,000 gallons of rated capacity. Site installation and startup can usually be accomplished in five working days or less. Few spare parts are needed. Purchase costs are competitive with other similar-sized treatment units. Unskilled operators have successfully followed infrequent, telephoned system adjustment instructions to maintain top performance.

The disadvantages are that most components are fabricated, assembled, and shipped from Colourado-few local materials would be used. The blower, although small, needs to be run continuously. The noise would have to be muffled, and a constant power cost would be involved.

The National Park Service has five of these plants in service in four widely varying environmental areas. One plant has been in service for several years.

Alternative Disinfection Methods

Traditionally, water from conventional treatment systems is disinfected with chlorine or chlorine compounds before being released back into the environment or reused. A side effect of this is that the chlorine or chlorine compounds are very reactive and sometimes produce highly persistent, toxic chemicals. Many environmentalists believe that there is no justification for use of chlorine and its compounds for disinfection.

Most public health codes call for disinfection with chlorine, and they would have to be changed to allow for either no disinfection or the use of other disinfectants. Less desirable possibilities include using less chlorine or removing the chlorine after the proper contact time. Dechlorination requires additional chemical feed and training.

The purpose of disinfection is to ensure that no virulent organisms are present after the water has been processed in one of the systems described earlier. The most common alternative disinfectants are ozone and ultraviolet light.

An entirely new treatment technology, introduced in September 1993, appears to provide excellent disinfection without the formation of environmentally harmful by-products. The National Park Service is arranging pilot studies at existing water and sewage treatment facilities to evaluate this emerging innovative technology.

❐

4

Environmental Microbiology

DEFINITION

The study of how microorganisms affect the earth and its atmosphere is called environmental microbiology or microbial ecology. Environmental microbiology is the study of microbes in the environment and their interactions with each other. Microbes are the tiniest creatures on Earth, yet despite their small size, they have a have a huge impact on us and on our environment.

Most types of microbes remain unknown. It is estimated that we know fewer than 1% of the microbial species on Earth. Yet microbes surround us everywhere—air, water, soil. An average gram of soil contains one billion (1,000,000,000) microbes representing probably several thousand species.

Environmental microbiology is the study of the relationships that exist between microorganisms and the environment.

Environmental microbiology is the relationship of microorganisms with themselves and with their surroundings. Microorganisms have special impact on the whole biosphere. They are the backbone of ecosystems of the zones where light cannot approach. In such zones, chemosynthetic bacteria are present which provide energy and carbon to the other organisms there. Some microbes are decomposers which have ability to recycle the nutrients. So, microbes have a special role in biogeochemical cycles. Microbes, especially bacteria, are of great importance in the sense that their symbiotic relationship (either positive or negative) have special effects on the ecosystem.

Bacteria and their enzymes, along with some fungi and critical nutrient additives are cost effective agents for in-situ remediation of hazardous wastes and subsurface pollution in soils, sediments and wastewaters. The ability of each bacteria strain to degrade toxic waste depends on the nature of each contaminant. Since most sites are typically comprised of multiple pollutant types, the most effective approach to bioremediation is to use a mixture of bacterial species/strains, each specific to the degradation of one or more types of contaminants. It is critical to monitor the composition of the indigenous and added bacterial consortium in order to evaluate the activity level of the bacteria, and to permit modifications of the nutrients and other conditions for optimizing the bioremediation process.

OVERVIEW

The pervasive and manifold positive and negative impacts of microbial activities on human and environmental health, and human activities, require that efforts to improve the human condition include efforts to modulate microbial activities. Attainment of this goal will be greatly facilitated by acquisition of a fundamental understanding of microbial functioning in their natural settings, and of parameters that regulate such activities. Our current fragmentary knowledge of microbial processes is based largely on information obtained from individual strains of microbes studied in the Labouratory as pure, exponentially-growing cultures in nutrient-rich, homogeneous aqueous solutions.

Sewage treatment is the process that removes the majority of the contaminants from waste-water or sewage and produces both a liquid effluent suitable for disposal to the natural environment and a sludge. To be effective, sewage must be conveyed to a treatment plant by appropriate pipes and infrastructure and the process itself must be subject to regulation and controls. Sewage is the liquid waste from toilets, baths, showers, kitchens, etc. that is disposed of via sewers. In many areas sewage also includes some liquid waste from industry and commerce. In the UK, the waste from toilets is termed foul waste, the waste from items such as basins, baths, kitchens is termed sullage water, and the industrial and commercial waste is termed trade waste.

The division of household water drains into Greywater and Black water is becoming more common in the United States, with greywater being permitted to be used for watering plants or recycled for flushing toilets. Much sewage also includes some surface water from roofs or hard-standing areas. Municipal wastewater therefore includes residential, commercial, and industrial liquid waste discharges, and may include stormwater run-off.

Sewerage systems that transport liquid waste discharges and stormwater together to a common treatment facility are called combined sewer systems. The construction of combined sewers is a less common practice in the U.S. and Canada than in the past and is no longer accepted within Building Regulations in the UK. Instead, liquid waste and stormwater are collected and conveyed in separate sewer systems, referred to as sanitary sewers and storm sewers in the U.S. and as foul sewers and surface water sewers in the UK. Overflows from foul sewers designed to relieve pressure from heavy rainfall are termed storm sewers or combined sewer overflows.

As rainfall runs over the surface of roofs and the ground, it may pick up various contaminants including soil particles (sediment), metals, organic compounds, animal waste, and oil and grease. Some jurisdictions, such as certain communities located in southern California, require stormwater to receive some level of treatment before being discharged to the environment. Examples of treatment processes used for stormwater include sedimentation basins, wetlands, and vortex separators (to remove coarse solids).

"Conventional wastewater treatment" is the process used in modern wastewater treatment plants that removes the majority of the contaminants from wastewater. This produces a liquid effluent suitable for disposal to the natural environment and also produces a sludge.

The treatment process typically involves the following three stages: Primary treatment Primary treatment is to reduce oils, grease, fats, sand, grit, and coarse (settleable) solids.

Grit removal This stage typically includes a grit channel where the velocity of the incoming wastewater is carefully controlled to allow grit and stones to settle but still maintain all organic material within the flow. Grit and stones need to be removed early in the process to avoid damage to pumps and equipment in the remaining treatment stages.

Screening or maceration The grit free liquid is then passed through fixed or rotating screens to remove larger material such as rags. Screenings are collected and may be returned to the sludge treatment plant or may be disposed of off site by landfilling or incineration. Maceration, in which solids are cut into small particles through the use of rotating knife edges mounted on a revolving cylinder, is used in plants that are able to process this particulate waste. Macerators are, however, more expensive to maintain and are less reliable than physical screens.

SEDIMENTATION

Primary sedimentation tank at a rural treatment plantIn almost all plants there is a sedimentation stage where the sewage is allowed to stand in large tanks so that faecal solids can settle and floating material such as grease and plastics can rise to the surface and be skimmed off. The main purpose of the primary stage is to produce a generally homogeneous liquid capable of being treated biologically together with a sludge that can be separately treated or processed. Primary settlement tanks are usually equipped with mechanically driven scrapers that continually drive the collected sludge towards a hopper in the base of the tank from where it can be pumped to further sludge treatment stages.

Secondary treatment Secondary treatment is designed to substantially degrade the biological content of the sewage. The majority of municipal and industrial plants treat the settled sewage liquor using aerobic biological processes. For this to be effective, the biota require both oxygen and a substrate on which to live. There are number of ways in which this is done. In all these methods, the bacteria and protozoa consume biodegradable soluble organic contaminants (e.g. sugars, fats, organic short-chain carbon molecules, etc.) and bind much of the less soluble fractions into floc particles.

Roughing Filters Roughing filters are intended to treat particularly strong or variable organic loads. They are typically tall, columnar filters filled with open synthetic filter media to which sewage is applied at a relatively high rate. The design of the filters allows high hydraulic loading and a high flow-through of air. The resultant liquor is usually within the normal range for conventional treatment processes.

Activated sludge Activated sludge plants use a variety of mechanisms and processes to use dissolved oxygen to generate a biological floc that substantially removes organic material. It also traps particulate material and can, under ideal conditions, convert ammonia to nitrite or nitrate.

FILTER BEDS

Trickling filter bed using plastic mediaIn older plants and plants receiving more variable loads, trickling filter beds are used where the settled sewage liquor is spread onto the surface of a deep bed made up of coke (carbonised coal), rocks or specially fabricated plastic media with high surface areas. The liquor is distributed through perforated rotating arms radiating from a central pivot.

The distributed liquor trickles through this bed and is collected in drains at the base. These drains also provide a source of air which percolates up through the bed, keeping it aerobic. Biological film comprising of bacteria, protozoa and fungi forms on all the available surfaces and this provides the required biological treatment capability to effect the reduction in organic content. Rotating Plates and Spirals In some smaller plants slowly revolving plates or spirals are used which are partially submerged in the liquor. A biotic floc is created which provides the required substrate.

Secondary sedimentation The final step in the secondary treatment stage is to settle out the biological floc or filter material and produce an effluent with very low levels of organic material and suspended matter.

Secondary Sedimentation tank at a rural treatment plantTertiary treatment Tertiary treatment provides a final stage to raise the effluent quality to the standard required before it is discharged to the receiving environment (sea, river, lake, ground, etc.) More than one tertiary treatment process may be used at any treament plant. If disinfection is practiced, it is always the final process.

Effluent polishing Sand filtration Sand filtration removes much of the residual suspended matter. Lagooning Lagoons provides settlement and further biological improvement through storage in large man-made ponds or lagoons.

Constructed wetlands These include engineered reed beds and a range of similar methodologies, all of which provide a high degree of aerobic biological improvement and can often be used instead of secondary treatment for small communities.

Nutrient removal Wastewater may also contain high levels of nutrients (nitrogen and phosphorus) that in certain forms may be toxic to fish and nvertebrates at very low concentrations(e.g. ammonia) or that can create nuisance conditions in the receiving environment (e.g. weed or algal growth). Although the growth of weeds and algae may seem to be primarily an aesthetic issue, algae can produce toxins, and in dying their decay and consumption by bacteria in the environment can result in the depletion of oxygen in the water and the possible consequential suffocation of fish. Where receiving rivers discharge to lakes or shallow seas, the added nutrients can cause severe and sometimes irreversible eutrophication with the loss of many sensitive clean water species. The removal of nitrogen and/or phosphorus from wastewater can be achieved either biologically or by chemical precipitation treatment processes.

Nitrogen removal Biological treatment of nitrogen generally involves creating conditions within the treatment process for bacteria to convert the ammonia to nitrate, and then allowing other bacteria to reducing the nitrate to nitrogen gas, which is released to the atmosphere. Sand filters, lagooning and the use of reed beds can all be used to reduce nitrogen. Sometimes the conversion of toxic ammonia to nitrate alone is referred to as tertiary treatment.

Phosphorous removal The biological treatment of wastewater to remove phosphorus also involves the design and creation of specific environmental conditions within a treatment plant to enable specific bacteria to bio- accumulate large quantities of phosphorus. When the bacteria containing the phosphorus are removed, the resulting bacterial biosolids often have a high fertilizer value. Phosphorus can also be removed by chemical precipitation using (commonly) salts of iron (i.e. ferric chloride) or aluminum (i.e. alum). The resulting chemical sludge, however, is difficult to dispose of, and the use of chemicals in the treatment process is expensive and makes operation difficult and often messy.

Disinfection Disinfecting substantially reduces the numbers of living organisms in the water to be discharged. The effectiveness of disinfection depends on the quality of the water being treated, the type of disinfection being used, the application rate, the contact time and environmental variables. Turbid water will be treated less successfully since solid matter can shield organisms, especially from Ultraviolet light or if contact times are low. Generally, short contact times, low doses and high flows all mitigate against effective disinfection. Common methods of disinfection include ozone, chlorine, or UV light. Chloramine, which is used for drinking water, is not used in waste water treatment because of its persistence.

A highly efficient dye-deColourizing bacterium, strain S12(T), was isolated from activated sludge of a textile-printing waste-water treatment plant in

Guangzhou, China. The cells were Gram-negative and motile by means of a single polar flagellum. The strain was capable of anaerobic growth either by fermentation of glucose or by anaerobic respiration and utilized a variety of electron acceptors, including nitrate, iron oxide and thiosulfate. The physiological properties, tested by using the Biolog GN2 system, were similar to those of the genus of Shewanella. Analysis of the nearly complete 16S rRNA gene sequence of strain S12(T) showed the highest similarity (98 and 97%, respectively) to Shewanella baltica and Shewanella putrefaciens. However, the level of gyrB similarity between strain S12(T) and S. putrefaciens was 87%. DNA from strain S12(T) showed 41.8 and 41.9% DNA relatedness, respectively, to the DNA of S. baltica DSM 9439(T) and S. putrefaciens DSM 6067(T). The DNA G+C content of strain S12(T) was 49.3 mol%. The predominant menaquinone was MK-7 and the predominant ubiquinones were Q-7 and Q-8. The dominant fatty acids were 15: 0, 16: 0, iso-15: 0 and 16: 1omega7c, similar to the profiles of other Shewanella species. On the basis of its physiological and molecular properties, strain S12(T) appears to represent a novel species of the genus Shewanella, for which the name Shewanella deColourationis sp.nov. is proposed. The type strain is S12(T) (=CCTCC M 203093(T)=IAM 15094(T)).

Hydrogenophaga palleronii (98.5%) and Hydrogen-ophaga taeniospiralis (98.0%). The fatty acid patterns and substrate-utilization profiles displayed similarity to the those of the five Hydrogenophaga species with validly published names, although clear differentiating characteristics were also observed. The two strains showed DNA-DNA hybridization values of 51% with respect to each other. No close similarities to any other Hydrogenophaga species were detected in hybridization experiments with the genomic DNAs.

On the basis of these results, two novel Hydrogenophaga species, Hydrogenophaga defluvii sp. nov. and Hydrogenophaga atypica sp. nov. are proposed, with BSB 9.5(T) (=DSM 15341(T)=CIP 108119(T)) and BSB 41.8(T) (=DSM 15342(T)=CIP 108118(T)) as the respective type strains.

Effect of ethoxylate number and alkyl chain length on the pathway and kinetics of linear alcohol ethoxylate biodegradation in activated sludge; Itrich NR et al.; Batch activated-sludge die-away studies were conducted with various pure homologs to determine the effect of ethoxylate number and alkyl chain length on the kinetics of primary and ultimate biodegradation of linear alcohol ethoxylates. The 14C-(ethoxylate) homologs C14E1, C14E3, C14E6, and C14E9 were used to investigate the effect of ethoxylate number, and 14C-(ethoxylate) homologs C12E6, C14E6, and C16E6 were used to examine the effect of chain length. Activated sludge was dosed with a trace concentration (0.2 microM) of each homolog, and the disappearance of parent, formation of metabolites, production of 14CO2, and uptake into solids were monitored with time. Ethoxylate number had little effect on the first-order decay rates for primary biodegradation, which ranged from 61 to 78 h(−1).

However, alkyl chain length had a larger effect, with the C16 chain-length homolog exhibiting a slower rate of parent decay (18 h(–1)) compared to its corresponding C12 and C14 homologs (61-69 h(–1)). Ethoxylate number affected the mechanism of biodegradation, with fission of the central ether bond to yield the corresponding fatty alcohol and (poly)ethylene glycol group increasing in dominance with increasing ethoxylate number. Based upon the measured rates of primary biodegradation, removal of parent during activated-sludge treatment was predicted to range between 99.7 and 99.8% for all homologs except C16E6, which had a predicted removal of 98.9%.

Based upon the measured rates of ultimate bio-degradation, removal of ethoxylate-containing metabolites was predicted to exceed 83% for all homologs. These predictions corresponded closely with previously published removal measurements in Labouratory continuous activated-sludge systems and actual treatment plants.

It was found that the estimated theoretical maximum growth yield of aerobic granules was as low as 0.2 g biomass g(-1) chemical oxygen demand (COD) and 10-40% of input substrate-COD was consumed through the maintenance metabolism, while experimental results further showed that the unit oxygen uptake by aerobic granules was 0.68 g oxygen g(–1) COD, which was much higher than that reported in activated sludge processes. Based on the growth yield and unit oxygen uptake determined, an oxidative assimilation equation of acetate-fed aerobic granules was derived; and this was confirmed by respirometric tests. In aerobic granular culture, about 74% of the input substrate-carbon was converted to carbon dioxide. The growth yield of aerobic granules was three times lower than that of activated sludge. It is likely that high carbon dioxide production is the main cause of the low growth yield of aerobic granules, indicating a possible energy uncoupling in aerobic

GRANULAR CULTURE

Comparing activated sludge and aerobic granules as microbial inocula for phenol biodegradation; Tay ST et al.; Activated sludge and acetate-fed granules were used as microbial inocula to start up two sequencing batch reactors (R1, R2) for phenol biodegradation. The reactors were operated in 4-h cycles at a phenol loading of 1.8 kg m(–3) day(–1). The biomass in R1 failed to remove phenol and completely washed out after 4 days. R2 experienced initial difficulty in removing phenol, but the biomass acclimated quickly and effluent phenol concentrations declined to 0.3 mg l(–1) from day 3.

The acetate-fed granules were covered with bacterial rods, but filamentous bacteria with sheaths, presumably to shield against toxicity, quickly emerged as the dominant morphotype upon phenol exposure. Bacterial adaptation to phenol also took the form of modifications in enzyme activity and increased production of extracellular polymers. 16S rRNA gene fingerprints revealed a slight decrease in bacterial diversity from day 0 to day 3 in R1, prior to process failure.In R2, a clear shift in community structure was observed as the seed evolved into phenol-degrading granules without losing species-richness. The results highlight the effectiveness of granules over activated sludge as seed for reactors treating toxic wastewaters.

Processes for managing pathogens; Godfree A et al.; Wastewater contains human, animal, and plant pathogens capable of causing viral, bacterial, or parasitic infections. There are several routes whereby sewage pathogens may affect human health, including direct contact, contamination of food crops, zoonoses, and vectors. The range and numbers of pathogens in municipal wastewater vary with the level of endemic disease in the community, discharges from commercial activities, and seasonal factors.

Regulations to control pathogen risk in the United States and Europe arising from land application of biosolids are based on the concept of multiple barriers to the prevention of transmission.
The barriers are:

- Treatment to reduce pathogen content and vector attraction.
- Restrictions on crops grown on land to which biosolids have been applied.
- Minimum intervals following application and grazing or harvesting.

Waste water treatment reduces number of pathogens in the wastewater by concentrating them with the solids in the sludge. Although some treatment processes are designed specifically to inactivate pathogens, many are not, and the actual mechanisms of microbial inactivation are not fully understood for all processes. Vector attraction is reduced by stabilization (reduction of readily biodegradable material) and/or incorporation immediately following application.

Concerns about health risks have renewed interest in the effects of treatment (on pathogens) and advanced treatment methods, and work performed in the United States suggests that Class A pathogen reduction can be achieved less expensively than previously thought. Effective pathogen risk management requires control to the complete chain of sludge treatment, biosolids handling and application, and post-application activities. This may be achieved by adherence to quality management systems based on hazard analysis critical control point (HACCP) principles.

Evaluation of microscopic techniques (epifluorescence microscopy, CLSM, TPE-LSM) as a basis for the quantitative image analysis of activated sludge; Lopez C et al.; Microscopic techniques ranging from epifluorescence microscopy to confocal laser scanning microscopy (CLSM) and two photon excitation laser scanning microscopy (TPE-LSM) combined with fluorescent stains can help to evaluate complex microbial aggregates such as activated sludge flocs. To determine the application limits of these microscopic techniques, activated sludge samples from three different sources were evaluated after staining with a fluorescent viability indicator (Baclight Bacterial Viability Kit, Molecular Probes).

Image analysis routines were developed to quantify overall amounts of red and green stained cells, location of stained cells within the flocs, and the spatial organization in clusters and filaments. It was found that the selection of the appropriate microscopic technique depends strongly on the type of microbial aggregates being Analysed.

For flocs with high cell density, the use of TPE-LSM is preferred, since it provides a clearer image of the internal structure of the aggregate. Epifluorescence microscopy did not allow to reliably quantify red stained cells in dense aggregates. CLSM did not adequately image the internal filamentous structure and the location of stained cells within dense flocs. However, for typical activated sludge flocs epifluorescence and CLSM proved adequate.

Parallel hybrid modeling methods for a full-scale cokes wastewater treatment plant; Lee DS et al.; Parallel hybrid modeling methods are applied to a full-scale cokes wastewater treatment plant. Within the hybrid model structure, a mechanistic model specifies the basic dynamics of the relevant process and a non-parametric model compensates for the inaccuracy of the mechanistic model. First, a simplified mechanistic model is developed based on Activated Sludge Model No. 1 and the specific process knowledge of the cokes wastewater treatment process.

Then, the mechanistic model is combined with five different non-parametric models - feedforward back-propagation neural network, radial basis function network, linear partial least squares (PLS), quadratic PLS and neural network PLS (NNPLS) - in parallel configuration. These models are identified with the same data obtained from the plant operation to predict dynamic Behaviour of the process. The performance of each parallel hybrid model is compared based on their ease of model building, prediction accuracy and interpretability. For this application, the parallel hybrid model with NNPLS as non-parametric model gives better performance than other parallel hybrid models. In addition, the NNPLS model is used to Analyse the Behaviour of the operation data in the reduced space and allows for fault detection and isolation.

Recovery of an environmental chlamydia strain from activated sludge by co-cultivation with Acanthamoeba sp; Collingro A et al.; Chlamydiae are a unique group of obligate intracellular bacteria comprising important pathogens of vertebrates as well as symbionts of free-living amoebae. Although there is ample molecular evidence for a huge diversity and wide distribution of chlamydiae in nature, environmental chlamydiae are currently represented by only few isolates. This chapter reports the recovery of a novel environmental chlamydia strain from activated sludge by co-cultivation with Acanthamoeba sp. The recovered environmental chlamydia strain UV-7 showed the characteristic morphology of chlamydial developmental stages as revealed by electron microscopy and was identified as a new member of the family Parachlamydiaceae (98.7% 16S rRNA sequence similarity to Parachlamydia acanthamoebae).

Infection studies suggested that Parachlamydia sp. UV-7 is not confined to amoeba hosts but is also able to invade mammalian cells. These findings outline a new straightforward approach to retrieving environmental chlamydiae from nature without prior, tedious isolation and cultivation of their natural host cells, and lend further support to suggested implications of environmental chlamydiae for public health.

Fenton peroxidation improves the drying performance of waste activated sludge; Dewil R et al.; Advanced sludge treatment processes (AST) reduce the mount of sludge produced and improve the dewaterability, thus probably also affecting the heat transfer properties and the drying characteristics of the sludge. This chapter studies the influence of the Fenton peroxidation on the thermal conductivity of the sludge.

Results demonstrate that the Fenton's peroxidation positively influences the sludge cake consistency and hence enhances the mechanical dewaterability and the drying characteristics of the dewatered sludge. For the two sludges used in this study, i.e. obtained from the wastewater treatment plants (WWTP) of Tienen and Sint-Niklaas - the dry solids content of the mechanically dewatered sludge increased from 22.5% to 40.3% and from 18.7% to 35.2%, respectively. The effective thermal conductivity k(e) of the untreated and the peroxidized sludges is measured and used to determine the heat transfer coefficient h(s).

An average improvement for k(e) of 16.7% (Tienen) and 5.8% (Sint-Niklaas) was observed. Consequently the value of h(s) increased with 15.6% (Tienen) and 5.0% (Sint-Niklaas). This increased heat transfer coefficient in combination with the increased dewaterability has direct implications on the design of sludge dryers. A plate-to-plate calculation of a multiple hearth dryer illustrates that the number of plates required to dry the peroxidized sludge to 90% DS is less than half the number of plates needed to dry untreated sludge. This results in reduced dryer dimensions or a higher capacity for an existing dryer of given dimensions.

Discrepancies in the widely applied GAM42a fluorescence in situ hybridisation probe for Gammaproteobacteria; Siyambalapitiya N et al.; A bacterial culture collection of 104 strains was obtained from an activated sludge wastewater treatment plant to pursue studies into microbial flocculation. Characterisation of the culture collection using a polyphasic approach indicated seven isolates, phylogenetically affiliated with the deep-branching Xanthomonas group of the class Gammaproteobacteria, were unable to hybridise the GAM42a fluorescence in situ hybridisation (FISH) probe for Gammaproteobacteria.

The sequence of the GAM42a probe target region in the 23S rRNA gene of these isolates was determined to have mismatches to GAM42a. Probes perfectly targeting the mismatches (GAM42a_T1038_G1031, and GAM42a_T1038 and GAM42a_A1041_A1040) were synthesised, and used in conjunction with GAM42a in FISH to study the Gammaproteobacteria community structure in one full-scale activated sludge plant. Several bacteria in the activated sludge biomass bound the modified probes demonstrating their presence and the fact that these Gammaproteobacteria have been overlooked in community structure analyses of activated sludge.

Toxicity of metals and organic chemicals evaluated with bioluminescence assays; Ren S et al.; The development of a bioluminescent sensor organism (Shk1) that was created for assessing wastewater toxicity was reported several years ago. In order to establish a test battery to better characterize wastewater toxicity, additional luminescent sensor organisms were later created. The present study focused on one promising candidate (PM6), a Pseudomonas spp. strain, because of its high level of luminescence compared to that of other newly created organisms. Using a batch toxicity testing protocol, the toxicity of 7 metals and 25 organic compounds was evaluated with the PM6 and Shk1 assays. Results indicated that the toxicity data of the PM6 and the Shk1 assays were correlated, and no assay appeared to be particularly more sensitive to a group of toxicants than the other assay. The results of the PM6 and Shk1 assays were further evaluated by comparing with the results of the Vibrio fischeri luminescence inhibition assay and activated sludge inhibition assays. Data suggested that PM6 and Shk1 more closely represented activated sludge organisms than V. fischeri. The suitability of using PM6 and Shk1 for assessing wastewater toxicity on activated sludge, both individually and in a test battery, was discussed.

Aerobic biodegradation of organotin compounds in activated sludge batch reactors; Stasinakis AS et al.; The biodegradation Behaviour of four organotin (OT) compounds, namely tributyltin (TBT), dibutyltin (DBT), monobutyltin (MBT) and triphenyltin (TPhT), was studied in lab-scale activated sludge batch reactors. The activated sludge was spiked with the OT compounds at a level of 100mugl(–1) as Sn.

Determination of the OT compounds by GC-FPD after ethylation in the dissolved and particulate phase revealed that 24h after the start of the experiments, almost the total of OT compounds has been removed from the dissolved phase and is associated with the suspended solids. Calculation of mass balance in batch reactors showed that OT compounds biodegradation was performed via a sequential dealkylation process.

Removals due to biodegradation were differentiated according to the parent compound. In experiments with non-acclimatized biomass, a percentage of 27.1, 8.3, 73.8 and 51.3 was still present as TBT, DBT, MBT and TPhT, respectively, at the end of the experiment (18th day). Half-lives (t(1/2)) of 10.2 and 5.1 days were calculated for TBT and DBT, respectively, whereas apparent t(1/2) values could not be determined for MBT and TPhT (t(1/2)>18 days).

The capacity of activated sludge to biodegrade OT compounds in the absence of supplemental substrate indicated that these compounds can be metabolized as single sources of carbon and energy in activated sludge systems. Excluding TBT, the presence of low concentrations of supplemental substrate did not affect the biodegradation potential of activated sludge.

The acclimatization of biomass on OT compounds enhanced significantly biodegradation, resulting in significant decreases of half-lives of OT compounds. As a result in the presence of acclimatized biomass, half-lives of 1.4, 3.6, 9.8 and 5.0 days were calculated for TBT, DBT, MBT and TPhT, respectively.

Genes Involved in Aniline Degradation by Delftia acidovorans Strain 7N and Its Distribution in the Natural Environment; Urata M et al.; Aniline-degraders were isolated from activated sludge and environmental samples and classified into eight phylogenetic groups. Seven groups were classified into Gram-negative bacteria, such as Acidovorax sp., Acinetobacter sp., Delftia sp., Comamonas sp., and Pseudomonas sp., suggesting the possible dominance of Gram-negative aniline-degraders in the environment. Aniline degradative genes were cloned from D. acidovorans strain 7N, and the nucleotide sequence of the 8,039-bp fragment containing eight open reading frames was determined.

Their deduced amino acid sequences showed homologies to glutamine synthetase (GS)-like protein, glutamine amidotransferase (GA)-like protein, large and small subunits of aniline dioxygenase, reductase, LysR-type regulator, small ferredoxin-like protein, and catechol 2,3-dioxygenase, suggesting a high similarity of this gene cluster to those in P. putida strain UCC22 and Acinetobacter sp. strain YAA. Polymerase chain reaction (PCR) and sequencing analyses of GS-like protein gene segments of other Gram-negative bacteria suggested that Gram-negative bacteria have aniline degradative gene that can be divided into two distinctive groups.

Leucobacter chromiireducens sp. nov, and Leucobacter aridicollis sp. nov., two new species isolated from a chromium contaminated environment; Morais PV et al.; Two strains designated strains L-1T and L-9T were isolated from activated sludge of a treatment plant that receives wastewater from the tannery industry contaminated with chromium.

Phylogenetic analysis showed that the organisms represented two new species of the genus Leucobacter. Strains L-1T and L-9T could be distinguished from the type strain of L. komagatae and from the type strain of "L. albus"by the B-type peptidoglycan composition, fatty acid composition, several phenotypic and physiological characteristics. The major fatty acids of the organisms were iso- and anteiso-branched C15:0 and C17:0, straight-chain C16:0 was also found in relatively high proportions.

The organisms were halotolerant, grew in medium containing 9% NaCl, and all strains, including the type strain of L. komagatae grew in medium containing 5 mM Cr(VI). On the basis of the distinct peptidoglycan composition, 16S ribosomal DNA sequence analysis, percentage of DNA-DNA reassociation values, and phenotypic characteristics we are of the opinion that strain L-1T represents a new species of the genus Leucobacter for which we propose the name Leucobacter chromiireducens and that strain L-9T represents an additional new species of the same genus for which we propose the name Leucobacter aridicollis.

The solids retention time-a suitable design parameter to evaluate the capacity of wastewater treatment plants to remove micropollutants; Clara M et al.; Micropollutants as endocrine disrupting compounds (EDC) or pharmaceuticals are of increased interest in water pollution control. Wastewater treatment plant (WWTP) effluents are relevant point sources for residues of these compounds in the aquatic environment. The solids retention time (SRT) is one important parameter for the design of WWTPs, relating to growth rate of microorganisms and to effluent concentrations.

If a specific substance is degraded in dependency on the SRT, a critical value for the sludge age can be determined. In WWTPs operating SRTs below this critical value, effluent concentrations in the range of influent concentrations or a distribution according to the adsorption equilibrium have to be expected, whereas in WWTPs operating at SRTs higher than the critical value degradation will occur. Critical SRTs were determined for different micropollutants, indicating that the design criteria based on the sludge age allows an estimation of emissions. Different treatment technologies as conventional activated sludge systems and a membrane bioreactor were considered and no significant differences in the treatment efficiency were detected when operated at comparable SRT.

The results of the investigations lead to the conclusion that low effluent concentrations can be achieved in WWTPs operating SRTs higher than 10 days (referred to a temperature of 10 degrees C). This corresponds to the requirements for WWTPs situated in sensitive areas according to the urban wastewater directive of the European Community (91/271/EEC) in moderate climatic zones.Reduction of bromate to bromide coupled to acetate oxidation by anaerobic mixed microbial cultures; van Ginkel CG et al.; Bromate, a weakly mutagenic oxidizing agent, exists in surface waters. The biodegradation of bromate was investigated by assessing the ability of mixed cultures of micro-organisms for utilization of bromate as electron acceptor and acetate as electron donor. Reduction of bromate was only observed at relatively low concentrations (<3.0mM) in the absence of molecular oxygen. Under these conditions bromate was reduced stoichiometrically to bromide.

Unadapted sludge from an activated sludge treatment plant and a digester reduced bromate without lag period at a constant rate. Using an enrichment culture adapted to bromate, it was demonstrated that bromate was a terminal electron acceptor for anaerobic growth. Approximately 50%of the acetate was utilized for growth with bromate by the enrichment culture. A doubling of 20h was estimated from a logarithmic growth curve. Other electron acceptors, like perchlorate, chlorate and nitrate, were not reduced or at negligible rates by bromate-utilizing microorganisms.

Fate of Ah receptor agonists during biological treatment of an industrial sludge containing explosives and pharmaceutical residues; Gustavsson LK et al.; Goal, Scope and Background: Sweden is meeting prohibition for deposition of organic waste from 2005. Since 1 million tons of sludge is produced every year in Sweden and the capacity for incineration does not fill the demands, other methods of sludge management have to be introduced to a higher degree. Two biological treatment alternatives are anaerobic digestion and composting. Different oxygen concentrations result in different microbial degradation pathways and, consequently, in a different quality of the digestion or composting residue, It is therefore necessary to study sludge treatment during different oxygen regimes in order to follow both degradation of compounds and change in toxicity.

In this study, an industrial sludge containing explosives and pharmaceutical residues was treated with anaerobic digestion or composting, and the change in toxicity was studied. Nitroaromatic compounds, which are the main ingredients of both pharmaceutical and explosives, are well known to cause cytotoxicity and genotoxicity.

However, little data are available concerning sludge with nitroaromatics and any associated dioxin-like activity. Therefore, we studied the sludge before and after the treatments in order to detect any changes in levels of Ah receptor (AhR) agonists using two bioassays for dioxin-like compounds. Methods: An industrial sludge was treated with anaerobic digestion or composting in small reactors in a semi-continuous manner.

The same volume as the feeding volume was taken out daily and stored at –20 degrees C. Sample preparation for the bioassays was done by extraction using organic solvents, followed by clean up with silica gel or sulphuric acid, yielding two fractions. The fractions were dissolved in DMSO and tested in the bioassays. The dioxin-like activity was measured using the DR-CALUX assay with transfected H4IIE rat hepatoma pGudluc cells and an EROD induction assay with RTL-W1 rainbow trout liver cells.

Reduction of excess sludge production by 3,3',4', 5-tetrachlorosalicylanilide in an activated sludge process; Ye FX et al.; The potential of 3,3',4',5-tetrachlorosalicylanilide (TCS) addition to an activated sludge continuous process to reduce excess sludge production by disrupting coupling between anabolism and catabolism was investigated. TCS was chosen as a metabolic uncoupler for continuous test in a lab-scale completely mixed activated sludge process. TCS reduced sludge yield by approximately 30% at a dosage of 40 mg/day. Substrate removal capability was not adversely affected by the presence of TCS, but effluent nitrogen concentration increased during the 60-day continuous operation. Sludge settleability of treated and control samples was qualitatively comparable and not significantly different. Microbial activities in terms of specific oxygen uptake rate were also enhanced, and the microbial population was altered. The results suggest that TCS is an effective chemical uncoupler that reduces sludge yield; process performance was not significantly affected by introduction of the uncoupler.

Stability of the fluorogenic enzyme substrates and pH optima of enzyme activities in different Finnish soils; Niemi RM et al.; Fluorogenic artificial substrates facilitate sensitive enzyme activity measurements for a variety of processes in soil and other environmental samples. It is possible to use in situ pH for measurements on condition that the substrates are chemically stable.

We studied the stability of 12 different methyl umbellipherone (MUF) and amino methyl coumarine (AMC) derivatives used as substrates for arylsulphatase, alpha-glucosidase, beta-glucosidase, beta-xylosidase, cellobiosidase, chitinase, phosphomonoesterase (PME), phoshodiesterase (PDE), esterase, lipase and alanine- and leucine aminopeptidases (AP) over the pH range from 4.0 to 8.0 in modified universal buffer (MUB).

Stability of the substrates for lipase (4-MUF-heptanoate) and esterase (4-MUF-acetate) measurements was poor, especially at the higher pH values. Chitinase substrate, 4-MUF-N-acetyl-beta-d-glucosamide, was unstable at high pH values whereas the substrate for PME activity measurement (4-MUF- phosphate) disintegrated at low pH.

The other substrates and MUF and AMC standard solutions were stable over the pH range studied. The optima between pH 4 and 8 of the 11 different enzyme activities were measured in three forest and two agricultural soil samples and in one activated sludge sample. In soil, for alanine and leucine AP the pH optima were usually 7.5 or higher, for arylsulphatase, beta-glucosidase, beta-xylosidase, esterase and PDE between 4 and 5.5, and for cellobiosidase between 4 and 5. alpha-Glucosidase had an optimum below 5.5 but also exhibited high activity at pH 7. Soil-dependent variation in pH optima were observed for chitinase, esterase, PDE and PME. Enzyme activities were also measured in 0.5 M acetate buffer at pH 5.5. This buffer yielded the highest activities in all soil samples for arylsulphatase, PDE and PME.

Persistent organic pollutants (POPs) in the conventional activated sludge treatment process: fate and mass balance; Katsoyiannis A et al.; The fate and the mass balance of persistent organic pollutants (POPs) during the conventional activated sludge treatment process were investigated in the wastewater treatment plant of the city of Thessaloniki, northern Greece. The POPs of interest were 7 polychlorinated biphenyls and 19 organochlorine pesticides. Target compounds were determined at six different points across the treatment system: the influent, the effluent of the primary sedimentation tank, the effluent of the secondary sedimentation tank, the primary sludge, the activated sludge from the recirculation stream, and the digested/dewatered sludge.

The distribution of POPs between the dissolved and the adsorbed phases of wastewater and sludge was investigated. A good linear relationship between the distribution coefficients, K(d), and the octanol-water partition coefficients, K(ow), of the solutes was observed only in raw wastewater, suggesting that other factors affect the phase distribution of organic compounds in treated wastewater. For all POPs, a significant increase in partitioning with a decreasing solids concentration was observed, revealing an effect from non- settling microparticles remaining in the “dissolved” phase during the separation procedure.

A good linear relationship was also revealed between logK(d) and the dissolved organic carbon (DOC) content of wastewater, suggesting that DOC Favours the advective transport of POPs in the dissolved phase. Almost all POPs showed good mass balance agreements at both the primary and the secondary treatment. The losses observed for some species could be attributed to biodegradation/ biotransformation rather than volatilization. The relative distribution between the treated effluent and the waste sludge streams varied largely among different compounds, with p-p'-DDE being highly accumulated in the waste sludge (98%) but almost 60% of alpha-HCH remaining in the treated effluent.

Sludge as source of energy and revenue; Onyeche TI; Sludge is a residue/ product from wastewater treatment plants and contains most of the contaminants released during human activities. Some stringent environmental regulations on sludge treatment and disposal exist in many countries. This has resulted in increasing interest in sludge treatment methods that encourage sludge reduction and improvement in biogas production during anaerobic digestion processes. This work demonstrates the first exploitation of valuable energy from homogenised sludge at technical scale with mass reduction. The optimal combination of sludge homogenisation at relatively low pressures using a modified high-pressure homogeniser led to the success of this unique project. Results showed that about 30% more energy could be obtained from thickened and disrupted sludge than from untreated samples. The energy produced was higher than that invested during disruption and digestion processes. About 23% sludge reduction was also observed with no increase in chemical oxygen demand.

This new process can produce extra energy for local electrification and heating the digester while the sludge reduction provides economic benefits. Concentration of sludge causes reduction in investment cost on digester as well as reduction in operational time for sludge dewatering.

Direct generation of electricity from sludges and other liquid wastes; Dentel SK et al.; Recently it has been shown that electrical energy can be harvested from marine sediments, simply by connection of an electrode (anode) in anaerobic marine sediments to an electrode in the aerobic zone above the sediments. We have now shown that similar applications are available in sludge treatment. Using a reactor with graphite foil electrodes in an aerated aerobic and anaerobic sludge zone, electrical current was generated, and enhanced when an additional organic substrate (acetate) was added. Electron microscopy, x-ray diffraction, and PCR examination of the anode surface showed no surface colonization and no increase in Geobacter relative to a control, indicating that microbial use of the anode as an electron acceptor wa indirect through the use of redox mediators. Given the demonstration of electricity generation from sludge, the potential for similar applications, using other organic waste sources, is evaluated.

Physical characteristics of a waste activated sludge: conditioning responses and correlations with a synthetic surrogate; Dursun D et al.; The efficient and economic management of waste activated sludge (WAS) requires a proper understanding of the sludge's material properties. Though there has been much study of WAS, an adequate linkage between its physical and chemical properties has been elusive. In particular, the conditioning and dewatering of WAS are expensive operations, and the addition of polymer to WAS leads to even more complex material Behaviour that is difficult to optimize. This chapter reports on an extensive characterization of WAS, both with and without polymer conditioning. We combined the classical "jar test" approach with less conventional rheometric and electrokinetic measurements. In addition to the use of sampled WAS, a synthetic surrogate sludge was formulated, attempting to duplicate properties of WAS and allow more extensive characterization of a reproducible surrogate to WAS. Results with both the synthetic surrogate and WAS indicated that the traditional, electrokinetic, and rheological properties were related. However, the dose optima by rheometry were somewhat higher in both cases. Minimization of sludge production and stable operational condition of a submerged membrane activated sludge process; Bhatta CP et al.; Modification of conventional activated sludge process by adding microfiltration type hollow fibre membrane module offers various advantages. The membrane module is immersed directly in the bioreactor to separate the treated water from the sludge and the modified process is called submerged membrane activated sludge process (SMASP). Bench scale SMASP units were operated at constant flux to study the possibility of minimizing the excess sludge production and stable operation at different SRT and BOD loading.

The long SRT developed high stabilized MLSS concentration in the bioreactor and decreased the sludge loading rate {kg-BOD/kg-MLSS-d}. In SMASP, very low sludge yield coefficients (0.04-0.09 kg-MLSS/kg-BOD) were obtained at long SRT (500 d) and small BOD loading (< or =0.5 kg/(m3 × d)) as compared to conventional activated sludge process and these coefficients showed considerable minimization in excess activated sludge production. At the same operation conditions, suction pressure was also maintained low (<25 kPa), and then SMASP was operated stably for long time (approximately 200 d).Co-conditioning and dewatering of alum sludge and waste activated sludge; Lai JY et al.; Co-conditioning and dewatering Behaviours of alum sludge and waste activated sludge were investigated. Two different sludges were mixed at various ratios (2:1; 1:1; 1:2; 1:4) for study. Capillary suction time (CST) and specific resistance to filtration (SRF) were utilized to assess sludge dewaterability. Relatively speaking, waste activated sludge, though of higher solid content, was more difficult to be dewatered than alum sludge.

It was found that sludge dewaterability and settlability became better with increasing fraction of alum sludge in the mixed sludge. Dosage required of the cationic polyelectrolyte (KP-201C) for dewatering was reduced as well. It is proposed that alum sludge acts as skeleton builder in the mixed sludge, and renders the mixed sludge more incompressible which is beneficial for sludge dewatering.

e, a, g. Implications of the results of the study to the sludge management plan for Taipei City that generates both alum sludge and waste activated sludge at significant amount are also discussed. The current sludge treatment and disposal plan in Metropolitan Taipei could be made more cost-effective.

Ultrasonic sludge treatment for enhanced anaerobic digestion; Hogan Fet al.; Ultrasound is the term used to describe sound energy at frequencies above 20 kHz. High-powered ultrasound can be applied to a waste stream via purpose-designed tools in order to induce cavitation. This effect results in the rupture of cellular material and reduction of particle size in the waste stream, making the cells more amenable to downstream processing. sonix is a new technology utilising high-powered, concentrated ultrasound for conditioning sludges prior to further treatment. This chapter presents recent results from a number of demonstration and full-scale plants treating thickened waste activated sludge (TWAS) prior to anaerobic digestion, therefore enhancing the process. The present studies have proved that the use of ultrasound to enhance anaerobic digestion can be achieved at full scale and effectively result in the TWAS (typically difficult to digest) behaving, after sonication, as if it were a "primary"sludge. The technology presents benefits in terms of increased biogas production, better solids reduction, improved dewatering characteristics of the digested sludge mixture and relatively short payback periods of two years or less subject to the site conditions and practices applicable at that time.

Three-dimensional dual-morphotype species modeling of activated sludge flocs; Martins AM et al.; An individual-based model, originally developed for a biofilm system, was adapted to simulate three-dimensional formation of activated sludge flocs. The model was extended to two different bacterial morphotypes (floc-forming and filamentous bacteria), allowing spatial development of the floc according to the bacterial morphology, diffusion, reaction, and growth processes.

The model needed also extension with a process for attachment of individual cells. Despite being in an early stage of development, the model is already a tool that enables us to obtain useful information about the microfloc environment. The model indicates that filamentous bacterial morphology and substrate microgradients are important aspects in the formation of bacterial structures.

In mass transport-limited regimes filamentous bacterial structures prevail, whereas in growth-limited regimes irregular shaped flocs with fingerlike structures are dominant. These modeling results suggest that activated sludge flocs and biofilms might be different manifestations of the same phenomena. The model results support the hypothesis that floc-macrogradients can be the most important parameter for development of bulking sludge.

The model suggests that attachment has a very strong effect on floc structure, leading to enhancement of the effect of substrate microgradients. Physicochemical properties and stability of activated sludge flocs under temperature upshifts from 30 to 45 degrees C; Morgan-Sagastume F et al.; The impacts of temperature shifts from 30 to 45 degrees C on the structural stability and surface charge of activated sludge flocs were assessed in four sequencing batch reactors (SBRs) treating pulp and paper mill effluent. The improvement in floc stability was tested by sludge magnesium enrichment in one SBR and by operating another reactor at a high sludge retention time (SRT) of 33 days. Floc stability was characterized by dissociation constants with solutions of CaCl(2), KCl, urea, and ethylenediamine tetraacetate (EDTA).

Surface charge was assessed by cationic-anionic titration and metals concentrations were also determined. The temperature shift consistently caused an increase in the negative sludge surface charge from approximately –0.180 to –0.300 meq/g MLSS.

Magnesium enrichment and a high SRT of 33 days promoted less negatively charged sludge, dampened the increase in negative sludge surface charge, and yielded structurally stronger flocs; however, sludge deflocculation still occurred. Manganese and iron appeared to be released by sludge under the temperature shift.

It was concluded that the temperature shift deteriorates the flocculating physicochemical properties of the sludge and that better floc stability achieved by magnesium enrichment and a high SRT is not enough to stop deflocculation. Further research is required to clarify the origin of the increase in negative sludge surface charge, the role of metals, and the governing factors in sludge deflocculation under such temperature shifts.

Optimized sequence retrieval from single bands of temperature gradient gel electrophoresis profiles of the amplified 16S rDNA fragments from an activated sludge system; Zhang X et al.; Sequence retrieval from single bands of polymerase chain reaction (PCR)-denaturing gel electrophoresis (DGE) profiles is an important but often difficult step for molecular diversity analysis of complex microbial communities such as activated sludge systems.

We Analysed the temperature gradient gel electrophoresis (TGGE) profiles of PCR-amplified 16S rDNA fragments from an activated sludge sample of a coking wastewater treatment plant. Single bands were excised, and a clone library was constructed for each.

Sequence heterogeneity in each single band was found to be significantly overestimated due to single-stranded DNA (ssDNA) contamination formed during the PCR amplification, since only 10-60% of library clones of each single TGGE band had identical migration Behaviour compared with the parent band.

Three methods, digestion with mung bean nuclease, optimization of PCR amplification, and purification via denatured polyacrylamide gel electrophoresis (d-PAGE), were compared for their ability to minimize ssDNA contamination, with the last one being the most efficient. After using d-PAGE to minimize ssDNA to a nearly nondetectable level, 70-100% of library clones for each single TGGE band had identical migration compared with the parent band.

Several sequences were found in each of six single bands, and this co-migration could be predicted with the Poland software. The predominant bacteria of the activated sludge were assessed via a combination of sequence retrieval from each single TGGE band and band intensity analysis. Only beta and alpha subclasses of the Proteobacteria were detected, 93.8% and 6.2%, respectively. Our work suggests that prior to constructing a clone library to retrieve the actual sequence diversity of a single DGE band, it is advisable to minimize ssDNA contamination to a nondetectable level.

High rate treatment by aerobic upflow sludge blanket (AUSB) with external oxygenation; Sharma KR et al.; A 3-year study was conducted using Encyclopaedia of Environmental Science an aerobic upflow sludge blanket (AUSB) reactor to achieve high-rate biological treatment through maintenance of a high biomass concentration (7-10 g VSS/L) together with a high oxygen flux. The AUSB reactor was not aerated directly; instead, oxygenation was provided in an external chamber with flow recirculation. The oxygenation was provided at four different pressurizations: 0, 15, 20 and 25 psig. The AUSB reactor was also pressurized to avoid the flotation of biomass. The flow recirculation rate was varied from 400% to 1,500% in order to ensure adequate oxygen supply. It was found that the AUSB system was able to handle a volumetric loading of as high as 10 kg COD/m3- day with a removal efficiency of 92%. Despite a high upflow velocity throughAUSB, the effluent suspended solid concentration was mostly below 60 mg/L. The active fraction of biomass in the AUSB sludge was about 3 times higher than that of the regular activated sludge. This was indicated by a very high specific oxygen uptake rate (SOUR), up to 180-250 mg O2/g VSS-hr. The sludge yield in the entire system was only 0.09 to 0.13 g VSS/g COD removed. This was mainly caused by additional auto-oxidation of biosolids in the oxygenation chamber due to flow recirculation.

Effect of chlorination bulking control on water quality and phosphate release/uptake in an anaerobic-oxic activated sludge system; Chang WC et al.; This study evaluates the effect of chlorination bulking control on water quality and phosphate release/uptake in an anaerobic-oxic activated sludge system. A series of batch experiments with different specific NaOCl mass dose were conducted to determine the sludge settling properties, supernatant water quality and phosphate metabolism Behaviour of filamentous bulking sludge.

The harvested sludge was from a continuous-flow anaerobic-oxic (A/O) activated sludge pilot-plant, i.e., enhanced biological phosphorus removal (EBPR) system, operated with 15 days of sludge retention time. The filamentous bacteria in the A/O pilot plant were identified to be Thiothrix according to Eikelboom's classification techniques, which was in accordance with the high influent sulfate concentration of this study (50 mg/L sulfate). Increasing NaOCI concentration, as revealed by experimental results, obviously decreased the sludge settling properties (SVI values and zone settling velocities) and meanwhile significantly reduced supernatant water quality (COD, SS, TP) mainly due to higher suspended solids caused by floc disruption. Moreover, the nine-hour batch experiments indicated that high NaOCI dosage (40 mg/ gMLSS) completely deteriorated phosphate metabolism of EBPR sludge. Such a high dosage of chlorination further confirmed overdosing through disappearance of intracellular PHB and death of protozoa by microscopic investigation.

Still, phosphate release/uptake Behaviour of EBPR sludge properly functions at low NaOCl dosage (5 mg/g MLSS). Besides, phosphate metabolism worsens rapidly before the SVI value reaches its lowest level. These findings imply that determining NaOCI requirement with merely SVI values can readily result in chlorination overdosing. Proper NaOCI dosage requires a delicately balanced consideration between sludge settling improvement, water quality demand and phosphate metabolism. Batch test of phosphate release/uptake is apparently a prerequisite to conclude an appropriate NaOCl dosage for bulking control.

Microbial degradation of estrogens using activated sludge and night soil-composting microorganisms; Shi JH et al.; In order to investigate the potential for microbial degradation of estrogens, and the products formed, activated sludge collected from Korea (ASK) and night soil-composting microorganisms (NSCM) were used to degrade estrogens. Results showed that both ASK and NSCM degraded almost 100% of the natural estrogens estrone (E1), 17beta-estradiol (E2), and estriol (E3) from initial concentrations of 20-25 mg/L, while synthetic estrogen, ethynylestradiol (EE2), was not degraded. Analysis of degradation products of E2 by using HPLC-ECD and a consecutive first-order reaction calculation confirmed that E2 was sequentially degraded to E1, which was further degraded to other unknown compounds by ASK and NSCM.

We then used the yeast two-hybrid assay to show that the unknown degradation products did not appear to possess estrogenic activity when E1, E2 or E3 were degraded to below the detection limit after 14 days of incubation, indicating that ASK and NSCM not only degrade natural estrogens, but also remove their estrogenic activities. Effects of phosphorus limitation and temperature on PHA production in activated sludge; Chinwetkitvanich S et al.; The study was designed to investigate the effects of temperature and phosphorus limitation on polyhydroxyalkanoate (PHA) production and storage by activated sludge biomass. The two-stage operation approach, i.e. a growth phase followed by a nutrient limitation phase, was applied to induce PHA accumulation.

The pre-selected temperatures of 10, 20 and 30 degrees C were investigated under phosphorus limitation conditions using three four-litre fully aerobic SBR systems operated at an SRT of 10 days with cycle time and HRT of 6 and 10 hours. PHA production was greater in the 10 degrees C system than in the 20 degrees C and 30 degrees C systems but there was little difference between the two higher temperatures. The maximum PHA fractions of the sludge were 52, 45 and 47%TSS for the three temperatures from low to high, and the maximum PHA concentrations in the mixed liquors were 1,491, 1,294 and 1,260 mg/l, respectively. However, it was observed that very low values of PHA yield per unit COD consumed were obtained, i.e., 0.05, 0.03 and 0.04 mgPHA/mgCODu, for the 10, 20 and 30 degrees C reactors, respectively. This was because all three systems required several days to reach maximum PHA accumulation in their mixed liquor biomasses. It is probable the bacteria still had some stored poly-P in their cells upon initiation of the phosphorus limited influent, and PHA accumulation was delayed until the stored phosphorus was depleted. Also,

PHA productivity was reduced by the large amounts of biomass lost from the systems because of sludge bulking.

Sludge thickening performance of mesh filtration process; Park MS et al.; Small-scale wastewater treatment facilities play an important role in improving the aquatic environment in many countries. Although sludge treatment is essential for overall wastewater treatment, it is difficult for small-scale facilities to use mechanical equipment or other facilities. As the first step of the sludge treatment, it is important to develop a convenient sludge thickening process for small-scale facilities.

In this work, we examined the sludge thickening performnce of a mesh filtration system: the mesh opening sizes of 100-500 microm, and the sludge (3,000-9,000 mg-SS/L) was obtained from a domestic wastewater treatment facility. The filtration was carried out only under the hydraulic pressure between the water level and the effluent port connected to the mesh filter module. The sludge reduction rates were in the range of 85-95% for 6-7 h; the initial filtration rate was very high, but the rate decreased with a decrease in hydraulic pressure due to the reduction of the water level in the vessel. In addition, the effluents (passed through the mesh) contained very low SS and could be directly discharged into the environment.

Biological treatabilty of raw and ozonated penicillin formulation effluent; Cokgor EU et al.; In the present study, oxidative pre-treatment of harmaceutical wastewater originating from the formulation of the penicillin Sultamycillin Tosylate Diydrate via ozonation at varying pH and ozone feed rates was investigated. Biological treatability studies were performed with a synthetic wastewater alone and supplemented with raw and ozonated penicillin formulation effluents.

The highest COD (34%) and TOC (24%) removal efficiencies were obtained at pH 11.0, whereas the BOD5 value increased from 16 mg l(–1) to 128 mg l(–1) after 40 min of ozonation, corresponding to an applied ozone dose of 1670mg l(-1) and 33% relative ozone absorption. The studies showed that no degradation of raw penicillin fraction (30% of total COD) occurred, and degradation of the synthetic wastewater being completely treatable without penicillin addition, was inhibited by 7%. Upon 40 min ozonation, the synthetic wastewater could be completely oxidized and at the same time 35% of ozonated penicillin wastewater removal was obtained.

Respirometric studies were conducted in parallel and produced results indicating a 22% decrease in the total oxygen consumption rate established for raw penicillin formulation effluent compared to the results obtained from the aerobic batch reactor. No inhibition of the synthetic fraction was observed for the 40 min-ozonated penicillin formulation effluent, biodegradability of the 60 min-ozonated penicillin effluent decreased possibly due to recalcitrant oxidation product accumulation.

The modeling study provided experimental support and information on inhibition kinetics in activated sludge model no. 3 (ASM3) by means of respirometric tests for the first time. Reduction of coagulant amount added to activated sludge for phosphorus removal; Nakajima J et al.; Adding coagulant to the activated sludge process is effective in maintaining the stability of phosphorus removal. However, the precise mechanisms of the reaction and Behaviour of coagulants and phosphorus are not well known. By introducing a new phosphorus removal model (PRM), the Behaviour of coagulant and phosphorus in the process could be described. The experimental data of the effluent phosphorus concentration and Fe content in the activated sludge agreed with the values calculated by PRM. The amount of coagulant additionto the activated sludge process for phosphorus removal is reduced with the enhanced biological phosphorus removal process. It is suggested that the amount of reduction is determined by using PRM.

Monitoring activated sludge settling properties using image analysis; Jenne R et al.; The goal of this study is to develop a monitoring system for activated sludge properties, as this is an essential tool in the battle against filamentous bulking. A fully automatic image analysis procedure for recognising and characterising flocs and filaments in activated sludge images has been optimised and subsequently used to monitor activated sludge properties in a lab-scale installation. The results of two experiments indicate that the image information correlates well with the Sludge Volume Index. It is shown that, at the onset of filamentous bulking, there is an increase in total filament length on the one hand, and a change in floc shape on the other hand.

Performances of three R-AN-D-N wastewater treatment plants in the Czech Republic; Bejvl Z et al.; The evolution of Czech standards requires higher efficiency of nutrient removal from municipal wastewaters. At the beginning of the last decade of 20th century, a new activated sludge configuration called R-AN-D-N process has been described, successfully tested and now largely used at several wastewater treatment plants (WWTP) in the Czech republic.

The main feature of the R-AN-D-N process is the introduction of a regeneration zone in sludge recycle, which enables to increase sludge age in the system without any substantial increase in WWTP volume. Performances of three Czech large WWTP with R-AN-D-N configuration have been monitored and compared within a period of one and a half years.

The results confirmed excellent nutrient removal efficiency for wastewaters with different proportion between sewage and industrial effluents. Two of three monitored WWTP received wastewaters from breweries (Budweiser and Pilsner Urquell). The settleability of activated sludge from all three WWTP was correct, with SVI values usually ranging from 50 to 150 ml/ g. Monitoring of sludge composition indicated proliferation of several filamentous bacteria, particularly types 0581, 0092 and M. parvicella. No severe bulking events were observed. Finally, the operational costs expressed in CZK (Czech crown: 1 CZK = {see symbol in text}0.0322) per cubic metre of treated sewage or per capita amounts respectively from 2.24 to 6.52, and from 285 to 342.

North American performance experience with anoxic and anaerobic selectors for activated sludge bulking control; Parker D et al.; Obtaining reliably high compaction characteristics and low SVI values has enormous economic consequences on activated sludge plant design and operation. Now that use of selectors has become more common for achieving low SVI values, the assessment of the full-scale performance characteristics of activated sludge plants incorporating selectors has high value for the design and operation of future plants.

Treatment plant effluent quality is seldom Analysed solely on an average basis and the impact peak events have on effluent quality must be considered in design and operation. Descriptive statistics should also be used to assess selector performance, so that the influence of infrequently occurring values on plant design and operation can be assessed. The performance of 21 activated sludge plants incorporating selectors was evaluated. All the selector plants for which post installation data were available showed that the operating SVI values were significantly improved. One plant studied in detail showed that final effluent quality was significantly better after the selector installation. As a group, activated sludge plants with anaerobic selectors outperform those with anoxic selectors. Dissolved oxygen control is just as important in activated sludge plants with selectors as in conventional activated sludge plants.

Influence of sludge properties and hydraulic loading on the performance of secondary settling tanks--full-scale operational results; Vestner RJ et al.; Full-scale investigations at a WWTP with a two-stage secondary settling tank process revealed relationships between significant operating parameters and performance in terms of effluent suspended solids concentration. Besides common parameters (e.g. surface overflow rate and sludge volume loading rate) feed SS concentration and flocculation time must be considered. Concentration of the return activated sludge may help to estimate the performance of existing secondary settling tanks.

Modeling of a large-scale wastewater treatment plant for efficient operation; Gokcay CF et al.; Environmental legislations in the Western world impose stringent effluent quality standards for ultimate protection of the environment. This is also observed in Turkey. The current paper presents efforts made to simulate an existing 0.77 million m3/day conventional activated sludge plant located at

Ankara, AWTP. The ASM1 model was used for simulation in this study. The model contains numerous stoichiometric and kinetic parameters, some of which need to be determined on case by case bases. The easily degradable COD (S(S)) was determined by two methods, physical-chemical and respirometric methods, namely. The latter method was deemed unreliable and rejected in the further study. Dynamic simulation with SSSP Programmeme predicted effluent COD and MLSS values successfully while overestimating OUR. A complete fit could only be obtained by introducing a dimensionless correction factor (etaO2 = 0.58) to the oxygen term in ASM1.

A control strategy for reducing aeration costs during low loading periods; Sahlmann C et al.; The efficiency of the aeration system in a full-scale activated sludge basin with 3 separately controlled aeration zones was improved for the low loading period in summer. The air flow rate to each aeration zone is currently regulated to hold a preset dissolved oxygen concentration (DO).

Four different DO setpoint combinations were tested, each one for a one week period, using dynamic off-gas testing to measure the standardised oxygen transfer efficiency (alphaSOTE). As the DO setpoints were lowered, the total air flow rate to the basin decreased initially.

A low DO in the first zones slowed biomass activity and pushed the load towards the end of the aeration basin. The relationship between alphaSOTE and the specific diffuser flow rate qD is different for each zone. In Zone 1 there was a strong decrease in alphaSOTE as qD increased, while Zones 2 and 3 were fairly independent of qD, Zone 2 at a higher level than Zone 3. Aeration costs were reduced by 15% for the most efficient combination. To achieve even more savings, a control strategy adjusting oxygen transfer rates over the aeration basin to the necessary oxygen transfer rates is suggested. It is based on changing the DO setpoints to reach the lowest total air flow rate while meeting the effluent requirements.

Nutrient removal process development and full scale implementation at the 4 million p.e. main treatment plant of Vienna, Austria; Kroiss H et al.; The Main Treatment Plant of Vienna is in extension for 4 million p.e. and very stringent nutrient removal requirements. The existing high rate BOD removal activated sludge plant (in operation since 1980) is extended by a second stage activated sludge plant and a newly developed flow scheme for nitrogen removal optimisation adaptable to the temperature variations over the year. For this plant pilot investigations have been performed for the development of a specific mathematical model (ASMV) and a specific aeration control strategy.

The civil work of the extension is already finished and the installation of the equipment has started. Operation should start in 2004. The whole project will cost about {see symbol in text 264 million of which about one half is for civil work. The effluent standards correspond to the requirements for sensitive areas in EU Directive for Municipal Waste Water. The raw primary and excess sludge are incinerated after thickening and dewatering. This chapter tries to condense the already existing literature with the construction progress and the cost situation.

Isolation and characterization of sulphur-oxidizing Thiomonas sp. and its potential application in biological deodorization; Chen XG et al.; AIMS: To isolate and characterize a sulphur-oxidizing bacterial strain from activated sludge and to evaluate its potential application in biological deodorization. Methods and Results: A dominant sulphur-oxidizing bacterial strain, designated as strain SS, was isolated from an enrichment culture using thiosulphate as a sole energy source and CO_2 as a sole carbon source.

The cells of this organism were aerobic, rod-shaped, Gram-negative and motile. Strain SS could grow autotrop-hically, heterotrophically as well as mixotrophically. Autotrophic growth was observed at pH values ranging from 2.3 to 9.0. Phylogenetic analyses revealed that strain SS belonged to Group 1 of the genus Thiomonas, closely related to Thiomonas perometabolis and Thiomonas intermedia. The thiosulphate oxidation rates of strain SS at different pH values were evaluated in terms of oxygen uptake using a Micro-Oxymax respirometer. The results showed that the maximum oxidation rate of 5.65 mg l(–1) h(–1) occurred at 56 h of growth and pH 6.0. Continuous H2S removal study demonstrated that strain SS could remove more than 99% of H2S when the inlet concentration was below 58.6 ppm.

Further increase of the inlet concentration to 118 ppm gave rise to a decline in the removal efficiency to ca 90%. Conclusions: The strong acidification of the culture medium during the later period could result in the deterioration of the growth activity and the metabolism activity of strain SS. In practical application, the problems caused by the end-product inhibition and the acidification can be alleviated by periodical replacement of culture medium with fresh medium. d, b. Given the physiological flexibility and the ability to remove H2S rapidly and efficiently, strain SS could be a good 'deodorizing' candidate. Significance and impact of the study: This is the first time that Thiomonas species has been reported for biological deodorization application. Start up of deammonification process in one single SBR system; Li X et al.; A process for autotrophic nitrogen removal named aerobic/anoxic deammonification wherein NH4+ is oxidized by nearly 50% to NO2– and subsequently the ammonia is converted together with the nitrite to molecular nitrogen (N2 gas), has come to full-scale application within the last few years. In this research, sludge from a biological rotation disk located at a landfil leachate plant at Mechernich, Germany, which is capable of performing the deammonification process, was used as seed sludge for acclimating deammonification activities in Labouratory scale batch-reactors.

In parallel, the same tests were performed with normal activated sludge. Research results indicated that deamm-onification activities could be obtained from the seeded reactor and also, with limited performance, from normal activated sludge in a single SBR system after several months acclimation. It was also seen that oxygen is an important factor that influences the deammonification from both the acclimatization process and process running.

Further results were approved that report an impact of nitrite as a process intermediate on the closely related process of anaerobic ammonia oxidation ("Anammox"). However, limiting concentrations on a bacteria population performing deammonification were found to be different to those reported for a pure Anammox-culture. Also the influence of another intermediate, hydrazine, was tested for speeding up the acclimating process by inducing the deammonification activities and recovering the activities of deammonification from nitrite inhibition. ❐

5

Wind and Climate Belts

This great, rotating ball that is Earth harbours within its atmosphere athousand varieties of wind and weather over the diverse oceans and continents of its expansive surface, even at any one instant of time; and all these weather vagaries are constantly changing, quickly or slowly, from day to day, week to week, month to month.

Nevertheless there is a grand and logical scheme underlying general world-wide weather—a pattern that unmistakably appears in the average weather, or climate, experienced at representative places spaced up and down the meridians of latitude; and in the prevailing winds that arise out of these belt-like, world-encircling regions of climatic unlikeness.

This average planetary wind-and-climate picture is seldom, if ever, duplicated by actual world-wide weather on any single day—particularly in north-temperate regions, where actual weather variability seems to discredit the existence of any such regular scheme. Yet over the oceans—where the sea surface is relatively flat and relatively frictionless to wandering breezes, where sameness of temperature discourages climatic oddities and local weather freaks, where the climate belts and prevailing winds are most evident, typical and regular—it is possible to glimpse in actual weather some prominent features of the grand planetary pattern. You can best sense certain of its gradual but definite and continuing changes, from latitude to latitude, by cruising north or south in a seagoing ship.

Here, for example, is the actual record of a March cruise from Panama to Philadelphia in the old but able American freighter Timber Rush—a slow two-thousandmile cruise from torrid to temperate latitudes at the comfortable ocean-going dog-trot of ten knots:

March 11th:	latitude 10°N, off Panama coast, temperature 79°F, broken clouds, wind ENE moderate (trade wind), caught barracuda on trolling line.
12th:	Lat 12°, in Caribbean, temp 79°, scattered to broken clouds, wind NE X N fresh to strong (norther), sea rough, water temperature 78°.
13th:	Lat 14½°, in Caribbean, temp 77°, scattered clouds, wind ENE fresh (intensified trade), water very rich blue.
14th:	Lat 18°, just southwest of Haiti, temp 78°, scattered, clouds to clear, wind E'ly light (in lee of Hispaniola).
15th:	Lat 22°, near Bahamas and Tropic of Cancer, temp 77°, scattered clouds, wind E'ly to variable light (horse latitudes) scattered clouds to clear.
16th:	Lat 26½°, in Atlantic, temp 73°, showers and thundershowers along a cold front, wind SW'ly moderate shifting to NW fresh with frontal passage.
17th:	Lat 30½°, in Atlantic, temp 67°, scattered to broken clouds, wind NW moderate (prevailing westerlies), water dark steely blue, water temperature 74°.
18th:	Lat 35°, off Cape Hatteras in Gulf Stream, temp 60°, broken clouds, wind N'ly fresh and increasing, water temperature 72°, passed school of whales.
19th:	Lat 38°, entering Delaware Bay, temp 35°, clear, calm to S'ly light, water temperature 50° after leaving Gulf Stream, 42° near coast, water dirty green.

Thus we passed, in the last four days of slow steaming at two hundred and forty easy sea miles a day, from a tropical climate to a temperate climate, from smiling summer skies and tufted cloudlets to cold, murky storms and driving showers, from moist, warm, enervating sameness to irritating and unexpected variability.

CLIMATES, TROPIC AND ARCTIC

Variability—that is the outstanding characteristic of temperate-zone weather. Even the brief sea log just quoted shows how weather in the tropics is much the same from place to place and day to day; temperate-zone weather, conversely, is diverse and changeable. But had our voyage extended far enough northward, it would have reached another region of comparative weather sameness—cold sameness instead of warm sameness -the arctic zone.

The temperate zone can have no stable and typical weather of its own. To southward (in the northern hemisphere) is the steady, warm, moist climate of the tropics. To northward is the fairly steady, mostly cold, fairly dry climate of the arctic. (This picture is of course repeated, with reverse directions, in the southern hemisphere.) The temperate zone is a buffer region between tropic and arctic climates, a battleground on which the gigantic atmospheric forces of heat and moisture advance and retreat, an unstable compromise between irreconcilable extremes, swayed this way and that in its evanescent weather as tropic forces (warm air masses) or arctic forces (cold air masses) momentarily gain ascendancy.

Tropic Climate

Tropic wind-and-climate belts migrate somewhat, north and south, with the apparent north-south seasonal swing of the sun. To get within-the-tropics weather in its most typical form we must go well south of the Tropic of Cancer, nearly down to the equator—say to Panama in about latitude 8°N. In these latitudes there are only two main seasons a winter dry season under the influence of the steady northeast trades, and a summer wet season brooded over by the fitful calms of the doldrums.

The outstanding features of tropical-marine climate are that the average temperature is practically constant from month to month (nor does it ever vary, by more than a few degrees, from day to day); and that the cloudiness and rainfall, always appreciable, reach high values during the wet-season doldrums invasion of summer and early fall.

Upper winds over Panama, at half-mile altitude, average northeast about 8 m.p.h. in the rainy season and north-northeast about 20 m.p.h. in the dry. At two miles altitude there is less seasonal change—the average wind only varies from east 10 m.p.h. (rainy) to cast-northeast 12 m.p.h. (dry).

Upper-air characteristics over the Caribbean Sea, quite typical of tropical-marine climate in general, were studied during the 1939 hurricane season by a United States Weather Bureau expedition to Swan Island. This four-mile- long islet, named after a nineteenth-century pirate, lies almost exactly in the middle of the Sea at about lat. 16°N, between Jamaica and British Honduras. Within four months, observers Rhamlow and Leech released some two hundred and fifty pilot balloons and one hundred and twenty radio-sounding balloons. They found that the trade wind extended well aloft, though with diminishing force, here as in Panama—prevailing easterlies all the way up to about two and a half miles altitude. But above this level, the upper winds are mostly out of the west and increase with altitude.

Surface temperatures at Swan Island, and all over the Caribbean area, average about +75° to 80°F the year round. But at the base of the stratosphere, ten miles up in these latitudes, it is cold indeed—about—105° to—110°F. Thus in the thick tropical troposphere there is an average lapse rate of something like—3½°F/1000 ft (about—6°C/Km). Together with plenty of warmth and moisture in the lower levels, this lapse rate means that conditional instability must extend through considerable air layers, particularly in the vertical-sun-warmed rainy season of summer and fall—and it is this conditional instability, together with inexhaustible supplies of moisture from the warm ocean, that produces the daily torrential showers and thundershowers, the regular afternoon downpours and the moist, misty mornings which can make equatorial regions such a continual sweat-bath for ill-adapted white men. On or close to the ocean this deep-tropical weather is bearable, or perhaps even comfortable. But the steaming jungles inland, where temperatures can rise unchecked and where all breeze is choked off by the tangled uprearing of dense vegetation, are sometimes aptly described as 'green hells.' As Julian Duguid wrote ' a rich eternal garment of green, dappled with golden sunspots a dense, fever-stricken thicket shimmering in the heat with a perpetual glassy haze.'

However, by no means all the land areas within the tropics are covered with jungle—some of them are sandy and arid in the extreme, and doubly hot though not so muggy. (The word 'jungle' itself, originally, meant only 'waste land' in India.) And even in the fecund jungle areas of the tropics, such as Panama and northern South America, vegetation waxes lush and green with the rainy season, wanes sparse and tawny with the dry season.

There are also irregular year-to-year variations in moisture and prevailing winds that change the face of tropical nature. I got off a ship on the Colombian coast in August, 1939, and walked inland for a while, expecting to wade through wetseason green tangles of vines and trees peopled with tropic creatures—but the soil proved to be dry and sandy, the jungle a dried-up shrinkage of scrub, the creatures hidden away from the glaring vertical sun.

Arctic Climate

Before Admiral Peary of the United States Navy marched five hundred icy miles on snowshoes from Ellesmere Land to the North Pole in the early spring of 1909, and since his time, many expeditions have entered the northern arctic regions. No latter-day explorer, incidentally, has ever surpassed Peary's record of intelligent planning and courageous, competent performance. But the airplane has rung startling changes in arctic travel.

In May, 1937, four large air transports set down four Russ an scientists and a dog, together with a little portable cabin and food and fuel for a year or more, on one of the uncertain drifting ice floes that cover the Arctic Ocean around the pole. Leader Papanin, astronomer-aerologist Federov, hydrologist Shirsov, and radioman Krenkel had hoped to remain at or near the earth's axis for a whole year of observations, but their ten-foot-thick ice floe soon began to drift at first very slowly, across the two-mile-deep polar basin towards Greenland. Then it moved, less gradually, southward along the east coast of that high continent, and finally sailed at twenty-five miles a day on southward.

Papanin's expedition found the summer temperatures of the polar basin hovering mostly around + 30° to 46°F; clear spells of a day or two alternated with dull, overcast, rainy-to-wetsnowy weather. Winds varied from calmate light or moderate, mostly out of the north or northeast. Much of the ice was melting in the warmer spells, adding to the troubles and perils of the ice-campers. There was always the danger that their floe would disintegrate and dump them unceremoniously into the Arctic Ocean.

When winter came they had drifted several hundred miles south of the pole itself, but winter climate around the polar basin is probably pretty much the same everywhere. Temperatures ranged all the way from 50°F below zero to 30° or more above, in accordance with the sweep of unlike air masses and cyclones, which occasionally extends well into the Arctic. For several months in midwinter the polar region is, of course, in continual darkness, with the sun visible at first in the spring only as a pale horizon glow around noon. On the thirty-first of December, strangely enough, the Russian wind-powered radio reported a temperature of + 43°F over the arctic ice! But within a few hours, the mercury had tobogganed to a more usual value of -31°F.

The general north-polar region is, of course, about midway on the most direct (great-circle) air routes between Europe and Asia, and between eastern United States and Asia. Some years hence, perhaps, great 500-mile-an-hour strato-transports will be winging daily across these icy-watery wastes. Compared with other climatic regions, the polar area offers good flying conditions—during at least two or three days out of a week, that is, and particularly at higher levels. Spring, when the polar high-pressure area lies symmetrically over the geographic pole, and not over the continental 'cold poles' in Greenland and northern Siberia, is the best flying season. A sealed- cabin, supercharged stratoliner, in any case, would be well up in the low arctic stratosphere (above thirty thousand feet) most of the time.

Aside from Ellesmere Land at the north tip of America, one of the best fixed bases for observing the climate of the north-polar basin is the small island of Franz Joseph Land in lat. 81°N long. 60°E, about six hundred land miles from the pole. In 1932-33 the Russian scientist Moltchanov (incidentally one of the pioneers in the development of radio-sounding balloons) made both surface and upper-air observations on Franz Joseph Land for a year.

Like the floe-campers, he found average summer surface temperatures around +30° to 40°F. Winter temperatures averaged around -10 F. Above the ground, for the first mile, there was usually a temperature inversion amounting to 5°F or so; above that height the lapse rate or vertical temperature gradient was always surprisingly uniform at -3° or -4°F per thousand feet up to the low polar 'tropopause,' marking the beginning of the stratosphere at about five miles altitude in winter, six miles in summer. Temperatures at the stratosphere base averaged about -75°F in winter, about -40°F in summer. July and August were the warmest months, as usual in the northern hemisphere; December was the coldest month at the surface, February coldest aloft. The southern hemisphere in general, being mostly water, shows more typical wind-and-climate belts than the northern hemisphere, which is all cluttered up with land masses that get too warm in summer and too cold in winter. By the same token, or a reverse one, the antarctic climate, founded upon a high, icecovered continent symmetric around the geographic pole, is more typically 'polar' than the arctic climate.

During the 1933-35 Byrd occupancy of 'Little America' in about lat. 79°S, long. 164°W, aerologists Haines and Grimminger recorded a great deal of antarctic weather. Mostly the average temperature was around + 20° to 25°F in summer (January), and around -35° to -40° in winter (July) considerably colder, that is, than the north-polar climate. The coldest summer temperature was +12° on January 1; the warmest, +37° on December 22. In winter the mildest reading was +23° on July 27; the coldest was -71° on July 21. During a roaring blizzard in July the temperature fell to -65° with the wind sixty miles an hour out of the southeast and wind, rather than temperature, is what really hurts in cold climates. These southeasterly blizzards in Antarctica correspond to the northeasterly gales around the north-polar basin.

High winds in Antarctica sometimes come from the north, just as north-polar gales can arise out of the south, but the colder blizzards from somewhere around the South Pole are more usual. Laurence M. Gould, Byrd's geologist and deputy in 1928-30, writes of one terrific southerly gale in the Rockefeller Mountains that first blew away the heavily packed snow wall in front of his firmly anchored airplane, and then finally blew the plane itself away and wrecked it. Some time before the maximum wind came, the airplane's air-speed meter indicated a steady velocity of ninety miles per hour, with gusts well over a hundred.

Conquest of Climate

The decade from 1900 to 1910 signalized an. amazing conquest of the tropics as a livable environment for civilized man. The years after 1940, likewise, may see a parallel conquest of the arctics. Within the tropics, except in certain areas plagued by extreme and wearisome heat, general living conditions were easy enough too easy for man's own good, in fact; and aside from air transport, radio communication, mechanical refrigeration, and other new conveniences, the conquest was nearly all medical.

Such dread warm-country diseases as typhoid, cholera, and yellow fever were finally vanquished at least in the case of selected, well policed areas. The work of Walter Reed, his assistant surgeons, and the heroic American soldiers who voluntarily entered the valley of 'yellow jack's' macabre shadow -all this really conquered the tropics and equatorial regions for white men, and contributed more than engineering and steam shovels to the triumphant completion, in 1913, of the world's greatest inter-ocean canal. Within the arctics and around the poles, on the contrary, disease is unknown unless brought to the scene by human carriers, but living conditions (without many bothersome precautions) are unduly hard and exacting. Aside from the extremes of dark and cold and gale and blinding snow, the chief hazards are: difficulty in finding your way around (particularly in the long, cold, dark arctic night); difficulty in surviving ordinary mischances such as falling into water, badly tearing clothes, and so on, particularly if alone; difficulty in getting adequately balanced diet and sufficient ultra-violet radiation.

Yet even now, with our immense progress in transportation and communication, some of these difficulties can be largely overcome at a properly organized arctic station. Clothing must, of course, be of arctic type—layers of wool or fur inside for warmth, layers of canvas or leather outside for wind-breaking. Adequate shelter, plenty of fuel, and fresh food can now be transported easily by airplane to almost any arctic site. On location, howling arctic gales will provide plenty of electric power for lighting and radio communication, and perhaps for heating as well.

Many of the dark terrors of the outer arctic night, which have heretofore held men prisoned for months in their base camps (except in the full of the moon) can perhaps be conquered by such devices as more powerful and efficient portable searchlights (using super-hot filaments, say, or possibly fluorescent light-sources); portable radio-telephone sets for constant two-way communication with and direction-finding on one or more base stations; more compact and accurate instruments, and simpler methods, for celestial navigation. Against the white, too-uniform snow-glare of the polar day, improved goggles (perhaps of polarizing type) may be employed.

Lighter, stronger and warmer tents and clothing can perhaps be devised. Much of the heavy local leg-work of arctic exploration can now be done with snow-mobiles, either tread-driven or propeller-driven. By means of vitamin diets, ultra-violet sunlamps and radio-diathermy, human beings near the poles will perhaps be enabled to maintain a high state of health during the long arctic night, and to overcome quickly any respiratory diseases that do develop.

TEMPERATE CLIMATE

We have said that the temperate zone is, in a climatic sense, nothing better than a battleground between tropic and arctic weather forces—between warm and cold air masses. Nevertheless the temperate zone (and in particular, the north-temperate zone) is that part of the earth in which civilized man evolved, and is probably the region most favourable to his continued progress and development.

On the average, of course, temperate temperatures are pleasantest and best for all human activities—neither too hot nor too cold—though this average may rarely be matched by the vagaries of daily weather. And even the extreme moodiness and changeability of temperate weather has its advantages. Variety in weather, at least as to the four- or five-day spells and sequences so common in middle latitudes, perhaps satisfies some deep, innate rhythmic urges in man which cannot be allowed to atrophy without intellectual and cultural and physical loss.

However such obscure causes may or may not operate, it is certain that most human progress up from the dumb-ape level—most human art and ingenuity and imagination, most human science and engineering and social organization—has been best nurtured so far, and probably will be best nurtured in the future, in temperate latitudes.

Among all the various meteorological factors that go to make up a climate, temperature is the one that most directly influences human health, happiness, and efficiency. How do various temperatures really feel to average human beings? Different degrees of heat of course feel quite different to different people, or to the same people under different conditions—in particular, under varying conditions of exertion. But it is possible to indicate average human reactions to temperature, and they are somewhat as follows:

TABLE: Outdoor Temperatures Feel to the Average Person

Temperature		**Human Effects**
(Fahrenheit)	**(Centrigrade)**	
Below 0° Below	-18°	Unpleasantly cold without heavy arctictype parkas, etc., unless exertion is continuous and heavy.
0°-32°	-18°-0°	Feets cold with anything but arctic-type clothing, if exertion is light and exposure long.
32°-50°	0°-10°	Under same conditions, feets comfortable with heavy overcoat, hat, etc., cold otherwise.
50°-65°	10°-18°	Feets cool with ordinary indoor clothing, comfortable with light overcoat, etc.
65°-75°	18°-23°	Feets hot under any conditions, unpleasantly hot with high humidity and light wind
90°-100°	32°-40°	Unpleasantly hot under all conditions.
Above 100°	Above 40°	

High humidity makes warm temperatures feel warmer, by retarding evaporation from the body. High humidity also makes cold temperatures feel colder, by making the clothing conduct heat better away from the body.

Wind makes cold temperatures feel colder, by removing the cushion of warm air that tends to form around the body in the absence of wind. This wind-cooling effect is particularly powerful at below-zero temperatures. On the other hand, wind makes warm temperatures up to 100°F feel less warm, by increasing evaporation from the body. But if the temperature is over 105°, wind increases the discomfort. The optimum temperature range, 68°-70°F, feels more comfortable if the air is normally moist.

For health and comfort, indoor relative humidity should be around sixty per cent. But in the average home or office, in winter, it is seldom more than twenty per cent. Evaporating pans on radiators help some, but not very much. Constant-duty electric evaporators, preferably combined with electric lights, might well be used in temperate-zone houses during the winter.

TABLE: Indoor Temperatures feel to the Average Person

Temperature		Human Effects
(Fahrenheit)	(Centrigrade)	
Below 04°	18°	Unpleasantly cool with ordinary clothing.
64° 68° 18°	-20°	Too cool for comfort if sitting still.
68°-70°20°	-21°	Optimum for health and comfort,
70o-74o21o	-230	Warm but not uncomfortable if clothing is light.
		Neediessly warm for healthy people.
		Unpleasantly warm for healthy people.
74°-78°	23°-25°	
Above 78°	Above 25°	

Today's temperature anywhere is indicated by the thermometer; tomorrow's (which in temperate or arctic latitudes may be shockingly different) is foretold by a good forecast. Beyond these limits, our best present guide to future temperatures is the 'normal.' To get the normal temperature for any particular day of the year at any particular place, the average temperatures for that day, observed over as long a span of years as possible, are all averaged together. The normal temperature is, of course, merely a fictitious quantity seldom or never duplicated by any actual daily average; for climatically speaking, we dwellers in the temperate zone are always traveling now south, now north with the ebb and flow of warm and cold air masses.

Is our North American temperate-zone climate changing? This question keeps recurring in weather talk, and most of the oft-quoted answers foist off something about 'colder winters in Grandfather's day' a notion perhaps promoted by oldsters in support of their claims to general superiority. The rising generation, contrariwise, have been inclined to charge the whole idea off to dotage and general boastfulness. A definitive answer has been given by J. B. Kincer of the United States Weather Bureau:

For a good many years up through 1936 there has been a decided tendency to warmer and drier weather, the trends being especially marked during the past quarter of a century, notwithstanding an occasional bad flood or a severely cold winter. Take the winter season, for example: With the exception of those for 1917-18 and 1935-36, the winters for the past twenty-five years or more, considering the country as a whole, have been rather uniformly warmer than normal, and, on an annual basis, every year since 1929 has had above normal temperature. Also, in the matter of rainfall, there has been an equally marked tendency to droughts in recent years. However, an examination of the longer weather records of the country, going back one hundred years or more, indicates that this does not represent a permanent change of climate, but rather a warm, dry phase of our normal climate, to be followed, doubtless, by a cooler, wetter phase, when there will be more rain in summer and lower temperatures in winter.

Since the turn of the century, Mr. Kincer goes on to say, there has been a distinct yearly trend towards warmer winters and drier summers all over the world and this in spite of a few exceptionally cold winters here and there.

In other words, there are long (say fifty-year) cycles in climate, just as there are short (say five-day) cycles in weather. If all these cycles were only considerate enough to maintain constant length, amplitude, and form, both weather forecasting and climate forecasting would be child's play. But the form, intensity, and length of weather and climate changes are all complexly variable.

In any case the semipermanent (say 500-year) picture of our American climate shows, apparently, no appreciable change. But the geologic-time picture, measured in thousands or millions of years, would again show marked climatic cycles the advance and retreat of the various ice ages.

THE PLANETARY PATTERN OF WINDS

Just as there are on this great globe fairly definite zones of climate, so there are even more definite belts of prevailing wind. Some of these round-the-world prevailing winds are quite steady and uniform, such as the trades and anti-trades. Others, such as the prevailing westerlies of temperate latitudes, are extremely variable from day to day and week to week. But the average wind directions observed in temperate latitudes prove that here, too, there is a great round-the-world wind system from west to east—winds that blow generally across continents and oceans, across desert sands, sawtooth mountains, fields of rippling grain, waving trees in virgin forests, and wide expanses of heaving, restless, white-capped blue water. The winds of the world are in essence a heat-pressure mechan- ism similar to that which causes a sea breeze, though enormously larger and more complex. The driving force behind all earthly winds, and all other earthly motions, is the enormous outpouring of radiant energy from the sun, our celestial ancestor and gigantic power-house ninetythree million miles out in space.

This radiant energy falls nearvertically on equatorial regions, and very obliquely on polar regions. Hence the equatorial regions are warmer, and are receiving more heat than they lose by earth-radiation. The poles, on the other hand, are radiating away more heat than they receive by sun-radiation. Hence, unless the tropics are to become unbearably hot and the arctics are to chill down to absolute zero (about -460°F. or -273°C), there must be some atmospheric circulation between the two. A heated region usually develops low pressure at the surface and high pressure aloft, that a cooled region develops high pressure at the surface and low pressure aloft. The result, where only short distances are involved, is that lower winds blow generally from cold to heat, and upper winds blow generally from heat to cold. Hence we might expect, as part of the planetary wind picture, some tendency towards an equatorward drift at the surface together with a poleward drift aloft.

However, that any largescale wind is gradually turned aside, to the right in the northern hemisphere and to the left in the southern, by the earth's slow rotation under the wind in question. In other words this 'earth's deflective force,' so called, tends to turn any equatorward-drifting air westward, and to turn any poleward-drifting air eastward.

In the upper troposphere, say at altitudes above two and a half miles, the average world wind pattern is a grandly simple thing that stems directly and understandably out of these elementary heat-pressure-wind principles. Upper pressure is relatively high over the equator, and relatively low over the poles, with a fairly uniform average horizontal pressure gradient between the two.

Hence the average upper-tropospheric winds of the northern hemisphere, directly driven by this pressure gradient and swayed by the earth's deflective force, are from southerly to southwesterly to westerly (drifting, that is,generally towards he northeast, or spirally towards and around the pole). In the southern hemisphere, driven by similar but oppositely directed forces, the average upper-tropospheric winds are mostly northwesterlies, drifting generally southeastward.

Either of these average drifts can be temporarily upset by regional vagaries, such as a southward surge of cold upper air in the north hemisphere. But as a norm or average condition, the two great upper drifts perennially move from the equator spirally eastward and poleward in each hemisphere.

The planetary pattern of lower winds (below the two-and-ahalf-mile level) is more complex, even in idealized form. For the lower winds are influenced by surface friction, and the earth's deflective force acts on them as a more drastic limitation on poleward or equatorward progress. Thus the planetary lowerwind pattern includes, not one wind belt in each hemisphere, but three. Moreover, these wind belts, and the pressure belts which separate them and regulate them, are in fact much modified and distorted (at least in the northern hemisphere) by the alternating surfaces of great continents and great oceans.

Perhaps the easiest and best introduction to them would be a voyage through ideally typical weather at just 5° of latitude (three hundred sea miles) each day down the whole 'length' of the world, along the ocean-parting meridian of long. 30°W from arctic to antarctic. Any such imaginary meridianal voyage would cut across the following idealized or average wind-andweather belts: a zone of arctic easterlies above lat. 60°N a stormy 'polar frontal' zone (actually variable in both width and location) around lat. 60°N or south of this; prevailing southwesterlies-to-westerlies (the familiar prevailing winds of temperate latitudes) between lat. 60°N and lat. 30°N a dry, clear calm belt called the 'horse latitudes' (probably because horses often died here on becalmed sailing ships) around lat. 30°N northeast 'trades' (so called by reason of their steady, perennial character and consequent importance to commercial sailing ships) from south of lat. 30°N down towards the equator: a humid, oppressive, and showery calm belt called the 'doldrums' (because of their dispiriting or dulling or 'dolting' effect on human sailors unwillingly becalmed in them) around the equator; southeast trades from near the equator down to somewhere near lat. 30°S the southern horse latitudes around lat. 30°S; the southern prevailing westerlies between lat. 30°S and lat. 60°S; the southern polar-frontal zone around lat. 60°S or north of here; and the antarctic easterlies south of lat. 60°S.

The steady poleward air transfer of the prevailing westerlies and southwesterlies in our latitudes is compensated by occasional large and violent southward outbreaks of polar air (cold waves), especially in winter, when the polar anti-cyclone extends down over Alaska and northern Canada. Southward outbreaks sometimes also occur far aloft, with 100-m.p.h. north winds in the stratosphere. Similar equatorward (northward) outbreaks occur in the southern hemisphere.

Considering first the northern hemisphere: There is an average high-pressure area, or anti-cyclone, around the 'cold pole' (despite the fact that many arctic observers have noted low pressures). This polar high-pressure area drives the cold, arctic lower air spirally outward (or southward and westward) as the arctic easterlies (mostly northeasterlies) down towards lat. 60°N. Along lat. 30°N (the horse latitudes) there is a subtropical high-pressure belt, the northern slope of which drives the prevailing westerlies (mostly southwesterlies) up towards lat. 60°N. Around lat. 60°N or south of there, these two great and conflicting wind systems meet along the low-pressure belt of the polar-frontal zone—a stormy region which may actually swing far north or south, and may extend as separate and vari-formed fronts, between various unlike air masses, through a wide sweep of latitude. The subtropical high-pressure belt in lat. 30°N (or its southern slope) also drives the northeasterly trades, as a general southwestward-to-westward drift around the earth, down into the equatorial low-pressure belt of the doldrums.

In the southern hemisphere, the subtropical high-pressure belt around lat. 30°S drives the southeast trades up into the doldrums, and also drives the southern prevailing westerlies (mostly northwesterlies) down into the low- pressure polar-frontal zone around lat. 60°S or north of there, where they meet and battle with the antarctic easterlies (mostly southeasterlies) from the south-polar high-pressure area. In temperate latitudes, both north and south, lower winds have the same general or average direction as upper winds, and the prevailing westerlies form a deep average pattern (seldom fully duplicated in actual existing winds) that extends with increasing force up through the troposphere, moderating again in the lower stratosphere, and then mostly reversing, in light to moderate intensity, above twelve miles or so. Within the tropics, on the 'winter side' of the equator, the easterly trades give way, at about two and a half miles altitude, to the planetary upper westerlies.

Called the counter-trades or anti-trades, these tropical upper winds blow mostly from southwesterly in the northern hemisphere and northwesterly in the southern. The same planetary upper westerlies are present, to some extent, in arctic regions above the arctic easterlies.

CENTERS OF ACTION

The whole idealized wind-belt picture described above is based, it will bear repeating, on average conditions. Daily departures from this average, in any one region, may be very wide. It is also based on the conception of a uniform earth—either all water, or all uniform land. As a matter of actual fact the southern hemisphere is so nearly all water that this idealized pattern actually holds there much of the time; and the pattern may also hold, though less consistently, well out in the Atlantic and Pacific Oceans of the northern hemisphere.

But in general the northern hemisphere is something like equally divided into great land surfaces and great water surfaces, which vary greatly in the frictional drag they oppose to surface winds.

Also, while oceans tend to remain equable in temperature at all hours and all seasons, continents tend to get too warm in summer and too cold in winter. As cold areas tend to develop high surface pressure, and warm areas low surface pressure, the idealized latitudinal pressure belts of the northern hemisphere are in fact somewhat broken up into continental and oceanic 'highs' and 'lows' which change considerably with the seasons.

These semipermanent, or seasonal, lows and highs are called centers of action, because their counter-clockwise or clockwise circulations control in somedegree the general movements and conflicts of air masses in their neighbourhoods, and influence greatly both the position and movement of various segments of the polar front and the development and motion of the traveling cyclones that are largely responsible for most temperatezone bad weather.The centers of action in and around North America—those that have most bearing on our weather in the United States -are as follows:

Winter	Summer
Canadian High	Hawaiian High
Aleutian low	Bermuda-Azores high
Icelandic low	(Rokies low - weak) (Aleutian and Icelandic lows - weak)

It will be well to recall these centers of action, and to remember the cardinal fact that (in the northern hemisphere) winds blow clockwise about a high- pressure area and counter-clockwise around a low-pressure area—particularly when we come to consider the traveling 'highs' that are usually associated with vagrant air masses, and the traveling 'lows' or cyclones out of which grow warm and cold fronts.

TABLE: Typical Climate of The Deep Tropics

Month	Mean Temperature	Mean Cloudiness (% of Sky)	Monthy Rainfall (inches)	Average Wind direction and force
Jan.	80	50	3-5	North, moderate
Feb.	80	52	1.5	North, moderate
Mar.	81	50	1.5	North, moderate
Apr.	81	61	4.2	North, light
May	81	75	12.6	Variable, light
June	81	01	13.7	Variable, light
July	81	81	15.6	Variable, light
August	80	80	15.2	Variable, light
Sep.	81	77	12.5	Variable, light
Oct.	80	77	15.5	Variable, light
Nov.	79	79	22.2	Variable, light
Dec.	80	61	11.1	Variable, light

What are the outdoor conditions of life and thought and feeling near the poles and within the arctics? As a brief and partial answer to this always interesting question, Frederick Crockett, who was also on the 1928-30 Byrd expedition, gave me the following word-picture:

Usually during the months-long arctic 'day' there are low-hanging clouds, but below them, along the ground, visibility may be very good. The view in all directions is extraordinarily uniform—no shadows, no dark objects, no horizon, all gray above blending with chalky white below, 'like walking in a bowl of milk.' On this white, featureless surface it is impossible to see hummocks before stumbling over them, and walking is consequently very tiresome without skis, which more or less 'feel their way' along.

Distances in this milky light are very deceiving. A piece of dog dung a hundred yards away may look like a far-distant house. Because of the uniform white glare from the surface below, from which men's eyes are not naturally protected, sun blindness is likely unless protective goggles are worn. A very painful experience once it is incurred, something like burning hot sand in your eyes. We had no trouble sleeping safely out on the trail at temperatures around 50°F below zero. Our sleeping bags were caribou hide with the hair turned inward, with thin woollen blankets inside. When crawling out in the morning one had to exhaust the bag of its warm air, lest the vaporous moisture condense and freeze, making a very uncomfortable sleeping place for the next night.

For clothing, we departed somewhat from previous arctic experience and wore separate thin layers of wool (successive sweaters). These sweaters we put on or took off, according to temperature and the degree of exertion needed at the time. On the trail we carried an outer natural-fur parka, but did not usually wear it (even at fifty-below temperatures) until we made a long halt, or when outside in the evening, and so on.

In general, wind was more important than temperature as a factor in comfort. On one winter day in July, 1929, two other men and I started out on skis for a short local exploration. It was 9 A.M., and still completely dark, when we left camp. The temperature was colder than 70 below-actually - 72°F, as I remember it. Fortunately, however, there was no wind. It wascompletely calm; the sky was clear, with the intensely bright southern-hemisphere stars (Canopus, the Southern Cross, and so on) swinging slowly around over our heads as the morning wore on, interspersed with the pale-green, flickering curtains of the aurora australis.

Starting off cool in thick clothing, we got well warmed up in a half-hour of skiing across the sand-like antarctic snow, which does not melt slightly under your skis like snow in warmer climates, providing a slippery surface. Hence skiing calls for heavy exertion in order to make any progress. Strange as it may seem, though at 9•30 A.M. the temperature was still 70 below and it was still dark, we were going forward barehanded and almost stripped to the waist. The only time we really felt the cold was on slight downslopes where we could slide along at six or seven miles an hour.

Here the lack of heavy exercise, combined with the slight wind of our forward speed, made us immediately feel the bitterness of the Antarctic cold. We continued thus in the darkness until around noon, when the sun appeared for an hour or so as a faint glow on the northern horizon.

To find the day-to-day march of normal temperature in your own particular American region, you must consult a Weather Bureau publication entitled 'Normals of Daily Temperature of the United States.' But by averaging the normals for Washington City with those of certain inland cities in the same latitude, we can obtain something fairly representative of a large (central and eastern) portion of the country. Atlantic seaboard temperatures average two or three degrees Fahrenheit warmer in winter than inland temperatures for the same latitude; in summer the seacoast may average a degree or two cooler.

Also, yearly maximums and minimums on the seacoast come several days later than corresponding maximums and minimums inland.

The northern United States (latitude of Boston) averages about five degrees Fahrenheit colder than this table. The southern United States (latitude of Savannah) averages about ten to fifteen degrees warmer than this table in winter and about three to five degrees warmer in summer. Temperatures in the Rockies and other high mountain regions vary widely with altitude and topography. Temperatures on the Pacific coast are largely conditioned by the water temperatures of the eastern Pacific Ocean, and by topography. In general, Pacific seaboard temperatures show far less annual change than inland temperatures, and yearly maximums and minimums are delayed a month or so longer. Summer is typically cool and foggy, because of cold sea water that wells up from the great depths along the west coast.

The normal-temperature table printed above shows many well-known but interesting facts. The time of greatest cold comes around January 25, lagging more than a month behind the sun. The temperature remains almost constantly cold for a while; then, with the coming of spring, it begins to rise, slowly at first, later at the rate ½°F per day or 10° to 15°F per month. But the month of April is nearly as cold as the month of November; though many people, rushing out scantily clothed into the brightening sun, fail to realise the fact, and return indoors to meditate on it while nursing a spring cold. In the northernmost parts of the United States winter lingers into April, perhaps into May.

A period of nearly uniform summer heat comes in July and August, the maximum (around July 25) again lagging more than a month behind the sun, which has already turned back southward; and this summer heat is surprisingly uniform, by and large, over the whole United States. Then the temperatures begin to fall, slowly at first, then more rapidly, until by midautumn the rate of decrease is about equal to the rate of increase in spring. Finally the winter snows begin, and the snow itself, during the ensuing months, tends to hold air temperatures below freezing by reason of its reluctance to melt and its high reflecting power.

Tabulated observations at various levels show the following prevailing directions and velocities: altitude 1 mile, mostly westerly, averaging about 15 m.p.h. in summer and about 25 m.p.h. in winter; 2 miles, mostly westerly, averaging around 18 to 28 m.p.h.; 4 miles, mostly westerly, averaging about 25 to 40 m.p.h.; 6 miles, westerly to west-northwesterly, averaging about 40 m.p.h.

❐

6

Climatic Effect

CHANGES IN CLIMATE

People vary greatly in their reactions to thunderstorms. Some few take to feather beds in the fear that Thor will personally seek them out for punishment and destruction (though a feather bed actually avails against a lightning bolt about as effectively as a garden gate stops a railroad train). Other people, perhaps equally foolish, rush to hilltops in the fond faith that Divine Providence will somehow ward any wandering electric bolt away from their exposed heads. Actually a thunderstorm is a beautiful and gigantic experiment in physical science, doubly interestng to those who know something of its causes and its mechanism, wildly beautiful to all men conscious of beauty in Nature. On what earthly mountains does a sunset glow compare with its pure rose radiance on the white cloud-cliffs of a towering and detached thunderstorm? What ground outlook compares with the spectacle of a distant towering cumulus in the clear night sky, flickeringly lit and veined by an infinite and never-ceasing variety of soundless lightnings? A great, wide-spreading, fast- traveling line squall, in its most violent and magnificent form, may be a towering range of such vapour-mountains, extending across the country for hundreds of miles and advancing in line like the fire-spitting front of some colossal and destructive army.

THOR'S APPROACH

Beautiful though thunderclouds are at a distance, either from the ground or from the air, on closer approach their still, turreted, snow-white serenity changes abruptly to a dark gray void of whirling winds, blinding firebolts, and torrential rain. If you (unwisely) fly into a thunderstorm in an airplane, this transition is very abrupt -- one moment you are admiring the infinite overhang of a thousand cloud cliffs all gleaming with every gradation of sunlight and shadow; the next you are plunged into turbulent blue-gray obscurity, split by recurring yellowish flashes, deluged with rain, and disrupted by updrafts and air bumps of unbelievable ferocity.

On the ground or the water, as the storm moves towards you at perhaps twenty or thirty miles an hour, the transition is slower and more obscure, yet spectacular and interesting enough.

It does not matter particularly whether you happen to be in the path of a single large summer 'air-mass' thunderstorm, or whether you are caught in the five-hundred-mile sweep of a spring or summer 'frontal' storm or line squall -- the general effects will be the same, except that the line squall may be more violent, and its squall wind, temperature drop, and pressure rise will be more pronounced and permanent. In either case, reduced to typical form, the preceding summer weather is usually sultry and oppressive, both temperature and dewpoint being high.

The sky is generally clear, but somewhat obscured by haze and cirriform clouds out of the west. The wind is southerly, and perhaps light or occasionally lacking altogether with a single thunderstorm in the offing, but likely to be strong and gusty if a line squall (cold front) is bearing down from the west or northwest.

If detached cumulus clouds are watched closely, they can often be seen to grow vertically upwards, sometimes at surpris-ing speed. Towards the middle of the afternoon a wall of thunderheads may be glimpsed, beyond obscuring haze and lower clouds (and are often plainly to be seen from an airplane at high altitude), rising slowly out of the west. But the thunderheads themselves often stream out into cirrus-like nebulosity at higher levels, and this, combined with ground haze, may hide them from view.

In any case, the westering sun is soon blotted out, and the sky to west and northwest slowly darkens, at least in its lower regions, to an ominous 'blue- black colour. Thunder begins to be heard perhaps half an hour to an hour before the storm strikes, and becomes louder and more frequent by the minute.

The wind, which has perhaps died down altogether, rises and begins to blow directly towards the storm. Perhaps fifteen minutes before the squall strikes, the 'squall head' or roll cloud, whitish against the black clouds above and behind it, may be seen advancing in the shape of a long and irregular roll across the front of the storm -- and this roll sometimes visibly revolves like a sort of horizontal tornado.

Perhaps fifteen minutes or more before the storm strikes, also, distinct lightning bolts come into view, most of them striking directly downward into the ground from the bottom and front of the black clouds just behind the roll cloud. You can check the narrowing distance away of these fiery ground- strikes, of course, by noting the flash-thunder time interval with a stop watch—the lag between sight and sound (as sound travels about eleven hundred feet per second in air) is about five seconds per mile.

After the lightning bolts approach within a mile or two, it is only a moment (in a severe storm) before they are striking all around you, and in this 'dry' front of the storm, before the rain begins, the lightning danger is greatest.

A few minutes after the roll cloud passes overhead, the squall wind sweepsout of the west or northwest, usually at something like thirty or forty miles an hour, but occasionally as a flattening blast of sixty or more. Along with the squall, or shortly after its onset, comes a wall of spattering, large-drop rain that may descend in torrents, and may be accompanied by lashing hailstones.

In a detached 'air-mass' storm the squall wind usually begins to moderate after a few minutes, and may have died away altogether in a half-hour or so. The rain is less heavy, and the lightning less frequent and less intense, in the rearward and less active parts of the storm. But after a line squall the high westerly wind may continue or even increase.

THUNDERSTORMS OCCUR

Thunderstorms occur most frequently in the warm, moist, and unstable air masses within the tropics, where (in the rainy season) they are a daily afternoon occurrence, and sometimes so regular that they serve as a convenient time indicator for human affairs. In arctic regions, where the air is cold and comparatively dry, thunderstorms may not occur oftener than once in several years. Temperate regions experience their share of storms in the late spring and summer. No part of the United States is free from thunderstorm activity, but different regions vary widely in thunderstorm frequency. In winter the center of electrical activity is over the lower Mississippi Valley; in summer there are two far more active centers, one over Florida and the eastern Gulf coast, the other over northern New Mexico. But these centers are only relative, of course, for many parts of the nation see nearly as many storms. And in some regions, such as the Central, North Central, and East Central States, thunderstorms are likely to be mostly of the frontal type, and hence more severe and destructive.

Thunderstorms are classed, according to their apparent severity, as 'mild,' involving few if any ground-strikes and relatively light winds; 'moderate,' involving fairly frequent groundstrikes, moderate to heavy rain, and moderate to strong winds; and 'severe,' attended by nearly incessant lightning, torrential rain, and severe squalls. Temperate-zone thunderstorms are also classed, according to the mechanism of their formation, as 'airmass' and 'frontal.' The older names for these same classes of storms, before air-mass-and-front analysis came into wide use, were 'convective' and 'cyclonic.'

Air-mass thunderstorms occur entirely within one air mass, and without aid from any frontal action between air masses. They arise where the characteristics of the air mass, aided by wind and topography, favour them.

They are most frequent in the United States in the warm, moist, and conditionally unstable air masses of tropical maritime origin. But moist polar air masses advancing southward rapidly, and hence remaining relatively cold in their upper levels, can occasionally be unstable enough to produce scattered air-mass thunderstorms.

The atmospheric instability that produces air-mass thunderstorms and keeps them active for hours over land areas in daytime is usually 'triggered off' by intense solar heating at the surface, which steepens the vertical temperature gradient; or perhaps by the importation of cold air aloft; or by the rising currents produced by upward deflection of horizontal winds (together with perpendicular solar heating) on the slopes of hills or mountains. Thus air-mass storms are frequent along the Gulf coast, where the sea air strikes the hot and slightly rising land; and thunderstorms of any type are more frequent over mountains than over flat country or at sea.

Frontal thunderstorms may occur along a warm front, where even gradual lifting on the warm-front slope is sufficient to trigger off the instability of very warm and very moist tropical air. They may also occur along the wave-like oscillations of a more or less stationary front that is, for the moment, alternately warm and cold. They may even arise, rather mysteriously for unaccustomed forecasters, out of upper fronts that give little or no other indication of their motivating presence. But the great majority of frontal thunderstorms occur either directly in front of, or actually ranged along, a vigourous cold front marked on the weather map by large temperature contrasts, a deep pressure trough, and strong shifting winds (that is to say, by a fairly strong pressure field having isobars sharply bent into V-shape where they cross the front). This whole formation is, of course, nothing more or less than a line squall.

Thunderstorms of all types are most frequent, on land, during the middle of the afternoon—that is, during the hours when solar heating at the surface reaches its maximum and the lapse rate (vertical temperature gradient) is consequently steepest. As the average life of a thunderstorm is six or seven hours, however, thunderstorms formed in the afternoon often persist into the early evening. Over the ocean, where the equable water warms up by only a degree or so during the day, and cools off by even less at night, he thunderstorm maximum comes during the late night and early morning hours, when radiational or airmass-transport cooling aloft is greatest, and, again, when the total lapse rate is steepest.

As the mechanics of frontal thunderstorms depend not so much on surface heating or upper cooling as on the violent upthrusting of warm and moist air masses by an undercutting wedge of cold air, frontal storms may occur at any time of the day or night. But even the frontal storms are most likely, and most severe, during the middle of the afternoon on land or during the early morning hours at sea. Plenty of active summer cold fronts, for example, do nothing much over land during the morning hours, yet develop into roaring line squalls by mid-afternoon.

FORMATION OF A THUNDERSTORM

What are the aerological conditions that favour the formation of a thunderstorm? First, there is the all-important factor of atmospheric instability, evidenced by a steep or steepening lapse rate (–3° F/1000 ft or steeper, or about –6° C/km or steeper). This steep temperature gradient is perhaps being increased by air-mass transfer below or aloft—lower warm winds from the south and southeast and/or upper cold winds from the north and northwest—and above all, by excessive solar heating at the surface.

In addition to the steep temperature gradient, high relative humidity at all levels (meaning that clouds will be formed by very little rise of air) also favours the formation of a storm. Finally the 'ice level,' or altitude of freezing temperature (+32° F. or 0° C.), should be located at considerable elevation, as it always is over the tropics. These upper-air characteristics, of course, can be fully known in advance only by means of airplane or balloon soundings, though they can sometimes be sensed by the rapid (and occasionally prodigious) vertical growth—the spectacular upward mushrooming—of detached cumulus clouds during the late morning.

There are also some surface thunderstorm portents not without value—high temperatures, say 80, 90, or 100° F. by day and 70° or 80° by night; high surface specific humidity as indicated by high dewpoint, say more than 6o° or 70°; and steadily decreasing surface pressure during the late morning and early after. noon. Another useful early-afternoon indication is increasing 'crash' static from a radio set—actually radio waves from lightning flashes beyond the horizon. Finally there are the definite evidences of frontal activity and advance towards one's position to be obtained from some sort of weather map—even a simple regional map that anyone can draw from airways radio weather reports in a few minutes' time.

One or more of the indications just listed can of course be present without resulting in thunderstorms, and thunderstorms can occur when one or more of them are absent; but in general the presence of most of them makes thunderstorms likely, and the agreement of all of them makes thunderstorms certain.

Given these favourable conditions, how does an air mass thunderstorm actually form and grow? The beginning of a thunderstorm is similar to the formation of an ordinary cumulus cloud or an ordinary shower. But in the case of the thunderstorm the air is more moist and more unstable, and this moist instability extends to greater heights. Hence the updrafts are more violent, and the cloudbuilding is more rapid and extensive.

As in ordinary cumulus-building, the rising currents of a thunderstorm cause the warm, moist air to cool by expansion to its dewpoint, forming minute liquid droplets (on commonly present condensation nuclei), perhaps 1/2500 inch in diameter, which droplets appear as the white or gray opacity of the cloud; and this condensation of water vapour, in a vicious cycle, releases latent heat which helps to drive and intensify the rising currents, which condense further moisture, and so on. The result is that under favourable conditions the thunderstorm grows fast and far—perhaps up to altitudes of six or seven miles.

Just as in the formation of all rain, so, while the minute cloud droplets of a thundercloud are carried up in the rising currents, there is some jostling between them, which jostling tends to coalesce many of them into much larger drops. These larger drops, if the updraft is not too strong, will begin to fall through the cloud as rain. If the upsurging cloud droplets reach a temperature as low as +5 °F. (–15 °C.), which they soon do in a by high dewpoint, say more than 60° or 70°; and steadily decreasing surface pressure during the late morning and early after. noon. Another useful early-afternoon indication is increasing 'crash' static from a radio set—actually radio waves from lightning flashes beyond the horizon. Finally there are the definite evidences of frontal activity and advance towards one's position to be obtained from some sort of weather map—even a simple regional map that anyone can draw from airways radio weather reports in a few minutes' time.

One or more of the indications just listed can of course be present without resulting in thunderstorms, and thunderstorms can occur when one or more of them are absent; but in general the presence of most of them makes thunderstorms likely, and the agreement of all of them makes thunderstorms certain.

Given these favourable conditions, how does an air mass thunderstorm actually form and grow? The beginning of a thunderstorm is similar to the formation of an ordinary cumulus cloud or an ordinary shower as described But in the case of the thunderstorm the air is more moist and more unstable, and this moist instability extends to greater heights. Hence the updrafts are more violent, and the cloudbuilding is more rapid and extensive.

As in ordinary cumulus-building, the rising currents of a thunderstorm cause the warm, moist air to cool by expansion to its dewpoint, forming minute liquid droplets (on commonly present condensation nuclei), perhaps 1/2500 inch in diameter, which droplets appear as the white or gray opacity of the cloud; and this condensation of water vapour, in a vicious cycle, releases latent heat which helps to drive and intensify the rising currents, which condense further moisture, and so on. The result is that under favourable conditions the thunderstorm grows fast and far—perhaps up to altitudes of six or seven miles.

Just as in the formation of all rain, so, while the minute cloud droplets of a thundercloud are carried up in the rising currents, there is some jostling between them, which jostli ng tends to coalesce many of them into much larger drops. These larger drops, if the updraft is not too strong, will begin to fall through the cloud as rain. If the upsurging cloud droplets reach a temperature as low as +5 °F. (–15 °C.), which they soon do in a towering cloud (at a height of four or five miles in temperatezone summer), some of them are jostled into ice particles; and ice particles are perhaps also subliming directly out of water vapour (onto sublimation nuclei already present at these cold heights), if the temperature and humidity conditions are favourable.

In a fully developed thunderstorm the ice-particle cloud is often visible as a cirrus-like, anvil-shaped top or cap that blows out somewhat ahead of the main cumulus turret on the upper winds.

In this fashion the production of rain continues until large drops perhaps a quarter of an inch in diameter are falling through the cloud; and the rising currents build up until, in the forward and most active part of the storm, they exceed twenty miles an hour. This updraft will just support a three-eighths-inch raindrop; and no raindrop larger than three-eighths inch can exist in nature, for it would be instantly blown into smaller drops by the wind of its fall.

Thus no rain can fall through the most active updrafts of the thundercloud, and the active front of the storm is marked by an abrupt stoppage of large-drop rain falling from above, by the deflection of this rain upward towards the rear of the storm, and by the breaking up of large drops into smaller drops.

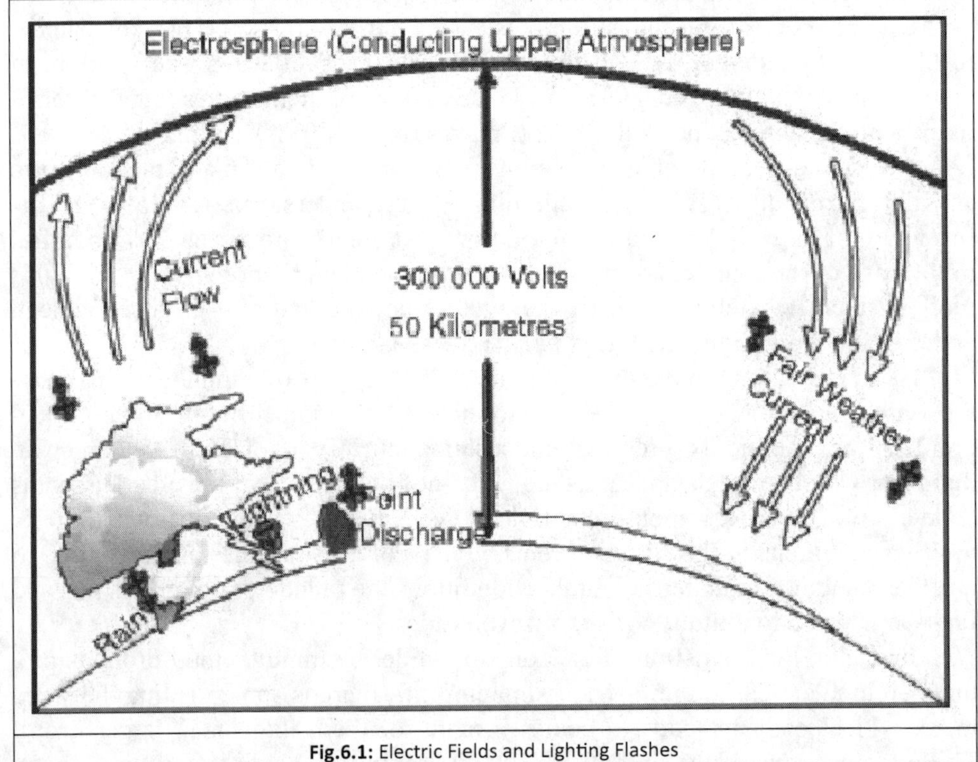

Fig.6.1: Electric Fields and Lighting Flashes

From the breaking up of these water drops, which can, of course, recombine above and descend to be broken again, come the electric charges that produce the lightning. At the instant of spattering, the to-be-broken drops are charged with positive electricity, and the air surrounding them is negatively charged. This negatively charged air is carried away into upper and rearward portions of the thundercloud, but the positively charged raindrops, breaking and recombining, tend to accumulate in the most active (forward and lower) portion of the storm, which hence becomes a region of high positive electrical charge.

From this positively charged region, usually just above and behind the roll cloud or 'squall head,' frequent heavy thunderbolts strike downward and forward directly into the ground (whose charge is neutral or somewhat negative), while other longer but less intense flashes strike upward and backward to lose themselves in the upper immensities of negatively charged cloud.

Towards the rear of the storm, however, where the earth is positive with respect to the negative clouds above, less intense bolts strike upward from the ground into the clouds.

Lightning may be only one feature of a great storm perhaps marked also by cloudbursts, ice showers, and destructive blasts of wind, but it is spectacular enough, and under some conditions hazardous enough, to warrant examination in some detail before considering the other aspects of the storm.

THOR'S HAMMER

The potentials set up by the cloud-formed electric charges of a thunderstorm are enormous—in the neighbourhood of one hundred million volts—and they discharge through dry air, ordinarily considered one of the best insulators, to usual distances of half a mile, or even occasionally to a mile or more. The lightning bolt itself, though similar in many ways to a small spark in the labouratory, yet shows additional and peculiar characteristics introduced by its enormous voltage and the great length and diversity of its path. There is nothing oscillatory or alternating about the lightning bolt—it is one or more surges of current in the same direction. Moreover the bolt does not jump instantly from a positive cloud, say, to the negative earth—it appears to grow or burrow through the air in a succession of electronic darts, building its own ionized (or electrically conducting) path as it goes.

Ordinaily this growth is frightfully rapid, and takes place in perhaps one millionth of a second. But occasionally it is slow enough to be seen by the naked eye in the form of 'rocket lightning,' in which a growing and perhaps upward-directed bolt burrows visibly at no greater speed than an ordinary rocket. Some of the veined lightning in a detached thunderhead, viewed from several miles away on a clear night, can often be seen to grow or burrow in this fashion progressively along the bulges of cloud. Usually lightning strikes from thundercloud to ground, from ground to cloud—or, even more commonly, between clouds. But on rare occasions it flickers or glows out of a clear sky, as charged masses of air pass near each other or the ground.The amperage or rate of electric current flow in a thunderbolt is likewise enormous, compared with most man-made currents. Its value ranges all the way from ten thousand amperes in weaker flashes up to perhaps two hundred thousand amperes in larger bolts. One hears of people being struck by lightnin and surviving the ordeal, but these fortunates are probably grazed by side flashes or forked branches while the main bolt goes elsewhere. Few if any animals, human or otherwise, have ever been hit by a heavy main bolt and lived to tell the tale. For like its voltage and amperage, lightning power is unbelievably large—perhaps ten million million kilowatts or thirteen trillion horsepower! But as an average single flash lasts only three or four millionths of a second, the total amount of electricity (or electrical energy) involved is not very large—perhaps three thousand kilowatthours or so, obtainable commercially within more usable limits of current and voltage for seventy to a hundred dollars.

Lightning punishment at one point is by no means limited to a single bolt, however, and the old adage that 'lightning never strikes twice in the same place' is about as untrue a statement as anyone could conveniently concoct. Many lightning flashes are of the 'multiple' variety; the bolt not only strikes twice, but may strike repeatedly—five times, ten times, or even forty times in the course of a second or two. This multiple lightning usually appears to the eye as a single, flickering, long-lived bolt, and it is the kind most likely to start fires in any dry and combustible material along its path.

The brilliant flash that accompanies a lightning discharge, formerly thought to be caused by sudden heating of air, is now known to result from ionization of the air by the terrifically strong electric field along the path, causing the very atoms of dense air to vibrate at such high frequencies that they radiate visible light in the same way that rarified neon in a glass tube glows under moderate electrification. The lightning flash is thus kindred (though once or twice removed) to the flickering aurora of polar skies, and to the glowing message of an advertising sign. Lightning is usually white in colour, combining the spectrum of nitrogen with that of oxygen. But it may appear bluish in contrast to yellowish artificial lights, and may appear yellowish if viewed through haze, fog and clouds. Occasionally a reddish or pinkish flash occurs, when air rich in water vapour is ionized so as to give the spectrum of hydrogen.

Thunder takes many audible forms, but they all originate in one underlying cause—the sudden increase of air pressure along the path of the bolt (for any sound wave is only a sudden increase in air pressure) resulting from the heating, molecular dissociation, and ionization of air by the bolt. Earlier vague assertions about 'air collapsing to fill up the hole hollowed by the bolt,' and such notions, are now in the discard. If lightning strikes close enough—within a few feet of you—the sound is simply one sharp and terrific 'bang!' like that of a large cannon fired in the same place; for the ear is too paralyzed by this first and nearest blast to notice the subtler and later gradations of reverberating thunder arriving from more distant portions of the bolt.

But if the bolt is farther away, and particularly if it is a long one, the thunder arriving successively (at eleven hundred feet per second) from its more and more distant portions has the sound of a long roll, which wavers considerably even as the characteristics of the bolt, and its direction with respect to the hearer, vary greatly along its path. Under certain conditions of air stratification, and often in the presence of steep mountains and cliffs that also serve to reflect and intensify sound, the roll of the thunder echoes back and forth; these overlapping echoes, together with simultaneous sound reinforcements from equidistant portions of the path, may build up into great booming crescendos of noise. Thunder is seldom heard over horizontal distances of more than fifteen miles; and if the flash occurs at altitudes of more than two miles, it may not reach the ground at all in audible form.

Several kinds of lightning have been observed, of which by far the most common is 'streak' lightning—the ordinary bolt. A streak of lightning is not a zigzag, as commonly drawn, but sinuous in shape, like a river in broken country. The streak may be single, but more commonly, part of it splits off into smaller downward branches that may or may not reach the ground. Occasionally a bolt splits into 'forked' lightning—two or more main branches that each strike home. Multiple streak lightning, drifted sideways by a strong wind, sometimes appears as 'ribbon' lightning. And if a single streak follows a very sinuous and varied path, it may present the illusion of glowing beads where it is either more intense, or else where part of the flash is seen end-on.

Another distinct variety, very rare but very dangerous, is 'ball' lightning. cientists long doubted altogether the existence of this variety, but so many accounts of its appearance were authenticated that it has come to be believed in, and more or less explained. It appears in the form of a luminous ball, perhaps the size of a clenched fist, which either falls through the air, floats horizontally, or rolls along some solid object such as the ground. Visible only for a second or two, it disappears either quietly, or with a terrific explosion. According to present theory, ball lightning is nothing more than a creeping corona discharge along a relatively intense portion of the electric field—or, one might say, a 'stalled thunderbolt,' similar in kind to rocket lightning but much slower. If the electric charge behind it dissipates, the ball fades away also; but if the original charge increases enough to break down the whole path with a large thunderbolt, as it well may—then look out! Another variey of Nature's electric illumination is 'sheet' lightning. This is not to be confused with so-called "heat' lightning—the diffuse and silent reflection of distant electric flashes hidden either by clouds or by the horizon.

True sheet lightning occurs as a momentary glow extending through fairly large masses of air in the cloud region (or troposphere), and is nearly always white in colour. It is probably an ionization effect similar in principle to the aurora, though the sheet-lightning glow occurs on a much smaller scale than the aurora, and at much lower levels in the atmospheric ocean.

Another and somewhat similar variety of thunderstorm light is ' Saint Elmo's fire,' a bluish, brush-like discharge that sometimes appears on rooftrees, tall steeples and chimneys, the mastheads of a ship, or any other high, sharp- pointed object during a thunderstorm. It is merely the well-known corona discharge, reproducible at will in the labouratory or on a high-voltage transmission line. Occurring in nature, it indicates intensification of the electric field, and may indicate the imminence of a thunderbolt. Such a corona discharge, of course, presents a very real fire hazard—particularly in the case of a hydrogen-filled balloon or airship.

During a great line squall over New Jersey, the German airship Hindenburg was valving explosive hydrogen in preparation for a landing at Lakehurst. Saint Elmo's fire ignited the escaping hydrogen-air mixture, and the great airship crashed in flames, killing thirty-six of its transatlantic passengers. This incident marked the tragic end of hydrogen-filled airship activity; and as the United States Government wisely declined to release helium to a nation arming to the teeth for a treacherous and unprovoked attack on its neighbours, it marked the effective end of all German airship activity for some time.

LIGHTNING HAZARDS

Just as lightning bolts themselves exhibit almost endless varieties of form and intensity, so do their earthly effects vary greatly, not only with the bolts' severity, but even more with the things and substances which a ground-strike encounters on its path to earth. Metallic conductors of sufficient size—say a bar of copper or other metal the size of a pencil—pass the bolt without harm to themselves or anything else. But if the con- ductor does not offer the shortest and easiest path to earth (from the viewpoint of the hop-along bolt itself), it may jump to a more tempting and destructive path; and even if the main bolt follows the conductor, side flashes may jump to neighbouring conductors, be they metal tanks or pipes, pools of water, or human beings.

Small, hollow conductors may be crushed by the intense electrostatic and electromagnetic fields accompanying lightning passage. Irregularly shaped, loose, or insecure rods or wires are likely to be moved, bent, burned, or melted; and insufficiently large conductors, such as a radio antenna, are invariably melted or vaporized instantly by a direct stroke. Non-conductors such as wood and stone fare much worse, at least during the 'dry' part of the storm before they are wetted over with a film of partly conducting water. Dry trees and poles are often split and splintered throughout their length by the explosive action of their moisture content suddenly converted into steam, and the pieces are blown forcibly outward in all directions. Dry or partly moist bricks and mortar may be violently exploded, and very rarely a small and flimsily built house is wrecked completely, as if by an earthquake.

Some of lightning's worst destructive effects are seen where large conducting masses are separated by non-conducting ones -such as a metal roof, ungrounded, on top of a wooden house. Here the conductor seems to collect and intensify the bolt, which wreaks havoc on the non-conducting structure beneath. No ordinary insulator such as a glass plate, of course, offers any effective resistance to a lightning bolt. If an insulating plate stands in the way and cannot easily be got around, the bolt simply blasts or burns a hole right through it. The best possible form of lightning protection for any object, animate or inanimate, is a 'Faraday cage'—in complete form a grounded shell of metal not less than one twentieth inch thick (Number 16 A.W.G.), completely surrounding the object in question and several feet away from it on all sides. But as such a metal box would be very expensive and inconvenient, it can be approximated, with nearly as good practical results, by a skeletonized cage of interconnected conductors—each conductor being at least one fourth inch thick, if of copper; or one third inch thick, if of iron.

Of such general form are the lightning-rod systems now in use on thousands of isolated houses and barns in thunderstorm areas—systems that prove their usefulness by greatly reducing fires and other lightning hazards. In general, these protective rods should extend at least a foot or two above all the higher projecting parts of a building; they should extend downward as directly as possible to thoroughly good ground connections; and they should be bonded to each other and to any other large metallic masses in the building, which latter should be independently grounded. Any high metallic conductor, such as a steel tower or a steel-frame city skyscraper, usually confers about it a 'cone of protection' having a radius equal to three or four times the height of the tower; occasionally, however, this cone of protection is violated by an erratic bolt.

But for all the unpleasantness of lightning in the open—the vivid flashes of uncomfortably close ground-strikes and the deafening crash of near-by thunder—the actual hazard is really very small. Out of all the millions of people exposed to lightning in the United States, only about five hundred are killed by it each year. Under the intense part of a very severe storm, however, and in various unwise locations beneath any storm, the risk may notably increase.

The Bareau of Standards has issued some rules of wise personal conduct for those caught out in the open by storms, which are as follows:

- Do not go out-of-doors or remain out during thunderstorms unless it is necessary. Stay inside of a building where it is dry, preferably away from fireplaces, stoves, and other metal objects.
- If there is any choice of shelter, choose in the following order:
  - Large metal or metal-frame buildings.
  - Dwellings or other buildings which are protected against lightning.
  - Large unprotected buildings.
  - Small unprotected buildings.
- If remaining out-of-doors is unavoidable, keep away from:
  - Small sheds and shelters if in an exposed location.
  - Isolated trees.
  - Wire fences.
  - Hilltops and wide open spaces.
- Seek shelter in a cave, a depression in the ground, a deep valley or canyon in dense woods, or a grove of trees.

Their cumulus clouds are merely prosaic banks of condensers, but their lightning is real and convincing, as anyone can testify who has seen one of their fivemillion-volt flashes strike downward in the dark through fifteen feet of air to shatter a three-inch fence post with the bang of a small cannon.

Safety of People in Ordinary Wood-Frame Houses

Much will depend upon the construction of the house and what is in it, and particularly the amount of metal in the walls, whether in the form of gas, water, or vent pipes, how the electric wiring is arranged—whether in rigid conduits or in BX, or knob-and-tube open wiring. Quite obviously, the only absolutely safe place to be is inside of a completely enclosed metal structure, which may or may not be grounded.

The degree of safety enjoyed by the occupants of an ordinary wood-frame house will depend upon the degree of approach to this completely shielded tructure. The wave front of current discharge in a direct stroke is steep enough, and the magnitude great enough, so that in many cases fairly high potentials will be built up in conductors of thirty- to forty-foot lengths—so that side-flashes may occur to other metal structures which may be connected to the same ground but at some distance away.

This was particularly exemplified in a case where a man had a lightning rod system properly installed, as far as the lightning rods were concerned, and yet it flashed through about six inches of wood and air to a conduit located in the peak of the attic roof underneath the lightning rod system. Of course if these two had been tied together at this point, the fire which resulted would have been avoided.

So far as my experience goes, the greatest danger from lightning in the ordinary wood-frame house is from fire, and of course this is particularly so in the rural communities where fire-fighting apparatus is not readily available. In cities the damage done to buildings as a result of lightning is not as a rule serious, although cases are on record where explosions have occurred during lightning storms, but apparently these explosions were due to the existence of gas or other explosive materials enclosed within the house at the time it was struck.

Wood-Frame Houses Protected by Lightning Rods

Insurance records for the State of Iowa indicate a very substantial improvement, which is also substantiated by those from the Province of Ontario, as reported by the Fire Marshal of Ontario.

These results have indicated that lightning protection is better than ninety per cent efficient, from the point of view of preventing fires in protected buildings. Of course not all of the lightning protection systems were properly installed. Many were, no doubt, inadequately grounded, which accounts for the imperfect record to some degree.

At the same time, we know today that lightning discharges do not always strike the highest object, and that now and then the cone of protection, which labouratory tests have demonstrated exist about a lightning rod, is violated. I have, for instance, a photograph showing lightning striking the top of the Empire State Building, in which the discharge failed to contact the very top of the building but struck about forty feet down. It is apparent that lightning discharges take a path which might well be called the best path, from the point of view of the discharge itself.

Safety of People in Closed Car with Steel Turret Top

Following the principle of the Faraday cage, there is no reason why people inside of a car with a steel top should not be very safe from lightning discharges, provided that they are not in the process of getting in or getting out of the car, or do not have their heads or arms out of windows, or are not standing on the ground changing a tire, and assuming that the gas tank is not struck and punctured, which is relatively unlikely. I believe the best advice to a man riding in an automobile during a lightning storm is to keep going and not to get out of the car, and the chances are that he will be perfectly safe.

Of course, if power wires fall on the car, which has happened during violent storms, he is wise to remain in the car, and not attempt to get out, since this procedure has in the past resulted in serious accidents. Safety of People in Closed Wooden Yacht

As a general thing, lightning is more likely to strike a wet mast on the water, if it is striking round about, than a similar object on land, because of the relative smoothness of the water and the relative height of the projecting mast. It is probable that suitable protection could be provided to the boat and its occupants by connecting a conductor following down the mast to the metal keel of the boat. In case it is an engine-driven boat, care should be taken to be sure that all the metal objects within the boat which are not connected together are electrically bonded together, and connected to some outside metallic conductor, such as the keel, which furnishes good contact with the water.

I don't know of any definite rule which has been set up concerning the protection of boats, but it seems to me that the general principle of the Faraday cage, or its equivalent, is one which can be applied to boats, coupled with the idea of the lightning rod and its usual cone of protection, and means for discharging the current of the lightning discharge itself into the water, without having either the boat or its occupants become a part of the circuit. It might be well at this point to inject an observation concerning people in swimming, to the effect that, in my estimation, the danger to the swimmer is not so much that he may be struck by lightning, as that a small current flowing through the water, even at considerable distance from the stroke itself, may be sufficient to render him unconscious, and he dies by drowning, rather than by electrical shock.

Safety of People in an Airplane

Once more the question is one of preventing the passenger or pilot from becoming a carrier of current of the lightning discharge by being in series with it. The only known way to prevent this is either to keep the lightning current from entering the plane by means of an all-metal plane, or its equivalent protection by some means of shunting the discharge which gets into the plane from the body of the occupants, and means also of preventing the lightning discharge from interfering with the mechanical and electrical operation of the plane.

The best method of doing this is probably the all-metal plane, which brings us to the Faraday cage, where if the joints in the plane are thoroughly bonded together, and the metal parts so connected that the measuring instruments and other electrical circuits of the plane are not likely to become a part of the lightning path, the possibility of damage is relatively small. The occupants of the plane ought to be perfectly safe provided they are entirely surrounded by metal having a reasonable current carrying capacity....

From the point of view of fire, the multiple stroke is undoubtedly far more serious than the single stroke, and I believe I am quite safe in stating, that the single stroke of large magnitude is probably responsible for the explosive effects frequently observed, whereas the multiple stroke, even though of many relatively small currents, may be responsible not only for explosive effects but for fires also, provided inflammable material is present. As many as forty strokes have been observed within one second of time, and data which we have been taking during the past two years seem to indicate by far the great majority of lightning strokes are multiple....

DOWNBLAST AND DELUGE

Earlier we saw how a typical thunderstorm forms out of nothing into a towering Everest of threatening cloud, how the terrific electric potentials of the lightning arise out of a mere spattering of raindrops. But what about the lashing hail, the scouring squall winds, and the other thunderstorm accompaniments that often prove more dangerous and destructive than the lightning itself?

We had earlier in mind the picture of a towering cumulus cloud marked by fast-rising air currents and fast-falling raindrops of near-maximum size, whose spattering in the zone of strongest updrafts produced positive electric charges large enough to forge the thunderbolts. As most of the warm, moist air contributing its latent heat to these rising currents is coming from in front of the storm, the most violent updrafts are usually in the front part of the main thundercloud; and they tend to slope somewhat towards the rear of the storm, carrying the rain rearward also. Just behind the main updrafts, where their speed has fallen off to less than twenty miles per hour, will be the region of heaviest rainfall. Much of this heavy rain, coming from upper and colder re- gions of the thundercloud, is cold; some of it may even be frozen. This cold deluge cools the air as it falls, partly by its own coldness, but even more by its partial evaporation. The result is that air in the heavy rainfall region is soon chilled well below the surrounding air, hence becomes denser than the surrounding air, hence begins to sink, and soon develops into a great downdraft of cold air perhaps nearly as violent as the updraft of warm air in front of it. This cold downdraft reaches the ground with the forward speed of the whole storm at upper levels (which may be twenty or thirty miles per hour or more), and at the ground level it sweeps forward as a violent outrushing squall wind, blowing in heavy gusts directly out of the storm.

But the squall wind may far exceed the speed of the storm itself. It is not impossible for a storm moving at thirty miles an hour, say, to produce a surface squall which (however shortlived) may blow at sixty or eighty miles per hour and sweep all before it. These excessive squall winds are perhaps partly explained by the crowding of the downdraft against the ground (which also contributes to the sharp barometer rise noted with the squall). But the chief reason for their ferocity is the action of a horizontal vortex. With a great updraft in one part of the cloud and a great downdraft behind it, all sorts of eddies and minor whirlwinds are bound to exist in thunderclouds, as airmen who have endured the blind bumpiness inside one can fervently testify. Most of these vortices, being well up in the cloud, are invisible (except on the gyrating instrument dials of an unhappy airplane); but one of them, being formed just at the cloud base, where the relative humidity is such that the slightest drop in pressure condenses moisture, may plainly appear as the long, light-coloured roll cloud or 'squall head' arching horizontally across the front of the storm.

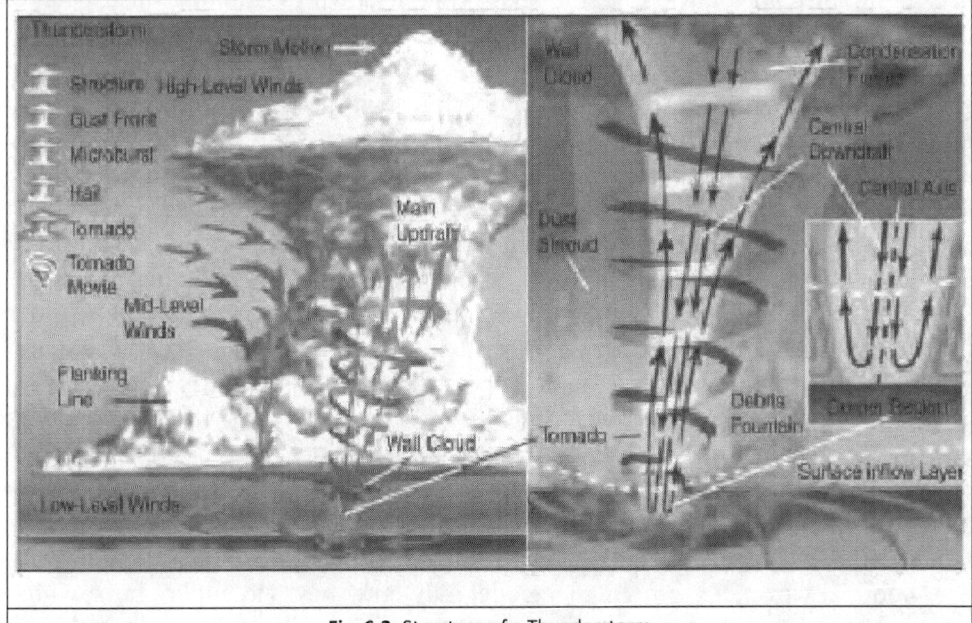

Fig. 6.2: Structure of a Thunderstorm

As the roll-cloud vortex revolves, it moves across land or water like the revolving brush of a gigantic carpet-sweeper, piling its speed of rotation on top of its speed of horizontal motion—all to the consternation of squall-blown humans in one case, or insects swept out of a carpet in the other. This horizontal vortex of the roll cloud is, in fact, a sort of horizontal tornado; and under certain conditions, parts of it can turn down to the ground and become real tornadoes. In any case, it is something for an aviator to avoid at all costs.

This whole thunderstorm wind-mechanism picture, has, of course, been snapped at a single and very typical instant. Actually, thunderstorms and line squalls vary greatly in size, form, and intensity. Some thunderstorms have no squalls; some squalls have no thunder; and any one storm is constantly changing, from moment to moment, its form and intensity. Updrafts and downdrafts alike are constantly changing their speed and their position in the storm.

Thundershowers are usually of the brief, heavy type, and may reach cloudburst intensity. Raindrops in the active parts of the storm are much larger than ordinary raindrops, being mostly of near-maximum size; and they are crowded together, in certain places, by the interplay of violent and ever-varying air currents. It is not at all uncommon for these thunderstorm deluges to fall at the rate of three inches or more an hour, flooding low places with a few feet of suddenly rising water. On occasion, the rate of water-fall for a brief interval has been known to run as high as one inch per minute! To a ground observer such a cloudburst seems to be falling in sheets of solid water, as if the heavens had indeed opened; to an aviator trying to fly through the downpour, it seems almost like boring through the depths of the ocean, and airplanes have actually been brought down by running unsuspecting into a heavy cloudburst at night. On the ground a cloudburst is most destructive in normally dry desert and mountain regions—among steep converging slopes bare of brush or timber—where any dry ravine may within a few minutes become a raging torrent, and walls of tumultuous water ten or twenty feet high may roar along the rock-walled bottoms of steep canyons.

Nor is torrential rain the last of possible thunderstorm visitations on men caught in or below the storm. On occasion, thousands of tons of ice may fall from the skies in the destructive form of hail. As we have seen earlier, the upper parts of a towering thundercloud harbour many slowly falling ice particles. Under favourable conditions of temperature and humidity, some of these may grow into sphero-crystals or snow balls (graupel) about the size of small peas. Alternatively, or in addition, raindrops carried by the storm's great updrafts into temperatures well below freezing, and sufficiently jostled by turbulence, may be frozen into ice balls (sleet) of similar size. When the updraft becomes less violent, or when these pellets reach the edge of it, they fall groundward again, perhaps gathering a coating of snow above the freezing level, and a further coating of slush or water below it.

Soon, however, the growing ice pellet may enter new violent updrafts, which carry it upward again into the arctic zone, where its acquisitions of snow, slush, and water are frozen into a new concentric shell of solid ice, increasing its diameter. In like manner the ice pellet may fall again through several thousand feet, be carried upward again through another great rise, and repeat the whole down-up cycle ten or twenty times, growing the while into an ice ball of considerable size.

The largest hailstones ever authentically observed, which fell on some Nebraska farms in 1928, were the size of small grapefruit. One of these stones weighed one and one half pounds and was about six inches in diameter! Three-inch hailstones, the size of ordinary oranges, probably shower down somewhere in the United States a few times every summer. Stones of golf-ball size are not at all uncommon. In 1936, nineteen Bushmen in South Africa were reported to have been killed by a fall of oversized hail that covered their bodies to a depth of three feet. Hail of some sort accompanies a large proportion of temperate.

One thunderstorms, but the hail zone of any one storm is usually not more than a few miles wide. But in tropical thunderstorms hail is very rare, however heavy the lightning and rain, because rain-making portions of the thunderclouds seldom extend far into a freezing zone perhaps five miles high.

The higher a temperate-zone thunderstorm extends, the blacker are its underparts, and the more likely it is to unleash severe lightning, torrential rain, destructive winds, and hail. The convergence of two or more storms tends to intensify these factors. Wind-tunnel tests with hail-sized balls have revealed some astounding probable speeds for the updrafts of great storms.

To sustain a one-inch hailstone an updraft of about sixty miles per hour is necessary; a three-inch stone requires a hundred-mile current; and a five-inch stone must have been held aloft at some time by an upblast of two hundred miles an hour! In severe thunderstorms, therefore, 60-mile updrafts (and commensurate adjacent downdrafts) are common, while 100-mile and 200-mile currents are not impossible. Which seems good enough reason for aviators to keep out of them if they can.

An average detached or 'air-mass' thunderstorm starts as a pile of cumulus clouds perhaps five miles or so in diameter, gradually spreading out horizontally and building up vertically as it sucks in warm surface air and gathers to itself near-by clouds. The altitude of its highest cumuliform turrets may be four or five miles in temperate latitudes (or considerably higher in the tropics); and the cirriform anvil or 'cap' may reach six or seven miles aloft. As the storm moves bodily along with the upper winds, it spreads out laterally to a width of perhaps twenty or thirty miles, so that its whole path resembles a pear in shape. After six or seven hours, with the decline and disappearance of the sun, an air-mass storm usually dies out, having raged in all through a distance of perhaps a hundred and fifty or two hundred miles.

The airplane has of course made it possible to survey the breath-taking topography of thunderstorm clouds, quickly and accurately. One of the TWA 'over-weather' research aviators made a detailed investigation of a large air-mass storm above Missouri. The general storm-top of cumuliform clouds was at twenty-two thousand feet.

The highest cumulo-nimbus turrets boiled up to twenty-five thousand feet, at which altitude the plane could just clear the turbulent wind-shift line. The cirriform anvil, ten miles wide, extended on up to about thirty-five thousand feet. Flying through the anvil at twenty-six thousand feet, the plane encountered fine snow and some subcooled water droplets which froze as light icing; the temperature was 5° F., compared with 100° F. at the ground.

Even an ordinary thunderstorm has within it a cold front in miniature—a rain-cooled wedge of squall-wind air surging outward and displacing the warm air in front of the storm. But the cold air behind a large-scale front usually comes from arctic or polar regions, whereas the cold air in a thunderstorm is, one might say, manufactured locally by the refrigerating process of the storm. Yet if a very large thunderstorm spreads out laterally beyond fifty miles or so, it begins to take on the characteristics of a line squall.

THE WINGS OF THE STORM

Reduced to typical form and simplest terms, a line squall is nothing more than a violent cold front often spotted with many thunderstorms, continuous or intermittent, along the hundreds of miles of its length. The front itself is necessarily marked by more than ordinarily large temperature contrast between the unlike air masses; by high moisture content and great depth of the warm air mass; and perhaps also by some moisture and instability in the cold air mass. Also in evidence is the rapid advance of the cold air mass, the enforced retreat (mostly sidewise) of the warm air mass, and conflicting impact between the two. This air-mass conflict shows up as high winds along and across the front, sharply contrasted in direction, and appears on the weather map as isobars packed close together and bending in a sharp V where they cross the front.

From this typical picture as a norm, individual line squalls of course vary widely. Few cold fronts are violent enough to be spectacular line squalls, and few of these, even, conform strictly to type. The one common characteristic of all surface cold fronts, at least over flat country or the ocean, is a marked shift in wind direction, usually from some southerly to some northerly quarter. The squall line itself usually advances from the northwest or west (less frequently from the southwest) at something like twenty to thirty miles per hour in summer and thirty to forty miles per hour in winter. (Occasionally, in some parts of the country, the squall may come out of the north or even the northeast.)

But rather than further generalities, often violated, here are accounts of two actual line squalls that struck the city of Washington. One came early in June:

At 2 P.M. the sky was still clear, but hazy; and a few clouds wer rising in the distant west. The wind was moderate out of the southwest, and some twenty or thirty small sailboats were racing on the wide Potomac River, which was also dotted with various other craft. Upper clouds gradually grew out of the west, gradually spread and darkened, obscuring the sun in a quarter of an hour. Then a storm collar or squall cloud appeared at lower levels out of the southwest. It consisted of dark gray stratocumulus clouds, whose violent agitation showed the ferocity of their inner winds.

The squall cloud came rapidly on until it passed overhead, but there was no lightning and no rain. The wind slackened somewhat directly under the squall cloud, becoming rather puffy, and the sailboats continued their gay progress all unheeding. But perhaps five minutes after the squall cloud had passed overhead, the surface squall struck very suddenly and fiercely out of the southwest. As the wind rose shrieking to fifty or sixty miles per hour the sailboats, one after another, were pressed down to the water and capsized, some with sails still set, some under bare poles. The extreme wind lasted only three or four minutes, then moderated to twenty miles an hour for another half-hour, with light rain; after which it fell nearly calm while the rain continued.

Another squll came in May, when the late spring contrasts between tropical and polar air were most intense: Up to mid-afternoon the sky was perfectly clear but very hazy, the wind moderate southwest. About 3 P.M. the forerunner clouds appeared out of the northwest—perhaps a sort of upper-level squall head—arching across the sky from northeast to southwest. They appeared as a solid wedge of high strato-cumulus or low altocumulus, advancing steadily at an altitude between one and two miles. There were moderate puffs of wind out of the northwest, followed by a gusty northwest wind of twenty-five or thirty miles per hour—but for the moment, no rain. The high clouds swept by overhead and out to eastward, shutting out the sky, and for twenty minutes nothing more happened, except that the pressure jumped up by five hundredths as the cloud edge passed, and the gloom steadily deepened in the northwest.

Then the gray rain curtain appeared, and with it came the real squall at 3.50 P.M. For a few minutes the northwest wind whistled around sixty-five miles an hour, driving torrential rain and rattling hail horizontally against windows and doors. Lightning flickered incessantly above and banged heavily to ground, trees and poles were borne to earth in exposed places, and the horizontal rain flooded over everything, cutting visibility to zero. The base of the black thunderclouds, when they could be seen at all, looked about one thousand or two thousand feet high. The temperature fell twenty-four degrees in an hour. The worst of the squall was over in a few minutes, but scattered showers and mild thunderstorms continued, interspersed with broken clouds, throughout most of the evening in the unstable invading cold air mass.

These are the picturesque and transient line squalls of spring and summer. In winter, marked cold fronts are less spectacular, but if anything more unpleasant. The worst weight and hazard of summer line squalls comes on the water, as the sailboat episode above indicates.

FLYING THROUGH A LINE SQUALL

Full-bodied line squalls are colourful enough, viewed from the ground or the water. But the way to see and feel one at its best—or its worst—is to get up into the air where you are almost a part of the gyrating squall winds. This is something that wise aviators seldom do, intentionally (except in big, scheduled transports, or in urgent military flying); but they sometimes get caught in squalls or plow through them deliberately in the course of an urgent mission. On the same afternoon that the airship Hindenburg crashed in flames at Lakehurst, an army pilot and I were flying from Boston to Washington in BT-8—a medium-fast, stubby, powerful pursuit-type training monoplane:... Towards Trenton the clouds were more scattered, and we dropped down through them to 2000, heading out westward so as to pass well to windward of Philadelphia's smoke pall. The scattered cumulus clouds, now above us, were becoming larger, with some scattered showers among them.

Some of these shining white cumulus heads were beginning to mushroom quite rapidly up into the blue—a sign of thundery weather.

Soon there appeared far ahead, somewhere over northeastern Maryland or southeastern Pennsylvania, a dark line or roll of cloud arching across the entire sky from southwest to northeast. This was the real front, the line squall itself. Great white thunderheads towered along it, in some places much higher than others.

Beneath these higher cloud-peaks were gray curtains of rain, and around them flickered the seeming-miniature sparks of lightning. For the time being, the controls were in my not very experienced hands.

Our course was about 240° magnetic (southwest-by-west true), and as we approached to within ten miles or so of the squall line, made out two large thunderheads towering far above the long, continuous roll cloud—turreted white mountains three or four miles high. Under each thunderhead a dark deluge of rain was falling from the squall cloud to the ground. It seemed wise to keep out of these active storms if possible, and we changed course 15° to the right so as to fly midway between them, just under the long, black roll cloud (which appeared fortunately to be well up off the ground, at a height of half a mile or more).

Beneath this rapidly approaching black roof of turbulent cloud the visibility was good—we could see far out ahead into patches of sunlight and shadow beneath other clouds behind the front. We were still cruising in level flight at 2000, and the black ceiling at about 3000, now rushing towards us overhead, seemed to offer plenty of clearance.

As we passed beneath the dark cloud-roof the BT-8 began to bounce around crazily. I had expected bumps, but these were so sharp and ferocious that it was all do to maintain oncourse control of a plane that was, at best, rather tricky to fly. Snatching a glance at the instruments, was alarmed to see the rate-of-climb meter swing to plus 1000 ft/min (12. m.p.h. straight upward), in spite of the fact that was already nosing downward in an effort to get lower and escape from the wildly gyrating black festoons of cloud that seemed now just over our heads.

This black cloud-roof, marked by turbulent downward bulges of mamato-cumulus and momentary 'cloud spouts' or incipient whirlwinds, was getting nearer by the second. It seemed as though we were about to get sucked bodily up into its black chaos. I had the nose shoved down as steeply as seemed prudent, for the air speed was now over 200 m.p.h. and the terrific bumps might strain the wings at this speed. Yet I hesitated to change the throttle setting without instructions from the pilot.

Finally I jammed the nose down in a steeper dive, taking a terrific beating from the bumps and still not losing altitude. At this point the great down-bulging cloud-pocks seemed to be almost couching our heads. Just then the pilot took over, and throttled back to case the bump-strains. And within another minute, we had passed out from under the roll cloud into a region of scattered sunlight and shadow, marked by detached storms in the cold air mass. These detached storms, of course, could be easily flown around.

Passing through a line squall in a controllable airplane may be exciting enough, but it is relatively easy and pleasant compared to navigating through one in an uncontrollable free balloon. Lieutenant Settle of the United States Navy had this unenviable experience over western Pennsylvania on the twentyeighth of May, 1938. Wrote he, in part: Dropped ballast and at 18.15 balloon was at 4500 feet altitude; the line cloud overtook and passed under us and we started up rapidly, the variometer going 'hard over' on the ascent side. At 8300 feet I valved two seconds, and balloon leveled off at 8400 feet and then started down at a violent rate, the variometer ink going out of sight on the descent side.

While falling the air was extremely turbulent; hail (size of peas), heavy rain, and snow flurries hit us; the basket swung violently from side to side like a pendulum and spun in azimuth; heavy gusts hit the envelope and basket; the motion of the balloon was like when attempting to hold one on the ground in very gusty surface wind. Lightning and thunder were heavy and apparently right on top of us. The rain and air were very cold and our hands numbed.

During the descent we dropped approximately fifteen bags of ballast; we came through the clouds at 3000 feet and leveled off at about 2000 feet. Found myself under a dense black cloud, the line boundary of which was several miles south of me; the line extended as far as see in either direction.

There were areas of heavy rain along the line on both sides of me and heavy rain to the northward of me, with lightning and thunder on all sides. The rain in my immediate vicinity had temporarily ceased; my air was NNW, about thirty miles per hour, and for a few minutes seemed relatively smooth. It was soon apparent that we were fast overhauling the line cloud.

When nearly up to the line, we again started violently up through the cloud on a second 'vertical circle,' repeating the history of the first one, and attaining a maximum altitude of 8200 feet. Some of the elements were more violent in the second than in the first excursion. Settle finally managed, by adroit maneuvering with gas valve and ballast release, to case his balloon out of the squall line. He landed at dusk on a farm near Parryopolis. He lived to write of his experience, but others who dared line squalls in free balloons, such as the brilliant American aerologist Meisinger, were less fortunate and plunged to lightning-lit deaths.

CAUSES OF AIR

Tornadoes are the smallest of all true storm whirls, but they are by long odds the most fearful and the most violent. The path of tornadic destruction may be only a few yards wide, and never spreads to more than a mile across— but within that path, winds roar at super-hurricane velocities approaching five hundred miles an hour.

Laths are driven like spears through large trees, and straws are driven like darts through inch-thick boards. Trees are uprooted and carried away, houses are exploded and laid in ruins, débris is strewn over the countryside, and the ground itself is harried and scoured. Cattle and men are carried aloft, cut in two or spared entirely, dropped yards away or miles away, gently or roughly, dead and mangled or alive and kicking.

The basic principle of a tornado can be glimpsed in the vortices that a vigourously plied canoe paddle creates in still water -or in the circular whirl of water around a washbowl drain. For the great air whirls likewise originate as vortices between conflicting currents—currents in the flow of the upper air. These strongly opposed upper winds arise oftenest in the more turbulent portions of a large and vigourous cyclone, either temperate or tropical—a cyclone that is usually some hundreds of miles wide. Sometimes the smaller whirls are themselves incorrectly called 'cyclones,' but 'tornado' is the true name. It comes from the Spanish and Portuguese words for 'thunder,' having been first applied to twisting thunder squalls experienced by early mariners off the northwest coast of Africa. Full-bodied atmospheric tornadoes are rare storms even in the geographic regions of their most frequent occurrence—parts of the American Midwest and Midsouth. The belt of maximum tornadic activity extends roughly northeast-southwest through Iowa and Kansas; but other midwestern and southern states also have their share.

No state cast of the Rockies, in fact, is free from occasional 'twisters,' and some few have been reported along the Pacific coast. All across America the yearly twister toll is about two hundred and fifty lives, some of which could doubtless be saved by a better application of aerological knowledge.

The storm waterspout, equally rare, is simply a tornado gone to sea, with all the sucking, rending violence of a land tornado—except that it usually has nothing to tear apart except sea water, unless some mariner is unlucky enough to fall afoul of it. But the commoner fair-weather waterspout of tropic seas is a milder creature, somewhat akin to the familiar hot-weather dust whirl ashore.

Basically the dust whirl is a sort of model tornado, formed in plain viewin the lower air levels on a small scale. Hence, it is worth examining as an introduction to the more awesome and confused mechanism of true tornadoes.

Growth of a Dust Devil

Almost everyone has noticed the small air whirls that frequent dusty roads and dry plowed fields on a hot summer afternoon. In subtropical desert country, where summer daytime temperatures are well above a hundred degrees under a blazing sun, where dry winds ebb and flow across the flat and sandy expanses, these dust whirls grow to rather imposing size and power.

Perhaps one overheated spot, like a hot stove, starts a rising current of air. The surrounding air flows inward to equalize the pressure difference, possibly from vagrant currents that are already somewhat in conflict—blowing in opposite or converging directions, or even blowing in the same direction at different speeds. Thus is started an eddy of whirling winds that ascends in the form of a spiral. Beginning with a sudden whirl on the ground, it grows with the ballooning of its overheated air to higher levels, perhaps reaching a thousand feet or so above the desert if conditions are favourable, and moving along with the general motion of the air.

The column of whirling dust is usually not more than a few feet wide, and its progress is slow enough so that anyone on foot or on horseback can easily avoid it. But even if it hits you squarely, no serious damage is done. It will buffet you with shifting winds, fill your mouth and ears with dust, and try to snatch off your hat. Papers and light trash lying loose are carried aloft for a few hundred feet.

A large tent fly may be burst upward like a paper bag. But an ordinary house stands firm. Thus the desert whirlwind is something like a tornado, at least in principle, though comparatively puny in effect. The dust whirl is limited in size and intensity by the limited amount of energy present in its atmospheric causes. Unstable air and conflicting air currents, in some form, are responsible for all whirlwinds—dust whirls, waterspouts, and tornadoes alike. But in the case of dust whirls both these causes operate on a small and limited scale.

The instability over a superheated deser may be very pronounced; the vertical temperature gradient or lapse rate may be more than about—19° F. per 1000 feet (about –34° C/km), so that the air overturns of its own accord; and it is certainly greater than –5° F. per 1000 feet (–10° C/km), which permits overturning once begun to continue without the aid of moisture. But this instability does not extend to very great heights, and there is practically no moisture present to contribute its latent heat of condensation to the building of large updrafts.

The conflicting winds that feed into the eddy are merely local winds, close to the ground and retarded by friction with the ground. As the earth's rotation is not effective on this small scale, dust whirls may spin either clockwise or counter-clockwise, in accord with chance.

Fig.6.3: Dust Devil

Nevertheless, a dust whirl has one quality that illustrates in miniature the underlying reason for a true tornado's destructiveness—the winds inside the dust column, near the center of the whirl, spin with a speed far greater than the speed of the outer air currents that feed them. This effect is due to a physical law—the 'conservation of angular momentum'—which governs planets in their courses no less than winds in their wanderings. In essence, this law requires that anything circling a center must tend to circle faster as it approaches that center.

So much for the small whirlwinds. Now for the portents that usher in the worst among all the varied freaks of the weather.

Tornado Weather

For many years aerologists had noticed that most tornadoes appeared in the southern quadrants of temperate cyclones (more rarely, in tropical hurricanes)—not all cyclones, of course, but certain ones. The reason, now coherently explained by airmass-and-front analysis, is that certain of these southern quadrants, or 'warm sectors," provide the atmospheric conditions out of which tornadoes are most readily born.

Conditions are most favourable in a 'thundery' warm sector where temperature and dewpoint are very high and where the air is conditionally unstable (having a vertical temperature gradient or 'lapse rate* greater than about –3° F. per 1000 feet [–6° C/km] with high relative humidity) to great heights. Moreover, most tornadoes seem to form somewhat (say fifty to a hundred miles) ahead of a marked and violent cold front or line squall that is invading the warm sector from the north or west.

All the conditions mentioned so far are equally favourable to violent thunderstorms, and violent thunderstorms usually (but not always) do occur along with tornadoes. The one essential condition that distinguishes a 'tornado clay' from a general thunderstorm day is strong, conflicting upper winds—blowing some forty or fifty miles an hour or more in converging or conflicting directions—at the lower cloud levels around half a mile high.

High southwest upper winds are always induced in the warm sector by the crowding of the oncoming cold front, and sometimes the cold squall head, pushing eastward at say halfmile altitude above the retarding effects of surface friction, may project outward fifty or one hundred miles beyond the cold front at the ground. If so, there is formed a steep upper front between the cold and warm air.

On the southeast side of this upper front the upper winds may be blowing a gale from the southwest; on the opposite side of the front they may be blowing equally hard from the north or the northeast. Under such conditions the slightest convection or updraft along the upper wind front will start a monstrous vortex between the two great slip currents.

Such updrafts are provided in gigantic form by the latent heat of condensing moisture in air that is conditionally unstable to great heights. The terrible vortex forms, perhaps a mile or so wide at the cloud level, drawing unlimited energy from the two great opposing upper air currents.

Conservation of angular momentum speeds the inblowing vortical winds, not at the original forty or fifty miles an hour, but up to perhaps four or five hundred miles an hour. The vortex grows downward from the cloud level to the earth, and where it touches the earth, there is destruction.

With our present aerological knowledge, tornado weather and tornado portents still cannot be surely or accurately appraised. Quantities of radio pilot balloons and radio sounding balloons might accomplish the task.

To our five senses and the old, reliable ground instruments, possible tornado weather is discerned as:

- A sticky, sultry oppressive day with southerly winds;
- Very high temperature and dewpoint;
- Pressure falling steadily and perhaps rapidly during the day.

There may be, an hour or two before the tornado, a topsy-turvy sky of mammato-cumulus clouds, crazily bulging down instead of up and indicating a distinct upper layer or front of colder air. (But these clouds do not necessarily mean tornadoes.)

In the west and northwest, perhaps (if we are in for a tornado), great thunderstorms grow nearer, with lightning both severe and continuous, and some hail usually in evidence. The towering thunderclouds have an ominous appearance—their colour is often described as a sickly greenish-black—and the lower clouds are perhaps visibly in rapid and confused or conflicting motion. Then, out of the base of a dark thundercloud, the rope-like funnel cloud appears, spinning counter-clockwise in the northern hemisphere—and if this funnel hangs near-by to the southwest or west of you, it is time to go underground or to move.

The favourable times of day for tornadoes are the same as those or thunderstorms—mid-afternoon to late afternoon—although frontal or pre- frontal twisters can occur at any time. the favourable times of year are spring and early summer, when air-mass contrasts and collisions are most intense. Professional aerologists have so far had very little luck forecasting tornadoes, and they are naturally hesitant to scare everyone in a whole state when the twister, even though it occurs, will mow only one narrow swath of destruction. But it is my guess that a keen amateur aerologist in the tornado belts (or a keen professional, for that matter) could do something along this line by drawing hourly regional weather maps from airway radiophone weather broadcasts. So far, indeed, regular meteorologists have sometimes not even been able to keep track of tornadoes already formed. One that struck Hutchinson, Kansas, several years ago had been traveling steadily towards the place for five hours, yet the people of the town were not warned. The usual height of a tornado funnel is the same as the 'ceiling' under the thundercloud base—say two thousand to three thousand feet. Since the funnel cloud results from condensation of moisture as the result of expansional cooling in the low pressure at the whirl center, the funnel varies greatly in appearance with atmospheric conditions. Oftenest it is a sort of thin, dangling rope, sometimes a gigantic 'elephant's trunk,' occasionally a fairly wide and solid-looking funnel.

Rarely (in very dry air) the funnel may be entirely invisible, though nevertheless destructive. Owing to peculiar light-refraction effects in its rarified center, the funnel sometimes looks almost black and very solid, 'as if made oftar.'

The funnel itself (or the air whirl around it) emits a roaring noise, when close by, that has been compared to the roar of hundreds of airplanes in flight or scores of freight trains going through a tunnel.

The narrow, sinuous path of destruction is usually twenty to forty miles long (though it may be as much as three hundred), and anywhere from fifty feet to a mile wide, averaging perhaps a thousand feet. The speed of advance continuously varies, being anywhere from twenty to sixty miles per hour, and the direction is usually towards the east or northeast.

In addition to its forward motion, the funnel often swings pendulously from side to side, changes form and shape, and bounces up and down—all of which gyrations serve to make matters interesting for frantic humans trying to escape the dread thing. Frequently the funnel lifts off the ground for a good distance, capriciously sparing one farmhouse and blasting the next.

Tornado Havoc

The earliest American colonists were introduced, here and there, to an occasional tornado. One was reported in New Haven in 1682, and another in Charleston in 1762. In the nineteenth century settlers penetrating into the plains states were oftener exposed, but what with all the other hazards of pioneer life, twisters do not seem to have been very seriously regarded until the seventies or the eighties, when tornado caves or dug- outs gradually came into use. At first there was some prejudice against these shelters, on the ground that it was undignified for a brave man to 'hunt his hole' at the sight of a dark cloud. But this idea disappeared, in time—and so did its proponents, if they tried to stay out in tornadoes.

Tornado destruction may start on the ground beneath the vortex forming at cloud level, even before the funnel cloud has appeared. People who have watched the near-by birth of a tornado tell of clouds racing along two or more opposite or conflicting paths, and then an inward, whirling motion of the clouds as the terrible vortex forms. Under the vortex trash and débris begin ascending spirally to the clouds.

Then, in one reported instance, explosions began on the ground beneath: 'the ascending wreckage looked like the explosion of sparks from a great fire'; these explosions were 'similar to blasting operations.' After three separate ground explosions the funnel cloud formed, resembling 'a misshapen cornucopia,' disappeared, formed again, and moved away to the northeastward.

When a tornado is large and well developed, almost every ground object touched by the lower end or spout of the funnel cloud is carried away or destroyed. The path of total destruction, a sinuous track that may extend for many miles across country, is usually some three or four hundred yards wide. But if the funnel lifts, ground havoc ceases; and sometimes the funnel spout sails along perhaps thirty feet up in the air, tearing the upper stories from buildings whose lower stories remain unscathed.

Ordinary buildings caught in the funnel spout seem to explode, partly as a result of the sudden drop in pressure—roofs fly upward and walls outward. If windows are open, releasing the internal pressure, a building may be spared. Heavy bricks and concrete blocks mean nothing to the five hundred-odd-mile winds of the tornado—they are strewn about like chaff. Modern, steel-frame buildings have apparently stood up fairly well, at least in smaller tornadoes. But not so with ordinary frame buildings.

Trees are uprooted and laid flat or carried away; windfall tracks in forest country are graphic and mournful records of tornado passages in bygone days. Curious projecti lelike effects occur in the fast-whirling winds—straws shot through boards, boards shot through logs and trees, and stones blasted through walls. Heavy timbers were carried completely away, in one instance, to varying distances between twelve and twenty miles. In another twister an eight- hundred-pound ice chest was carried several miles.

Another whirl whisked aloft a large bull and dropped him to earth forty yards away, unhurt, but frantically bellowing. A horse was carried half a mile without being killed; another was flung two hundred yards and cut in two. A child was carried half a mile and merely scratched. A thirteen-year-old boy flew three quarters of a mile to land again on solid ground, bruised, stunned, and stripped of his clothing. A woman was carried seven miles and killed. Tornado wounds, incidentally, are difficult to cleanse and heal, being deeply scored with gravel and débris.

No vehicle can continue its course through a real tornado. Airplanes must avoid them at all costs. On a main highway a speeding sedan automobile, struck by a rather small twister, was lifted bodily, carried eighty-five feet, and dropped undamaged on its wheels in the opposite traffic lane. Some years ago the 'Empire Builder,' fast limited train of the Northwest, was struck by a sizeable tornado while rolling at a mile a minute. The tornado moved in from the side nearly as fast, struck the train fairly, and wrecked it. Five seventy-ton cars were lifted bodily from the rails, one being moved eighty feet. The onehundred and forty-ton locomotive and one hundred-ton tender remained on the track, but the cab windows were blown out and the engineer's goggles snatched from his head. The heavy steel cars saved most of the passengers, as in other train wrecks, from serious injury.

Fig. Path of Tornado

As for tornado protection, the lower inside rooms of a modern steel-frame office building are reasonably safe from twisters, as from all earth storms and earth shocks. There is also the old, reliable tornado cellar (alliteratively but incorrectly called 'cyclone cellar'), or dugout. After a dugout, the next safest refuge is probably the southwest corner of the cellar of a woodframe building (which will most likely be carried bodily away to the northeast). As in earthquakes, unframed buildings of brick and stone are often worse than none. For people caught outside, the first requisite is quick but careful observation of the tornado's movement and probable path, together with an equally hurried orientation of compass directions. Most tornadoes advance out of the west or southwest towards the east or northeast. Plans for flight must also take account of the sinuous swaying back and forth of the funnel cloud, like that of a snake advancing on its victim. From directly in front of the tornado, it is best to run towards the north or northwest (as the northern side is likely to be slightly less violent).

At any rate, never retreat to the eastward or northeastward unless you are in a vehicle capable of outracing the funnel. If flight is impractical, the best outside refuge is a deep, narrow trench or ditch, or the best hole or gully that comes to hand, or the eastern side of a large boulder mostly embedded in the ground. If there is nothing for it but an open field, lie face down, head to the east, and arms over head.

Seeing the Inside of a Tornado

As most human observers caught near a tornado are (to say the least) excited, and weather instruments caught within one are usually destroyed, full and accurate tornado observations are rare. By long odds the calmest and most penetrating tornado account that has come to my knowledge is the following one, written some years ago by Alonzo A. Justice for the United States Monthly Weather Review from the eyewitness story of Will Keller, a farmer of southern Kansas:

An umbrella-shaped cloud in the west and southwest and from its appearance suspected that there was a tornado in it. The air had that peculiar oppressiveness which nearly always precedes the coming of a tornado. But my attention being on other matters, I did not watch the approach of the cloud.

However, its nearness soon caused me to take another look at it. I saw at once that my suspicions were correct, for hanging from the greenish-black base of the cloud was not just one tornado, but three. One of the tornadoes was already perilously near and apparently headed directly for our place. I lost no time therefore in hurrying with my family to our cyclone cellar. The surrounding country is level and there was nothing to obstruct the view. There was little or no rain falling from the cloud. Two of the tornadoes were some distance away and looked to me like great ropes dangling from the clouds, but the near one was shaped more like a funnel with ragged clouds surrounding it. It appeared to be much larger and more energetic than the others and it occupied the central position of the cloud, the great cumulus dome being directly over it.

Around the lower rim of the great vortex small tornadoes were constantly forming and breaking away. These looked like tails as they writhed their way around the end of the funnel. It was these that made the hissing noise. The opening was entirely hollow except for something which could not exactly make out, but suppose that it was a detached wind cloud. This thing was in the center and was moving up and down. After it passed my place it again dipped and struck and demolished the house and barn of a farmer by the name of Evans.

The Evans family, like ourselves, had been out looking over their hailed-out wheat and saw the tornado coming. Not having time to reach their cellar they took refuge under a small bluff that faced to the leeward of the approaching tornado. They lay down flat on the ground and caught hold of some plum bushes which fortunately grew within their reach. As it was, they felt themselves lifted from the ground. Mr. Evans said that he could see the wreckage of his house, among it being the cook stove, going round and round over his head. The eldest child, a girl of seventeen, being the most exposed, had her clothing completely torn off. But none of the family were hurt.

WATERSPOUTS

Waterspouts are sea or lake whirlwinds, and occur in many parts of the world. At sea off North America they are most numerous in the southern Pacific, across the Atlantic doldrums, and up along the warm waters of the Gulf Stream. These 'great horn spouts' can occur at any time of year, in warm weather or cold. They can also form at any time of the day or night, though most likely (as are sea thunderstorms) around dawn or around noon—at which times over-ocean lapse rates are steepest. There is great variety of form and intensity between different spouts, and in any single spout, even, at different stages of its life cycle. But by considering the way in which they are formed, we can divide all of them into two general classes—fair-weather spouts and storm spouts.

Similar in general mechanism to a dust whirl on land is the fair-weather waterspout. In place of desert intensive surface heating and absolutely unstable air, however, the sea affords a large supply of moisture which, by condensation, provides large quantities of latent heat. The over-ocean lapse rate may be moderate—say –3° or –4 °F. per 1000 feet (about –6° or –7 °C/km)—the air only conditionally unstable; yet the temperature and humidity are both high enough to insure that this conditional instability shall be realised in the form of updrafts and clouds. Here again the whirl itself originates in a rising current fed by conflicting winds, and may materialize at the start as nothing more than one or two rings of spray on the water surface. As its feeder winds are neither very strong nor very extensive, and the earth's deflective force is weak near the equator, the fair-weather waterspout, like the dust devil, spins either clockwise or counter-clockwise by chance. True, a fairweather sea spout is usually somewhat bigger and more violent than a desert dust whirl. Yet in ferocity it falls far short of either a true tornado or a large storm waterspout formed in tornado fashion—that is, by the conflict of large and rapid and conflicting upper-air currents at cloud levels.

A waterspout at sea is always a fascinating sight that brings all hands on deck. While doing some marine weather work aboard the American freighter Timber Rush in April, 1939, it was my good fortune to observe several fair-weather spouts in about lat. 8° N., long. 84° W.—that is to say, about fifty sea miles off the south coast of Costa Rica. Records my journal: At this time the sun had advanced well north of the equator, swinging three or four degrees northward of our zenith at noon; the rainy season was imminent; and the hot, moist, fitful weather belonged to that shifting equatorial belt of calms and sweats and squalls known as the 'doldrums.' Daily great white thunderheads towered upward. There came a day hotter than usual (eighty-five degrees or so), and more humid (dewpoint around eighty degrees), when the thunderheads towered white against the blue sky and black against the vertical sun.

Off the starboard bow, to leeward in the light breeze, there developed gradually a towering dark heap of cumulus cloud—a typical thunderhead. Its level but ragged base was perhaps about fifteen hundred feet above the ocean; its mushrooming top had reached perhaps ten thousand feet. Suddenly a narrow black funnel, a thin, horn-shaped cornucopia of cloud, grew downward from the cloud base. The ocean surface beneath the funnel erupted into an extraordinary fountain of whirling white spray as a long and whitish whirling tube, sinuous like the trunk of an enormous elephant, reached downward and bridged the gap between funnel and water. It was a waterspout, the ocean counterpart of the justly dreaded tornado.

The waterspout first formed about three miles to the northward of our course, and moved slowly towards us, traveling against the light surface breeze (in accordance with the similar drift of its parent cloud), even as we steamed ahead towards the track of the spout. There was no visible evidence, in the parent cloud, of pronounced, conflicting air currents; both surface and upper winds were light. The larger of these sea tornadoes are sudden death to any small craft they happen to strike, and have been known to strip lighter rigging and other loose top-hamper from a full-sized steamer. The black cloud-cone, showing a lighter core, was now very plain; the lower, whitish tube writhed about on the ocean, picking up a fountain of white spray perhaps a hundred feet high and two hundred across. But suddenly, at about a mile's distance, now almost abeam to starboard, the spout came apart in the middle; the whirling spray fountain quickly subsided; and the upper funnel very slowly drew up into the parent cloud above. Presently another funnel grew downward from the cloud, but never developed into a complete spout. Later, when the thunderhead was far astern, another spout grew downward, very long and very thin, and swayed slowly from side to side for several minutes before it broke up.

While an ordinary waterspout should do no real damage to a steel steamship, it is something to be avoided by any small-boat mariner. As a fair-weather spout usually moves rather slowly, and can be seen at long distances across the water, it is not at all hard to steer clear of them with a powered craft—except at night, or when, as rarely happens, a dozen or so may be wandering around at one time. Sailing-craft mariners without power, becalmed in front of an oncoming spout, can of course do nothing but make the best of whatever comes.

An account of a storm waterspout encountered in the seventeenth century appears in William Dampier's Voyages. The ship was of three hundred tons (larger than most yachts, fishing schooners and sailing coasters of the present day); yet it was manhandled and dismasted by a spout, while en route from London to the Guinea Coast in 1674: When we came into latitude seven or eight degrees north, we saw several spouts one of which came directly towards the ship, and we having no wind to get out of the way of the spout, made ready to receive it by furling the sails. It came on very swift, and broke a little before it reached the ship, making a great noise, and raising the sea round it, as if a great house, or some such thing, had been cast into the sea.

The fury of the wind still lasted, and took the ship on the starboard bow with such violence, that it snapt off the boltsprit and foremast both at once, and blew the ship all along, ready to overset it; but the ship did presently right again, and the wind whirling round, took the ship a second time with the like fury as before, but on the contrary side, and was again like to overset her the other way. The mizzenmast felt the fury of the second blast, and was snapt short off, as the foremast and boltsprit had been before. The mainmast and maintop-mast received no damage, for the fury of the wind (which was presently over) did not reach them. Three men were in the foretop when the foremast broke, and one on the boltsprit, and fell with them into the sea, but all were saved. The danger area for boats (or airplanes) around a waterspout is quite narrow—usually not more than a few hundred feet wide. The Navy airplane flew within fifteen hundred feet of a small spout near Pensacola,

Florida. Even at this distance the pilot noted no undue turbulence, and the observer was able to take a good photograph of the spout and its spray fountain.

Unlike the widely distributed waterspouts of tropic seas, temperate-zone storm spouts usually form along or in front of a marked cold front that is advancing eastward or southward over the sea, much as tornadoes form along a marked front on land. In tidewater regions, such as southeastern Virginia, a vigourous twister may change from tornado to waterspout and back again many times as it crosses bays and rivers.

Some years ago a whirl near Norfolk began by destroying trees and sheds on a point. Then it crossed a creek, sucking up the water so that the bottom was plainly visible and gouging out the exposed mud. Anchored small boats were hurled up onto the shore. Then it ripped off part of a heavy pier, destroyed some buildings, became a waterspout again in Hampton Roads, changed back to a tornado and dumped a railroad gondola car and some refrigerator cars off the tracks of a railroad yard, crossed and sucked up another creek, damaged some airplane hangars on land, and finally headed up Chesapeake Bay as a waterspout.

A storm waterspout may damage even a steel steamship. The S.S. Hestia passed through a fair-sized one, in 1902, at about sunset off Cape Hatteras. During heavy and widespread thunderstorms several spouts were seen forming, the largest of which headed directly for the ship in such a way that it could not be avoided. The captain ordered all hands below decks, remaining out himself until the last moment. This oncoming spout was about fifty feet wide, with an almost black core about two feet across. As it struck, the captain dived below. There was a deafening roar, strong wind gusts, and a sudden shock as the spout passed across the ship. Bobbing out again onto the bridge, the captain saw two great hatch tarpaulins and a large plank high overhead; and the log line from the taffrail, with the weighted spinner at its far end, extended straight up into the air like a large-sized Indian rope trick. In appearance a waterspout is quite similar to a tornado -the funnel itself is a hollow tube of spiraling cloud, formed by dynamic condensation.

The usual spout height is one thousand to two thousand feet, though a specimen five thousand feet high was seen off Australia. Any large spout moves with the general drift of the cloud-level upper air—which upper drift may be across or against the surface wind. Around the base of the spout is the 'cascade,' a circular whirl of spray considerably wider than the spout itself. A small spout cascade may be only fifty feet wide by ten feet high; but a storm-spout cascade runs larger than this—say two hundred feet wide by fifty feet or more high. Waterspout damage to ships is occasioned partly by tornadic winds, partly by suddenly reduced pressure, and partly by the deluge of water sometimes released. While an ordinary spout consists entirely of fresh-water droplets produced by dynamic pressure-fall condensation, there is good evidence that larger spouts carry considerable quantities of sea water up and away for some distance.

Probably the largest waterspout ever carefully observed by many intelligent people appeared between Martha's Vineyard and the Massachusetts mainland one sultry August day in 1896. Heavy, towering thunderheads were forming all about, and particularly banked up in the north and west, by noon of that oppressive day. A marked cold front (though unrecognized as such in those days) was advancing out of the northwest—it struck Cottage City on Martha's Vineyard later that afternoon, knocking the temperature from 72° down to 56°. The great waterspout was first seen at 12.45 P.M., just as all the summer boarders were rising from a good old-fashioned dinner. The monstrous apparition formed and disappeared three times during the next half-hour. A long spiral column seemed to fall from the inky clouds, according to eye-witness accounts, increasing in size as it lowered and changing its colour from gray to black.

This record-breaking spout was seen by thousands of people (including a party on the yacht Avalon becalmed uncomfortably close to its cascade); and it was recorded by scores of photographers both amateur and professional—though not with the modern films and lenses and filters that would have captured a striking likeness. The ocean was flat calm, except immediately around the spout, as the immense black funnel slowly and majestically moved southeast towards Nantucket Sound. The recorded dimensions of this great spout may strain credulity, but they were checked by too many reliable and scientificminded people to be lightly disregarded. The height was estimated at 3600 feet; the column was 840 feet wide at its top, 140 feet at its middle, and 240 feet at its base. The cascade was 720 feet across and 420 feet high.

Shortly after 1.20 P.M. the ponderous column drew slowly up into the clouds, and the aerological spectacle was ended. But salty rain fell on the Vineyard that afternoon, proving that this extraordinary spout had scooped up plenty of water from the Sound in the form of spray.

❐

7

Introduction

ATMOSPHERE

Some of the atmosphere's profoundest mysteries are far above any possible ascent of man-carrying balloons or propellerdriven airplanes—beyond reach, even, of the twenty-five-mile sounding balloon. In the upper stratosphere there are levels warmed by the action of sun rays on ozone. Beyond the stratosphere, in the ionosphere, there are levels where solar radiation separates innumerable free ions and electrons out of the extremely rarefied air. Some traces of our earth's atmosphere, becoming ever more and more tenuous with increasing distance, probably extend out to cosmic depths of two thousand miles.

Our present knowledge of the upper stratosphere and the ionosphere beyond is mostly indirect painstakingly gleaned, that is, from such intangible phenomena as sound reflection, light reflection, radio reflection, semi-cosmic clouds, meteors, polar lights, and so on. Or, to put it plainly, our present picture of the upper atmosphere is largely limned with guesswork; and subject to constant correction as Doctor X announces a new and inexplicable observation, as Doctor Y propounds a new and scarcely intelligible theory, or as Doctor Z obs up to confound Doctor Y. This state of affairs is likely to continue until the sounding rocket at last carries recording instruments, or possibly human observers, above the lower stratosphere into black depths beyond the blue.

Nevertheless, innumerable passive observations, as well as certain active experiments with sound waves and radio waves, have already revealed many interesting facts about the upper atmosphere.

ABOVE THE STRATOSPHERE

The colour of the sky, of course, comes from the blue light which the air molecules scatter, and at sunset the light slanting through miles and miles of dense air is predominantly of penetrating red. The duration of twilight is a key, imperfect but not entirely valueless, to the density of the upper air; for the twilight afterglow, so infused with simple and reposeful beauty, is nothing but the shining of air particles high overhead in the rays of the sunken sun. Knowing the time after sunset through which twilight persists, perhaps an hour or two, and the consequent dip of the sun below the horizon, perhaps eighteen degrees or so, scientists have calculated that air sufficiently dense for this reflection extends to a height of about fifty miles.

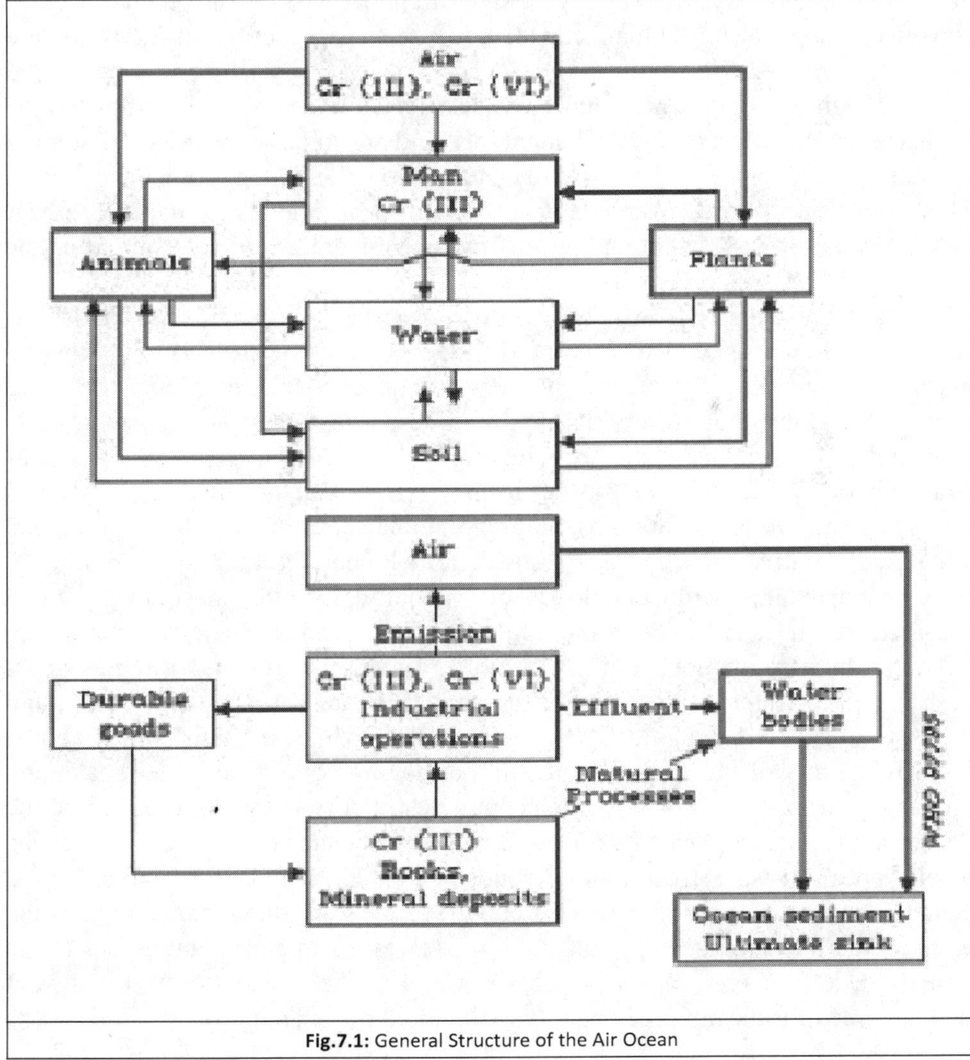

Fig.7.1: General Structure of the Air Ocean

This fifty-mile altitude is perhaps the upper limit of air as we can perceive it and imagine it the end of the stratosphere and the beginning of the ionosphere. Also to be seen at dawn and dusk under favourable conditions is the great domed shadow of the earth itself on the upper atmosphere, rising or sinking in that quarter of the sky opposite from the sun a few minutes after sunset or before sunrise.

It is seen best in crystal-clear weather from high altitudes. The curving shadow-are the sun lights the upper atmosphere with pale rose; below it, there is only dull-gray darkness. Another sunset phenomenon that tells somewhat of the nature of the upper air is the rose-tinted glow known as the 'purple light,' seen after sunset on very clear evenings above the point where the sun disappeared. It is caused by the refraction and reflection of the longer-wave (red) light through the upper atmospheric layers around the bulge of the earth. An observer out in space (say on the moon during a lunar eclipse) would see it as a continuous ruddy aureole around the whole globe.

Though the stratosphere and the black depths above it are cloudless in the ordinary sense, clouds of a very diaphanous kind are sometimes observed floating at these upper levels. At heights of fifteen to twenty miles patches of parti-coloured filmy iridescence, called nacreous or 'mother-of-pearl' clouds, sometimes appear in the crystal-clear skies attending an invasion of cold, dry polar air after the passage of a cyclonic rain area.

These are distinct from, and much higher than, the cirriform clouds that sometimes show iridescence around their edges; in a true nacreous cloud, floating ten miles above the troposphere, the play of colour extends clear into the cloud centre. Nacreous clouds are perhaps composed of microscopic ice crystals or minute, super-cooled water droplets; and they show the existence of mid-stratospheric winds that are mostly from westerly and northwesterly in north temperate and sub-arctic latitudes, ranging in violence from flat calm to more than a hundred and fifty miles an hour. Considerably higher, marking the beginning of the ionosphere, are noctilucent clouds, occasionally seen after sunset in arctic and sub-arctic latitudes at heights around fifty to fiftyfive miles -- heights determined, of course, by triangulation from two or more ground stations situated along known base lines. Noctilucent clouds are excessively thin and wispy in appearance, and their colour shades from golden near the horizon to white and bluish-white above.

Unlike any of the lower clouds, they probably come from outer space and probably consist of excessively fine cosmic dust. Perhaps unable to penetrate the denser atmosphere below fifty miles, they always seem to hang just above this level. Appearing as faintly luminous clouds in the night sky, they are, nevertheless, actually lit by sun rays, and perhaps energized by solar ultra-violet, beyond the earth's shadow. Noctilucent clouds give evidence of superhurricane winds at the base of the ionosphere -- drifts from easterly and northerly that sometimes blow at two hundred or three hundred or even six hundred miles an hour!

In 1879, Josef Stefan discovered i old Vienna an important law of physics that might seem at first sight far removed from any usefulness to aerologists. Stefan's law is, in brief, that both the sort and intensity of light rays (both visible and invisible) emitted by any hot body are related to its temperature. The temperature of the sun is known; therefore, physicists can say how far its radiations extend into the ultra-violet. But it was found that, as we measure them on earth, solar ultra-violet rays stop short of this theoretical spectrum (which extends, with decreasing intensity, even down to X-rays). Hence something in the atmosphere must be absorbing them. That something has turned out to be ozone—the rare triatomic form of familiar oxygen.

Some slight traces of ozone occur near the earth's surface -popularlyassociated with pine woods and sea breezes. Measurable amounts are formed by electric sparks, and particularly by lightning flashes. But most of the atmosphere's ozone (still only a minute percentage of the other aerial gases) occurs at heights around fifteen to twenty miles—the zone of the nacreous clouds. At heights of twenty-five to thirty miles the ozone layer is more tenuous, but more important because it receives the full force of solar ultra-violet. It is a fortunate thing for all God's creatures that this great natural screen of ozone hangs up there thirty miles aloft, incredibly thin and diffused, transparent and invisible, and until recently unknown. For without it, the far ultra-violet and near X-rays shot forth from the sun would destroy all life on the sunlit earth within a few seconds.

But the high-altitude ozone screen has an important effect on the upper air itself. In the words of Haurwitz, 'ozone as a powerful absorber of solar radiation determines to a large extent the temperature of the upper atmosphere.' We have already seen that temperature may increase somewhat with elevation in the lower stratosphere. In the upper stratosphere, where ozone enters powerfully into the picture, temperatures may be much higher—say up to summer heat (around +85°F) or even near boiling (around +160°F). These apparent high temperatures have been discovered by nothing more tangible than sound waves.

SKY-REFLECTED SOUND

It is no new fact that powerful noises, such as heavy gunfire, are sometimes distinctly heard at distant places, though entirely unheard at nearer points. Back in 1666, Londoners heard the British Fleet cannonading the Dutch line-of-battle in the Channel, while the sound apparently skipped nearer places along the English coast. This circumstance puzzled Samuel Pepys and other philosophers of that day; and was not explained, in fact, until 1903, by a German named von dem Borne.

Since von dem Borne's time it has often been noticed that the sound of a loud explosion, though heard at great distances, passes completely over intervening points. This peculiar circumstance is best accounted for by supposing that there is in the upper atmosphere some sort of layer which acts in effect as a great sound reflector, and directs outgoing sound waves downward again towards the earth.

In order to fit in with all the other known facts about the upper air, this reflecting 'roof' must take the form of a very warm and very diffuse upper-air layer—it is, in fact, the upper part of the ozone layer. As the velocity of sound is greater in warmer air, the outgoing sound waves travel faster in their upper portions (which encounter high temperatures first), and are hence gradually bent back to earth. The net result is the same as if they had been reflected back from the warm upper layer.

Many experiments to check the existence and height of this sound-reflecting layer were performed in Germany and England soon after the First World War Whipple, for example, set up some super-sensitive microphones while British artillery was firing on an island in the lower Thames. He heard the gunfire at Birmingham, two hundred and thirty miles away; the sounds had skipped intervening places entirely and seemed to come from the upper air. Studies of the skip-distance patterns of sound from dynamite explosions, extending out to a distance of six hundred miles, point to the same general conclusion. The only way they can well be accounted for is by assuming high atmospheric temperatures at twenty to thirty miles altitude. The actual reflection summit, at about twenty-five miles, corresponds closely with what is known and assumed about the ozone layer.

The actual pattern of audibility of some great explosion, such as an ammunition dump or a shipload of TNT, is somewhat as follows: The blast is of course heard, and almost felt, out to a distance of perhaps twenty-five or fifty miles. From there to around a hundred miles is a 'skip zone,' or ring of comparative silence. From perhaps a hundred to a hundred and fifty miles extends the first zone of reflected audibility; then a second skip zone; then a second audibility zone from perhaps two hundred and fifty to three hundred miles; also, if the explosion is powerful enough, third and perhaps fourth or fifth skip and audibility zones. The sound of this battle practice was plainly audible at Santa Barbara, one hundred and twenty miles from the ships. It came in as a dull, jarring, almost soundless rumble that rattled house walls and windows like a mild earthquake.

These upper-air sound reflections are subject to curious vagaries. There are differences between eastward and westward transmission, varying with seasonal effects on the upper-wind distribution. The reflecting layer is warmer and contains more ozone in winter; it is colder and less effective in summer. The curved reflecting layer sometimes serves to intensify the downcoming sound in a peculiar way. In one case, firing with sixteen inch guns actually broke windows in a city nearly two hundred miles away!

The noisiest sound ever heard by civilized men blasted forth when a volcanic island called Krakatoa, between Sumatra and Java, blew up in 1883. The actual blasted-out material, pumice and rock, smashed and covered whole forests on islands miles away. The volcano eruption shot to a height of seventeen miles, and its black pall of concentrated dust extended out one hundred and fifty miles. An appalling ocean tidal wave rolled out, a hundred feet high, for thousands of miles, swamping hundreds of ships and drowning thirty-six thousand people.

The actual sound of this prodigious blast was faintly heard at Rodriguez Island, nearly three thousand miles away, and powerfully audible at many places more than a thousand miles distant! The atmospheric wave-disturbance, in sub-audible form, travelled several times around the earth. Also over the whole earth, carried by tropospheric and stratospheric winds during the next few months, spread thin brownish clouds of Krakatoa's dust.

LIGHTS OF THE OUTER AIR

Definite and useful indications about the upper atmosphere are given by meteors, those swift points of light, erroneously called 'shooting stars,' that streak across the starry sky every few minutes on a clear, moonless night. As these bits of cosmic débris bore into the atmosphere at speeds of ten to forty miles per second, it is no wonder that they are intensely heated by friction and soon burst into flame. If they are observed or photographed simultaneously against a fixed-star background from two or more known locations, their altitude at successive instants can be determined, and it is found that most of them begin to flame about one hundred miles or more above the ground and burn completely out before reaching the thirty-mile level. The larger and slower ones, that is, may continue down to thirty miles, but most of the smaller and faster ones burn out around fifty miles. This knowledge, of course, gives a rough idea of atmospheric density at these heights.

Meters are seen at their best and brightest in the small hours of the morning; for at that time one is on the 'forward' side of the earth as it swings at a comfortable eighteen miles a second along its orbit, and the fiery collisions of the meteors with our outer atmosphere are more or less head-on. Perhaps ten or twenty million meteors enter the earth's atmosphere every day. When we meet a definite swarm of them, this number is vastly increased; during the Leonid swarm of 1833 they appeared as a night-long shower of celestial snow.

Most visible meteors are not much larger than pinheads (ranging, perhaps, from birdshot to buckshot in size) and are completely burned up in the higher atmosphere. But occasionally a larger meteorite, or 'fireball,' weighing several pounds or even several tons and burning brightly enough momentarily to rival the sun, crashes through to the earth's surface. Larger meteors leave trails in the upper air that gleam phosphorescently at night, and resemble lines of white smoke by day. These trails have been seen to drift rapidly before dissolving, and their motion confirms the idea that enormously high winds can occur at heights above thirty miles eteors also confirm another idea about the far upper atmosphere; for the calculated heights at which they should blaze agree with observed heights only if one assumes high temperatures at altitudes around thirty to fifty miles. This result, of course, strengthens the testimony of sound waves and ozone.

Despite their tenuous mystery, introducing new problems, the northern polar lights and their southern counterpart also offer some evidence about the upper air. In northern countries the active aurora is a sight never to be forgotten; widespreading curtains of flame appear and vanish like the draperies of an enormous stage, while arcs and bands of red and green merge into higher rays of violet and blue marching like searchlight beams across the depths of space. Carl Störmer has observed and photographed the aurora borealis many times in Norway. He finds that it occurs in two general varieties: the lower and more common types, mostly apple-green arcs and bands, flame in the earth's shadow zone at heights around sixty or seventy to one hundred miles.

The higher types, mostly blue, violet and gray-violet rays, flicker at around two hundred miles, or even occasionally as high as four hundred miles, perhaps far up where the sun is shining across the pole above the blackness of arctic night. These heights correspond roughly, with the observed heights of the ionized layers that influence radio transmission.

The spectrum of the aurora shows that the rarefied gases of these far heights are mainly oxygen and nitrogen, rather than the helium and hydrogen formerly thought to predominate in the upper air. There is, in particular, one prominent green line that results when rarefied monatomic oxygen is electrically excited. Moreover, it has been found that this diffuse green glow is always present in the night sky, over all the earth, even above the equator.

It appears to come from atomic oxygen superheated by solar ultra-violet at heights around one hundred to two hundred miles. So the natural human conception of is somewhat of an illusion. Actually about one third of the moonless night sky's luminosity comes from the stars; more than one half comes from this highaltitude oxygen glow.

The primary cause of the aurora is probably 'searchlight' beams of electrons and ions shot out of the sun—beams that rotate with the sun's rotation, somewhat after the fashion of light beams from a rotating beacon, and sweep past the earth. Sometimes, when a brilliant aurora extends visibly to temperate latitudes, this solar 'beam-excitation' can almost be seen. Thus at midnight the northern aurora may appear from the south as successive outward pulses of light, originating at the left (west) of the sky and flashing progressively towards the right (east) of the sky. These pulses were very apparent in an aurora I witnessed some years ago in New York State. It began with great hanging curtains, but changed in an hour or so to a broad, white band along the northern horizon, surmounted by radial beams of white and pale green. From observed angles and the known distance of the magnetic pole, this aurora appeared to be about a thousand miles away from me, about a thousand miles wide, and several hundred miles high. The actual speed of the light pulses in the heavens appeared to be around one hundred or two hundred miles a second; and at the earth's distance a rotating solar beam should, in fact, sweep past at about two hundred and fifty miles a second. Probing the Ionosphere

Within recent years the aurora-haunted depths of the ionosphere have been probed daily and almost hourly by means of upward-sounding radio signals. Radio waves of course speed outward in all directions from the transmitter, along the ground and through the air, at the velocity of light. The ground part of the wave, however, usually dies out within a few score miles, leaving only the 'sky' wave for long-distance transmission. And the sky wave, to reach any distant place on earth, must somehow continue to bend around the bulge of our great globe. This bending of the outward-speeding sky wave is accomplished by what amounts to reflection in the ionized layers—layers where solar ultra-violet has disassociated free electrons and ions, increasing the electrical conductivity—of the earth's upper atmosphere.

Actually, just as we saw in the case of the warm ozone layer and sound, an ionized layer does not reflect radio waves; it refracts them by virtue of the fact that the outspeeding electromagnetic vibrations tend to move faster in the more intense ionization of greater heights within the 'reflecting' layer, and hence bend gradually downward again towards the ground. Nevertheless, this refraction process has the same net effect as reflection, so that one may think of the ionized layers as radioreflecting shells or 'roofs' in the upper atmosphere.

Within the last few years it has become possible to measure the height and other characteristics of the ionized layers almost at will. By means of a special transmitter, a radio 'tick,' or signal lasting less than one five-thousandth of a second, is shot upward. In something like one thousandth of a second it plunges a hundred-odd miles into space and returns to a special receiver on the ground, accurately recording the height from which it was in effect reflected—the height of an ionized layer. By this method the main reflecting layers have been probed and checked time and time again. These changes explain why a powerful broadcasting station is heard only a hundred miles or so by day, yet reaches out a thousand miles or more by night. When solar disturbances are active, some of the layer heights vary in erratic and spectacular fashion, flopping up and down by several miles in a few seconds and playing havoc with long-distance reception.

The General Picture

Before leaving the upper atmosphere it might be well to gather up some of the essential facts—the aerological elements, as a scientist would say—into one unified picture. First of all there is the matter of pressure in the air ocean above us. Pressure decreases very uniformly with increasing height. This rule never varies; pressure cannot possibly increase with height. A long and complicated formula has been deduced, called the hypsometric formula, which shows the exact variation of pressure with altitude. The pressure at a given altitude is not always the same; it is influenced somewhat by the air temperature, and also very slightly by the amount of moisture in the air.

The complete hypsometric formula must therefore include temperature and moisture terms. In general, however, pressure decreases in the lower levels, by about one inch of mercury per thousand feet of altitude (or, in the metric system, by about one hundred millibars per kilometre of altitude). At upper Introduction

9 levels, of course, the pressure decrease with altitude is much less. At sea level the average pressure is about one thousand millibars, or about thirty inches of mercury. At about three and one half miles altitude, pressure has fallen to five hundred millibars or fifteen inches, and half of the mass of the air ocean is therefore below one. This accounts for the extraordinarily clear, transparent air characteristic of high mountain summits and high altitudes in general; it also accounts for the difficulty in breathing that most humans begin to feel at ten or fifteen thousand feet altitude. At ten miles altitude the pressure fraction has fallen to a mere tenth; at twenty miles, it is something like a hundredth; at thirty miles, something like a thousandth.

Temperature, unlike pressure, varies with height in very erratic fashion. Even in the troposphere, temperature may decrease with increasing altitude, it may remain constant, or it may even increase as one goes higher. On the average, though temperature does decrease with height in the troposphere, by the amount of about 3°F per 1000 feet or about 6°C per kilometre. But above the troposphere, throughout the lower stratosphere, temperature is fairly constant regardless of altitude. This isothermal layer extends upward to about fifteen or twenty miles in winter; and perhaps extends higher, to about twenty-five or thirty miles, in summer.

Above this isothermal layer, temperature increases until, at heights around thirty miles (maximum absorption of ultra-violet by ozone), the almost vacuous air is probably as warm as the tropics at sea level (say around + 85°F or + 30°C), or perhaps up to near-boiling heat (say around + 160°F or + 70°C).

Above this warm layer there is perhaps a cooler one. Then, at still greater heights above sixty miles (maximum absorption of ultra-violet by atomic oxygen), temperature may increase to several hundred degrees—the heat of a roasting oven—at least in the daytime. But at night these far heights are probably much colder—say below zero F. And above this upper warm zone, at altitudes of two or three hundred miles, for all we know, the temperature may decrease again to something near absolute zero.

The composition of the upper air has long been a bone of contention. Most meteorological texts printed prior to 1930 or so show the atmosphere above sixty miles (one hundred kilometers) as consisting largely of hydrogen and helium. More recent theoretical views of the matter, such as those of Chapman and Milne, Gutenberg and Haurwitz, have disproved this idea completely. In the light of present knowledge it appears that the composition of the atmosphere up to at least thirty miles, and probably up to sixty miles—so far as the proportions of various gases are concerned—must be very much like that of air in the lower altitudes.

At lower levels, in any case, the air consists on the average of about 78 per cent of nitrogen, 21 per cent of oxygen, nearly 1 per cent of argon, 0.03 per cent of carbon dioxide, 0.01 per cent of hydrogen, and even more minute fractions of neon and helium. At lower levels there is also, of course, a variable amount of water vapour, ranging up to about four per cent of the air in volume—a factor of great practical importance in tropospheric weather which will be dealt with at some length. In general this water vapour decreases with height, being almost absent from the stratosphere; but even up into the ionosphere, there are probably minute traces of it.

In the upper stratosphere the proportion of helium may increase somewhat. In the ionosphere the oxygen percentage may decrease, most of what oxygen there is takes the monatomic form, and the percentage of nitrogen probably increases. Helium and hydrogen are probably not present in these black depths above the blue, warmed to roasting heat by the super-glaring sun—for if present, they would soon diffuse away into cosmic space.

COMPOSITION OF ATMOSPHERE

The most primitive man—the wildest savage—recognizes a wind when out in it, but it took a Greek philosopher to tell us what it is: air in motion. What motion is we know, at least well enough for practical purposes, but what is air? What is that invisible and odourless something we breathe and therefore call atmosphere, the thing that affords us all our weather perceptions and whose states and conditions we have learned to measure and even to foretell, and what was its origin?

ORIGIN OF THE ATMOSPHERE

To start as nearly as possible at the beginning, how and when did the earth ever get an atmosphere—its gaseous envelope? Well, there are two, and so far as we know, only two basic substances, the electron, a certain extremely minute quantity of negative electricity, plus something else, maybe; and the proton, an equally small quantity of positive electricity, plus something else, also maybe. The neutron or chargeless mass is omitted as so little is known about it.

Of the origin of these entities no one has the slightest idea. They can exist separately, in which case they are electrically very active; or variously grouped together in equal numbers, with increase of inertia but almost total loss of electric force, at least on things external.

Every such group that is stable, or even measurably durable, is a chemical element. There are no other elements, and this number is limited. Furthermore, though occurring in unequal quantities, all these elements appear to be distributed throughout the universe.

Most of those known on the earth have been found in the sun, for instance, and we believe the others are there, too, even if in such relatively small amounts as to be difficult of detection. Hence, when that other star, according to one cosmic theory, some three billion four hundred million years ago, passed so near our sun as to drag off from it by tidal action the masses that coalesced into the planets, there were present, and came off together, all the possible elements.

That is, the primordial material of the earth, as it was pulled off from, or out of, the sun consisted of all the elements that now make it up, so that at the very beginning of the independent existence of the earth it had, if not an actual atmosphere, at least the makings of one.

Fig.7.2: Kilauea, an Air Factory

It is believed, by those who hold to this theory that as the earth mass drew together in a molten sphere the heavier and more refractory substances formed mainly the inner core, while the lighter and more volatile elements and compounds formed the rocky shell and the gaseous envelope. In this way, they claim, the earth soon had oceans and an atmosphere of some kind. However, even if all the gases, hydrogen, oxygen, nitrogen, and others, that can combine with various elements and compounds and form solids had then so com- bined, leaving neither water nor air, it is certain that before long the earth would have begun accumulating both.

It is doing so now through every volcano, every fumarole, and every bubbling spring, and must always have done so since before even the first crust began to form. But it seems unlikely that our present atmosphere actually did come entirely from the molten interior, whether by way of volcanic activity or otherwise, because volcanoes do not, so far as we know, give off free oxygen—volcanic gases captured before there has been any chance for admixture with the air show no trace of it. Furthermore this element could not exist uncombined in the presence of hydrogen and sulphur at high temperatures, both of which are abundant in volcanic vapors.

But we have oxygen; where did it come from? It is known that highly developed green plants extract, under the stimulus of light, a great deal of oxygen from carbon dioxide, a gas abundantly emitted by volcanos. But most primitive plants do not, they consume it; hence this promising source of free oxygen appears, on close examination, to be quite uncertain. However, some forms of lower life thrive in the absence of free oxygen and yet, in the presence of light, evolve it from certain of its compounds, especially water.

Again, lightning, which must have occurred from the beginning, frees a little oxygen from water; and so also may ultra-violet light. There have been, then, continuously active means of obtaining free oxygen from its compounds since the beginning of the world. Nevertheless, it seems most likely that at the beginning there was more oxygen present than was necessary to use up the free hydrogen and to combine with the available surface materials—enough to do all this and to have a goodly amount left over as free oxygen of the atmosphere.

Presumably, too, there were present other primitive gases, especially nitrogen, argon, carbon dioxide and water vapour. But whatever the primitive state of the atmosphere when the earth first formed, it may be regarded as practically certain that during the whole of the three billion four hundred million years since that time it has been continuously depleted by combination with many things in and of the crust, and also as continuously repleted by their decomposition. It is always changing, but, except in respect to water vapour, the change is so small in comparison with the whole that we are not ordinarily aware of it.

It has been argued that the could not have retained an atmosphere when molten, or even when dull red. But this is true of only the lightest two gases, hydrogen and helium, and of them only if no other gases were present in large amounts. A deep atmosphere of water vapour, for instance, would catch any light gas that might leave the earth beneath with an escaping velocity. It could escape only if the outer portions of the atmosphere also were quite hot.

At any rate, as soon as a crust, however thin, formed over the earth the supply of heat from beneath was so reduced (the crust being a good insulator) that the upper air necessarily became cool enough to retain the lightest gases. Also water must soon have begun gathering on the surface. Even if, up to this stage, all helium and free hydrogen had been driven wholly away from the earth, a condition that seems unlikely, there has been since then, and still is, abundant opportunity to accumulate both of them—hydrogen from volcanos, and helium from radioactive materials everywhere.

Presumably, therefore, the atmosphere is primitive in part—pulled off from the sun with all the other elements—and in part, at least, certainly regenerated inasmuch as every volcano is an active air factory.

Apparent Simplicity of air

Presumably every one usually thinks of air as being just air, a homogeneous and single thing. Many of us always think of it that way, as the ancients did, when we think of it at all. Indeed so far as its behaviour and most of its physical properties are concerned it shows no obvious complexity.

It is just air in motion that is responsible for a thousand familiar things from the stir of a leaf to the wreck of a house; and just air that floats the balloon and sustains the aeroplane. In all these matters it usually is quite satisfactory to regard the air as the single substance it ordinarily seems to be.

Evidence of Complexity

This apparent oneness of air does not, however, extend to all physical processes. When we try to liquefy it, for instance, evidences of its complexity soon become amazingly conspicuous. If untreated air is forced through the cooling coils they quickly become choked with ice; and they still clog up when even the driest air is used, if nothing but the water vapour has been removed—this time with solid carbon dioxide. Then, too, the liquid air itself shows abundant evidences that it is a mixture and not a simple substance like water. What then are the known constituents of the atmosphere, and how and when were they discovered?

Water Vapour

The earliest considerations of the composition of the air that have come down to us are those of the Greeks. In their speculative philosophy on the composition of objects, they considered air to be one of the four elements (fire, earth and water being the other three) that, singly or variously combined, make up all substances.

From this it might seem that these Greek philosophers regarded the atmosphere as strictly a single thing—the "element" air. Yet it appears that by "air" they meant anything gaseous, and not necessarily the atmosphere. At any rate, Aristotle 250 years B.C., says very distinctly, in his work on meteorology, that cloud and rain are caused by condensation from the atmosphere of water vapour that had gotten there by the evaporation of water at the surface of the earth. He thus makes it very clear that the air consists of at least two things, and that water vapour is one of them.

Water vapour, then, was the first constituent of thc atmosphere to be explicitly recognized. Aristotle mentions it, but it is not certain that this discovery was original with him. However, his is the earliest record we have of it, and for that reason, there being no evidence to the contrary, we regard him as one of the discoverers the earliest one of the constituents of the air.

DELAY OF FURTHER DISCOVERIES

For more than 22 centuries, therefore, and perhaps for much longer, it has been known that the air we breathe consists of at least two things, water vapour and whatever is left after the water is removed. And for more than two thousand years after the days of Aristotle this is all that was known about its composition. Indeed it was practically impossible to push our knowledge of the atmosphere any farther without something of the facilities and methods of the modern laboratory, nor before there had been acquired—very slowly and tediously it was—a fair concept of chemical elements and pure substances.

Not until the beginning, then, of the 18th century was it reasonably possible for any constituent of the air to be discovered in addition to water vapour, nor indeed was any discovered until long after that. The chief obstacles that prevented such discovery for more than a hundred years after enough advance for that purpose had been made in laboratory technique, for that had been adequate from the beginning of the 17th century, were:

- The fixed idea that all gases are alike, all just air, and that any differences between various samples are due only to greater or less modifications of one and the same thing.
- The completely misleading and faulty concept that flame or combustion is the escape of something, phlogiston they called it, from within the burning object.
- The failure to recognize that change of weight incident to strong heating, during combustion, or under any other circumstances, was a matter of importance or had any scientific significance whatever.

A century before any constituent of the atmosphere, except of course water vapour, was recognized and collected in an approximately pure state, and while the faulty notions just listed were still prevalent, two people, working entirely independently, came near to finding one or more of its elements. The first of these was Robert Boyle a wealthy bachelor, chemist and theologian; discoverer of the fact, known as "Boyle's Law," that doubling the pressure on a gas reduces its volume by one-half, of course for the same temperature. It seems very probable that Boyle would have discovered some of the constituents of the air if he had carried to completion certain experiments that he definitely listed. But there is nothing in his voluminous writings to show that he ever got them beyond the paper stage, despite the fact that, so long as his health permitted, he was a persistent worker.

After many and well-devised experiments Mayow concluded that the air consists of at least two portions, one that supports combustion and sustains life, and another part that does neither. The former he called "fire-air," because it keeps a flame going. He also said that it consists of "nitro-aerial particles," that is, particles in a gaseous form of the kind that makes a mixture of niter and charcoal, or other combustible, burn, when lighted, in the absence of air—even under water.

All this is true enough, but his proof that the air consists in part of a special constituent, different from all the rest, was not complete. He did not collect "fire-air," the gas we now call oxygen, in a practically pure form and show that it is identical with the "fire-air" of the atmosphere. However, he recognized the incompleteness of some of his arguments, and it seems likely that if he had lived a few years longer his proofs would have been perfected,and our knowledge of the composition of the atmosphere set forward almost a hundred years. But this near attainment to the goal, and also even the direct route to it, appear to have been lost sight of for nearly a century. Indeed "fire-air," that came so near to being the first constituent of the atmosphere to be discovered, except, of course, water-vapour, and whose properties make it the most conspicuous, turned out to be the very last of all the major ones, save only argon.

Carbon Dioxide

The first of the permanent gases of the air of nearly constant quantity to be clearly discovered was not, as would seem most likely, either oxygen or nitrogen that together make up nearly the whole, but carbon dioxide, that is present as scarcely more than a trace—three parts in ten thousand. At that time alkalies were given to persons suffering with urinal calculi, and certain physicians recommended lime water for the same purpose. Which the view of finding something still better Black undertook investigations with magnesia alba, a form of carbonate, as we now know, or, more exactly, basic carbonate, of magnesium.

On strongly heating this substance a gas is given off, and Black turned his attention particularly to that gas, or air, as all gases were then called. He tried calcining, or burning to a powder, various substances, such as limestone, that we now know to be carbonates, and studying the gas thus obtained. In the end he found that this gas is much heavier than ordinary air, that it will not support life or combustion, that it will recombine with the calx, or powder, produced by the strong heating of the original substance, and that heat will again expel it as before. This gas seems to be fixed, or somehow fastened, in the objects from which it may be obtained.

Nitrogen

The composition of the air in a closed vessel after it no longer would support combustion or life, and to determine the cause of its unwholesomeness. After burning charcoal, phosphorus, or other combustible, in a closed volume of air, as long as possible, he removed the "fixed air" (carbon dioxide), if any had been formed, by means of lime, or an alkali, all in accordance with the previous investigations of Black, and then examined the remaining gas. He showed that this residue is not ordinary air because it supports neither life nor combustion, and that it is not fixed air for the alkalies do not absorb it. He called this residue "mephitic air," because it does not support life.

We now know and say that this residue had been obtained by burning out the oxygen of the confined air and then absorbing the carbon dioxide thus produced. And we know, too, that this residue, this "mephitic air," was nearly pure nitrogen. But at that time oxygen had not yet been discovered, and of course Rutherford could not talk in terms of things and chemical reactions then unknown. He did know, however, that it was obtained by combustion in an inclosed or limited volume of ordinary air, and therefore concluded, after the philosophy of his day, that it was atmospheric air combined with, or modified by the addition of, phlogiston—a mysterious fire substance whose escape from an object commonly is manifested by flame. Nevertheless, and no matter what his ideas as to its nature, Rutherford did obtain reasonably pure nitrogen and did record some of its properties, and therefore may be regarded as the discoverer of this constituent of the air, the most abundant of all.

Oxygen

Almost immediately after the discovery of nitrogen, the other major constituent of the atmosphere, oxygen, was independently found by Joseph Priestly. Priestley was a preacher by occupation and a chemist for recreation. At one time, while waiting for the building of the parsonage to be finished, he had the good fortune to live next door to a brewery—good fortune, because he was induced thereby to take up the chemistry of gases, his chief avocation for many years, on which he published several volumes, and which made his name famous. In the course of this work he obtained a red powder by heating mercury in the presence of air, and then on more strongly heating this powder with a burning lens he got a gas which supported combustion much better than ordinary air. He had gotten oxygen, as we now know, by first burning mercury to an oxide and then decomposing that oxide by raising it to a high temperature.

This element of the atmosphere he called dephlogisticated air. Combustible things burned in it readily; their phlogiston, or fire principle, rushed into it as air into a vacuum. It was air, that is, a gas, but air deprived of phlogiston. But regardless of what he called it Priestley had discovered a new constituent of the atmosphere, the one we now call oxygen, the one that sustains life and supports combustion. On the other hand, Scheele was a professional chemist, so completely absorbed in his subject that he had little time for anything else, whether occupation or diversion.

His incentive to study the air was his wish to solve the riddle of fire, a thing essential to so many chemical processes. He found substances, such as phosphorus, which on burning reduced the volume of the confined air (at the same temperature and pres sure) in which they were burned by about one-fifth. In all cases the remaining air would not support combustion.

He also found many ways of getting a gas that would support combustion far better than ordinary air, that was heavier than the air left after combustion, and that made burnt air indistinguishable from ordinary air when mixed with it in the proportion of one volume to four, or thereabouts. The residual or burnt air, the nitrogen, essentially, as we now name it, he called "vitiated air." The other portion, the part that supports combustion, he called "fire-air," just as Mayow had called it a century before.

Scheele, like Priestley and every other chemist of his day, except the immortal and tragic Lavoisier, interpreted fire phenomena in terms of phlogiston. He therefore considered "fire-air," which we call oxygen, to be a combination of phlogiston and a subtle acid substance. But whatever his opinions may have been as to its possible composition, he did find this constituent of the atmosphere, and that too even before it was found by Priestley.

Another student of the atmosphere that must be mentioned in connection with the discoveries of its composition was Henry Cavendish. He may not generally be credited with the discovery of any one of these constituents, and yet he appears to have found nitrogen about the same time that Rutherford did, if not earlier, by the simple process of passing the same confined air back and forth over red hot charcoal and then removing the fixed air (carbon dioxide) with an alkali. He did not publish this—he appeared always to be indifferent about publishing his investigations—but described it in a letter to Priestley in 1772, the year Rutherford published his discovery of mephitic air.

Lord Rayleigh had been making careful measurements of the densities of different gases, and thus came upon the fact that the nitrogen of the atmosphere, or rather that residual then regarded as pure nitrogen, was a little denser than thc nitrogen obtained from any one of several chemical compounds.

This was capable of three or four different interpretations, one of which, then regarded as the least likely, was the presence in the atmosphere of an unknown gas denser than nitrogen. At that stage Sir William Ramsay joined Lord Rayleigh in a common attack on this problem. Each tried removing from the air the then known constituents, using entirely different methods for getting rid of the nitrogen, and each found conclusive evidence of the presence in the atmosphere of an unknown gas to the extent of about one per cent. It proved to be chemically inert, and for that reason was called argon.

Other Inert Gases

During the years 1895-1898, that is, immediately after the discovery of argon, four additional inert gases, helium, neon, krypton and xenon, were found by Sir William Ramsay and his assistant. Travers, to be constituents of the atmosphere.

Helium was found as that portion of the atmosphere that is still gaseous while the rest has been chilled to liquids or solids; and the others, neon, krypton and xenon by fractional distillations of large quantities of liquid air and liquid argon, impure, of course.

HYDROGEN

Lord Rayleigh and others also found free hydrogen always present in the air, but in amounts that varied from, roughly, one part in 1,000,000 to one 5,000. Perhaps this gas can not be regarded as a permanent component of the atmosphere in the same sense that oxygen and nitrogen are, but rather as an accidental and variable impurity. At any rate it is irregularly added to the air, at least by volcanoes, and more or less irregularly removed from it.

Traces and Impurities

Varying traces of ammonia, nitric and nitrous acids and their compounds, sulphuric and sulphurous acids and their compounds, oxides of nitrogen, hydrogen dioxide, and ozone are among the innumerable things always in the air if not of it. Minute particles of sea salt from evaporating spray, fine earth dust caught up by winds, pollen of every description and spores of many kinds are among the coarser and ever present pollutions of the atmosphere from the earth beneath. Shooting stars furnish a continuous, though invisible, shower of dust the world over from outer space. Also radioactive products of radium, thorium and other elements are pouring into the atmosphere continuously and contributing to the maintenance of its electrical state. This state means, in part, around 20,000 electrified particles, or ions, per cubic inch in the lower air. The number of ions 60 miles or so above the surface is far greater, at least 1,600,000 per cubic inch, as we infer from the phenomena of radio communication.

In short, while the things of the atmosphere are but few the things in it are innumerable. All these latter are relatively very small in amo t, but some of them are exceedingly important. The salt particles and some others are essential to condensation and rainfall; the ammonia and other nitrogen compounds, brought down by precipitation, add much to the fertility of the soil; the ions of the high atmosphere make distant radio communication possible; and ozone, also in the upper air, shields us from that portion of the ultra violet radiation that would destroy our eyesight, as we are now constituted, and otherwise do us irreparable harm.

PERCENTAGE COMPOSITION OF PURE DRY AIR

If the atmosphere is freed from all its numerous impurities and also freed from water vapour, there will remain pure dry air whose percentage composition at the surface of the earth is, very closely:

The amount of water vapour in the atmosphere, or better, perhaps, the amount of the water vapour constituent of the atmosphere, varies from scarcely more than a trace, at extremely low temperatures, to at least 5 per cent by volume on the hottest and most humid days. At and above the height of 5 miles, say, the amount of water vapour always is small, even when saturation obtains, owning to the very low temperatures at these levels. That is, water vapour is confined almost wholly to the lower atmosphere. Its average value, the world over, is such that if it were all condensed it would be the equivalent of a layer of water about one inch deep over the entire earth.

Nitrogen	78.03
Oxygen	20.99
Argpm	0.9323
Carbpm doxide	0.03
Hydrogen	0.01
Neon	0.00018
Helium	0.0005
Krypton	0.0001
Xenon	0.000009

FORMS OF OXYGEN AND THEIR DISTRIBUTION

Oxygen also is peculiar in its distribution, and it occurs in three different forms. All but about one part in 400,000 is ordinary oxygen, or, in the language of the chemist, diatomic oxygen. Triatomic oxygen, or ozone, occurs chiefly beyond the highest clouds, its greatest density being, apparently, at the levels of 20 to 30 miles. It occurs in the lower air only as a trace. In the highest air, 60 miles and more above the surface, the oxygen appears to be at least partly monatomic, according to the spectrum of the aurora.

Importance of Ozone

Every one knows that life would be impossible if there were no ordinary or diatomic oxygen in the atmosphere, and that without it nearly if not quite all vegetable life also would be impossible. But it is not so well known, though equally true, that animals, including the human species, could not exist as now constituted if the air did not contain a small amount of triatomic oxygen or ozone.

And yet, paradoxical as it may if seem, ozone were just a little more effective in its goodness, again life, as now constituted, could not last. These surprising facts come about in this way: The radiation from the sun includes not only every colour, that is, the whole of the visible spectrum, but also extends indefinitely beyond the red into the long-wave-length invisible region, and likewise, in the opposite direction, well beyond the limit of the violet.

But in this ultra violet portion of the spectrum the radiation from the sun ceases far short of the limit to which that from an electric arc, for instance, can be followed; and ceases, not because no radiation beyond that limit is given out by the sun, but because it is absorbed by the ozone in the upper atmosphere.

Now, much of that particular radiation which ozone absorbs is destructive to the eye, and when intense probably injurious to other tissues as well.

On the other hand, it also absorbs a large part of the radiation that is effective in preventing rickets, but, and this is of the utmost importance, it does not absorb quite all of this antirachitic portion. Enough is let through to keep us in proper health.

This particular limiting, then, of the solar radiation that reaches the earth is amazingly located. If it was a little farther out in the ultra violet, eyes, as now constituted, could not have developed; and if a little nearer the visible, again, animals, as they now are, could not have come into being. Of course ozone, in moderate amount, presumably was in the air long before there were animals of any kind to be affected by its action on solar radiation. This radiation, therefore, was not adapted to them, but they developed in adaptation to it; nevertheless the fit is so close and of such a surprising nature as to give us decided pause for thought.

Origin of Ozone

Ozone is produced by the action of extreme ultra violet light on oxygen, and herefore at great heights; a little where the oxygen is very rare, then more and more with decrease of height and increase of the oxygen supply until by absorption the ozonizing rays are considerably enfeebled. As the density of the oxygen increases the intensity of the effective radiation decreases, hence the rate of production of ozone by this process must be very slow in the outermost air, increase to a maximum with decrease of height and then rapidly fall off at still lower and lower levels.

The total range of ozone, top to bottom, probably is not less than 100 miles, with its maximum concentration around 25 miles, perhaps, above the surface, and yet all told, and despite its great importance, it is the equivalent of a layer of this gas only about one-tenth of an inch thick at atmospheric pressure and room temperature. Ozone is produced also by electric discharges through oxygen. Lightning certainly produces some ozone, and it may be that the auroras form it too, though the great heights at which they occur, 60 miles and more above the earth, seem to render this conclusion doubtful.

Distribution of Ozone

As already stated, we know that at most there is only a trace of ozone in the atmosphere up to the level of the highest clouds, and that it exists to an appreciable extent at considerably greater heights. But this is not all. The total amount of ozone vertically over one to the limit of the air appears to increase, in general, with increase of latitude; to be greater during winter and spring than summer and fall; and to be greater in winds from high latitudes than in those of the opposite direction. This complicates the problem of the origin of ozone. If it is produced by ultra violet radiation alone why should it not be most abundant in tropical regions, and elsewhere in the summer time? If we surmise that it is produced largely by auroral discharges, how, we ask, can these discharges, 60 miles up and more, reach a sufficient supply of oxygen? And if there is enough oxygen at these great heights how can the ozone subsequently become concentrated at the intermediate levels? These are some of the questions to be answered by future observations and studies.

UPPER AND LOWER ATMOSPHERE

The composition of the lower atmosphere up to at least 6 or 7 miles in temperate regions, and 8 or 9 within the Tropics, is well known, except in\ respect to condensation nuclei and certain impurities.

Throughout this region too the percentages of the several gases, except water vapour, are practically constant owing to their continual mixing incident to convection and turbulence. Somewhere in the upper air, however, the percentages of the lighter gases must increase and those of the heavier decrease under the action of gravity since at these levels vertical convection is very feeble.

But as the upper air grows thinner the more and more rapidly does our knowledge of it become less. We know that its outermost portion is the very extensive region of the aurora, hundreds of miles thick; that near the base of the auroral region there is enough ionization to make wireless communication around the world entirely practicable; and that far below this Kennelly-Heaviside layer, in turn, and yet well within the upper air occurs most of the ozone, the triatomic oxygen that indirectly is so vital to all terrestrial life. We seldom give any thought to this upper air, but it is so important that we really must know more about it.

Mass of the Atmosphere and of its Constituents

From the known percentages of the several constituents of dry air, given above, their molecular weights, and various other pertinent facts such as the amount of water vapour present, height of the barometer, volume of land above sea level, and distribution of temperature with height, it is easy to compute the approximate mass of the atmosphere as a whole and of each of its several gases.

The results are given in the following table, in which the factor 108 means: Add eight ciphers, or multiply by 100,000,000. Since the values in this table were determined more or less independently it could not be expected that the percentages found of the constituents would add up to exactly 100, nor that the sum of the computed masses of the several parts would precisely equal the mass of the whole. These deviations, however, are very small—probably within the present limits of experimental errors.

TABLE: Mass of the Atmosphere and of its Constituents in Tons (2000 Pounds)

Substance	Volume Percent dry air, at surface	Total Mass
Total atmosphere		56,328,000×108 tons
Dry Air	100.00	56.181,850” ” ”
Nitrogen	78.03	42.684, 725 ” ” ”
Oxygen	20.99	12.782.647 ” ” ”
Argon	0.9323	982, 125 ” ” ”
Water vapor		146, 150 ” ” ”
Carbon dioxide		23, 874 ” ” ”
Hydrogen		1, 423 ” ” ”
Neon	0.03	759” ” ”
Krypton	0.01	141” ” ”
Helium	0.0018	88” ” ”
Ozone	0.0001	33” ” ”
Xenon	0.0005	19” ” ”
	0.00006	
	0.000009	

The numbers here given that express the masses of the atmosphere and its several constituents are useful as quantitative values and for exact comparisons, but so great, even though in terms of tons, that we can form no distinct conceptions of them—they are just awfully big! A clearer idea may be gotten from the fact that the total mass of the atmosphere is the equivalent, roughly, of that of a block of granite a thousand miles long, a thousand miles broad and half mile thick; while the least abundant of the constituent, xenon, if loaded on cars, 19 tons to the car, would freight a train reaching 40 times around the earth along a great circle, and which, traveling 20 miles all hour, would be 6 years in passing any point on the road.

❐

8

Survey of Natural Resources

ENVIRONMENTAL problems and issues received special attention of the Government of India during the beginning of the Fourth Five Year Plan. As a follow-up step, a National Committee of Environmental Planning and Co-ordination (NCEPC) was set up in 1972 under the Department of Science and Technology. A separate Empowered Committee was set up in 1980 for reviewing the existing legislative measures and administrative machinery for ensuring environmental protection and for recommending ways to strengthen them.

On the recommendations of this Empowered Committee, a separate Department of Environment was set up in 1980 which was subsequently upgraded to a full-fledged Ministry of Environment and Forests in 1985 to serve as the focal point in the administrative structure of the Government of India for the planning, promotion and co-ordination of environmental and forestry programmes. The State departments of environment, Central and State Pollution Control Boards, the Botanical and Zoological Survey of India, the Forest Survey of India, the National River Conservation Authority (formerly Central Ganga Authority), the National Afforestation and Ecodevelopment Board, the Indian Council for Forestry Research and Education, the Wildlife Institute of India, the National Museum for Natural History, etc. , are the Ministry's partners in carrying out environmental protection activities.

SURVEY OF NATURAL RESOURCES

The Botanical Survey of India (BSI), established in 1980, is responsible for surveying and identifying the plant resources of the country. With its headquarters at Calcutta and nine circles located in different regions of the country, the BSI undertakes exploration tours of the country regularly and the results of such tours are published in the form of national, state and district flora. Established in 1916, the Zoological Survey of India (ZSI) is responsible for carrying out surveys of the faunal resources of the country. While the headquarters of the Zoological Survey of India is at Calcutta, it has 16 regional stations located in different parts of the country. ZSI also undertakes regular faunistic survey tours of the entire country.

The Forest Survey of India (FSI), established in 1981,is entrusted with the task of surveying the forest resources of the country. Besides the headquarters at Dehra Dun, FSI has four regional offices located at Bangalore, Calcutta, Nagpur and Shimla. It prepares thematic maps on 1: 50,000 scale and forest vegetation maps on 1: 2,50,000 scale of the country.

The thematic maps are prepared for the entire country on a ten-year cycle. The vegetation maps are prepared on a two-year cycle for the entire country. The State of Forest Report, 1997 is the sixth report in the series of biennial reports published since 1987 and presents a comprehensive account of forest cover.

NATIONAL CONSERVATION STRATEGY

The National Conservation Strategy and Policy Statement on Environment and Development, adopted by the Government of India in June 1992, lays down strategies and actions for integration of environmental considerations in the development activities of various sectors of the country, thus paving the way for achieving sustainable development. The action points pertaining to individual ministries/departments ensure that they take action for reorienting their policies and programmes in conformity with the strategy.

BIOSPHERE RESERVES

Biosphere reserves are multi-purpose protected areas to preserve the genetic diversity in representative eco-systems. The major objectives of biosphere reserves are:

- To conserve diversity and integrity of plants, animals and micro-organisms;
- To promote research on ecological conservation and other environmental aspects
- To provide facilities for education, awareness and training.

So far eleven biosphere reserves have been set up.

They are:

- Nilgiri
- Nanda Devi
- Nokrek
- Great Nicobar
- Gulf of Mannar
- Encyclopaedia of Environmental Science
- Manas
- Sunderbans

- Similipal
- Dibru Saikhowa
- Dehong Deband
- Pachmarhi.

Comprehensive guidelines for them emphasize formulation of eco-development and demonstration projects, development of data-base, conservation plans of key species, establishment of research stations and implementation of social welfare activities. Non-governmental organisations are being involved in the biosphere reserve programmes for creation of public awareness.

WETLANDS, MANGROVES AND CORAL REEFS

India has a wealth of wetland eco-systems distributed in different geographical regions from the cold arid zone of Ladakh in the North to the wet humid climate of Imphal in the East, the warm arid zone of Rajasthan in the West to the tropical monsoonic Central India and the wet and humid zone of Southern Peninsula. Most of the wetlands in India are directly or indirectly linked with major river systems such as Ganga, Brahmaputra, Narmada, Tapti, Godavari, Krishna, Cauveri, etc.

A National Level Committee constituted to advise the Government on appropriate policies and measures to be taken for conservation and management of the wetlands, has so far identified 20 wetlands for conservation and management on priority basis. Steering Committees have been set up by the concerned State governments in which representatives of State government departments, universities and research institutions are included. Nodal research/academic institutions have been identified for each of the selected wetlands.

Management Action Plans have been drawn up for most of the identified wetlands. A directory on wetlands in India has been published which gives information on location, area and ecological categorisation of wetlands in different parts of the country. India is a signatory to the Convention on Wetlands of international importance, especially as Waterfowl Habitat (Ramsar Convention) and six Indian Wetlands, viz., Keoladeo National Park, Bharatpur and Sambar (Rajasthan), Chilka(Orissa), Loktak (Manipur), Wullar (Jammu and Kashmir), and Harike (Punjab) have been designated under this Convention. Mangroves are salt-tolerant forest ecosystems found mainly in the tropical and sub-tropical inter-tidal regions of the world. They are reservoirs of a large number of plant and animal species associated together over a long evolutionary period and exhibiting remarkable capacity for salt tolerance. They stabilize the shoreline and act as a bulwark against encroachments by the sea. India harbours some of the best mangroves in the world and these occur all along the Indian coastline in sheltered estuary, tidal creeks, backwaters, salt marshes and mud flats covering a total area of 4,827 sq km.

Under the scheme on Conservation and Management of Mangroves, 15 mangrove areas have been identified for intensive conservation and management purposes: Northern Andaman and Nicobar, Sunderbans (West Bengal), Bhitarkanika (Orissa), Coringa, Godavari Delta and Krishna Estuary (Andhra Pradesh), Mahanadi Delta (Orissa), Pitchavaram and Point Calimer (Tamil Nadu), Goa, Gulf of Kutch (Gujarat), Coondapur (Karnataka), Achra/ Ratnagiri (Maharashtra) and Vembanad (Kerala). Management action plans for all the 15 mangrove areas have been sanctioned. Coral reefs are shallow-water tropical marine ecosystems, characterised by high biomass production and rich floral and faunal diversity. Four coral areas, Gulf of Mannar, Andaman and Nicobar Islands, Lakshadweep Islands and Gulf of Kuchch have been identified for conservation and management. State-level steering committees have been constituted for the formulation and implementation of management action plans.

Such action plans have been sanctioned for Andaman and Nicobar and Gulf of Mannar coral reefs so far. The objectives of the Convention on Biological Diversity (CBD) are: the conservation of Biological Diversity; the sustainable use of its component and the fair and equitable sharing of the benefits arising out of the utilization of genetic resources. Following the ratification of the Convention on Biological Diversity by India in 1994, several steps have been initiated both to meet the commitments under the Convention and to realise the opportunities offered by the CBD. A draft National Action Plan on Biodiversity has been finalised which seeks to consolidate the on-going efforts of conservation and sustainable use of biological diversity and to establish a policy and programmes regime for the purpose. A legislation on bio-diversity is also under finalisation. India regularly participates in important international conventions on Biological Diversity.

FORESTS

Forests are a renewable source and contribute substantially to economic development. They play a major role in enhancing the quality of environment. The country has an area of 633.4 lakh hectares notified as forests which represents 19.27 per cent of geographic area. Out of this, dense forest (crown density more than 40 per cent) accounts for 11 per cent, open forest (crown density 10-40 per cent) represents eight per cent, while mangrove forest occupies 0.15 per cent.

FOREST POLICY AND LAW

India is one of the few countries which have a forest policy since 1894. It was revised in 1952 and again in 1988. The main plank of the Forest Policy of 1988 is protection, conservation and development of forest.

Its aims are:

- Maintenance of environmental stability through preservation and restoration of ecological balance;
- Conservation of natural heritage;
- Check on soil erosion and denudation in catchment area of rivers, lakes and reservoirs;

- Check on extension of sand dunes in desert areas of Rajasthan and along coastal tracts;
- Substantial increase in forest tree cover through massive afforestation and social forestry programmes;
- Steps to meet requirements of fuelwood, fodder, minor forest produce and soil timber of rural and tribal populations;
- Increase in productivity of forest to meet the national needs;
- Encouragement of efficient utilisation of forest produce and optimum substitution of wood
- Steps to create massive people's movement with involvement of women to achieve the objectives and minimise pressure on existing forests.

The entire gamut of forest activities are being given a new orientation in the light of the National Forest Policy of 1988. In order to operationalise the National Forest Policy 1988, a National Forestry Action Programmes (NFAP) is being prepared. As a part of this exercise State Forestry Action Programmes are also being prepared for each State.

Under the provisions of the Forest (Conservation) Act, 1980,prior approval of the Central government is required for diversion of forest lands for non-forest purposes. Since the enactment of the Act, the rate of diversion of forest land has come down to around 25,000 hectares per annum from 1.43 lakh hectares per annum, before1980.During 1998, 851 proposals from various State and UT governments were processed under this Act. A scheme titled 'Association of Scheduled Tribes and Rural Poor in Regeneration of Degraded Forests on Usufruct Sharing Basis' is under implementation in nine States of the country. Besides improving the forest cover, the scheme also aims at providing wage employment and usufructs to the tribal people. Joint Forest Management (JFM) is being practised in 21 States of the country. About 7 million hectares of degraded forests in the country are being managed and protected through approximately 35,000 village Forest Protection Committees.

FOREST FIRE CONTROL

Fire is one of the major factors responsible for destruction of forests in the country. In India, most of the fires are man-made and are generally deliberate and on rare occasions accidental. Some of the reasons for intentional fire are grazing, mahua seeds and flower collection, tendu leaf collection, poaching, shifting cultivation, etc.

In order to reduce the incidents of fire in forests, the following steps have been undertaken by the forest department:

- Development of firelines
- Establishment of watchtowers, besides employing firewatchersduring fire seasons.

The UNDP-assisted Modern Forest Fire Control Project, initiated in 1984 in Chandrapur (Maharastra) and Haldwani/Nainital (Uttar Pradesh), is in operation in 11 States of the country. Financial assistance is provided to the state governments for hand tools, fire-resistant clothing, wireless communication sets, fire-fighting instruments, construction of watch-towers, creation of fire lines, etc. The project is being revised to include the monitoring of forest fire incidents in the country.

WILDLIFE

The National Wildlife Action Plan provides the framework of strategy as well as programmes for conservation of wildlife. The Indian Board of Wildlife, headed by the Prime Minister, is the apex advisory body overseeing and guiding the implementation of various schemes for wildlife conservation. At present the protected area network comprises 84 national parks and 447 sanctuaries covering 4.5 per cent of the total geographical area of the country. The Wildlife (Protection) Act, 1972 adopted by all states except Jammu and Kashmir (which has its own Act), governs wildlife conservation and protection of endangered species. The Act prohibits trade in rare and endangered species. An Inter-State Committee has been set up to review the Wildlife (Protection) Act, 1972 and other laws. India is a signatory to the Convention on International Trade in Endangered Species of wild flora and fauna (CITES). Under this, export or import of endangered species and their products is subject to strict control. India is also a signatory to the Memorandum of Understanding concerning the conservation of Siberian Cranes. The Centre provides financial and technical assistance to states for development and improved management of national parks; protection of wildlife and control of poaching and illegal trade in wildlife products; eco-development in areas around national parks and sanctuaries; conservation of elephant and its habitat; and conservation of rhinos in Assam.

Under Project Tiger, launched in April 1973, 25 Tiger Reserves have been set up in 14 States covering an area of about 33,875 sq km. The tiger population now reported by 14 States is 3,435 only. Financial assistance is available for voluntary relocation of tribal families from the Protected Areas. Under the Project Elephant, States having free-ranging population of wild elephants are being given financial as well as technical and scientific assistance to ensure long-term survival of identified viable population of elephants in their natural habitats. A Central Zoo Authority has been set up to look after the management of zoological parks in the country. It coordinates the activities of over 200 existing zoos and also supervises the exchange of animals on a scientific basis.

A national policy on zoos prepared by the Authority has been adopted which provides appropriate directions to the Government and other zoo operators. The export of the shed antlers of deer (Chital and Sambhar) and articles made of these have been banned. Export of 29 medicinal plants, parts thereof, and pure extracts of these plants have been banned.

ANIMAL WELFARE BOARD

The Animal Welfare Board of India, established in 1962 under the provisions of the Prevention of Cruelty to Animals Act, 1960 is an autonomous organisation of the Ministry of Environment and Forests working for the cause of animal welfare in the country.

More than 9,000 Honorary Animal Welfare Officers have been appointed by the Board so far to check cruelty to animals and to protect them. A series of camps for training the members of the Society for Prevention of Cruelty to Animals and animal welfare organisations are being held all over the country. Animal Welfare Fortnight is celebrated from 14 January every year.

ENVIRONMENTAL IMPACT ASSESSMENT

Impact assessment is a pointer to the environmental compatibility of the projects in terms of their location, suitability of technology, efficiency in resource utilisation, recycling and so on.

Impact assessment was introduced in India in 1978 and now covers projects such as:

- River valley;
- Thermal power;
- Mining;
- Industries;
- Atomic power;
- Rail, road, highways, bridges;
- Ports and harbours;
- Airports;
- New towns and
- Communication projects.

- Those which require the approval of the Public Investment Board/ Planning Commission/Central Electricity Authority.
- Those referred to the Ministry of Environment and Forests by other ministries.
- Those which are sensitive and located in environmentally degraded areas.
- Public sector undertakings of the Centre where the project cost is more than `50 crore.

A notification issued in January 1994 makes Environmental Impact Assessment statutory for 29 categories of developmental projects under various sectors such as industrial, mining, irrigation, power, transport, tourism, communication, etc. The Environmental Impact Assessment (EIA) Notification was amended in 1997 to provide for public hearing as well as for empowering State governments for according environmental clearance in respect of certain Thermal Power Projects. As per the provisions of the EIA Notification 1994, mining projects are subject to environmental clearance.

Until now, all proposals for prospecting and exploration were required to obtain the clearance. However, it has been now decided to exempt prospecting through aerial survey and/or test drilling in smaller areas from the clearances. Applications complete in every material aspect are normally examined and decision conveyed to the applicants within 30 days in cases of site clearance and 120 days in cases of environmental clearance of projects. Once an application has been submitted by the project authority, the preliminary scrutiny of the project is done by the respective technical divisions and the overall appraisal of the projects is undertaken by specially constituted environmental appraisal committees of experts.

In addition, special groups/committees and task forces are constituted as and when needed for expert inputs on major projects. After detailed scrutiny and assessment, the appraisal committee makes its recommendations for approval or rejection of the project. To ensure transparency, the position of forest and environmental clearance has been brought out on web site: httpc/ www. nic. in/ envfor since February 1999.Depending on the nature of the project, certain safeguards are recommended. For monitoring and timely implementation of safeguards suggested, six regional offices of the Ministry have been set up at Shillong, Bhubaneswar, Chandigarh, Bangalore, Lucknow and Bhopal.

AUTHORITIES UNDER ENVIRONMENT PROTECTION ACT, 1986

A National Environmental Appellate Authority has been constituted to hear appeals with respect to rejection of proposals from the environmental angle. The objective is to bring in transparency in the process and accountability, and to ensure the smooth and expeditious implementation of developmental schemes and projects.

An Environmental Impact Assessment Authority for the National Capital Region has been constituted to deal with environmental protection problems arising out of projects planned in the National Capital Region (NCR). An Aquaculture Authority has been constituted to deal with the situation created by the Shrimp Culture industry in the coastal States and Union Territories.

The Central Ground Water Authority for regulation and control of ground water management has initiated action regarding registrations for ground ater pollution/depletion. It has also initiated mass awareness programmes. Besides this, different authorities have been created for dealing with specific problems in the States of Tamil Nadu and Maharashtra.

PREVENTION AND CONTROL OF POLLUTION

The policy statement on Abatement of Pollution, adopted in 1992, provides instruments in the form of legislation and regulation, fiscal incentives, voluntary agreements, educational programmes and information campaigns to prevent and control pollution of water, air and land. Since the adoption of the policy statement, the focus of activities has been on issues such as promotion of clean and low waste technologies, waste water minimisation, reuse/ recycling, improvement of water quality, environment audit, natural resource accounting, development of mass-based standards, institutional and human resource development, etc.

The whole issue of pollution prevention and control is dealt with by a combination of command and control methods as well as voluntary and egulations, fiscal measures, promotion of awareness and involvement of public. To facilitate industries in preparing environmental statements, sectorspecific environmental audit manuals have been prepared.

A software package, Paryavaran, along with a user manual has been prepared for analysis of information submitted in these environmental statements and distributed to all the State Pollution Control Boards. Training programmes are also being organised for officials of CPCB and SPCB to enable them to use the software. An "Eco-mark" label has been introduced to label consumer products that are environment-friendly.

Nineteen products have been identified for labelling and 18 notifications have been issued so far on different products criteria. The Bureau of India Standards (BIS)/Directorate of Marketing and Inspection (DMI) is the implementing agency for this scheme. So far one license has been granted by the BIS to a product under the soaps category. Under the scheme for adoption of clean technology in small-scale industries and for extending necessary technical support, training and awareness programmes for personnel in Small Industry Development Organisation and for entrepreneurs are being organised.

'From Waste to Profits', a manual giving guidelines for waste minimisation, has been prepared. Sector-specific manuals on waste minimisation in the areas of pulp and paper, pesticides formulations and textiles, dyeing and printing and electroplating, have also been prepared.

Waste Minimisation Circles (WMCs) are being established to promote group efforts in increasing productivity and improving the environmental conditions in small and medium-scale industries through adoption of waste minimisation techniques. Fifteen Waste Minimisation Circles have been established so far in different industrial clusters across the country. An Indian Centre for the Promotion of Cleaner Technologies (ICPC) with a network of institutions including industries, academic institutions and other user agencies is being set up for which the World Bank has provided $2 million as grant-in-aid.

The Centre will provide evaluated and ranked technology options to entrepreneurs. A major programmes for improvement of automotive fuels such as motor gasoline and diesel is being introduced in the country in a phased manner, with the objective of improving the air quality of the cities of our country. Low-lead petrol (0.15 gms/litre lead content) has been introduced at all retail outlets of the country from 1 April 1996. Unleaded petrol has been made available in selected retail outlets of the four metro cities of Delhi, Mumbai, Calcutta and

Chennai from 1 April 1995. From 1 June 1998 unleaded petrol is also available in all state capitals and major towns. From 1 September 1998 only unleaded petrol has been available at all the outlets in Delhi and it is extended to National Capital Territory of Delhi from 1 January 1999.

Ban on 2T oil has been imposed in the National Capital Territory of Delhi from 31 December 1998. High-speed diesel containing 0.5 per cent sulphur as been introduced in the four metro cities from 1 April 1996. HSD with 0.25 per cent sulphur is being supplied in the Taj Trapezium area from 1 September 1996. It is also being supplied in Delhi from 15 August 1997. Stricter emission norms for new vehicles effective from 1 April 2000 have been notified.

CENTRAL POLLUTION CONTROL BOARD

The Central Pollution Control Board (CPCB) is the national apex body for assessment, monitoring and control of water and air pollution. The executive responsibilities for enforcement of the Acts for Prevention and Control of Pollution of Water (1974) and Air (1981) and also of the Water (Cess) Act, 1977 are carried out through the Board. The CPCB advises the Central Government in all matters concerning the prevention and control of air, water and noise pollution and provides technical services to the Ministry for implementing the provisions of the Environment (Protection) Act, 1986.

Under this Act, effluent and emission standards in respect of 61 categories of industries have been notified. Seventeen categories of heavily polluting industries have been identified. They are: cement, thermal power plant, distilleries, sugar, fertilizer, integrated iron and steel, oil refineries, pulp and paper, petrochemicals, pesticides, tanneries, basic drugs and pharmaceuticals, dye and dye intermediates, caustic soda, zinc smelter, copper smelter and aluminium smelter. Out of a total of 1,551 units identified under these 17 categories, 1,266 units have installed adequate facilities for pollution control and 130 units have been closed down. The Central Pollution Control Board, in consultation with State Pollution Control Boards, has identified critically polluted areas in the country which need special attention for control of pollution.

These are:

- Vapi (Gujarat)
- Singrauli (Uttar Pradesh)
- Korba, Ratlam, Nagda (Madhya Pradesh)
- Digboi (Assam)
- Talcher (Orissa)
- Bhadravati (Karnataka)
- Howrah (West Bengal)
- Dhanbad (Bihar), Pali and Jodhpur (Rajasthan)
- Manali and North Arcot (Tamil Nadu)
- Visakhapatnam and Patancheru, (AndhraPradesh)

- Chembur (Maharashtra), Najafgarh (Delhi)
- Govindgarh (Punjab)
- Udyog Mandal (Kerala)
- Parwanoo and Kala Amb (Himachal Pradesh)

Thc CPCB in collaboration with the SPCBs monitors the quality of fresh water resources of the country through a network of 480 monitoring stations located all over the country. Based on such monitoring, 13 heavily polluted and 26 medium-polluted river stretches have been identified. Under the National Ambient Air Quality Monitoring programmes, 290 stations covering over 90 towns/cities monitor the ambient air quality of the country.

The Central and State Pollution Control Boards regularly conduct surveys in different cities of the country pertaining to vehicular and noise pollution, sanitation status, status of solid waste, etc. A survey on the status of solid waste conducted in 299 Class I cities of the country indicates that 62 per cent of the total solid waste generated in the country comes from the 23 metro cities of the country. The average per capita generation of solid waste for Class I cities is about 376 gms per person per day. A total of 1,532 grossly polluting industries in 24 States/Union Territories have been identified under th National River Action Plan. Comprehensive River Basin Documents for the rivers Ulhas, Brahmaputra, Pennar, Indus Part II, Rishkulya and Chaliyar are being prepared by the Board. The Central Pollution Control Board has a NGO Cell for interacting with NGOs.

Simple water-testing kits are distributed free of cost to selected NGOs and financial assistance provided to them for conducting mass awareness programmes relating to prevention and control of pollution. The White Paper on status of pollution in Delhi with an Action Plan for its control prepared earlier is being implemented. The Action Plan contains specific measures for control of pollution relating to vehicular pollution, water pollution, industrial air pollution, solid waste, hospital wastes, industrial hazardous wastes, noise pollution and people's participation in making Delhi a cleaner city. Directions have been issued by the National Capital Territory of Delhi for imposing restrictions on all commercial vehicles in Delhi in a time-bound programmes beginning from April 1998. Based on the recommendations of a National Level Committee on Noise Pollution, directions have been issued to the State governments to check noise pollution from bursting of crackers. Apart from introducing unleaded petrol in the National Capital Territory of Delhi from 1 September 1998 and the National Capital Region from 1 January 1999, there is a proposal to promote the use of CNG in buses, private vehicles and taxies.

There is also a proposal to introduce a certain percentage of ethanol/methanol in gasoline so as to reduce the emission of carbonmonoxide and hydrocarbons. It was also proposed to set up a high level body to look into issues concerning vehicle emissions. There is also a plan to introduce higher parking fees, to augment parking spaces in the city of Delhi and to introduce no traffic zones.

MANAGEMENT OF HAZARDOUS SUBSTANCES

The Ministry of Environment and Forests is the nodal agency for the management and control of hazardous substances which include hazardous chemicals, waste and micro-organisms.

The following rules have been notified under the Environment (Protection) Act, 1986:

- Manufacture, Storage and Import of Hazardous Chemicals, 1989;
- Hazardous Wastes (Management and Handling) Rules, 1989;
- Manufacture, Use, Import, Export and Storage of Hazardous Micro-organisms/Genetically Engineered Organisms or Cell, 1989
- Biomedical Waste Rules, 1998.

A central control room has been set up in the Ministry to deal with emergencies arising from hazardous chemicals and a Crisis Alert System has been established. Guidelines for preparation of crisis management plans have been issued to the state governments and financial support is being provided to them to strengthen infrastructure for the purpose. Emergency Response Centres have been set up at Bhopal, Baroda, Manali and Khapali.

The Red Book entitled 'Central Crisis Group Alert System', which includes names, addresses and telephone numbers of the Central and state authorities and experts to be contacted in case of emergency, has been updated and circulated to all concerned. A sub-scheme entitled 'Industrial Pocket-wise Hazard Analysis' has been in operation since the Eighth Five Year Plan.

Hazard analysis studies of 37 industrial pockets have since been completed. The remaining 8 Hazard analysis studies are being offered to the consultants. A Public Liability Insurance Act has been enacted to provide immediate relief to the victims of accidents by hazardous chemical industries. India is a signatory to the UNEP-sponsored Convention on Control of Transboundary Movement of Hazardous Wastes which was adopted at Basel, Switzerland, by 126 governments of the world in 1989. The Convention aims at checking the reported illegal traffic in hazardous wastes from one country to another. A scheme on National Register of Potentially Toxic Chemicals (NRPTC) has been started for setting up the basic infrastructure for implementing the London guidelines for thc exchange of information on chemicals in international trade, including the procedure for prior informed consent. Nine Regional Registers (RRPTC) have been set up in Uttar Pradesh, Madhya Pradesh, Andhra Pradesh, Gujarat, Orissa, Kerala, Himachal Pradesh, Punjab and Maharashtra.

National Poison Information Centre has been set up at the Department of Pharmacology in the All India Institute of Medical Sciences, New Delhi. An antidote data-bank is being developed at the Centre. The Ministry has constituted a National Waste Management Council to suggest ways and means for effective utilisation of wastes generated in the country. Three sub-groups have been set up by the Council to deal with the major categories of wastes, viz., industrial, urban and rural.

The sub-groups have been entrusted with the task of identifying wastes, suggesting technological action points, including legislation, taxes and incentives. A Pilot Project on Municipal Solid Waste Management has been sanctioned to the Hyderabad Municipal Corporation. Special emphasis is being laid on promoting the use of flyash in various fields, and the State governments have been asked to prepare a perspective action plan for 50 per cent utilisation of flyash by the turn of the century. A high-powered committee on hazardous wastes has been constituted to oversee the strict implementation of the rules and regulations on management of these wastes.

For the disposal of hazardous wastes, 15 sites have been identified in the country for preparation of secured land fills. Erring industries of the village Bichchri, Rajasthan, have been ordered to undertake remedial work on the 320-hectare land contaminated due to indiscriminate disposal of hazardous wastes. The Hazardous Wastes (Management and Handling) Rules, 1989 and Manufacture, Storage and Import of Hazardous Chemical Rules, 1989 have been amended. Final Rules on Bio-medical Wastes have been notified. A new set of rules, 'Emergency Planning Preparedness and Response to Chemical Accidents', has been notified. A notification has been issued banning import of wastes containing Beryllium, Selenium, Chromium (Hexavalent), Thallium, pesticides, herbicides and insecticides as well as Waste Asbestos (dust and fibre), and waste containing PCB, PCT and PBB.

NATIONAL RIVER CONSERVATION AUTHORITY

The Central Ganga Authority (CGA) established in 1985, lays down the policies for works to be taken up under the Ganga Action Plan (GAP). With the approval of the National River Conservation Plan (NRCP) in 1995, the CGA has been reconstituted as the National River Conservation Authority (NRCA) and the Ganga Project Directorate has been redesignated as the National River Conservation Directorate (NRCD).

The NRCA is assisted by a Steering Committee and a Monitoring Committee. The NRCD of the Ministry services these committees and coordinates the implementation of the schemes under the Ganga and other Action Plans. The main component of the Ganga Action Plan is diversion of sewage flowing into the river to other locations for treatment and conversion into valuable energy sources. As against the target of creating infrastructure capable of intercepting, diverting and treating 873 million litres per day (mld) of domestic waste, infrastructure capable of treating 728 mld of municipal sewage has been created.

Out of 261 schemes sanctioned, 254 have been completed and almost all low-cost sanitation schemes and construction of electric crematoria have been completed, leading to significant reduction in the pollution load of the river

Ganga. The Ganga Action Plan, Phase II has been merged with the National River Conservation Plan (NRCP) through a Government Resolution. The expanded NRCP now covers 141 towns located along 22 inter-state rivers in 14 States.

The total cost of the schemes is ` 2,014.39 crore. Pollution abatement works in 46 towns located along 18 inter-state rivers have been taken up. Works have also been taken up at 59 towns on the Main Stem of River Ganga. The scheme of cleaning the river Yamuna covers 21 towns in three states and its estimated cost is `509.45 crore.

Under the Gomti river scheme, works are proposed in three towns at an estimated cost of `64 crore. The scheme of pollution abatement of the river Damodar was approved in October 1996 at a cost of `24.54 crore. Works are proposed in 12 towns in two States. A National Lake Conservation Plan has also been initiated under which 11 urban lakes have been identified for immediate conservation. The World Bank has agreed to provide financial assistance for Conservation and Management of Lakes and Wetlands in Rajasthan and Punjab.

NATIONAL AFFORESTATION AND ECO-DEVELOPMENT BOARD

The National Wastelands Development Board (NWDB) established in May 1985, was bifurcated into a new Department of Wastelands Development and a National Afforestation and Eco-Development Board (NAEB) in 1992. While the NAEB is a part of the Ministry of Environment and Forests, the Department of Wastelands Development has been transferred to the Ministry of Rural Development.

The responsibilities of the NAEB include promotion of afforestation, tree plantation, ecological restoration and eco-development activities in the country with special attention to the degraded forest areas and lands adjoining the forest areas, national parks, sanctuaries and other protected areas, as well as the ecologically fragile areas like the Western Himalayas, Aravallis, Western Ghats, etc. The NAEB seeks to carry out its responsibilities through several schemes such as the integrated afforestation and eco-development project, area-oriented fuelwood and fodder projects, non-timber forest produce (including medicinal plants) scheme, aerial seeding, seed development, technology extension, grants-in-aid to voluntary agencies, etc.

During the Ninth Five Year Plan, four key elements, viz., Promotion of Joint Forest Management, Micro planning, Incorporation of improved technologies, and Monitoring and Evaluation of projects, will be emphasized while implementing these schemes. Eco-task forces of exservicemen have been deployed in the states of Uttar Pradesh, Rajasthan and Jammu and Kashmir for undertaking eco-restoration work through afforestation, pasture development and other water and soil conservation measures in selected highly degraded and inhospitable areas of these states.

To commemorate the 50th Anniversary of India's Independence, the Board provided financial assistance to State Forest Departments for distributing seedlings to all Village Panchayats to enable every village to raise a cluster of 50 trees. Wastelands maps on 1: 50,000 scale have been prepared for 237 districts of the country using satellite data. The NAEB has seven Regional Centres located at different universities and national-level institutions of the country. These centres help the NAEB in promoting extension of replicable technologies and in disseminating research findings to a wider audience. These centres also carry out problemspecific studies and undertake evaluation of NAEB's programmes in the field.

RESEARCH PROMOTION

ENVIRONMENTAL RESEARCH PROGRAMMES

The environmental research programmes aims at developing strategies for sustainable resource management and creating facilities and trained manpower to conduct research in priority areas.

Research projects are supported under three main schemes:

- Man and Biosphere Programmes;
- Environment Research Programmes;
- Action-oriented Research Programmes for Eastern and Western Ghats.

Many small, specific research projects are also being supported. Expert Committees and Expert Working Groups make recommendations for undertaking research in priority areas, scrutinise proposals, evaluate progress and recommend suitable mechanisms for implementation of research projects. Priority areas of research have been identified in five fundamental elements, viz., land, water, air, space and energy. During the year, 45 new projects were sanctioned, 45 were completed and 112 projects were serviced. The entire research programmes is being reviewed under a special project funded by the World Bank. The Framework Convention in Climate Change ratified by 127 countries of the world was ratified by India during 1993.

India participated in the 3rd Conference of Parties to this Convention held in Kyoto, Japan in December 1997 during which a protocol was agreed upon which calls for an average cut in the greenhouse gas emission, of 5.2 per cent below the 1990 levels, to be achieved between 2008 and 2012 AD. Specific research projects are being sponsored by the Ministry in the area of climate change for taking precautionary measures to anticipate, prevent and minimise the causes of climate change and mitigate its adverse effects. The Asia Least Environment Facility has been completed.

GB PANT INSTITUTE OF HIMALAYAN ENVIRONMENT AND DEVELOPMENT

The GB Pant Institute of Himalayan Environment, established in 1988, is an autonomous organisation of the Ministry of Environment and Forests for developing strategies, technologies and knowledge base for ecologically sound development of the Himalayan region. Besides undertaking actionoriented research projects, the Institute has also installed three weathermonitoring stations for collecting and analysing climatic data of the Himalayan region.

Units of the Institute are located in Uttar Pradesh, Sikkim, Nagaland and Himachal Pradesh. The Ministry also sanctions specific research projects under various programmes such as Biosphere Reserves, Wetlands, Mangroves and Coral Reefs, etc. The research activities of the National River Action Plan focus on identifying economically viable and technically feasible solutions for controlling the microbial pollution in the rivers. During the current year, it is proposed to initiate research studies in estimation of non-point pollution, heavy-metals uptake in vegetables and river fauna, etc.

FORESTRY RESEARCH

Forestry research is the primary responsibility of the Indian Council of Forestry Research and Education located at Dehra Dun. The following forestry research institutes and centres under the Council are responsible for undertaking research in their respective eco-climatic zones:

- Encyclopaedia of Environmental Science
- Institute of Arid Zone Forestry Research, Jodhpur
- Institute of Rain and Moist Deciduous Forests, Jorhat
- Institute of Wood Sciences and Technology, Bangalore
- Tropical Forestry Research Institute, Jabalpur
- Institute of Forest Genetics and Tree Breeding, Coimbatore
- Temperate Forest Research Centre, Shimla
- Centre for Forest Productivity, Ranchi
- Institute of Social Forestry and Eco-rehabilitation, Allahabad
- Institute of Forestry Research and Human Resources Development, Chhindwara
- Advanced Centre for Bio-technology and Mangrove Forests, Hyderabad.

In addition, the Indian Plywood Industries Research and Training Institute, Bangalore, has been transferred to the Ministry of Environment and Forests from the Ministry of Industry.

Besides forestry research in wood panel products, this Institute imparts training to workers in optimum utilisation of timber. The Indian Institute of Forest Management, Bhopal, also undertakes education, training, research and consultancy in forest management.

WILDLIFE RESEARCH

Research programmes in wildlife are carried out by the Wildlife Institute of India, Dehra Dun, and the Salim Ali Centre for Ornithology and Natural History, Coimbatore. Several projects on habitat evolution, elephant movement, ecology of gharials and turtles, status of endangered species, behavioural ecology, bio-diversity, resource study conservation, ecology and management of specific animals, etc. , are being carried out by both these Institutes.

NATIONAL NATURAL RESOURCES MANAGEMENT SYSTEM

The basic objective of the Natural Resources Management System (NRMS) is the utilisation of Remote Sensing Technology with the conventional methods for optimal use and management of the natural resources of the country. A Standing Committee on Bio-resources and Environment advises the Ministry on various aspects relating to the implementation of this scheme. Specific research projects are also supported under this scheme. Two Sub-Committees, the Technical and Financial Sub-Committee and the Forest Sub-Committee, help the Standing Committee in its task.

EDUCATION, AWARENESS AND INFORMATION

Priority is accorded by the Ministry of Environment and Forests to promote environmental education, create environmental awareness among various age-groups and to disseminate information through Environmental Information System (ENVIS) network to all concerned. A major initiative to include environment education as a separate and compulsory subject in the educational curricula has been taken by the Ministry at all levels of formal education, i.e., secondary, senior secondary and tertiary levels.

A discussion paper prepared on strengthening environment education was presented by the Minister for Environment and Forests at the State Education Ministers' Conference held from 22 to 24 October 1998. The paper was adopted by the Conference. The Chief Ministers were urged to introduce environment education in the school curricula from the 1999-2000 academic session.

Maharashtra is the first State to introduce the subject in the school curriculum. Special emphasis is given to non-formal environmental education through seminars/symposia/workshops, training programmes, eco-camps, audiovisual shows, etc. The Ministry has been organising a National Environmental Awareness Campaign (NEAC) since July 1986. As a part of this campaign, 19 November to 18 December every year is observed as the National Environment Month. The main theme for the 1998 campaign was 'Keep Our Water Resources Clean'. A large number of organisations have been granted financial assistance by the Ministry to organise various activities for creating environmental awareness. The Ministry also provides financial assistance for setting up eco-clubs in schools.

A new scheme, Paryavaran Vahini, was launched in 1992-93 to create environmental awareness and to ensure active public participation by involving the local people in activities relating to environmental protection.

Paryavaran Vahinis are proposed to be constituted in 194 selected districts all over the country which have a high incidence of pollution and density of tribal and forest population.

The Vahinis also play a watch-dog role by reporting instances of environmental pollution, deforestation, poaching, etc. They function under the charge of District Collectors, with the active cooperation of the State/Union Territory governments. This scheme is entirely financed by the Ministry of Environment and Forests. Seven Centres of Excellence have been set up by the Ministry to strengthen awareness, research and training in priority areas of Environmental Science and Management. These are: Centre for Ecological Sciences, Bangalore; Centre for Mining Environment, Dhanbad; Centre for Environmental Education, Ahmedabad; CPR Environmental Education Centre, Chennai; Salim Ali Centre for Ornithology and Natural History, Coimbatore; the Centre for Environmental Management of Degraded Ecosystems, Delhi, and the Tropical Botanical Garden and Research Institute, Thiruvananthapuram, Kerala. The National Museum of Natural History (NMNH) set up in New Delhi in 1978, is concerned with the promotion of non-formal education in the area of environment and conservation.

Besides permanent exhibit galleries on various aspects of environment, the museum also conducts temporary exhibitions and a large number of educational programmes and activities for school children, college youth and the general public. Three Regional Museums of Natural History have been established at Mysore, Bhopal and Bhubaneswar. The Indian Council for Forestry Research and Education is the focal point for forestry education and extension development in the country.

The Indira Gandhi National Forest Academy, Dehra Dun, imparts in-service professional training to Indian Forest Service (IFS) professionals. State forest service colleges provide training to the officers of the State Forest Service (SFS). The Indian Plywood Industries Research and Training Institute, Bangalore, organises short-term courses in the area of wood science. The Indian Institute of Forest Management, Bhopal, also provides training in forest management and allied subjects to persons from the Indian Forest Service, forest development corporations, and forest-related industries to develop forestry programmes. The Wildlife Institute of India, Dehradun, provides in- service training to forest officers, wildlife ecologists and other professionals for conservation and management of the wildlife resources of the country.

ENVIRONMENTAL INFORMATION SYSTEM

An Environmental Information System (ENVIS) was set up by the Ministry of Environment and Forests in 1982 as a decentralised information network for collection, storage, retrieval and dissemination of environmental information. Besides the Focal Point in the Ministry, ENVIS network presently consists of 25 subject-oriented centres known as ENVIS Centres, set up in various institutions/organisations of the country in priority areas of environment such as environmentally sound and appropriate technology, bio-degradation of wastes, desertification, estuary, mangroves, coral and lagoons, media and environment, environmental education, solid wastes disposal, animal ecology, Himalayan ecology, etc. The Focal Point brings out a quarterly abstracting journal, Paryavaran Abstracts, containing information about environmental research in the Indian context. ENVIS has been designated as National Focal Point (NFP) and Regional Service Centre (RSC) for South Asia sub-regional countries by INFOTERRA (an international referral system for sources of information on environment) of United Nations Environment Programmes (UNEP). The ENVIS network responded to a total of 12,756 queries during 1998-99 about a variety of environmental concerns.

FELLOWSHIPS AND AWARDS

Instituted in 1987, the Indira Gandhi Paryavaran Puraskar, consisting of a cash component of rupees one lakh, a silver trophy and a citation, is awarded every year to an organisation and to an individual for significant contributions in the field of environment. The Indira Priyadarshini Vrikshamitra Awards were constituted by the Ministry of Environment and Forests in 1986 to recognise outstanding contributions of individuals and organisations in the field of afforestation and wastelands development. Starting in 1993, 12 annual awards are given to individuals, educational institutions, panchayats, voluntary agencies, government agencies and the corporate sectors. Each award carries a medallion, a citation and a cash component of `50,000.

The Mahavriksha Puraskar, instituted by the National Afforestation and Eco-development Board (NAEB) during 1993-94 is given every year to individuals/organisations for trees of notified species having the largest girth and height and in good health and vigour. The award consists of a cash prize of `25,000, a plaque and a citation. Rajiv Gandhi Environment Award for Clean Technology instituted in 1993, is given to industrial units that have made significant and measurable contributions towards development of clean technologies and innovative solutions to environmental problems created by industrial operations. With the objective of encouraging original and applied research among Group 'A' scientists in the Ministry of Environment and Forests and its associated offices and autonomous bodies, Paryavaran Evam Van Mantralaya Vishisht Vaigyanik Puraskar was instituted in 1992-93. The scheme provides every year for two awards worth `20,000 each.

The Pitambar Pant National Environment Fellowship Award instituted in 1978, is awarded annually to recognise, encourage and support excellence in any branch of research related to environmental science. In recognition of the Bishnoi community's contribution to nature conservation and to encourage studies on desert ecology, the Ministry has instituted a Desert Ecology Fellowship at the University of Jodhpur.

Instituted in 1995, the B. P. Pal National Environment Fellowship Award for Biodiversity is in recognition of significantly important research and development contributions and is intended to encourage talented individuals to devote themselves whole time to R&D pursuits in the field of bio-diversity.

The Rajiv Gandhi National Wildlife Conservation Award and Salim Ali/ Kailash Sankhla Fellowships were instituted recently by the Ministry for recognising eminent officers and field workers for exemplary work in the field of Wildlife Conservation and Research. Award in the name of Amrita Devi Vishnoi has been instituted for showing valour and courage in protection of wildlife.

ENVIRONMENTAL LEGISLATION

Major legislations directly dealing with the protection of environment are: the Wildlife (Protection) Act, 1972, the Forest (Conservation) Act, 1980, the Water (Prevention and Control of Pollution) Act, 1974, the Water (Cess)t, 1977, the National Environment Appellate Authority Act, 1977, the Air (Prevention and Control of Pollution) Act, 1981, the Environment (Protection) Act, 1986, the Public Liability Insurance Act, 1991 and the National Environment Tribunal Act, 1995. The Wildlife (Protection) Act, 1972 provides for rational and modern wildlife management while the Forest (Conservation) Act, 1980 has been enacted to check indiscriminate deforestation/diversion of forest land for non-forest purposes. The Water and Air Acts are the major instruments for the control of water and air pollution and these have provided for the establishment of the Central and State Pollution Control Boards.

The Environment (Protection) Act, 1986 is a landmark legislation which provides for a single focus in the country for the protection of environment and aims at plugging the loopholes in the existing legislation. The Public Liability Insurance Act, 1991, provides for mandatory insurance for the purpose of providing immediate relief to the persons affected by accidents occurring while handling any hazardous substances. The National Environment Tribunal Act, 1995, seeks to constitute a Tribunal with Benches to award compensation for damage to persons, property and the environment arising out of any activity involving hazardous substances. The National Environment Appellate Authority Act, 1977 provides for the establishment of a National Environment Appellate Authority to hear appeals against environmental clearance given by the Ministry.

All these Acts are amended from time to time to rationalise and expand their scope, coverage and penal provisions. The Ministry is preparing a comprehensive

National Policy on Environment. The Water (Prevention and Control of Pollution) Cess Act, 1977 is also being amended.

INTERNATIONAL COOPERATION

The Ministry of Environment and Forests functions as a nodal agency for United Nations Environment Programmes (UNEP), South Asia Cooperation Environment Programmes (SACEP), and International Centre for Integrated Mountain and Development (ICIMOD), International Union for Conservation of Nature and Natural Resources (IUCN) and various international agencies, regional bodies and multilateral institutions.

- India is signatory to the following important international treaties/ agreements in the field of environment:
- International Convention for the regulation of Whaling
- International Plant Protection Convention tarctic Treaty
- Convention on Wetlands of international importance
- Convention on International trade in Endangered Species of Wild Flora and Fauna
- Protocol of 1978 relating to the international convention for the prevention of pollution from ships
- Vienna Convention for the protection of the Ozone Layer
- Convention on Migratory Species
- Basel Convention on Trans-boundary movement of hazardous substances
- Framework Convention on Climate Change
- Convention on conservation of bio-diversity
- Montreal Protocol on the substances that deplete the ozone layer and International Convention for Combating Desertification

The Ministry and its agencies receive assistance from various countries such as Sweden, Netherlands, Norway, Denmark, Australia, U.K., U.S.A, Canada, Japan, FRG, etc. , on bilateral basis and from several UN and other multilateral agencies such as UNDP, World Bank, Asian Development Bank, OECF (Japan) and ODA (U.K.) for various environmental and forestry projects.

OZONE CELL

Global efforts to protect the ozone layer started in the early seventies leading to the adoption of the Vienna Convention on Ozone Depleting Substances (ODS) in 1985 and the Montreal Protocol in 1987. India acceded to the Montreal Protocol, along with its London Amendment in 1992. To meet the country's commitment on ODS phase-out under the protocol and to disseminate information on ozone and ODS, the Ministry has established an Ozone Cell.

A newsletter on ozone issues is being published by the Cell every two months and a number of workshops and seminars are also conducted to create awareness about ozone among industries. UNEP-IE Ozone Action Programmes, Paris held its first south-Asia ODS officers Network Meeting in Delhi. Swiss Development Corporation organised a strategy workshop on service skill development in ecological refrigeration. The Ministry provides custom/excise duty exemption for ODS phase-out projects and detailed guidelines/procedures have been finalised to grant duty exemption for new investments with non-ODS technologies. A policy to issue licences for import of ODS has been implemented and the Reserve Bank of India, on the recommendation of the Ministry, has issued instructions to all commercial banks prohibiting finance or refinance of new investments with ODS technologies. One hundred seventy seven projects worth about US $ 49 million have been approved for India by the Multilateral Fund under the convention.

Trade in ODS with non-Parties has been banned. Imports and exports of ODS have been licensed. Exports of CFCs to developed countries has been stopped. Draft rules on Ozone Depleting Substances (ODS) phaseout were published on 9 May 1998 in the Gazette of India. Policy to issue licences for imports of ODS has been implemented.

CHRONOLOGICAL HIGHLIGHTS

- 1951 National Forest Policy of 1894 replaced Van Mahotsava (national tree planting festival launched)
- 1952 Indian Wildlife Board Constituted
- 1953 Central Soil Conservation Board set up
- 1962 Animal Welfare Board of India established
- 1965 Central Forestry Commission set up to study the implementation of the National Forest Policy by State Governments and Union Territories
- 1966 Indian Forest Service constituted
- 1973 Project Tiger initiated
- 1980 Separate Department of Environment
- 1981 Forest Survey of India (FSI) set up
- 1983 National Wildlife Action Plan adopted
- 1985 Department of Environment and Forest created Central Ganga Authority set up
- 1987 Indira Gandhi Paryavaran Puraskar instituted

❐

9

Fogs and Clouds

Such ordinary aerological processes as the formation of rain are now understood almost as well as any laboratory experiment. First causes in the sun or elsewhere may still be obscure, but the general earthly mechanism of fogs and clouds and showers is reasonably clear.

MOISTURE VISIBLE AND INVISIBLE

Invisible water vapour is always present in the air, varying in amount from a mere trace up to about four per cent by volume or two and one half per cent by weight (the water vapour being somewhat lighter than dry air). 'Specific humidity'—the actual mass of water vapour present in a given mass of air, measured usually in grams of water vapour per kilogram of air, and varying from less than 1 g/kg in very dry air up to perhaps 20 g/kg in very moist air—is the most conservative and generally useful measure of atmospheric moistures.

For some purposes, though, dryness or moistness of air is more conveniently expressed as 'relative humidity'—the ratio of the actual amount of water vapour present (the existing specific humidity) to the greatest possible amount of water vapour that could exist at the existing temperature (the saturation specific humidity).

And for many practical weather purposes, the best measure of atmospheric moisture is the 'dewpoint'-that temperature which has a saturation specific humidity equal to the present, existing specific humidity. Finally, we saw how any one of three conditions indicates that condensation of water vapour into liquid form is imminent:

- The nearness of the existing specific humidity to the saturation specific humidity;

- The nearness of the dewpoint to the temperature;
- The nearness of the relative humidity to one hundred per cent.

When water vapour in the air comes out of its invisible form into the visible form of liquid water, however finely divided, it must have some solid nucleus on which to form. Now, on a summer night, say, if the lower air is quite still as the dark earth radiates away its heat, the cooling ground cools only that layer of air immediately next to it.

When the temperature of this still, surface air reaches its dewpoint, invisible moisture condenses into liquid dew on sticks and stones, blades of grass, or any other cool and convenient objects—nuclei of condensation, we might call them, albeit large ones. Thus the fairly common idea that 'dew falls' is wrong. If the temperature happens to be below freezing (less than +32°F or 0°C), the invisible water vapour sublimes directly into feathery ice crystals, forming hoar-frost.

Invisible water vapour that condenses into minute droplets in the free air, forming fog or clouds, also needs solid nuclei on which to form. But whereas the nuclei for dew and frost are large and evident, nuclei for fog and cloud droplets are of infinitesimal size. Nevertheless these nuclei—minute particles of sea salt, smoke, hygroscopic dust, and the like—are always present in the troposphere. Given these nuclei, rapid condensation of invisible moisture into minute droplets of liquid water (visible in the aggregate as fog or cloud) begins when, by reason of evaporation, dewpoint increases to existing temperature; or when (as is far more usual), by reason of atmospheric cooling, temperature reaches existing dewpoint. In any case, condensation begins when and where dewpoint and temperature come together, for any reason, at any level in the air ocean.

FOG

Fog can be at best a creation of elfin beauty, is on the average an inconvenience scarcely ever welcomed by honest people, and becomes at worst, in transportation, a great hazard. It may stall land vehicles for hours, force ships along the coast to anchor for days, and even more sorely bedevil airplanes that must keep moving at two or three miles per minute if they are to keep the air at all. In temperate latitudes the gray, creeping fog-mass is usually formed entirely of myriads of minute water droplets, each something like one thousandth of an inch in diameter.

Damp haze, which in its denser forms may approach the murkiness of light fog, can (unlike fog) begin forming when temperature and dewpoint are far apart. This denser form of haze sometimes extends for several thousand feet above the earth's surface, and the well-defined top of the haze layer usually marks the base of a temperature inversion. To an aviator above the haze it appears almost as a level, pearly-white sea brilliantly reflecting the sunlight; only as he descends into the haze do ground patterns begin to loom murkily into view. But at night, lights shine upward clearly through this haze, whereas true fog of any considerable thickness blots them entirely from view. The reason for this is that fog droplets are much larger than damp-haze droplets (though still incredibly small)—large enough, in relation to the wavelength of visible light, to block off almost completely any direct rays. Yet the transition from dense haze to light fog may be gradual, with gradually increasing relative humidity. As dense fog requires both plenty of vaporous moisture to start with, and then considerable cooling, for its formation, it is found mostly in temperate latitudes. In arctic regions, or at high altitudes, fog (or cloud) may be composed of myriads of minute ice-needles—'frost smoke' that glitters and sparkles in the sun. In the tropics fog occurs but rarely, because the temperature contrasts that favour it are usually lacking.

Fog is classified, according to the mode of its formation, into three main types: (1) radiation fog, also called 'land fog' or 'summer fog,' in which air already stagnant on the scene cools until its dewpoint is reached; (2) advection fog or transportation fog, also called 'winter fog' or 'sea fog,' in which air is brought in bodily from outside regions to the scene of its fogginess; and (3) frontal fogs, usually temporary and transient, that occur with the passage of fronts between unlike air masses.

Radiation Fog

Radiation fog is most frequent during the long, clear nights of autumn, though it may occur at any season, particularly in rather stagnant air conditions near the centre of a high-pressure area. The fog sometimes appears to rise from the ground; but actually, the only thing that rises is a zone of temperatures lower than the dewpoint of the air in question. As the ground and the moist air immediately above the ground slowly radiate their warmth away into space, the temperature of this lower air falls, approaching its dewpoint. When the two coincide, fog appears in visible form. As the top of the fog layer itself radiates away heat much faster than dry air, the fog usually continues to deepen until sunrise.

On a flat plain, if the wind is very light (say only one or two miles per hour), the fog layer may be very shallow. This 'ground fog,' perhaps only a few feet thick, results from very gentle mixing in the air immediately above the ground; there is not enough eddying and turbulence to raise the surface coldness higher. But in rolling or mountainous country, even on a calm night, cooled surface air drains into the hollows (as anyone can notice by driving over darkened hill and dale in an automobile, or even walking across broken terrain, with a thermometer). As a result, large pools of valley air may be cooled sufficiently to produce fog 'lakes' many fathoms deep.

If the breeze is somewhat stronger—say two to six miles per hour—the surface coldness is mixed by eddying and turbulence through a considerable layer of the lower air, causing true fog that blots out hillocks and hollows alike. This sort of fog usually begins at or near the surface, working gradually upward, though to an observer on the ground it seems to materialize rather suddenly out of nothing more tangible than a few stray wisps of vapour. The maximum height of the fog depends on many aerological factors, but it is seldom more than a few hundred feet. Temperature in the fog layer usually decreases with altitude at something like the moist adiabatic rate (-2½°/1000 ft); and just above the top of the fog there is usually, if not always, a temperature inversion.

If the breeze is much stronger than six miles an hour or so, eddying and turbulence diffuse the surface coldness through such an extensive layer of overlying air, and bring down so much potential warmth from aloft, that the air temperature never reaches its dewpoint and fog is entirely prevented. The conditions favourable to deep and bothersome radiation fog are, therefore: to begin with a fairly high dewpoint for the season (plenty of actual water present in vaporous form, that is), and not much difference (say less than 15°F at sunset) between temperature and dewpoint (or a corresponding small difference, say less than 7°F, between wet- and dry-bulb temperatures); second, a clear or nearly clear sky, for maximum radiational cooling; and third, through at least part of the night, light breezes registering two to six miles per hour at the surface.

Once begun, radiation fog continues to deepen through the hours of darkness and, curiously enough, may increase for a brief time after sunrise. This last effect may be due to either of two factors: the power of sunlight to make floating particles attract water; or the increase of lower-level turbulence and mixing caused by solar heating at the surface. It is a rather frightening experience to take off from a clear airport at dawn, and see it fog in under you a few minutes later.

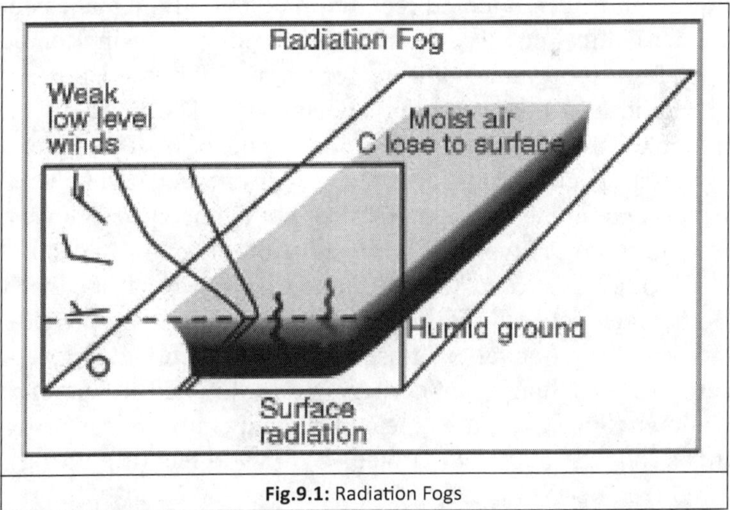

Fig.9.1: Radiation Fogs

Within a few hours after sunrise, however, an ordinary radiation fog begins to 'burn off.' This burning off always proceeds from the ground upward as the solar radiation, penetrating through the fog layer, warms first the ground itself and then, progressively, the air above it. The lower fog does not actually 'lift' in this case any more than it 'rises' out of the ground at night; what actually lifts at first under the morning sun is a zone of temperatures higher than the dewpoint. But before long rising currents from the warmed ground break through the fog layer here and there as the solar radiation continues. And the sun, shining through these breaks with redoubled energy, soon produces enough warm updrafts to raise the whole fog layer slowly upward and sweep it into scattered drifts of cloud.

The worst radiation fogs are found around great cities, where innumerable smoke, oil, and soot particles impair visibility on their own account, serve as ready nuclei for condensation, and hinder the evaporation of fog once formed. Thus the infamous London fogs, which greatly increased during the late nineties with the growth of the great city, may hang on in yellowish 'pea-soup' form, perhaps for days, until the advent of a new air mass and freshening winds.

Though less tenacious, radiation fogs are nearly as common in many regions of the open country. They can also occur over limited areas of calm ocean, but the usual sea fog arises out of different causes. An extreme example of radiation fog occurs in the Great Valley of California, where murky-gray obscurity may hang on for days beneath the 'lid' of a temperature inversion.

A particular type of radiation fog, called 'high fog' or 'stratus fog,' starts in mid-air and builds downward. This high fog, also, is particularly prevalent in the walled-in valleys of California, but I have noticed it occasionally over the eastern seaboard, and it probably occurs widely, though rather rarely, in other regions.

Once we had a formation of army airplanes operating all night from an air base in the District of Columbia. This high fog formed suddenly, around midnight, at a height of about fifteen hundred feet, trapping one of the planes above it. The predisposing aerological condition here is a zone of decreasing temperature and high humidity from the ground upward for the first thousand feet or so, with a pronounced and lid-like temperature inversion above. The air below the inversion is thoroughly mixed, equally moist, and coldest at the base of the inversion in mid-air. In other words, specific humidity is fairly constant (dewpoint decreasing about 1°F/1000 ft) throughout the moist lower layer, but temperature is lowest at the top.

As the whole moist layer cools by radiation at night, the top temperature reaches its dewpoint first, and the fog builds downward from the base of the nversion. Sometimes you can see it beginning in wisps along the side of a steep mountain. Its downward advance is usually rather sudden at first—perhaps five hundred feet or so in an hour; then it lowers more slowly. But within a few hours, it may reach down to the ground. Some typical temperatures, in a California fog of this sort, are as follows: 600 feet (surface), 57°F; 900 feet (top of fog), 54°; 1200 feet, 64°; 1500 feet, 66°.

Advection Fog

Whereas radiation fog is due to simple cooling of the air in place, advection fog may be due to either of two causes. One is the horizontal advance of warm, moist air over a cold land or water surface. The other is the advance of cold air over a moist, warm surface. Of these two, the first cause is by long odds the most common.

On land, a warm-air advection fog often results when gentle, moist southerly winds blow in from the ocean, particularly if the land surface is covered with snow. The warm, moist air is chilled by the cold ground below it until its falling tempera- ture, uniformly distributed through several hundred feet of depth by gentle turbulence, reaches its dewpoint. In the United States any large northward outbreak of warm, moist (tropical maritime) air in wintertime may produce widespread advection fog.

The warm sector of a marked temperate cyclone may be completely blanketed by such a fog, and if the pressure gradients are steep, this may be one of the rather exceptional cases where fog accompanies a high wind, usually from some southerly quarter. These warm-air advection fogs are far less common on land than radiation fogs; but the advection fogs are likely to be more widespread and prolonged, forming or advancing at any hour of the day or night.

On the ocean, warm-air advection fogs are by far the most common type, and more frequent in summer than in winter. They are likely to form wherever a cold ocean current or area is surrounded by considerably warmer water—such as off Newfoundland, where the Labrador current projects southward; off Nantucket, where the colder water from the northern New England coast meets the warmer waters from south of Cape Cod; or along the whole line of the Pacific coast, where water upwelling inshore from chill depths below is much colder than the Pacific waters offshore.

The cold-water and fog area may be small and distinct—such as the five-mile-wide cold current in the middle of Santa Barbara Channel. The whole fog area moves with the prevailing winds, which are usually light -say fifteen miles per hour or less on the open water—though in a great temperate cyclone they may be much higher.

I remember one case of moderately thick fog on a steamship off the Atlantic coast, with the wind blowing a whole gale out of the southeast and a mountainous sea running. If the wind is much above twenty miles per hour, however, the minute fog droplets tend to coalesce into the larger droplets of mist or drizzle, with somewhat improved visibility.

These sea fogs vary widely in height—sometimes they are so shallow that the mastheads of ships project into clear air above. At other times they are more than a thousand feet deep. Their average depth is perhaps around five hundred feet. In any actual case, of course, this height can be determined by a kite or captive-balloon sounding from a ship in the fog—or from shore elevations along a mountainous coast like California. Sometimes the sea murk may take the form of 'high fog' or low stratus, with good visibility along the water surface.

The second type of advection fog occurs when cold air flows over a moist, relatively warm surface such as ocean or lake water, or possibly land wet by recent rain or melting snow. Here the invisible water vapour, evaporating from the water surface, is chilled below its dewpoint by the cold air above so that it visibly 'steams.' On land, the steaming of lakes on autumn mornings, as radiation-cooled land air sweeps across water areas less prodigal with their summertime heat, shows the mechanism of cold-air advection fogs in miniature form. Similar steaming fogs occur in arctic regions when cold, belowzero polar air sweeps across leads of open water in the pack ice; though here, of course, the fog is formed of ice crystals.

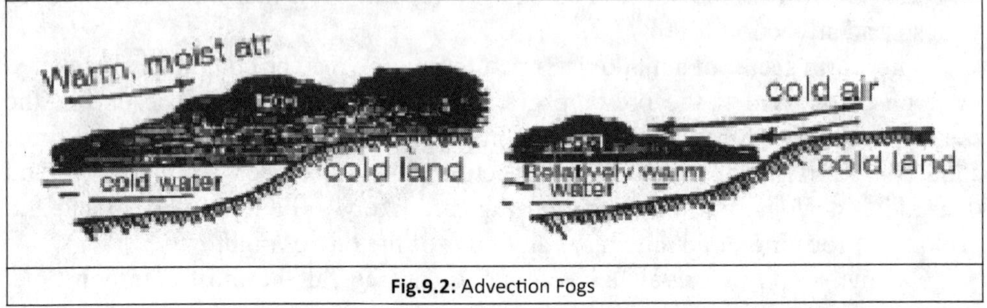

Fig.9.2: Advection Fogs

Cold-air advection fog on the ocean is quite rare, but notably hazardous when it does occur. Near the United States it is confined mostly to the cast coast, and arises from the rapid southward and eastward outbreak of a large, cold (polarcontinental) air mass. When the fresh polar air, at near-zero temperature, rushes southward or southeastward behind a marked cold front over water at temperatures around 50°F (or even 70° in the Gulf Stream), steaming fog arises on a widespread scale and perhaps masthead high. What with the viewless murk, the piercing cold, and a howling gale out of the north or northwest, even a steel steamship may do well to anchor inshore, dcspite every modern aid to navigation.

Frontal Fogs

Somewhat different in mechanism from radiation fogs and advection fogs are the fogs that occur near the marked 'fronts' between unlike air masses. These frontal and pre-frontal fogs, however, are neither very common nor very persistent, and only two varieties of them are worth mentioning here.

First there is a type of pre-frontal fog that occurs in advance of a warm front, in a zone perhaps fifty or a hundred miles wide where heavy rain from the warm-front sloping cloud-surface above has saturated the cooler air of the underlying cold wedge, and falling pressure ahead of the front causes further expansional cooling of the cold, moist air. (This pre-frontal fog is distinct, of course, from the warm air mass [advection-type] fog sometimes present in the warm sector behind the warm front.) The other type of pre-frontal fog may occur a few miles ahead of an advancing cold front, where a rapid fall in pressure encourages the fog-forming proclivities of the warm air mass.

Any fog, however formed, may drift bodily elsewhere and engulf regions well removed from its origin, provided aerological conditions do not favour its rapid evaporation. The front of the fog, particularly at sea, may take the well-defined form of a steeply sloping 'bank,' with a distinct and clear-cut edge. When a sea fog comes ashore, it remains on the ground if the land is colder than the water. If the land is relatively warmer, however, the lower surface of the fog lifts somewhat above the ground.

FORMATION OF CLOUDS

Fogs and clouds are of the same texture, and their different appearances are mainly a matter of viewpoint. High cloudturrets may gleam as white, as apparently solid, as alabaster in the bright upper sunlight; but explore the insides of those seeming-solid vapour-peaks in an airplane, and they become just as ragged, gray-murky, cold, and damp as any fog. Having seen how cooling of humid air near the ground produces fog, we may well guess that cooling of some sort in the upper air produces cloud.

How is this large-scale free-air cooling brought about? Not by conduction (for air is a very poor conductor of heat), not usually by radiation (for dry air neither absorbs nor radiates heat readily), but mainly by the expansion cooling of air forced in some way to rise. In the preceding we examined the underlying mechanismofthisrising-refrigerationprocess we followed the salient facts of dry-adiabatic cooling, of moist-adiabatic cooling, of dry-air and moist-air stability and instability. It remains now to see how these various aerological conditions play their part in forming clouds out of thin air.

Forced Cloud Formation

If an entire layer of moist air is driven by prevailing winds up the slope of a mountain range, or drifts up the inclined plane of an underlying wedge of colder air, it cools throughout its whole extent at something like the dry adiabatic rate (about -5°F/1000 ft—or 5.5°, to be exact) so long as moisture remains entirely and invisibly vaporous. But at some moderate altitude, if its humidity is high, the rising air layer reaches its dewpoint. Then moisture begins to condense on myriads of infinitesimal nuclei that are usually present in air—salt particles, haze particles, and so on—into the form of innumerable small water droplets visible in the aggregate as cloud. The individual droplets forming the cloud are about the same size as fog droplets or somewhat smaller—say 1/2500 of an inch in diameter. After this cloud condensation begins, of course, released latent heat retards the further cooling of the rising air to the moist adiabatic rate (say -3°F/1000 ft at moderate altitudes).

If the moist, rising air layer is ill-disposed towards updrafts and downdrafts within itself—if, in a word, it is stable—it forms merely one great, shapeless horizontal mass of flat, quiescent, layer-like cloud in the same way that a groundcooled air layer forms fog. A large proportion of all clouds are formed, as simply as fog, within such moist but stable air layers lifted in some way by external causes. If all clouds were so formed, the science of aerology would be much simplified. But more complicated mechanisms of cloud formation produce some of the most splendid clouds and spectacular storms.

Free Cloud Formation

It is worth recalling from the stability in the preceding that, in 'absolutely unstable' air (lapse rate steeper than about -5°F/1000 ft), vertical currents tend to continue and to grow, once they are triggered off by commonplace causes such as surface heating, mountain updrafts, and so on. The same may be true of 'conditionally unstable' air (lapse rate between -5° and -21/2°F/1000 ft), which behaves like absolutely unstable air once condensation has begun. By the 'beginning of condensation,' of course, we mean the formation of the minute water droplets that constitute visible cloud. The ultimate result of instability in any air layer, dry or moist, is updrafts (together with compensating downdrafts elsewhere). And the result of updrafts, in any moist air, is bound to be clouds.

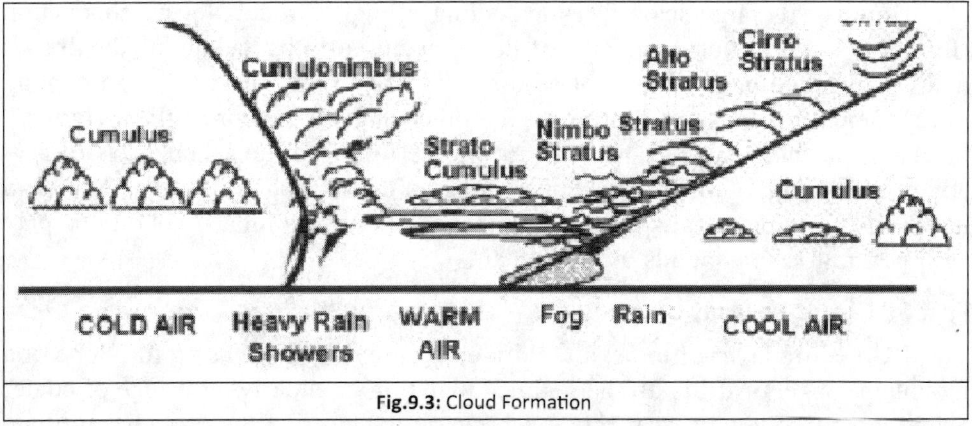

Fig.9.3: Cloud Formation

Unlike the flat, quiescent layer-like clouds formed by general air lifting under external forces, these cloud-children of instability are usually lumpy, heaped-up, puffy, and cauliflower-like in appearance. The instability clouds themselves may be very large, occasionally towering higher than the highest mountains, yet they tend towards localized lumps, turrets, and masses, called 'cumulus,' rather than widespread and shapeless layers, called 'stratus.'

In freely rising air, instability cloud-birth always begins at a certain definite level—the condensation level where falling temperature meets existing dewpoint. Under some conditions this condensation level is very definite, and readily apparent as a flat cloud base or 'ceiling.' The typical woolly, cream-puff clouds of fair-weather summer skies, for example, all have flat bases (and not the rounded bases that some artists like to paint into them); and these bases are actually all at the same elevation, though more distant clouds may appear lower to ground observers by reason of perspective. Each detached, puffy cloud in this idealized summer picture represents, of course, the visible crown of an updraft (invisible below the condensation level) that is sustaining the cloud. In the clear spaces between these detached woolpacks there are usually downdrafts more gentle and more extensive than the updrafts that they balance.

There is a rough but useful relationship between 'ceiling,' or height of cloud base, and surface temperature and dewpoint. At moderate temperatures, each degree of temperature-dewpoint difference means about two hundred and fifty feet of ceiling. Thus if the surface temperature is 70°F and the surface dewpoint 66°F, the ceiling should be around one thousand feet.

Occasionally the mechanism of updraft-builds-cloud is revealed in detached and clearly apparent form by a large fire. The smoke from the fire rises, of course, along with a great column of heated air. Just as the smoke begins to thin out, perhaps, at some moderate elevation, a new and thicker turruted murkiness appears and builds upward. This upper murkiness is not smoke, but cloud formed by the updraft, and the elevation where it appears is the condensation level. If a large and deep air layer is sufficiently unstable and sufficiently moist—conditions that exist in tropical or summertime thunderstorm weather—both the updraft and the consequent instability cloud may grow out of all proportion until the final result is vertical currents ballooning upward at possibly one or two hundred miles an hour, and a towering cloud mountain as high and wide as Everest.

For the wholesale condensation of water vapour into cloud droplets releases wholesale latent heat, which intensifies the updraft, which condenses still more moisture and releases still more latent heat, and so on. You can sometimes watch these thunderheads visibly growing against the blue sky—the faster they grow, the stronger the updraft and the more unstable the air to begin with. The upward momentum of rising air also helps in this cloud-building, which is ultimately checked, in extreme cases, by the great upper temperature inversion that is the stratosphere.

Instability clouds are not limited to hot weather. It is all a matter of relative warmth and coldness alow and aloft—of lapsed rate, provided enough moisture for cloud-building is available. A strong wintertime outbreak of frigid polar air over a relatively warm water surface (say along the Atlantic coast of the United States, or across the Great Lakes) can produce as puffy and squally clouds as anyone could wish. And in winter, with the sun low in the sky, these instability clouds and snow squalls look blacker and more forbidding than most summer storms.

Combined Cloud Formation

Not all clouds are distinctly of layer-like, quiescent stratus type or indisputably of puffy, instability cumulus type. It is possible to have both types present in the same section of sky, perhaps at different levels. Hybrid or combination types, such as strato-cumulus, are also possible.

The kind and prodigality of cloud-building depends on the relative strength and interplay of the three great cloud-building forces—moisture, instability, and external lifting. Any moist air layer bordering on instability tends to become more unstable, the more it is lifted by external forces. Such a layer, externally lifted, may at first form quiescent stratus clouds which, as the lifting goes on, may bulge upward here and there into growing instability cumulus turrets. Nevertheless, every cloud that we see in the sky must result from some chance combination of the three cloud-building forces. The more unstable the air is, the more moist it is, and the more violently it is lifted by external forces, the greater will be the bad-weather effects in clouds, rain, snow, hail, and the like.

We have said earlier that radiation is usually an unimportant factor in cloud formation. But moist air absorbs and radiates heat better than dry air, and it is possible for a high, moist layer (having a relative humidity close to one hundred per cent) to radiate away enough heat at night so that it forms a layer of high stratus cloud. Possible—but not common.

This ever-changing interplay of cloud-building forces in Nature's air ocean produces what, at first sight, seems to be an endless and bewildering variety of cloud forms. But these cloud forms (or the more common ones, at any rate) can be grouped into a few types and classes that are quite easy to remember and identify. The most basic cloud distinction of all, between layer-like stratus and heaped-up cumulus, has indeed already been made.

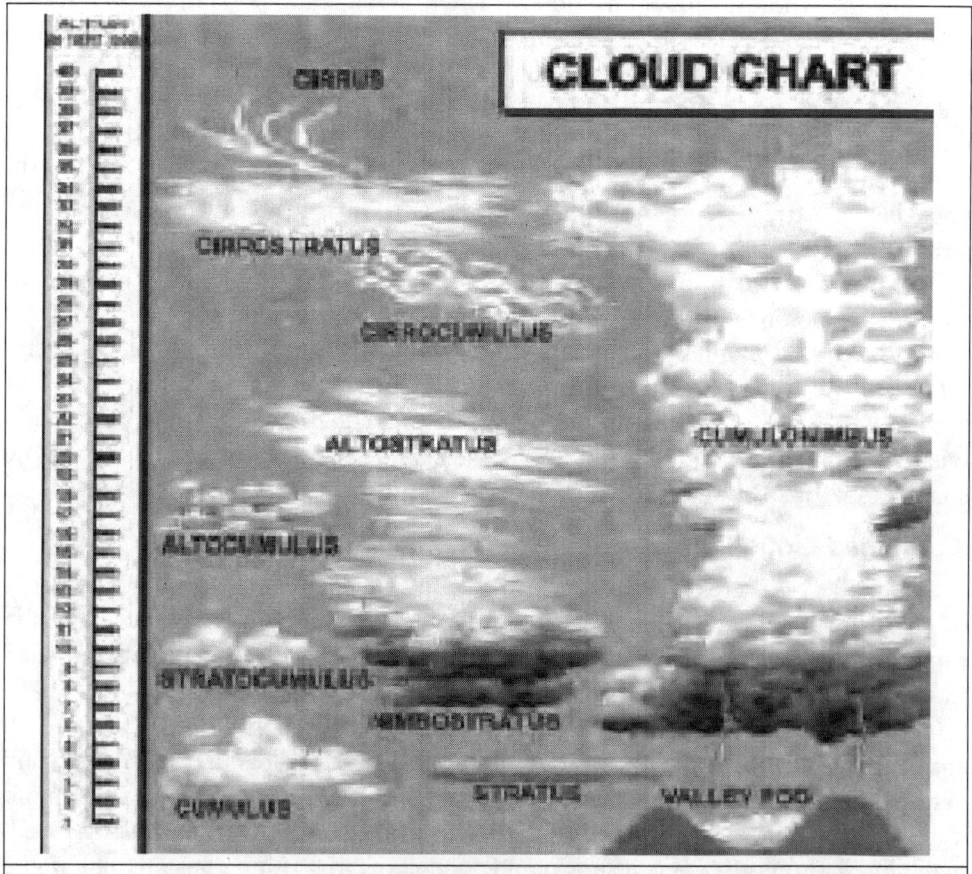

Fig.9.4: Clouds by Type and Altitude

PARADE OF THE CLOUDS

The first attempt at naming clouds systematically and completely was made by Luke Howard, an Englishman, in 1803. He classified all clouds by shape or appearance into three general types—cirrus or 'curl' clouds, cumulus or 'heap' clouds, and stratus or 'layer' clouds. Other types of clouds, such as nimbus or rain clouds, Howard associated in some way with his three basic types. During the nineteenth century other cloud classifications were proposed, but Howard's scheme, from time to time elaborated and improved, gradually grew into a standard system increasingly accepted throughout the world. We are concerned only with the commoner clouds, such as any amateur can learn to name and recognize. And most amateur difficulties in the past have come from too-exhaustive or illogical systems of classification.

So we shall keep our cloud charts very simple, and as logical as possible. The most fundamental cloud differentiation, already hinted at in the preceding section, is based, like Howard's, on shape or form:

CLOUD CLASSIFICATION BY FORM OR SHAPES

Two Fundamental Shop, Classes

Layer class (from stable air)	Stratus (= 'a spreading out') Cumulus	A quiescent cloud layer, rather flat and formless, extending horizontally for the most part.
Puffy class (from unstable air)	(='a heap')	A piled-up heap of cloud. usually having definite three dimensional shape, often growing vertically.

Two Special Shape Types

Wispy type	Cirrus (= 'curl')	White, fibrous mares ' tails or feathery curls at very high altitude.
Broken type	Such as Fracto-stratus Fracto-cumulus	Heat-evapoorated or windom into separate smaller cloud masses with ragged, wispy edges,

Mixed and Storm Typer

Combination type	Strato-cumulus	Combining some charicteristics of two different classes
Precipitation types Nimbus (= 'rainstorm') Clouds of strafform or cumulo-nimbusfrom which rain, snow, or ice is actually failing.		
Hyphens in cloud names have been dropped by professional serologists, but they are retained here for the sake of simplicity and clearness,		

Another way of classifying clouds, and perhaps a better one for understanding how they fit into the pageant of tropospheric weather, depends on their height above the earth's surface -or, in a more general sense, their altitude above sea level. Beginning with the highest and thinnest clouds, and ending with the lowest and thickest, the more common forms are:

TABLE: Cloud Classification By Altitude

Upper or citrus-like types	Cirrus Cirro-stratus Cirro-cumulus	Usually occur in the upper troposphere, often not fair below the stratosphere, at heights of about 4 to 7 miles (6-10km).
Middle or high ('ralto') types	Alto-stratus Alto-cumulus	Usually occur in the middle torposphere at heights of 2 to 4 miles (3-6 km).
Lower types	Stratus Cumulus	Usually occur in the lower troposphere, from the surface up to a height of nearly 2 miles (3km.).
Verticle Growth Type	Strato-cumulus Nimbo-stratus Cumulo-nimbus (thunderstorm)	Single cloud mass may extend from near the surface up to heights of 4 to 7 miles.

This table shows at a glance how various types of clouds fit into the general picture of the troposphere. From it one can appreciate why towering cumulo-nimbus clouds can be seen hundreds of miles away in clear air; one can sense why a spreading high deck of cirro-cumulus stands out in the red light of a clear sunset for perhaps three hundred miles to westward, where the sun-rays, receding at ten miles per minute, still light the cloud deck's edges half an hour after they have left the ceiling directly overhead.

These high decks of upper clouds (cirro-cumulus or alto-cumulus), together with clear lower air to westward, are the one unfailing recipe for sunset splendor—were long before Meredith, and always will be: 'Ay, but shows the southwest a ripple-feathered bosom, Blown to silver while the clouds are shaken and ascend, Scaling the mid-heavens as they stream, there comes a sunset Rich, deep like love in beauty without end.'

What do all these cloud forms really look like, and how do they arise out of the vast emptiness of the air ocean? The cloud photographs. And in further elaboration, we review briefly the various outstanding peculiarities of each cloud form in turn:

CIRRUS

True cirrus clouds are fibrous and delicate in appearance, showing as white wisps against the upper blue. Sometimes they arc combed out into mares' tails; sometimes they are formed more thickly into curls on feathery plumes or parallel 'plowed-field' bands. But always they are the most tenuous clouds of the sky, scarcely dimming the sun's disk, and the highest, floating at five to seven miles above the earth (8 to 11 kilometers). As the temperatures at these levels are always far below zero, cirrus clouds are formed from minute, free-floating crystals of ice (simple hexagonal prisms). In fair weather they may result from general lifting of the upper troposphere above the active, 'cold-dome' portion of a high-pressure area. As foul-weather harbingers they are 'locks of the approaching storm,' blown out by substratospheric gales far in advance of the general cloud area of a cyclonic warm front.

Sometimes broken or wispy clouds fracto-stratus or (fractocumulus) in the lower levels have a superficial resemblance to cirrus clouds, and are mistaken for them by laymen. But the true cirrus clouds, carefully watched, are very evidently much higher than these fracto-forms.

Cirrus clouds are always interesting to watch because they show the motion of air near the top of the troposphere. Sometimes the long, fibrous tails seem to radiate from a cyclone centre below the horizon, and actually are combed out along the direction of air motion by the upper winds. At other times cirrus plumes or rays can be seen to move, not along their length but across it, like successive ranks of marching soldiers. Here the cirrus plumes make visible a colossal wave motion in the substratosphere—white crests of upsurging condensation and blue troughs of downdropping evaporation—great, silent aerial surges that may measure ten or twenty miles from crest to crest. Some cirrus clouds, again, are probably nothing more than evaporating wisps of what was originally cirro-stratus -a sort of 'fracto-cirro-stratus,' one might say. After sunset or before sunrise, cirrus clouds often glow with beautiful tints of flame-yellow and rose-pink.

Cirro-stratus

Cirrus-like cloud smoothed into one continuous, thin layer, resembling a whitish veil, is called ciffo-stratus. It may be smooth and milky, or again uneven and somewhat fibrous, forming out of thickening cirrus. Cirro-stratus never casts shadows, nor does it greatly dim either the sun or the moon, though in thickening and lowering form (approaching alto-stratus) it begins to blur their disks somewhat. Cirro-stratus is almost as high as cirrus (3½ to 7 miles, or 5½ to 11 kilometers), and is likewise formed of minute hexagonal ice crystals. Inall citrus-like cloud-forms, but most noticeably in cirro-stratus, these ice crystals may refract and reflect light in such a way as to form halos or luminous rings around the sun or the moon. The most common halo is a single ring of about twenty-two degrees radius, coloured faintly if at all; but one or even two additional outer rings may be seen occasionally, together with rarer 'mock suns,' 'sun pillars,' tangent arcs, and the like. ('Coronas,' or rare, coloured sun-andmoon rings that may superficially resemble halos, are usually caused by water-droplet clouds at much lower altitudes.) Cirro-stratus cloud is most frequently formed in the upper and outer warm-front portions of a cyclonic cloud area, thickening to alto-stratus towards the storm centre, thinning out to true cirrus wisps in advance. Hence while quiescent detached cirrus is a frequent fair-weather cloud, cirrus thickening into a cirro-stratus veil is a more definite portent of possible cyclonic approach. The portent is still more definite if cirro-stratus thickens to alto-stratus—but by that time, the rain is not far behind.

Cirro-Cumulus

Cirro-cumulus clouds are small, fleecy, white flakes or very small cloud balls—a 'mackerel sky'—perhaps merging into cirrus or cirro-stratus, and composed of supercooled droplets or ice crystals floating at great heights (3½ to 5½ miles or 5½ to 9 kilometers). True cirro-cumulus clouds are not very common, and are easily confused with thinning forms of alto-cumulus—which latter are always well below the true citrus-like cloud levels. The distinguishing feature is, however, that in cirro-cumulus the individual flakes usually appear to be smaller by reason of their great height, and they are never thick enough to cast shadows or show dark undersides.

Cirro-cumulus clouds may sometimes be formed by small up-and-down currents along a wide but shallow unstable layer in the upper air. But more often they are arranged, like some types of cirrus, in parallel rows or lines, showing wave-motion between overrunning and underrunning air currents somewhat similar to ocean waves along the sea surface where water meets differently moving air. All these high, cirriform clouds can best be seen, incidentally, through dark (preferably purplish) glasses; and they can best be photographed, like most other clouds, through red filters.

Alto-Stratus

This cloud usually takes the form of a thick grayish or bluish veil, always somewhat fibrous in texture. It is thicker and less translucent than cirro-stratus, which often merges gradually into it with the approach of a cyclonic storm. Through typical alto-stratus the sun and the moon gleam only weakly and vaguely, as through ground glass; they have a watery and diffused look, or may be almost wholly obscured, usually casting no shadows. Unlike cirriform clouds, altostratus is composed mostly of minute water droplets plus ice crystals scattered here and there.

The usual height of alto-stratus is 2 to 3½ miles (3 to 6 kilometers). It occurs most frequently along the slightly sloping surface of a warm front, where the overrunning warmer, moister air, ascending the inclined plane of the cold-air wedge, has lost by condensation a good deal of its original warmth and moisture. This frontal alto-stratus often marks a zone where light rain or snow is just beginning to fall from the high clouds, though this outermost precipitation (except cold-weather snow) may evaporate again before reaching the ground. Aside from the typical warm-front cloud-sheet, local alto-stratus of a more level and quiescent type can occasionally result from the radiational cooling of a high and very humid air layer.

Alto-cumulus

Alto-cumulus is quite similar in general appearance to cirro-cumulus, but the individual flakes or cloud-balls of alto-cumulus are always larger and grayer. Usually they cast shadows and show dark undersides. The lower the alto-cumulus is, the larger are its opaque lumps. Certain forms of altocumulus ('castellatus') display little cloud-turrets growing upward from the main moisture-supplying layer.

In common with alto-stratus and most of the lower clouds, alto-cumulus is formed of minute water droplets, which may, however, be sub-cooled well below freezing in the liquid state. The usual height of alto-cumulus is 2 to 3 miles (3 to 5 kilometers). Like cirro-cumulus, alto-cumulus is often arranged in parallel rows along a wide and level deck, resulting from wave motion between differently moving layers of air. Alto-cumulus usually results from upper-air wave motion of some sort, or instability; it may grow into alto-stratus or nimbo-stratus.

Stratus

Any flat, shapeless, more or less uniform layer of cloud in the lower levels, usually dull gray in colour when viewed from below and whitish when viewed from above, is called stratus. Stratus resting on the ground is simply fog, formed and dissipated in the same way as fog; and high fog is simply low stratus. More usually, however, stratus occurs at heights of 1500 to 3000 feet (500 to 1000 meters). Low stratus is often a local cloud, gradually breaking away into shreds and smaller masses called fracto-stratus or 'scud.' Stratus can produce a fine drizzle, but not appreciable rain or snow.

CUMULUS

Cumulus clouds are thick, dense, detached 'woolpacks' or 'puffballs,' usually as high as they are wide. Their bases, as we have seen, are usually flat and uniformly fixed at the condensation level, but their upper parts are dome-shaped and variously bulbed and rounded, like cauliflowers. Their colour ranges through every nuance of sunlight and shadow from dazzling white above to dark gray below.

It will bear repeating that a detached cumulus cloud always marks the top of a rising air current, which perhaps originated in the first place from an overheated portion of the earth's surface, or from some updraft in the general air flow. In any growing cumulus cloud conditional instability is being realized and condensation of moisture is contributing its latent heat to make the updrafts more violent. If moisture is plentiful and the instability extends upward to great heights, the cumulus grows into a cumulo-nimbus, or thunderstorm. But if an ordinary growing cumulus bumps into a very stable layer such as a deep temperature inversion, its top flattens and its growth stops.

Most ordinary fair-weather cumulus clouds, formed under clear skies at levels between 2000 and 8000 feet (600 and 2500 meters), wax and wane daily with the sun (or with surface temperature). Foul-weather cumulus clouds, on the other hand, occur at any time of day along cold fronts and within moist, unstable air masses. A torn, immature, or partly evaporated cumulus is sometimes called fracto-cumulus. Sometimes cumulus clouds on the lee side of a mountain range evaporate in the dynamically heated downdraft so rapidly that they seem to disappear as if by magic.

Strato-cumulus

Strato-cumulus usually occurs in the form of a wide and fairly level layer marked by thick rolls and dark, rounded masses of cloud. It somewhat esembles alto-cumulus, but its cloud masses are usually larger, thicker, and darker. Sometimes there are small or large open breaks between the rolls and masses; sometimes the layer is continuous but steeply undulating, especially along its upper surface as seen from an airplane. Strato-cumulus clouds usually occur in the same levels as fair-weather cumulus—say 3000 to 8000 feet (10002500 meters)—and may be formed by the flattening of cumulus clouds, instability in a stratus layer, or wave motion between two differently flowing layers of air. Strato-cumulus can change into, or develop from, stratus.

This cloud type, formerly called plain 'nimbus,' is in general any low, thick, formless, dark-gray cloud layer from which rain, snow or other precipitation is actually falling. Even if the rain momentarily stops, this cloud has a ragged 'rainy' look, and is still called 'nimbo-stratus.' Below the main cloud deck there may be various ragged 'scud' clouds (fracto-stratus or fracto-cumulus) that occasionally merge into nearly solid lower layers or into the main cloud deck above. Nimbo-stratus transmits a feeble illumination that seems to come from within the cloud itself.

Usually nimbo-stratus develops out of alto-stratus and rainformed lower fracto-stratus or 'scud,' marking the near approach of a warm front in the central region of a cyclone. Less frequently, nimbo-stratus may develop out of heavy and threatening strato-cumulus. In either case the nimbo-stratus cloud, at heights of perhaps 1000 to 5000 feet (300 to 1500 meters), results from widespread condensation in air forced to rise by reason of its movement around or towards the centre of a cyclone. Scattered areas of nimbo-stratus may develop from cumulonimbus, as in the rear parts of a thunderstorm.

Cumulo-Nimbus

This is a great cumulus cloud mass in which condensation of moisture, release of latent heat, and increasing updrafts (fed by a deep layer of moist and unstable air) have grown to gigantic proportions. Its colour varies all the way from alabaster whiteness on the upper turrets to dark gray, black, or dark greenish below. The whole cloud mountain may tower from some ordinary cumulus level, such as half a mile, up to a height of five, six or even seven miles.

In a broad sense any large cumulus mass producing a shower can be called cumulonimbus, and this type is the common shower cloud, with or without thunder, of the tropics. But in temperate latitudes, cumulo-nimbus connotes more specifically a large, towering thunderstorm cloud whose top, blown out into anvil shape and cirrus-like texture by substratospheric winds, consists mostly of ice crystals. Hoods of alto-stratus may mark temperature inversions that the growing cloud mountain breaks through.

Special Cloud Forms

One special cloud form worth mentioning is mammato-cumulus or 'rain balls,' whose cumulus heads bulge downward instead of upward from a high cloud deck. It usually appears as a multitude of small, dark downward bulges in a turbulent layer of alto-stratus, alto-cumulus, or stratocumulus. The pendulous cloud pocks are evidently caused by local cold downdrafts perhaps started by detached showers of falling hail or snow. Mammato-cumulus always indicates very disturbed and unstable conditions in the upper air, seldom occurs far from a thunderstorm, and sometimes precedes tornadoes.

Another interesting cloud form is the detached, lens-shaped ovals that are sometimes formed at considerable heights out of quiescent or slowly dissolving cloud masses (alto-stratus 'lenticalus,' and so on). Sometimes the crest cloud on a mountain summit, caused by the steady flow of air up over the peak, takes this lenticular shape. The lens shape is produced, in any case, by symmetrical air flow and selective evaporation.

When high winds are blowing in mountain country, any tall detached peak is likely to fly a 'banner cloud' that streams out, pennant-like, far to leeward. Sometimes the great plume is formed of blowing snow from the peak itself. But if the humidity is just right, a true banner cloud can form on a perfectly bare peak such as Yosemite's Half Dome in summer, caused by the dynamic cooling effect of lowered pressure in the lee of the peak.

RAIN AND SNOW AND ICE

At ordinary temperatures all lower clouds are composed of minute water droplets perhaps 1/2500 of an inch across. Left to themselves in still air, these cloud droplets would sink slowly, at a speed less than one mile per hour. Yet as the droplets jostled together, somewhat larger droplets might be formed here and there. From a cloud mass (low stratus or high fog) near the ground in very moist air, these larger droplets, perhaps ten times as large as the original cloud droplets, might sprinkle downward in the form of a gentle drizzle.

But such a drizzle does not even approach the wetting power of real rain. Ordinary raindrops are ten times as big again as the drizzle droplets say 1/25 of an inch across. And rain clouds, considerably less common than ordinary clouds, are distinguished from them by two characteristics: fairly strong rising currents, and considerable depth of cloud. Many rain clouds indeed probably extend, summer or winter, well above the freezing level, and ice particles probably play an important part in their mechanism.

In any case the cloud droplets of any active rain cloud are being borne steadily upward, though not quite as fast as the air surrounding them. In the upper parts of the cloud the droplets are probably farther apart, and somewhat larger by reason of longer and freer condensation. Here and there they are combining into still larger drizzle droplets. Well up in the clouds the rising currents, not very uniform anyway, must somewhere begin to diminish.

As a result the larger droplets begin to fall downward, or at least rise slower than most of the cloud droplets, and in either case begin to sweep through the smaller droplets and the moist air surrounding them, gathering unto themselves more and more moisture as they fall. Finally these growing drops perhaps become large enough to continue earthward, even through moderate rising currents, in the form of rain. Much tropical rain and some temperate-zone rain probably originates in this simple way.

But in temperate and arctic latitudes probably most rain, and perhaps all heavy rain and certainly all snow, evolve originally out of ice particles. These ice particles can sublime directly out of water vapour onto sublimation nuclei (dust particles and the like, always present in the atmosphere) very easily at low temperatures around 5°F (-15°C) or colder; or, perhaps less frequently, they can be frozen out of ascending water (cloud) droplets that are likewise cooled well below freezing and jostled sufficiently by turbulence.

The level at which the ice particles are first formed varies with latitude and the season and the general weather situation from as low as a mile or less in winter to as high as four miles or more in summer. The upper parts of a typical heavy rainstorm, therefore, are mostly iceparticle clouds cirro-stratus, alto-stratus, and the like rather thin-textured clouds with indistinct edges. The lower parts of the same typical rainstorm, at least during its earlier phases, are mostly water-droplet clouds cumulus, stratus, and the like opaque-looking and well defined.

As the ice particles from the upper clouds fall slowly (at perhaps two or three miles an hour) through the lower (waterdroplet) clouds, a curious transfer of moisture occurs. By reason of physical laws that need not be elaborated here, at certain temperatures and humidities water vapour is evaporated from liquid droplets, yet sublimed on solid particles. Thus the falling ice particles constantly gain in bulk at the expense of the water droplets around them, and with increasing size fall gradually faster.

And in addition to their sublimation-growth, they freeze to themselves any subcooled water droplets that they happen to hit. The ultimate form of the ice crystals depends on the humidity, and varies from simple ice needles to skeleton stars (snow) to snow balls (graupel). As these ice crystals fall, the ice-particle cloud itself thickens and lowers, perhaps encroaching on the water-droplet clouds below—or possibly engulfing them entirely, as in cold-weather nimbo-stratus. But after falling below the freezing level (provided it is well up in the air), the ice crystals soon melt into raindrops.

This is the idealized picture. But in a violent rainstorm -and most particularly, in a turbulent thunderstorm—cloud droplets, ice particles, raindrops, and snowflakes may all be milling around in some parts of the cloud.

RAIN

In 'light' rain the drops are very small, averaging perhaps 1/50 of an inch in diameter. Ordinary or 'moderate' raindrops are about twice as big (1/25 inch). Drops of 'heavy' rain are larger still, perhaps 1/16 inch wide. The very largest raindrops in a cloudburst may 1/8 inch or more across. Raindrops splashing on the ground of course look much bigger than the effective spherical sizes just given, and even while falling through the air they are somewhat flattened out.

The largest possible raindrop size is about 1 inch, which falls through still air at about twenty miles per hour; any larger drop, falling faster, would be broken into smaller drops by the wind of its fall. Though all rain results from rising moist air, the gentleness and uniformity or violence and variability of the formative updrafts go far to determine what form the precipitation will take. Rain out of stable (stratus) cloud layers, gently rising, say, along a warm front, is light, steady, and continuous. But rain out of unstable cumulus along a cold front (where updrafts are patchy but powerful) is likely to be heavy, variable, and showery.

In the heaviest cloudbursts rain falls at the rate of three or four inches of solid water per hour. (This excessive rainfall rate cannot continue for long, but even an hour of it may cause floods.) A fall of nine or ten inches in a few hours, by no means rare, brings creeks and rivers raging out of their banks and inundates the countryside. The heaviest rainfall so far ever recorded was at Baguio, Philippine Islands, where forty-six inches—nearly four feet of solid water on a level surface—fell in twenty-four hours!

Rain is most beautiful when it falls through clear lower air in the near-level rays of a low sun, glistening and sparkling in all directions, and perhaps painting the spectral arch of the rainbow. The primary bow, and perhaps a secondary are, are always centered about the observer's shadow. Hence they can never form more than half a complete circle for anyone on level ground or at sea. The cause of the rainbow is selective reflection and refraction of the various colours or wavelengths composing white light (at certain critical angles that vary slightly with the wavelength) by the small crystal-clear spheres of the falling raindrops. The primary bow, ranging from a sharp outer edge of red to a diffused band of inner blue or violet, has a radius of about forty degrees from the observer's shadow. The rarer secondary bow, if visible at all, begins rather sharply with red about nine degrees outside the primary and shades away outward into a nebulous greenish or bluish glow.

The larger the raindrops and the clearer the lower air, the more striking and beautiful are the rainbow colours. Ground observers, of course, see the largest bows when the sun is low. But from a high mountain-peak, or better still from an airplane flying between or below showery clouds, it is sometimes possible under a high sun to see a strangely unfamiliar but doubly miraculous and brilliant rainbow rounding into more than half a circle and seemingly laid on its side.

Less spectacular and less rare are the faintly coloured aureoles, called 'glories' or 'mountain-bows,' that anyone on a sharp peak or in an airplane can see around his shadow on the white surface of a sunlit cloudbank. If the cloud droplets are near at hand, these glories may resemble faint but orthodox rainbows. Smaller aureoles that appear on more distant cloud-floors—faint colour-rings only two degrees or so in diameter—are probably due to selective diffraction of light rays around the sharp opaque edge of the peak or airplane.

SNOW

Few places on the earth are not visited by rain, occasional or frequent, light or heavy, steady or showery. Over the whole earth, indeed, it has been estimated that rain is always falling at the stupendous rate of sixteen million tons per second! But in northern latitudes, through a large part of the year, this never-ending deluge takes the form of feathery-white showers of snow— snow that, unlike rain, accumulates everywhere, in the valleys and on the hills, to store moisture against a future melting. This snow blanket brings to sight general whiteness, and to sound general silence. Snow only four inches deep absorbs something like nine tenths of ordinary sound energy, being more efficient as a noise muffler than heavy draperies of velvet. A feather-like snowflake floating slowly downward seems to have little in common with a piece of ice. Nevertheless, that is what the snowflake really is, a piece of ice usually crystallized into a very intricate and beautiful pattern. Snowflake patterns are so elaborate, consisting mainly of outgrowths at sixty-degree angles, that each flake presents a large air-catching surface compared with its weight, and falls slowly like a parachute. Snowflakes can be caught on a dark cloth and examined under a magnifying glass, and they have often been photographed with microscopic cameras.

Snow, as already indicated, is probably formed by the same general causative factors as rain, operating however at temperatures entirely below freezing. As the nuclear ice crystals fall, they freeze to themselves as crystalline extensions any subcooled minute water droplets in their path, and grow also by direct sublimation out of invisible water vapour.

If snowflakes fall into temperatures above freezing on their downward path, however, each elaborate crystal melts into a drop of rain. In a climbing airplane it is not uncommon to sound through a region of rain up into a zone of snow (though the exploration is not particularly to be recommended, owing to icing hazards), and much winter rain undoubtedly begins as snow in the higher levels.

Although snow can melt into rain, rain cannot freeze back into snow. If rain from an upper, warmer air layer passes through a colder-than-freezing layer close to the ground, the result is frozen raindrops, or round pellets of clear ice, properly called sleet. Sometimes, however, snow falls in the form of round, compacted pellets which resemble sleet, but are more correctly called graupel; and occasionally these pellets come through to the ground. The remaining form of aerial ice is hail -pellets formed by repeated up-and-down rides in the towering thunderstorm clouds.

If rain falls in liquid drops on ground, houses, trees, or wires that have been previously cooled below freezing, it forms a layer of clear ice called glaze. 'Icing' of a somewhat similar sort can occur (to the embarrassment of pilots) on the wings or other parts of an airplane in flight through cold clouds or rain. Another type of ground icing (also noticed on airplanes) is rime, where subcooled fog or cloud droplets freeze on exposed projections, gradually building out into the wind. Widespread and continued glaze, which may bring down trees limbs, telephone wires, and power lines by the hundreds, is called an 'ice storm.'

The formation of clouds out of thin air, the birth of precipitation out of clouds—both these processes are physical experiments conducted by Nature in her open-air laboratory on a gigantic scale. Even a local summer shower involves millions of tons of air, hundreds of thousands of tons of water, and energy equivalent to the burning of thousands of tons of coal. A widespread storm, such as occurs along the path of a cyclone, may involve materials and forces many times greater.

❒

10

Biogeochemical Cycle

ECOSYSTEM

An ecosystem is a group of living and non-living components interacting together on a given physical landscape. The size of an ecosystem is arbitrary and could be as small as a few square centimeters if you are looking at a soil microbial ecosystem; as large as thousands of square kilometers if you are looking a biome like the Great Plains ecosystem; or a few hectares if you are looking at a single forest stand ecosystem.

FORESTED ECOSYSTEM WORKS IS TO BUILD A MODEL

An ecosystem model is an accurate but simplified representation of an ecosystem that can be very useful in thinking about or simulating the actions of a real ecosystem.

Because any ecosystem has many different but interrelated components, the best way to understand the system is to break it down into its component parts. To get an introduction to a very simplified forest model, which gives participants and introduction to how a hardwood forest ecosystem works before and after exotic earthworms invade.

Step One: The first step in building a graphical model of a hardwood forest ecosystem is to identify its major components. The components of any ecosystem are those physical things that contain energy and nutrients.

A forested ecosystem, by definition contains trees, so that is our first component. In addition to the various species and layers of trees in a forest, there are other distinct ecosystem components. The understory contains most of the visible plant life found between the sapling layer and forest floor. The forest floor is where one would find most of the plant roots, bulbs, fungi, seeds, years of accumulated leaves and twigs.

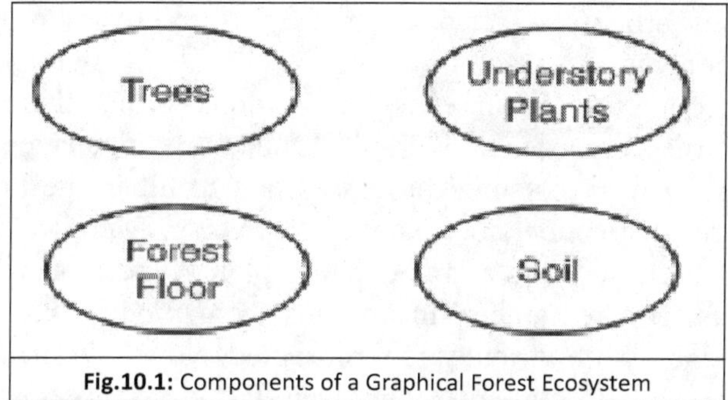

Fig.10.1: Components of a Graphical Forest Ecosystem

The soil is the "dirt" under the forest floor and is composed largely of minerals of various grain size (very small grain size = clay very large grain size = sand) and organic material that has been mixed in with the mineral component. In addition, there are numerous animals that live in the forest, and of course we cannot forget people. We will add those two components to our forest ecosystem model later

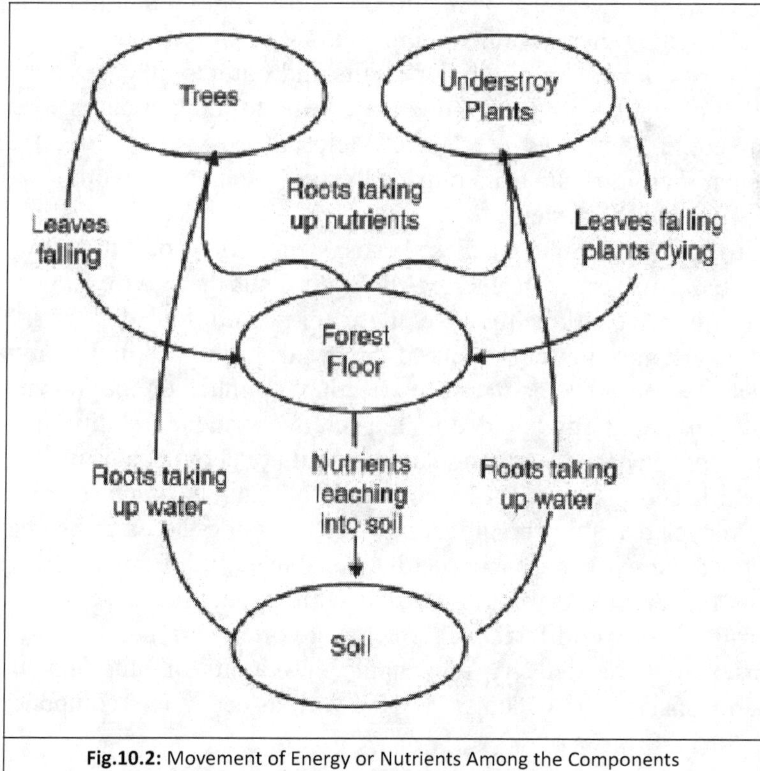

Fig.10.2: Movement of Energy or Nutrients Among the Components

Step Two: Once you have identified the components of your ecosystem model, you need to define the processes that connect the components. This is graphically done by using arrows to indicate the flow of nutrients or energy among the different ecosystem components. The components of our ecosystem model are now connected by processes that result in the movement of energy or nutrients among the components.

One thing to notice in our ecosystem model is that there is no process connecting the trees component with the understory plants component. This is because there are no substantial processes that result in the flow of nutrients directly from a tree to an understory plant or visa-versa. There would be important relationships between the trees and understory plants. Trees provide shade to the understory plants. But remember, in an ecosystem model, only processes that result in flow of energy or nutrients are represented. Now let's add the animals and the people components to our ecosystem. energy & nutrients flow from the trees and understory plants to the animals when they eat the leaves, twigs and buds of trees or graze on understory plants; and when the animal excrete waste products or die, energy & nutrients are returned to the forest floor component. Since people are really just a special kind of animal, energy & nutrients flow from the trees to people when they eat something from a tree, like maple syrup

We have added two components to our ecosystem model, along with some processes connecting them to other components.

Step Three: Determine the major inputs and outputs of your ecosystem. As you are building your ecosystem model, one thing to think about is whether your ecosystems could be opened or closed. A closed ecosystem is one that has no inputs of energy or nutrients from outside the ecosystem and no outputs of energy or nutrients leaving the system.

The earth is an example of a closed ecosystem with respect to nutrients and an open ecosystem with respect to energy All the nutrients that have ever been on earth are here and simply continue to cycle, there are no additions or losses. However, the earth is constantly getting inputs of energy from the sun and simultaneously radiating energy back. The earth doesn't heat up too much or cool down too much because the earth's energy balance is in a relatively stable equilibrium, meaning that the amount of energy being input and output are about equal.

The earth ecosystem and has no inputs or outputs of nutrients which are constantly recycled within the global ecosystem, while the earth has both inputs and outputs of energy that are in a relatively stable equilibrium.

Just as the Earth ecosystem is closed with respect to nutrients, unmanaged earthworm-free hardwood forest ecosystems are often very nearly closed nutrient ecosystems that there are very few inputs or outputs of nutrients. Rather the nutrients are constantly recycled among the various ecosystem components.

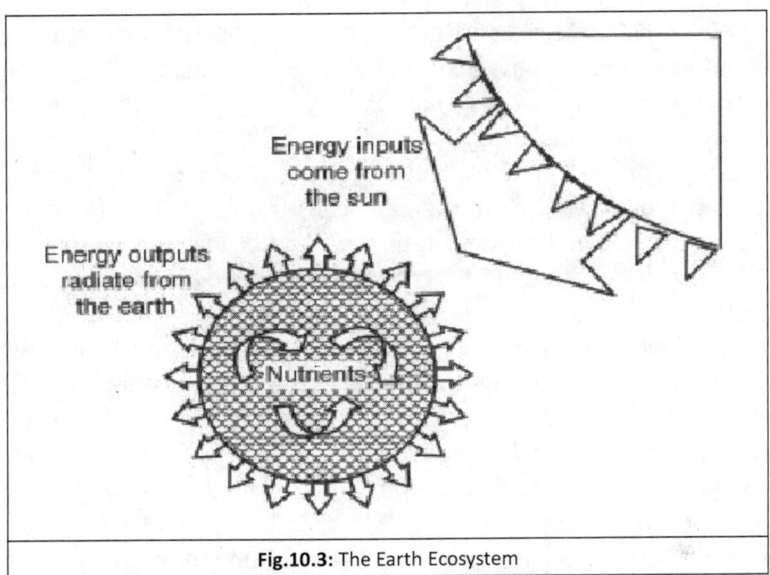

Fig.10.3: The Earth Ecosystem

In contrast, most agricultural ecosystems require nutrient inputs from outside to function properly some typical inputs and outputs of nutrients and energy for forested ecosystems include evapotranspiration, nutrient leaching, sunlight and rain.

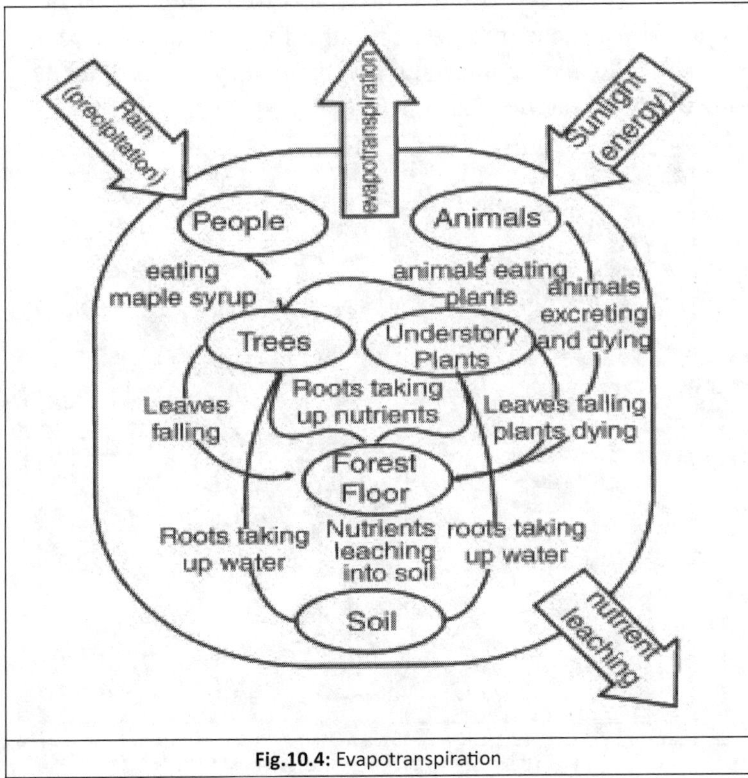

Fig.10.4: Evapotranspiration

Step Four: Once you have identified the components, processes and major inputs and outputs in your ecosystem model, then you can begin to add the actual values to these parts of your ecosystem by measuring them. The amount of litter that falls to the forest floor each year (a process), what the biomass of trees is in a given forest (a component), how much light reaches the forest over a growing season (an input), or how much nitrogen leaches from the forest (an output).

Needless to say, some of these things are easier to measure than others and for most of these things it would be very hard to directly measure the value for a whole forest.

It would be hard to catch every single leaf that fell from the trees in a given year and weigh them all! So, researchers estimate these values taking samples of the given measurement they want to know. In the case of leaf litter, you can put out trays in the forest and after all the leaves have fallen for the year, dry and weight the leaf litre in your trays. They you can use that value to calculate an estimate of the total leaf litter for your forest.

Step Five: Use your ecosystem model to think about how changes can cascade through an ecosystem or to ask specific questions that can be answered with further research. When the major components, processes and inputs and outputs of an ecosystem are understood, then you can use the model how changing one part of the ecosystem affects other parts. If you harvest trees from your forest, that will decrease the amount of leaf litter reaching the forest floor each year which may lead to decreases in available nutrients for understory plants. This is the type of thing forest ecology researchers often study.

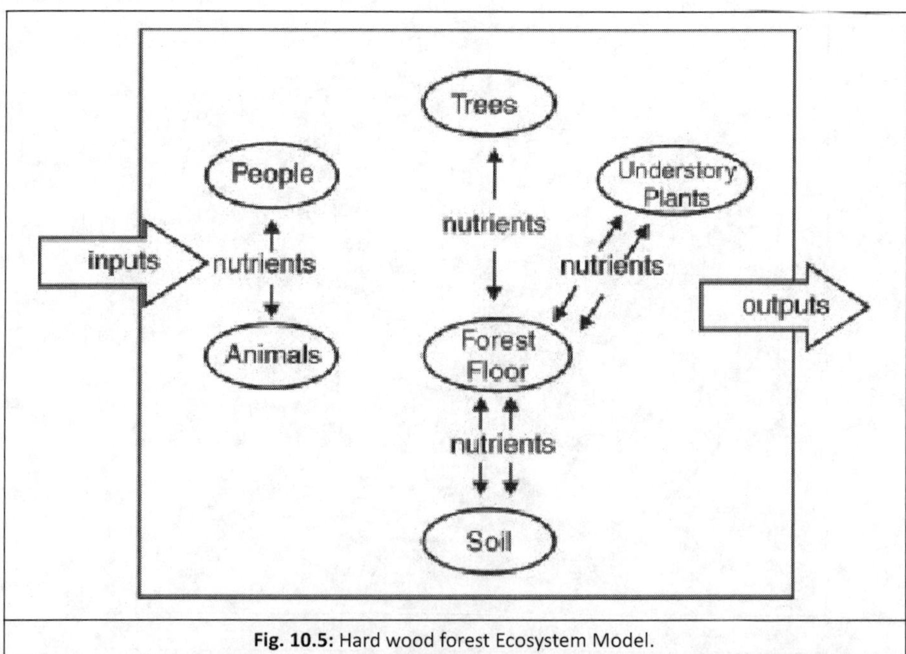

Fig. 10.5: Hard wood forest Ecosystem Model.

An ecosystem that is in equilibrium doesn't gain or lose nutrients. Researchers may monitor soil nutrient levels for many years after trees have been harvested how the real forest behaves compared to what they thought might happen based on their forest model, their understanding of how the forest works.

If the results in the real forest are very different than those predicted by their model, then they know that they don't have full understanding of how their forest works and they may go back to try to improve their model.

In general, when building a model, you start simple and add more detailas eeded. To get an introduction to a very simplified forest model which gives participants and introduction to how a hardwood forest ecosystem works before and after exotic earthworms invade.

Energy Flow Through the Ecosystem

Both energy and inorganic nutrients flow through the ecosystem. We need to define some terminology first. Energy "flows" through the ecosystem in the form of carbon-carbon bonds. When respiration occurs, the carbon-carbon bonds are broken and the carbon is combined with oxygen to form carbon dioxide. This process releases the energy, which is either used by the organism or the energy may be lost as heat. The dark arrows represent the movement of this energy.

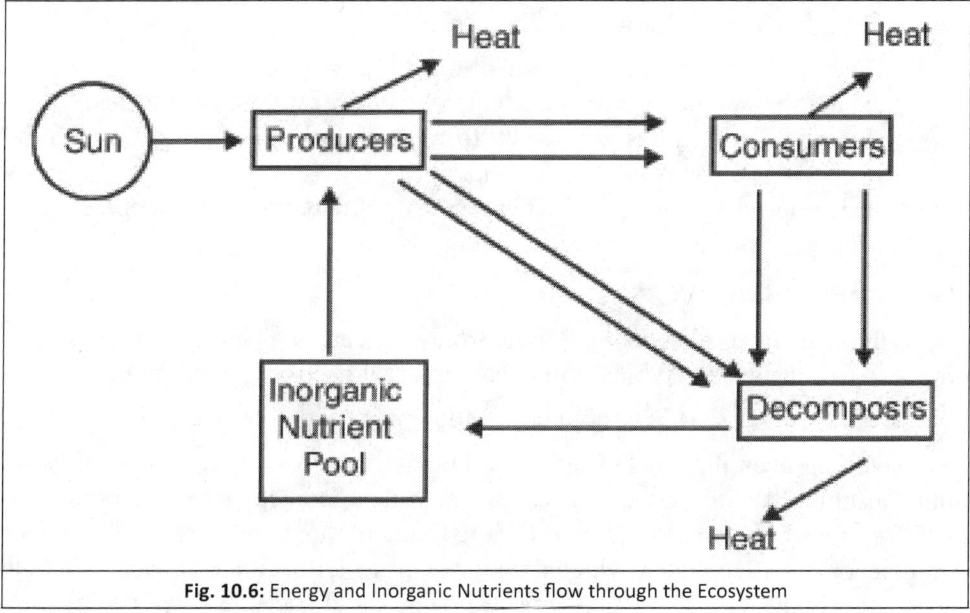

Fig. 10.6: Energy and Inorganic Nutrients flow through the Ecosystem

The other component are the inorganic nutrients. They are inorganic because they do not contain carbon-carbon bonds. These inorganic nutrients include the phosphorous in your teeth, bones, and cellular membranes; the nitrogen in your amino acids; and the iron in your blood.

The movement of the inorganic nutrients is represented by the open arrows. Note that the autotrophs obtain these inorganic nutrients from the inorganic nutrient pool, which is usually the soil or water surrounding the plants or algae.

These inorganic nutrients are passed from organism to organism as one organism is consumed by another. Ultimately, all organisms die and become detritus, food for the decomposers. At this stage, the last of the energy is extracted and the inorganic nutrients are returned to the soil or water to be taken up again. The inorganic nutrients are recycled, the energy is not.

Many of us, when we hear the word "nutrient" immediately think of calories and the carbon-carbon bonds that hold the caloric energy. When writing about energy flow and inorganic nutrient flow in an ecosystem, you must be clear as to what you are referring. Unmodified by "inorganic" or "organic", the word "nutrient" can leave your reader unsure of what you mean. This is one case in which the scientific meaning of a word is very dependent on its context. Another example would be the word "respiration", which to the layperson usually refers to "breathing", but which means "the extraction of energy from carbon-carbon bonds at the cellular level" to most scientists (except those scientists studying breathing, who use respiration in the lay sense).

To summarize: In the flow of energy and inorganic nutrients through the ecosystem, a few generalizations can be made:

- The ultimate source of energy (for most ecosystems) is the sun.
- The ultimate fate of energy in ecosystems is for it to be lost as heat.
- Energy and nutrients are passed from organism to organism through the food chain as one organism eats another.
- Decomposers remove the last energy from the remains of organisms.
- Inorganic nutrients are cycled, energy is not.

FOOD CHAINS AND WEBS

A food chain is the path of food from a given final consumer back to a producer. For instance, a typical food chain in a field ecosystem might be:

Grass →Grasshopper →Mouse →Snake →Hawk

Note that even though I said the food chain is the path of food from a given final consumer back to a producer we typically list a food chain from producer on the left to final consumer on the right. Note to international readers: In Hebrew or Aramaic, or other languages which are read right-to-left, is it customary to list the food chains in the reverse order? By the way, you should be able to look at the food chain above and identify the autotrophs and heterotrophs, and classify each as a herbivore, carnivore, etc. You should also be able to determine that the hawk is a quaternary consumer.

The real world, of course, is more complicated than a simple food chain. While many organisms do specialize in their diets (anteaters come to mind as a specialist), other organisms do not. Hawks don't limit their diets to snakes, snakes eat things other than mice, mice eat grass as well as grasshoppers, and so on. A more realistic depiction of who eats whom is called a food web. It is when we have a picture of a food web in front of us that the definition of food chain makes more sense. We can now a food web consists of interlocking food chains, and that the only way to untangle the chains is to trace back along a given food chain to its source.

The food webs are grazing food chains since at their base are producers which the herbivores then graze on. While grazing food chains are important, in nature they are outnumbered by detritus-based food chains. In detritus-based food chains, decomposers are at the base of the food chain, and sustain the carnivores which feed on them. In terms of the weight (or biomass) of animals in many ecosystems, more of their body mass can be traced back to detritus than to living producers.

Pyramids

The concept of biomass is important. It is a general principle that the further removed a trophic level is from its source (detritus or producer), the less biomass it will contain (biomass here would refer to the combined weight of all the organisms in the trophic level). This reduction in biomass occurs for several reasons:
Not everything in the lower levels gets eaten

- Not everything that is eaten is digested
- Energy is always being lost as heat

It is important to remember that the decrease in number is best detected in terms or biomass. Numbers of organisms are unreliable in this case because of the great variation in the biomass of individual organisms. For instance, squirrels feed on acorns. The oak trees in a forest will always outnumber the squirrels in terms of combined weight, but there may actually be more squirrels than oak trees. Remember that an individual oak tree is huge, weighing thousands of kilograms, while an individual squirrel weighs perhaps 1 kilogram at best. There are few exceptions to the pyramid of biomass scheme. One occurs in aquatic systems where the algae may be both outnumbered and outweighed by the organisms that feed on the algae. The algae can support the greater biomass of the next trophic level only because they can reproduce as fast as they are eaten. In this way, they are never completely consumed. It is interesting to note that this exception to the rule of the pyramid of biomass also is a partial exception to at least 2 of the 3 reasons for the pyramid of biomass. While not all the algae are consumed, a greater proportion of them are, and while not completely digestible, algae are far more nutritious overall than the average woody plant is (most organisms cannot digest wood and extract energy from it).

Biological Magnification

Biological magnification is the tendency of pollutants to become concentrated in successive trophic levels. Often, this is to the detriment of the organisms in which these materials concentrate, since the pollutants are often toxic.

Biomagnification occurs when organisms at the bottom of the food chain concentrate the material above its concentration in the surrounding soil or water. Producers, as we saw earlier, take in inorganic nutrients from their surroundings. Since a lack of these nutrients can limit the growth of the producer, producers will go to great lengths to obtain the nutrients. They will spend considerable energy to pump them into their bodies. They will even take up more than they need immediately and store it, since they can't be "sure" of when the nutrient will be available again (of course, plants don't think about such things, but, as it turns out, those plants, which, for whatever reason, tended to concentrate inorganic nutrients have done better over the years). The problem comes up when a pollutant, such as DDT or mercury, is present in the environment. Chemically, these pollutants resemble essential inorganic nutrients and are brought into the producer's body and stored "by mistake". This is the first step in biomagnification; the pollutant is at a higher concentration inside the producer than it is in the environment.

The second stage of biomagnification occurs when the producer is eaten. Remember from our discussion of a pyramid of biomass that relatively little energy is available from one trophic level to the next. This means that a consumer (of any level) has to consume a lot of biomass from the lower trophic level. If that biomass contains the pollutant, the pollutant will be taken up in large quantities by the consumer. Pollutants that biomagnify have another characteristic. Not only are they taken up by the producers, but they are absorbed and stored in the bodies of the consumers. This often occurs with pollutants soluble in fat such as DDT or PCB's. These materials are digested from the producer and move into the fat of the consumer. If the consumer is caught and eaten, its fat is digested and the pollutant moves to the fat of the new consumer. In this way, the pollutant builds up in the fatty tissues of the consumers. Water-soluble pollutants usually cannot biomagnify in this way because they would dissolve in the bodily fluids of the consumer. Since every organism loses water to the environment, as the water is lost the pollutant would leave as well. Alas, fat simply does not leave the body.

The "best" example of biomagnification comes from DDT. This long-lived pesticide (insecticide) has improved human health in many countries by killing insects such as mosquitoes that spread disease. On the other hand, DDT is effective in part because it does not break down in the environment. It is picked up by organisms in the environment and incorporated into fat. Even here, it does no real damage in many organisms (including humans). In others, however, DDT is deadly or may have more insidious, long-term effects. In birds, for instance, DDT interferes with the deposition of calcium in the shells of the bird's eggs. The eggs laid are very soft and easily broken; birds so afflicted are rarely able to raise young and this causes a decline in their numbers. This was so apparent in the early 1960's that it led the scientist Rachel Carson to postulate a "silent spring" without the sound of bird calls. The banning of DDT, the search for pesticides that would not biomagnify, and the birth of the "modern" environmental movement in the 1960's. Birds such as the bald eagle have made comebacks in response to the banning of DDT in the US. Ironically, many of the pesticides which replaced DDT are more dangerous to humans, and, without DDT, disease (primarily in the tropics) claims more human lives.

HUMAN VS. NATURAL FOOD CHAINS

Human civilization is dependent on agriculture. Only with agriculture can a few people feed the rest of the population; the part of the population freed from raising food can then go on to do all the things we associate with civilization. Agriculture means manipulating the environment to favour plant species that we can eat. In essence, humans manipulate competition, allowing favoured species (crops) to thrive and thwarting species which might otherwise crowd them out (weeds). In essence, with agriculture we are creating a very simple ecosystem. At most, it has only three levels-producers (crops), primary consumers (livestock, humans) and secondary consumers (humans). This means that little energy is lost between tropic levels, since there are fewer trophic levels present.

This is good for humans, but what type of "ecosystem" have we created? Agricultural ecosystems have several problems. First, we create monocultures, or fields with only one crop. This is simplest for planting, weeding, and harvesting, but it also packs many similar plants into a small area, creating a situation ideal for disease and insect pests. In natural ecosystems, plants of one species are often scattered. Insects, which often specialize on feeding on a particular plant species, have a hard time finding the scattered plants. Without food, the insect populations are kept in check. In a field of corn however, even the most inept insect can find a new host plant with a jump in any direction. Likewise, disease is more easily spread if the plants are in close proximity. It takes lots of chemicals (pesticides) to keep a monoculture going.

Another problem with human agriculture is that we rely on relatively few plants for food. If the corn and rice crops failed worldwide in the same year, we would be hard-pressed to feed everyone. tural ecosystems usually have alternate sources of food available if one fails.

A final problem associated with agroecosystems is the problem of inorganic nutrient recycling. In a natural ecosystem, when a plant dies it fall to the ground and rots, and its inorganic nutrients are returned to the soil from which they were taken. In human agriculture, however, we harvest the crop, truck it away, and flush it down the toilet to be run off in the rivers to the ocean. Aside from the water pollution problems this causes, it should be obvious to you that the nutrients are not returned to the fields. They have to be replaced with chemical fertilizers, and that means mining, transportation, electricity, etc. Also, the chemical fertilizers tend to run off the fields (along with soil disrupted by cultivation) and further pollute the water.

Some solutions are at hand, but they bring on new problems, too. No-till farming uses herbicides to kill plants in a field; the crop is then planted through the dead plants without plowing up the soil. This reduced soil and fertilizer erosion, but the herbicides themselves may damage ecosystems. In many areas, sewage sludge is returned to fields to act as a fertilizer. This reduces the need for chemical fertilizers, but still requires a lot of energy to haul the sludge around. Further, if one is not careful, things such as household chemicals and heavy metals may contaminate the sewage sludge and biomagnify in the crops which we would then eat.

BIOGEOCHEMICAL CYCLES

We have already seen that while energy does not cycle through an ecosystem, chemicals do. The inorganic nutrients cycle through more than the organisms, however, they also enter into the atmosphere, the oceans, and even rocks. Since these chemicals cycle through both the biological and the geological world, we call the overall cycles biogeochemical cycles. Each chemical has its own unique cycle, but all of the cycles do have some things in common. Reservoirs are those parts of the cycle where the chemical is held in large quantities for long periods of time. In exchange pools, on the other hand, the chemical is held for only a short time.

The length of time a chemical is held in an exchange pool or a reservoir is termed its residence time. The oceans are a reservoir for water, while a cloud is an exchange pool. Water may reside in an ocean for thousands of years, but in a cloud for a few days at best. The biotic community includes all living organisms. This community may serve as an exchange pool (although for some chemicals like carbon, bound in a sequoia for a thousand years, it may seem more like a reservoir), and also serve to move chemicals from one stage of the cycle to another.

For instance, the trees of the tropical rain forest bring water up from the forest floor to be evapourated into the atmosphere. Likewise, coral endosymbionts take carbon from the water and turn it into limestone rock. The energy for most of the transportation of chemicals from one place to another is provided either by the sun or by the heat released from the mantle and core of the Earth. While all inorganic nutrients cycle, we will focus on only 4 of the most important cycles-water, carbon (and oxygen), nitrogen, and phosphorous.

The Water Cycle

In the water cycle, energy is supplied by the sun, which drives evapouration whether it be from ocean surfaces or from treetops. The sun also provides the energy which drives the weather systems which move the water vapour (clouds) from one place to another (otherwise, it would only rain over the oceans). Precipitation occurs when water condenses from a gaseous state in the atmosphere and falls to earth. Evapouration is the reverse process in which liquid water becomes gaseous. Once water condenses, gravity takes over and the water is pulled to the ground.

Gravity continues to operate, either pulling the water underground (groundwater) or across the surface (runoff). In either event, gravity continues to pull water lower and lower until it reaches the oceans (in most cases; the Great Salt Lake, Dead Sea, Caspian Sea, and other such depressions may also serve as the lowest basin into which water can be drawn).

Frozen water may be trapped in cooler regions of the Earth (the poles, glaciers on mountaintops, etc.) as snow or ice, and may remain as such for very long periods of time. Lakes, ponds, and wetlands form where water is temporarily trapped.

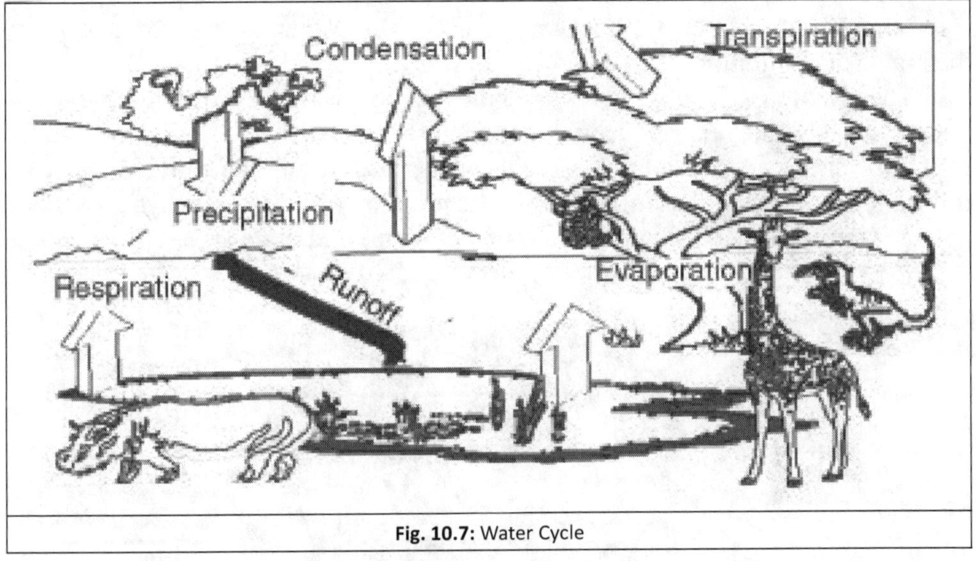

Fig. 10.7: Water Cycle

The oceans are salty because any weathering of minerals that occurs as the water runs to the ocean will add to the mineral content of the water, but water cannot leave the oceans except by evapouration, and evapouration leaves the minerals behind. Thus, rainfall and snowfall are comprised of relatively clean water, with the exception of pollutants (such as acids) picked up as the waster falls through the atmosphere. Organisms play an important role in the water cycle.

As you know, most organisms contain a significant amount of water. This water is not held for any length of time and moves out of the organism rather quickly in most cases. Animals and plants lose water through evapouration from the body surfaces, and through evapouration from the gas exchange structures (such as lungs). In plants, water is drawn in at the roots and moves to the gas exchange organs, the leaves, where it evapourates quickly. This special case is called transpiration because it is responsible for so much of the water that enters the atmosphere. In both plants and animals, the breakdown of carbohydrates (sugars) to produce energy (respiration) produces both carbon dioxide and water as waste products. Photosynthesis reverses this reaction, and water and carbon dioxide are combined to form carbohydrates. Now you understand the relevance of the term carbohydrate; it refers to the combination of carbon and water in the sugars we call carbohydrates.

Carbon Cycle

Once you understand the water cycle, the carbon cycle is relatively simple. From a biological perspective, the key events here are the complementary reactions of respiration and photosynthesis. Respiration takes carbohydrates and oxygen and combines them to produce carbon dioxide, water, and energy. Photosynthesis takes carbon dioxide and water and produces carbohydrates and oxygen. The outputs of respiration are the inputs of photosynthesis, and the outputs of photosynthesis are the inputs of respiration.

The reactions are also complementary in the way they deal with energy. Photosynthesis takes energy from the sun and stores it in the carbon-carbon bonds of carbohydrates; respiration releases that energy. Both plants and animals carry on respiration, but only plants (and other producers) can carry on photosynthesis. The chief reservoirs for carbon dioxide are in the oceans and in rock.

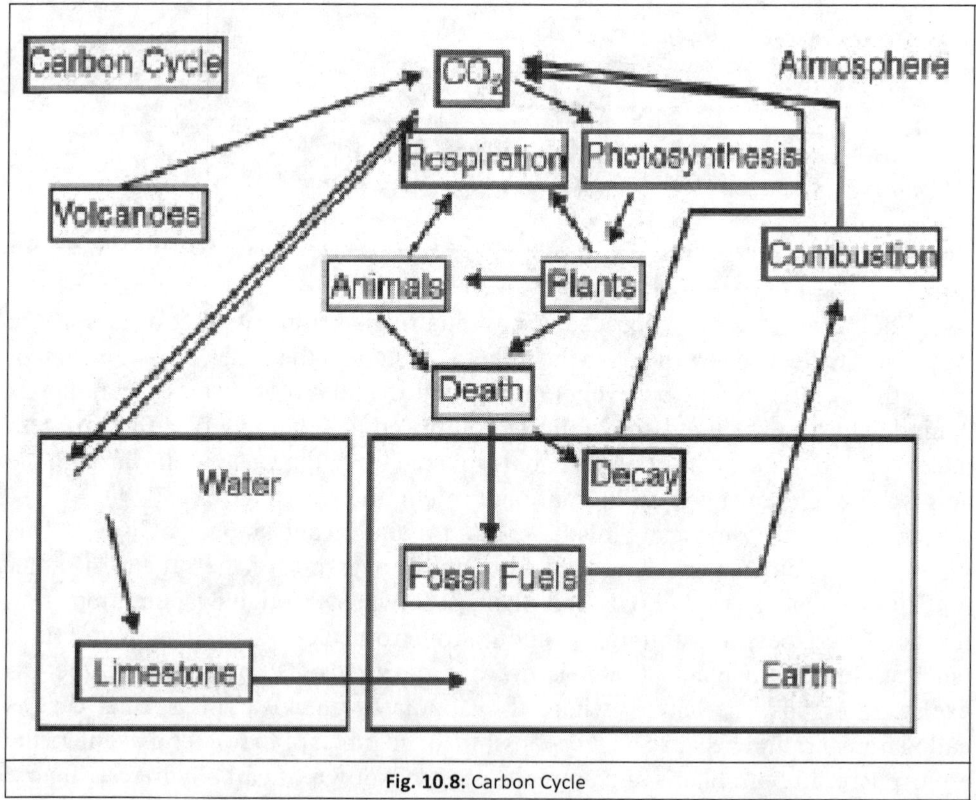

Fig. 10.8: Carbon Cycle

Carbon dioxide dissolves readily in water. Once there, it may precipitate (fall out of solution) as a solid rock known as calcium carbonate (limestone). Corals and algae encourage this reaction and build up limestone reefs in the process. On land and in the water, plants take up carbon dioxide and convert it into carbohydrates through photosynthesis.

This carbon in the plants now has 3 possible fates. It can be liberated to the atmosphere by the plant through respiration; it can be eaten by an animal, or it can be present in the plant when the plant dies. Animals obtain all their carbon in their food, and, thus, all carbon in biological systems ultimately comes from plants (autotrophs).

In the animal, the carbon also has the same 3 possible fates. Carbon from plants or animals that is released to the atmosphere through respiration will either be taken up by a plant in photosynthesis or dissolved in the oceans. When an animal or a plant dies, 2 things can happen to the carbon in it.

It can either be respired by decomposers (and released to the atmosphere), or it can be buried intact and ultimately form coal, oil, or natural gas (fossil fuels). The fossil fuels can be mined and burned in the future; releasing carbon dioxide to the atmosphere. Otherwise, the carbon in limestone or other sediments can only be released to the atmosphere when they are subducted and brought to volcanoes, or when they are pushed to the surface and slowly weathered away. Humans have a great impact on the carbon cycle because when we burn fossil fuels we release excess carbon dioxide into the atmosphere. This means that more carbon dioxide goes into the oceans, and more is present in the atmosphere. The latter condition causes global warming, because the carbon dioxide in the atmosphere allows more energy to reach the Earth from the sun than it allows to escape from the Earth into space.

The Oxygen Cycle

If you look back at the carbon cycle, we have also described the oxygen cycle, since these atoms often are combined. Oxygen is present in the carbon dioxide, in the carbohydrates, in water, and as a molecule of two oxygen atoms.

Oxygen is released to the atmosphere by autotrophs during photosynthesis and taken up by both autotrophs and heterotrophs during respiration.

In fact, all of the oxygen in the atmosphere is biogenic; that is, it was released from water through photosynthesis by autotrophs. It took about 2 billion years for autotrophs (mostly cyanobacteria) to raise the oxygen content of the atmosphere to the 21% that it is today; this opened the door for complex organisms such as multicellular animals, which need a lot of oxygen. The Nitrogen Cycle

The nitrogen cycle is one of the most difficult of the cycles to learn, simply because there are so many important forms of nitrogen, and because organisms are responsible for each of the interconversions. Remember that nitrogen is critically important in forming the amino portions of the amino acids which in turn form the proteins of your body. Proteins make up skin and muscle, among other important structural portions of your body, and all enzymes are proteins. Since enzymes carry out almost all of the chemical reactions in your body, it's easy how important nitrogen is. The chief reservoir of nitrogen is the atmosphere, which is about 78% nitrogen.

It is a pretty non-reactive gas; it takes a lot of energy to get nitrogen gas to break up and combine with other things, such as carbon or oxygen. Nitrogen gas can be taken from the atmosphere (fixed) in two basic ways. First, lightning provides enough energy to "burn" the nitrogen and fix it in the form of nitrate, which is a nitrogen with three oxygens attached.

This process is duplicated in fertilizer factories to produce nitrogen fertilizers. The other form of nitrogen fixation is by nitrogen fixing bacteria, who use special enzymes instead of the extreme amount of energy found in lightning to fix nitrogen.

These nitrogen-fixing bacteria come in three form: some are free-living in the soil; some form symbiotic, mutualistic associations with the roots of bean plants and other legumes (rhizobial bacteria); and the third form of nitrogen-fixing bacteria are the photosynthetic cyanobacteria (blue-green algae) which are found most commonly in water. All of these fix nitrogen, either in the form of nitrate or in the form of ammonia (nitrogen with 3 hydrogens attached).

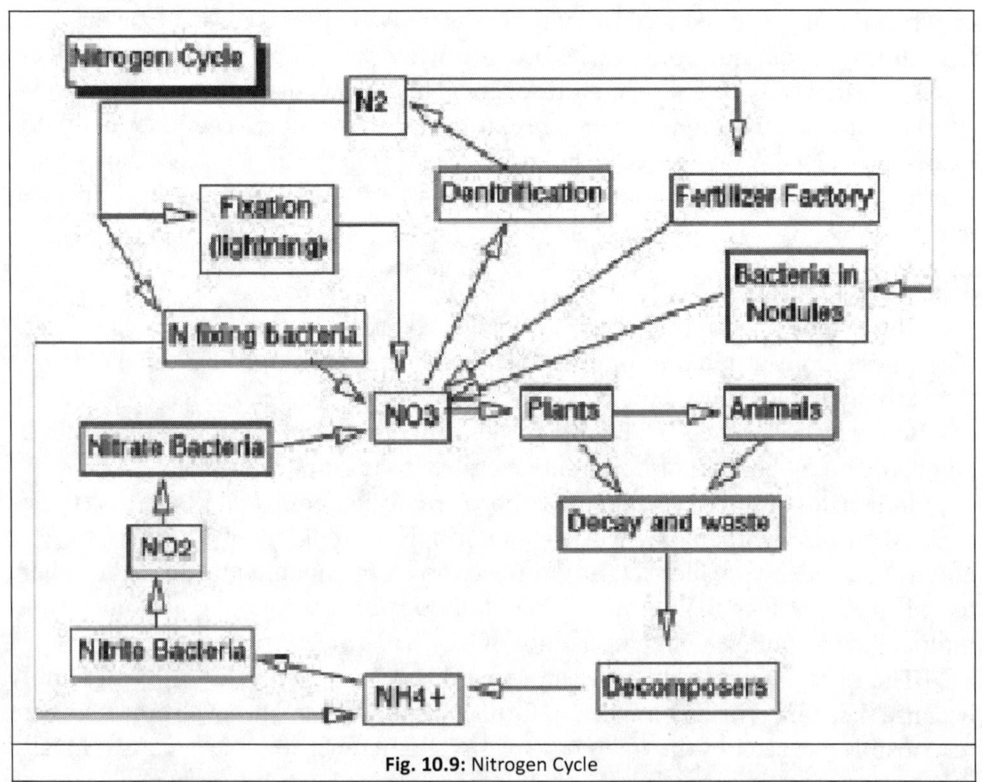

Fig. 10.9: Nitrogen Cycle

Most plants can take up nitrate and convert it to amino acids. Animals acquire all of their amino acids when they eat plants (or other animals). When plants or animals die (or release waste) the nitrogen is returned to the soil. The usual form of nitrogen returned to the soil in animal wastes or in the output of the decomposers, is ammonia. Ammonia is rather toxic, but, fortunately there are nitrite bacteria in the soil and in the water which take up ammonia and convert it to nitrite, which is nitrogen with two oxygens. Nitrite is also somewhat toxic, but another type of bacteria, nitrate bacteria, take nitrite and convert it to nitrate, which can be taken up by plants to continue the cycle. We now have a cycle set up in the soil (or water), but what returns nitrogen to the air? It turns out that there are denitrifying bacteria which take the nitrate and combine the nitrogen back into nitrogen gas.

The nitrogen cycle has some important practical considerations, as anyone who has ever set up a saltwater fish tank has found out. It takes several weeks to set up such a tank, because you must have sufficient numbers of nitrite and nitrate bacteria present to detoxify the ammonia produced by the fish and decomposers in the tank. Otherwise, the ammonia levels in the tank will build up and kill the fish.

This is usually not a problem in freshwater tanks for two reasons. One, the pH in a freshwater tank is at a different level than in a saltwater tank. At the pH of a freshwater tank, ammonia is not as toxic. Second, there are more multicellular plant forms that can grow in freshwater, and these plants remove the ammonia from the water very efficiently. It is hard to get enough plants growing in a saltwater tank to detoxify the water in the same way.

The Phosphorous Cycle

The phosphorous cycle is the simplest of the cycles. Phosphorous has only one form, phosphate, which is a phosphorous atom with 4 oxygen atoms. This heavy molecule never makes its way into the atmosphere, it is always part of an organism, dissolved in water, or in the form of rock.

When rock with phosphate is exposed to water (especially water with a little acid in it), the rock is weathered out and goes into solution. Autotrophs take this phosphorous up and use it in a variety of ways. It is an important constituent of cell membranes, DNA, RNA, and, of course ATP, which, after all, stands for adenosine triphosphate. Heterotrophs (animals) obtain their phosphorous from the plants they eat, although one type of heterotroph, the fungi, excel at taking up phosphorous and may form mutualistic symbiotic relationships with plant roots.

These relationships are called mycorrhizae; the plant gets phosphate from the fungus and gives the fungus sugars in return. Animals, by the way, may also use phosphorous as a component of bones, teeth and shells.

When animals or plants die (or when animals defecate), the phoshate may be returned to the soil or water by the decomposers. There, it can be taken up by another plant and used again. This cycle will occur over and over until at last the phosphorous is lost at the bottom of the deepest parts of the ocean, where it becomes part of the sedimentary rocks forming there. Ultimately, this phosphorous will be released if the rock is brought to the surface and weathered.

Two types of animals play a unique role in the phosphorous cycle. Humans often mine rock rich in phosphorous. For instance, in Florida, which was once sea floor, there are extensive phosphate mines.

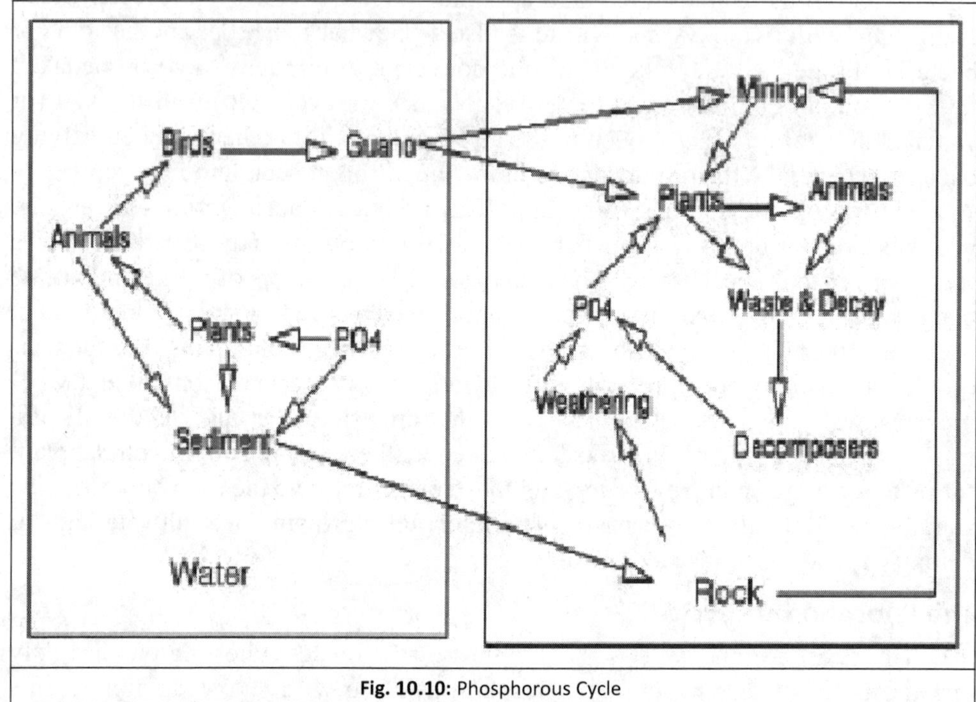

Fig. 10.10: Phosphorous Cycle

The phosphate is then used as fertilizer. This mining of phosphate and use of the phosphate as fertilizer greatly accelerates the phosphorous cycle and may cause local overabundance of phosphorous, particularly in coastal regions, at the mouths of rivers, and anyplace where there is a lot of sewage released into the water (the phosphate placed on crops finds its way into our stomachs and from there to our toilets). Local abundance of phosphate can cause overgrowth of algae in the water; the algae can use up all the oxygen in the water and kill other aquatic life.

This is called eutrophication. The other animals that play a unique role in the phosphorous cycle are marine birds. These birds take phosphorous containing fish out of the ocean and return to land, where they defecate. Their guano contains high levels of phosphorous and in this way marine birds return phosphorous from the ocean to the land. The guano is often mined and may form the basis of the economy in some areas.

Energy Transformations and Biogeochemical Cycling

The study of ecosystems mainly consists of the study of certain processes that link the living, or biotic, components to the non-living, or abiotic, components.

Energy transformations and biogeochemical cycling are the main processes that comprise the field of ecosystem ecology. As we learned earlier, ecology generally is defined as the interactions of organisms with one another and with the environment in which they occur. We can study ecology at the level of the individual, the population, the community, and the ecosystem.

Studies of individuals are concerned mostly about physiology, reproduction, development or behaviour, and studies of populations usually focus on the habitat and resource needs of individual species, their group behaviours, population growth, and what limits their abundance or causes extinction. Studies of communities examine how populations of many species interact with one another, such as predators and their prey, or competitors that share common needs or resources.

In ecosystem ecology we put all of this together and, insofar as we can, we try to understand how the system operates as a whole. This means that, rather than worrying mainly about particular species, we try to focus on major functional aspects of the system. These functional aspects include such things as the amount of energy that is produced by photosynthesis, how energy or materials flow along the many steps in a food chain, or what controls the rate of decomposition of materials or the rate at which nutrients are recycled in the system.

Components of an Ecosystem

A basic understanding of the diversity of plants and animals, and how plants and animals and microbes obtain water, nutrients, and food. We can clarify the parts of an ecosystem by listing them under the headings "abiotic" and "biotic".

Abiotic Components	Biotic Components
Sunlight	Primary producers
Temperature	Herbivores
Precipitation	Camivores
Water or Moisture	Omnivores
	Detiritivores
Chemistry (e.g., P. NH_4)	$_{+}$Soil or Water

All of these vary over space/time

By and large, this set of environmental factors is important almost everywhere, in all ecosystems. A functional group is a biological category composed of organisms that perform mostly the same kind of function in the system; all the photosynthetic plants or primary producers form a functional group. Membership in the functional group does not depend very much on who the actual players happen to be, only on what function they perform in the ecosystem.

PROCESSES OF ECOSYSTEMS

This figure with the plants, zebra, lion, and so forth illustrates the two main ideas about how ecosystems function: ecosystems have energy flows and ecosystems cycle materials.

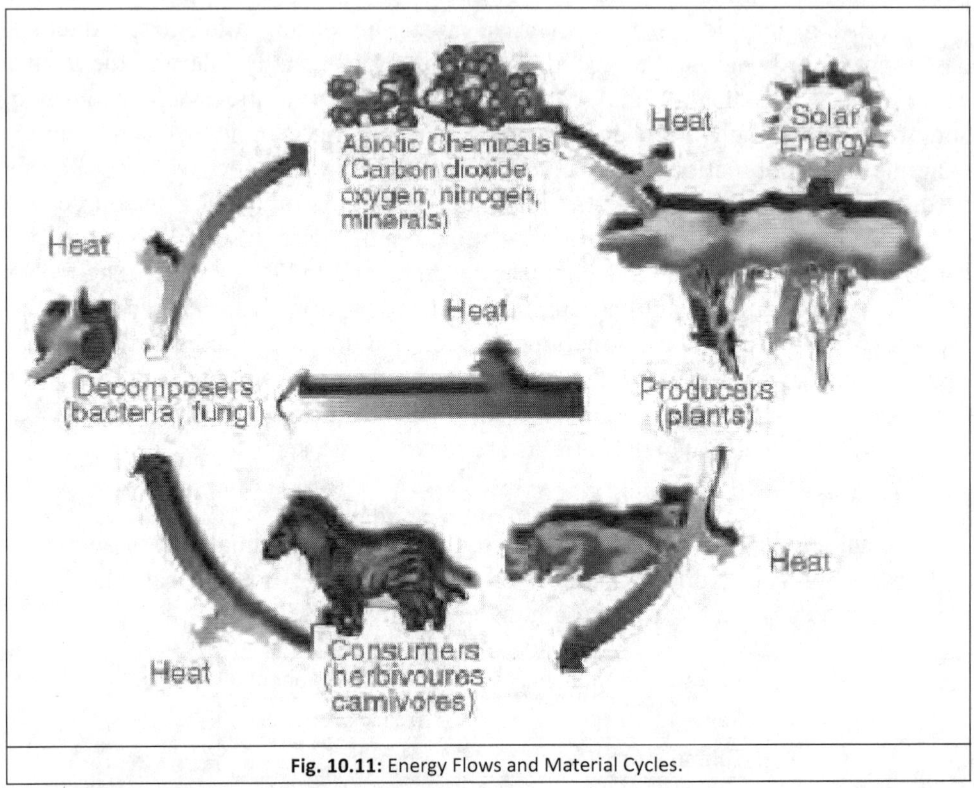

Fig. 10.11: Energy Flows and Material Cycles.

These two processes are linked, but they are not quite the same energy enters the biological system as light energy, or photons, is transformed into chemical energy in organic molecules by cellular processes including photosynthesis and respiration, and ultimately is converted to heat energy. This energy is dissipated, meaning it is lost to the system as heat; once it is lost it cannot be recycled.

Without the continued input of solar energy, biological systems would quickly shut down. Thus the earth is an open system with respect to energy. Elements such as carbon, nitrogen, or phosphorus enter living organisms in a variety of ways. Plants obtain elements from the surrounding atmosphere, water, or soils. Animals may also obtain elements directly from the physical environment, but usually they obtain these mainly as a consequence of consuming other organisms.

These materials are transformed biochemically within the bodies of organisms, but sooner or later, due to excretion or decomposition, they are returned to an inorganic state. Often bacteria complete this process, through the process called decomposition or mineralization During decomposition these materials are not destroyed or lost, so the earth is a closed system with respect to elements (with the exception of a meteorite entering the system now and then). The elements are cycled endlessly between their biotic and abiotic states within ecosystems. Those elements whose supply tends to limit biological activity are called nutrients.

The Transformation of Energy

The transformations of energy in an ecosystem begin first with the input of energy from the sun. Energy from the sun is captured by the process of photosynthesis. Carbon dioxide is combined with hydrogen (derived from the splitting of water molecules) to produce carbohydrates (CHO). Energy is stored in the high energy bonds of adenosine triphosphate.

The prophet Isaah said "all flesh is grass", earning him the title of first ecologist, because virtually all energy available to organisms originates in plants. Because it is the first step in the production of energy for living things, it is called primary production Herbivores obtain their energy by consuming plants or plant products, carnivores eat herbivores, and detritivores consume the droppings and carcasses of us all. portrays a simple food chain, in which energy from the sun, captured by plant photosynthesis, flows from trophic level to trophic level via the food chain.

A trophic level is composed of organisms that make a living in the same way, that is they are all primary producers (plants), primary consumers (herbivores) or secondary consumers (carnivores). Dead tissue and waste products are produced at all levels.

Scavengers, detritivores, and decomposers collectively account for the use of all such "waste"—consumers of carcasses and fallen leaves may be other animals, such as crows and beetles, but ultimately it is the microbes that finish the job of decomposition. Not surprisingly, the amount of primary production varies a great deal from place to place, due to differences in the amount of solar radiation and the availability of nutrients and water.

Energy transfer through the food chain is inefficient. This means that less energy is available at the herbivore level than at the primary producer level, less yet at the carnivore level, and so on. The result is a pyramid of energy, with important implications for understanding the quantity of life that can be supported. Usually when we think of food chains we visualize green plants, herbivores, and so on. These are referred to as grazer food chains, because living plants are directly consumed. In many circumstances the principal energy input is not green plants but dead organic matter.

These are called detritus food chains. Examples include the forest floor or awoodland stream in a forested area, a salt marsh, and most obviously, the ocean floor in very deep areas where all sunlight is extinguished 1000's of meters above. In subsequent lectures we shall return to these important issues concerning energy flow. Finally, although we have been talking about food chains, in reality the organization of biological systems is much more complicated than can be represented by a simple "chain". There are many food links and chains in an ecosystem, and we refer to all of these linkages as a food web. Food webs can be very complicated, where it appears that "everything is connected to everything else", and it is important to understand what are the most important linkages in any particular food web.

Biogeochemistry

One obvious way is to study the flow of energy or the cycling of elements. The cycling of elements is controlled in part by organisms, which store or transform elements, and in part by the chemistry and geology of the natural world. The term Biogeochemistry is defined as the study of how living systems influence, and are controlled by, the geology and chemistry of the earth. Thus biogeochemistry encompasses many aspects of the abiotic and biotic world that we live in.

There are several main principles and tools that biogeochemists use to study earth systems. Most of the major environmental problems that we face in our world toady can be analysed using biogeochemical principles and tools. These problems include global warming, acid rain, environmental pollution, and increasing greenhouse gases. The principles and tools that we use can be broken down into 3 major components: element ratios, mass balance, and element cycling.

Element Ratios

In biological systems, we refer to important elements as "conservative". These elements are often nutrients. By "conservative" we mean that an rganism can change only slightly the amount of these elements in their tissues if they are to remain in good health. It is easiest to think of these conservative elements in relation to other important elements in the organism. In healthy algae the elements C, N, P, and Fe have the following ratio, called the Redfield ratio after the oceanographer who discovered it:

$$C: N: P: Fe = 106: 16: 1: 0.01$$

Once we know these ratios, we can compare them to the ratios that we measure in a sample of algae to determine if the algae are lacking in one of these limiting nutrients. Mass Balance Another important tool that biogeochemists use is a simple mass balance equation to describe the state of a system. The system could be a snake, a tree, a lake, or the entire globe. Using a mass balance approach we can determine whether the system is changing and how fast it is changing. The equation is:

$$\text{Net Change} = \text{Input} + \text{Output} + \text{Internal Change}$$

In this equation the net change in the system from one time period to another is determined by what the inputs are, what the outputs are, and what the internal change in the system was. The example given in class is of the acidification of a lake, considering the inputs and outputs and internal change of acid in the lake.

Element Cycling

Element cycling describes where and how fast elements move in a system. There are two general classes of systems that we can analyse. A closed system refers to a system where the inputs and outputs are negligible compared to the internal changes. Examples of such systems would include a bottle, or our entire globe. There are two ways we can describe the cycling of materials within this closed system, either by looking at the rate of movement or at the pathways of movement.

- Rate = number of cycles/time * as rate increases, productivity increases
- Pathways-important because of different reactions that may occur

In an open system there are inputs and outputs as well as the internal cycling. Thus we can describe the rates of movement and the pathways, just as we did for the closed system, but we can also define a new concept called the residence time. The residence time indicates how long on average an element remains within the system before leaving the system.

- Rate
- Pathways
- Residence time, Rt
- Rt = Total amount of matter/output rate of matter

CONTROLS ON ECOSYSTEM FUNCTION

Now that we have learned something about how ecosystems are put together and how materials and energy flow through ecosystems, we can better address the question of "what controls ecosystem function"? There are two dominant theories of the control of ecosystems. The first, called bottom-up control, states that it is the nutrient supply to the primary producers that ultimately controls how ecosystems function.

If the nutrient supply is increased, the resulting increase in production of autotrophs is propagated through the food web and all of the other trophic levels will respond to the increased availability of food (energy and materials will cycle faster). The second theory, called top-down control, states that predation and grazing by higher trophic levels on lower trophic levels ultimately controls ecosystem function.

If you have an increase in predators, that increase will result in fewer grazers, and that decrease in grazers will result in turn in more primary producers because fewer of them are being eaten by the grazers. Thus the control of population numbers and overall productivity "cascades" from the top levels of the food chain down to the bottom trophic levels. So, which theory is correct? Well, as is often the case when there is a clear dichotomy to choose from, the answer lies somewhere in the middle.

There is evidence from many ecosystem studies that both controls are operating to some degree, but that neither control is complete. The "top-down" effect is often very strong at trophic levels near to the top predators, but the control weakens as you move further down the food chain. Similarly, the "bottom-up" effect of adding nutrients usually stimulates primary production, but the stimulation of secondary production further up the food chain is less strong or is absent. Thus we find that both of these controls are operating in any system at any time, and we must understand the relative importance of each control in order to help us to predict how an ecosystem will behave or change under different circumstances, such as in the face of a changing climate.

THE GEOGRAPHY OF ECOSYSTEMS

There are many different ecosystems: rain forests and tundra, coral reefs and ponds, grasslands and deserts. Climate differences from place to place largely determine the types of ecosystems. How terrestrial ecosystems appear to us is influenced mainly by the dominant vegetation. The word "biome" is used to describe a major vegetation type such as tropical rain forest, grassland, tundra, etc., extending over a large geographic area. It is never used for aquatic systems, such as ponds or coral reefs. It always refers to a vegetation category that is dominant over a very large geographic scale, and so is somewhat broader than an ecosystem. Every place on earth gets the same total number of hours of sunlight each year, but not the same amount of heat.

The sun's rays strike low latitudes directly but high latitudes obliquely.This neven distribution of heat sets up not just temperature differences, but global wind and ocean currents that in turn have a great deal to do with where rainfall occurs. Add in the cooling effects of elevation and the effects of land masses on temperature and rainfall, and we get a complicated global pattern of climate.

A schematic view of the earth shows that, complicated though climate may be, many aspects are predictable High solar energy striking near the equator ensures nearly constant high temperatures and high rates of evapouration and plant transpiration. Warm air rises, cools, and sheds its moisture, creating just the conditions for a tropical rain forest. Contrast the stable temperature but varying rainfall of a site in Panama with the relatively constant precipitation but seasonally changing temperature of a site in New York State. Every location has a rainfall-temperature graph that is typical of a broader region.

We can draw upon plant physiology to know that certain plants are distinctive of certain climates, creating the vegetation appearance that we call biomes. Note how well the distribution of biomes plots on the distribution of climates Note also that some climates are impossible, at least on our planet. High precipitation is not possible at low temperatures—there is not enough solar energy to power the water cycle, and most water is frozen and thus biologically unavailable throughout the year. The high tundra is as much a desert as is the Sahara.

- Ecosystems are made up of abiotic (non-living, environmental) andbiotic components, and these basic components are important tonearly all types of ecosystems. Ecosystem Ecology looks at energy transformations and biogeochemical cycling within ecosystems.
- Energy is continually input into an ecosystem in the form of light energy, and some energy is lost with each transfer to a higher trophic level. Nutrients, on the other hand, are recycled within an ecosystem, and their supply normally limits biological activity. So,"energy flows, elements cycle".
- Energy is moved through an ecosystem via a food web, which is made up of interlocking food chains. Energy is first captured by photosynthesis (primary production). The amount of primary production determines the amount of energy available to higher trophic levels.
- The study of how chemical elements cycle through an ecosystem is termed biogeochemistry. A biogeochemical cycle can be expressed as a set of stores (pools) and transfers, and can be studied using the concepts of "stoichiometry", "mass balance", and "residence time".
- Ecosystem function is controlled mainly by two processes, "top-down" and "bottom-up" controls.
- A biome is a major vegetation type extending over a large area.
- Biome distributions are determined largely by temperature and precipitation patterns on the Earth's surface.

BIOGEOCHEMISTRY

As the terminology implies, Biogeochemistry is a borderline specialisation involving three basic branches of natural, physical and earth sciences. On one hand living organisms are involved and on the other substances such as nutrients, are made available to the living organisms either through soil (the pedosphere) or through water (the hydrosphere). Of course there are certain organisms which can get their food through air (atmosphere) also. The living organisms, while differing vastly among themselves in size, morphology and physiology, share, all the same a basic feature common to them all: ability to metabolise their requirements in the habitats they live. An individual organism is perishable but life process, sustained by successive generation, is perpetual. Biogeochemistry is all about impact of such living processes on our environment. The bulk of living matter is concentrated in the landmass of the world though at the continent-ocean interface and at the bottom of the sea they are present in abundance. The physio-chemical interaction between organisms and the land that sustain them have been a subject of study for over a century now.

More than two hundred years ago, The turnover of elements on the surface of the globe in which he postulated on the cyclic exchange of elements among the three kingdoms of nature, namely mineral, vegetable and animal. Lavoisier paid a heavy price for stating so: he was executed after his work was published. The interaction among the above three kingdoms essentially constitute the broad field of Biogeochemistry as we know today. The original kingdoms have been replaced by reservoirs of different kind and atmosphere is brought into the picture because of gases involved in many biogeochemical processes. Since life is cyclic, all processes where living matter is involved has also got to be cyclic. Hence, in all the reservoirs are linked to each other in both directions:at the inlet and outlet point for each reservoir. The interconnection of all the reservoirs leads to a overall global cycle.

Such a cycle is known by various names depending on the user and the sector of interest: biogeochemical cycle, hydrological cycle, geochemical cycle and rock cyclé. The emphasis in each cycle varies: in hydrological cycle, one will be interested in transfer of water; in the geochemical cycle, one will be interested in transfer from the crust to other segments and in rock cycle, the interest will be transformation of large volume of crustal materials across various interfaces. Whereas in the biogeochemical cycle the focus is on nutrients essential for the life processes and at the toxic substances determinant to such living things. Thus biogeochemical processes can be defined in three stages:

Stage-1: Elements to be grouped according to the needs: Essential, such as Carbon (C), Nitrogen (N) and Phosphorous (P) Minor, such as Sodium (Na), Potassium(K) etc. Trace, such as Iron (Fe), Manganese (Mn), Copper (Cu) etc.

Toxic, such as Mercury (Hg), Selenium (Se), Cadmium (Cd) (all the above groupings according to the requirements of biological system in the biological world)-the bio part of the topic.

Stage-2: The availability of the above from the earth, generally from the top few metres of the earth (the geocomponent, more specifically the pedosphere the geo part of the topic and now.

Stage-3: The physio-chemical mechanisms that make the needs of the biological systems from the pedospheric layers of the earth in response to micro-environmental functions such as pH (acidity), weathering and release of elements by their breakdown due to chemical interaction of individual minerals with water. This constitutes the chemical part of the topic.

Let this linkages using an essential element Carbon as an example. Referring to carbon is present in the atmosphere mainly as carbon dioxide (CO_2) and in small quantities as carbon monoxides (CO). In the hydrosphere, carbon is present as bicarbonate and carbonate the major factor in the hardness of water. In the lithosphere- pedosphere, it is present as calcium carbonate ($CaCO_3$) or some combination thereof. In the biosphere, carbon is present as the building block of the plants and animals in the form of variety of carbon compounds such as carbohydrates, proteins and amino-acids. Carbon enters the biological system either through geological process (weathering of carbon-bearing minerals such as limestone ($CaCO_3$) and /or chemical process directly from the atmosphere (CO_2)

by photosynthesis. When the organism (including plant) dies out, the carbon is released back either directly to the atmosphere as CO_2 or is mineralised as coal, limestone etc. So that it is returned to the lithosphere. Thus the carbon cycle is complete. The oldest carbon-bearing organic world remains known to man is about 3.5 billion years whereas the age of the earth is at present estimated to be about 4.5 billion years. In spite of the complexity of the biological evolution through geological time scale, carbon has been and is being cycled through various reservoirs relatively unaffected and un- disturbed till recent times.

Among the three essential nutrients for life phosphorous (P) is extremely interesting. Unlike carbon and nitrogen, it is predominately derived from the crustal top of the solid earth and mobilised through the biosphere by the transport behaviour in the hydrosphereP ignites in air but glows when left in darkness. Thus it is a carrier of energy in all living systems. The homologous elements P and N has complimentary properties in relation to Hydrogen(H) and Oxygen(O_2)- the other important elements for most living matter on the surface of the earth. P in association with O_2 introduces some structural order in the molecules at cellular levels and thus the P-O association in the form of PO_4 units separates life from ordinary inanimate minerals represented by similar association of Silicon with O_2 (SiO4)

From a single cell amoeba some 3.5 billion years ago (3500 million), organisms progressively went through complex biological evolution as a function of geological time scale. The link between time and evolution is established through changing environmental parameters such the carbon dioxide (CO_2) levels in atmosphere prevailing at a given time, the O_2 levels, nutrients such as P and minor and trace elements etc. Geochemical evidences through rock records show that about 2.5 billion years ago, the atmospheric CO_2 was at least 100 times lower as that of today and O_2 levels negligible compared to today s environment. have been continuously interacting among each other through such a vast time scale to maintain the cyclic behaviour of nutrients and other elements. Excessive constituents in any one segment was removed over a period of time to other segments so that its biogeochemical cyclic behaviour was not affected. During Gondwana times (about 400 to 500 million years before) when India, Australia and Antarctica were neighbours, excess CO_2 in the atmosphere was neutralised by enhanced photosynthesis and subsequently the biota decayed and gave raise to most of the coal deposits which we are exploiting today.

Thus due to the biogeochemical behaviour, nature was able to adjust itself at its own rate in any parameters. On the other hand in modern times, man introduces any one variable to a segment in very abruptly and in such large dose that nature is unable to adjust to that change. A case in point the emission of green house gas CO_2 by the burning of coal. The essential nutrient element C is transferred from the lithosphere to the atmosphere so fast that the excess CO_2 is unable to be removed equally faster by the hydrosphere leading to enhanced levels of CO_2

in the atmosphere and hence we hear everyday about the global warming On the other hand, let us take a element such as Mercury (Hg) considered to be very toxic to the biological world. The movement of Hg in cyclic fashion through our environment. Hg generally occurs in minerals in association with sulphur in the form of mercury sulphide (HgS) systems.

Even in rocks, it is considered to be trace element since its abundance in the crustal earth is very low. However, during the exploitation of base metals resources, such as that of copper, lead and Zinc, mercury is extracted as a buy-product. Modern civilization uses mercury for a variety of products such as chemicals, coatings on seeds etc. Even though it does not react with water in the hydrosphere easily, mercury combines with any organic substances such decayed plant debris etc and immediately becomes toxic to the living biota. The well known minemata decease in Japan was the result of such affinity of mercury for many biotic materials. Being the only metal known to man to be in liquid state at room conditions, it easily vapourises and hence get also into the atmospheric segment Hence due to man s use, what was once locked up safely in inert geological materials in rocks, have been mobilised to the biosphere, hydrosphere and atmosphere.

The story is basically same for a few other toxic metals such as Cadmium though the details may individually vary due to their differing biogeochemical behaviour in our present environment.

It is clear that processes in our environment are better understood by focussing on the interfacing areas. Thus Biogeochemistry is the ideal such frontier area of study and research. It is however a relatively new field and only in the last two decades various researchers across the world have been focussing their attention on this inter- disciplinary subject. Outside India there are a few places where separate institutes exists for this purpose. At about the same time the Ministry of Environment and Forests also approved the setting up of a ENVIS centre in JNU on the same subject under his direction. Though there no other specific institutions in India in this subject, there are individual researchers in many universities working on topics of interest to Biogeochemistry.

Thus, the area of Biogeochemistry is an emerging branch of multi-diciplinary approach involving all major branches of Sciences. From a humble beginning two decades ago, now this subject is well established in the international field and in India is fast catching up as an important tool of research for understanding processes in our environment. Sufficient infra-structure has now developed in India through a variety of individual and institutional efforts so that in the years to come the importance work on Biogeochemistry will be better understood and appreciated.

The involvement of a number of specialists is obvious when ones looks into the above three stages: for understanding the biotic behaviour, one should have a good understanding of ecological principles; the requirements for the ecosystem is provided by basic geochemical processes of weathering helped by microbiological and biochemical reactions in soils and over rocks; farther, from the source to the receptor, chemical and hydrological principles operate as a middleman. Hence within the broad field of Biogeochemistry itself, there are sub-specialisations depending on the background of the researchers- be a ecologist, geochemist or chemical hydrologist. Hence the word Biogeochemistry is truly an integrating field to better understand our present, past and future environmental problems.

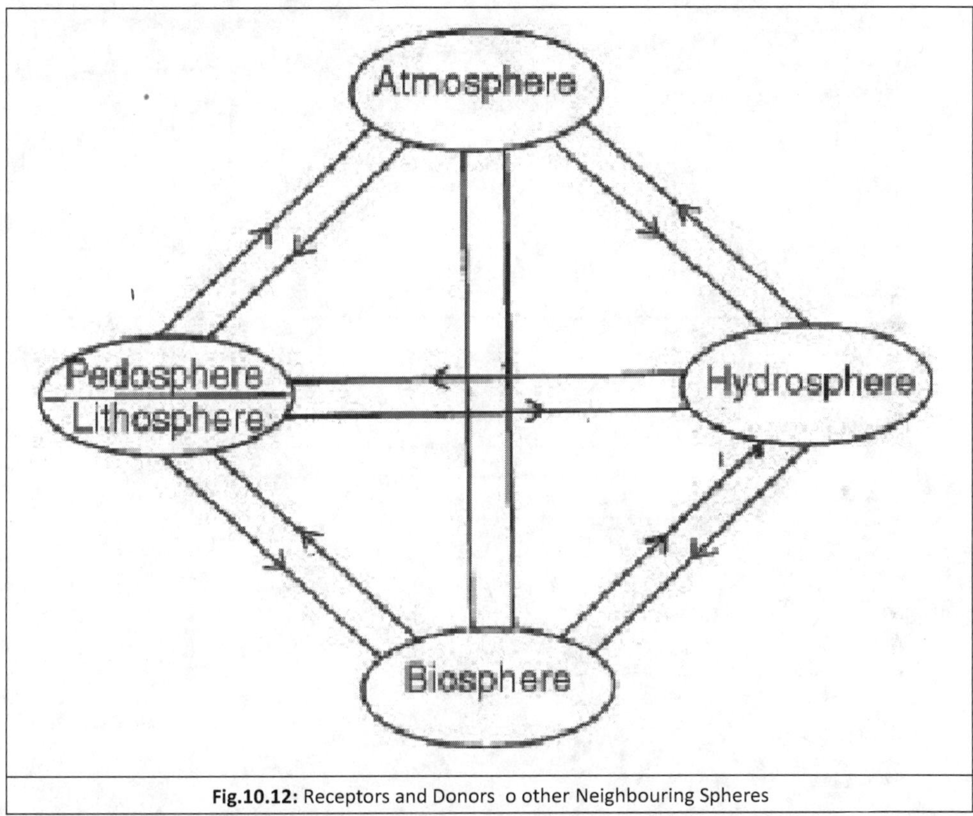

Fig.10.12: Receptors and Donors o other Neighbouring Spheres

Among the more recent sub-specialisations within Biogeochemistry that have emerged in the international research is the application of certain biochemical and molecular biochemical techniques to understand our present and past environment. A case in point is the study about the Alpine-man.

A full human body was found in tact in the frozen alps and several investigations on the body using most modern biochemical techniques have thrown some light about the habitat of the community some 5000 years ago, then the prevailing eco- system, palaeo-environment, anthropological lineage etc. DNA based study thus offers excellent scope even in environment related matter. In India, a student from our labouratory is working under the supervision of Prof. Laljit Singh, Centre for DNA Finger-printing at Hyderabad to apply DNA techniques in fossils bones of geological and anthropological importance to understand the habitat and palaeo-environment. DNA techniques to ancient biological specimens. Farther, characterisations of amino-acids in ancient sediments have been successfully used to understand palaeo-environment and also stresses acting on such systems. In India, however, this has not been attempted so far.

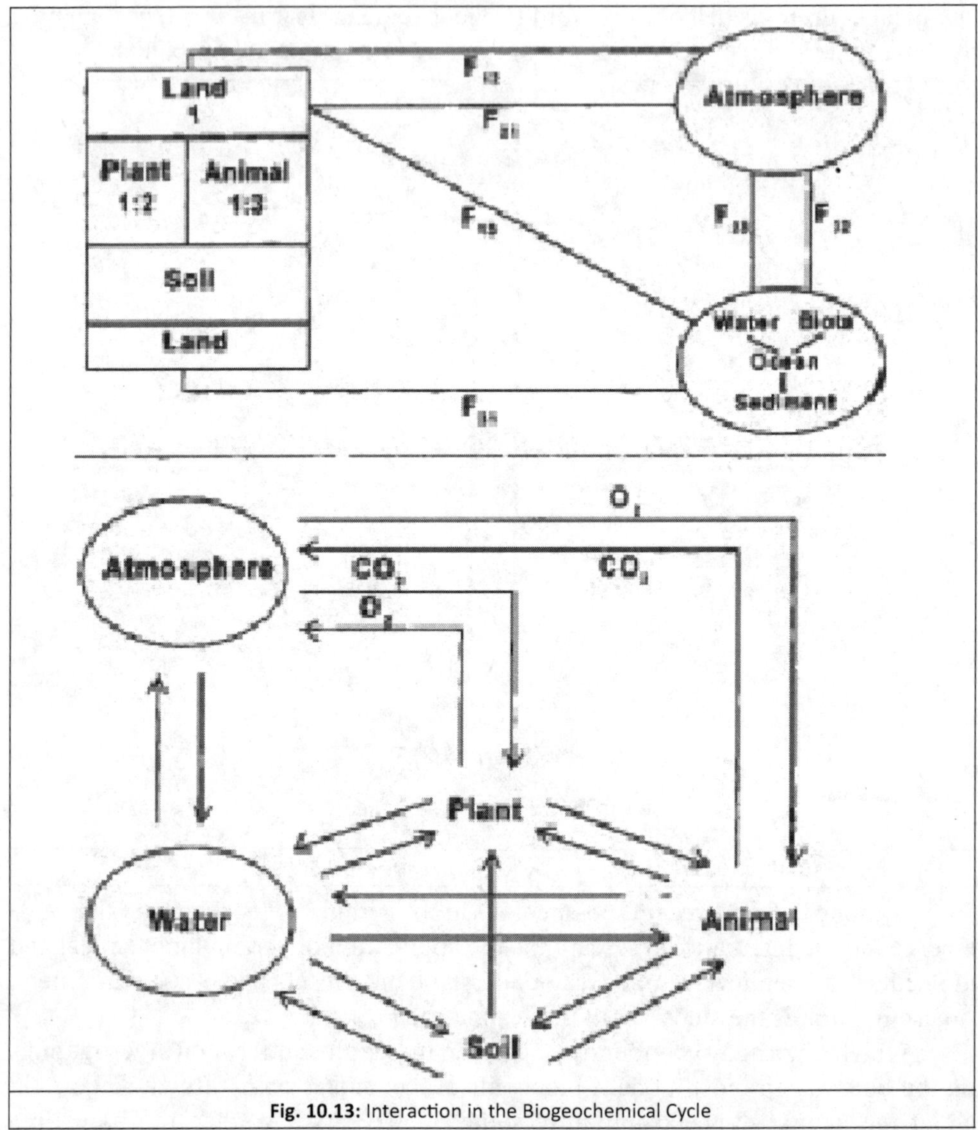

Fig. 10.13: Interaction in the Biogeochemical Cycle

Thus, Biogeochemistry can be viewed as a very challenging area of study involving a variety of basic sciences so that it is ideally suited to understand our environment better. Research should not go in the direction of reinventing the wheel but improve and if possible, replace it. Hopefully the 21st century consolidation of the field of Biogeochemistry as an established brach of science.

Enlarged version of specific component. whenevr possible. Important components in each segment and direction of transfer F means flux that is concentration for a given parameter the volume of water/air transferred in that direction per unit time. This, F has unit of wt/vol/time. F is likely to have volumes since the time involved is relatively shorter for some of the sectors. For eg. some of the variables transferred involving soil and plants/animals and plants and atmosphere etc. may have a time scale of a few seconds only.

The carbon transfer, the following fluxes have been incorporated: 1 to 34 point all have same magnitude, time scale and importance. Between river and young soils sediments, the transfer 26 and 27 represent deposition and erosion processes. Faster compared to deposition or accumulation so that net transfer is loss of carbon from soil (Pedosphere) to river (hydrosphere). Similarly, due to deforestation, evapouration may be higher than the net river volume.

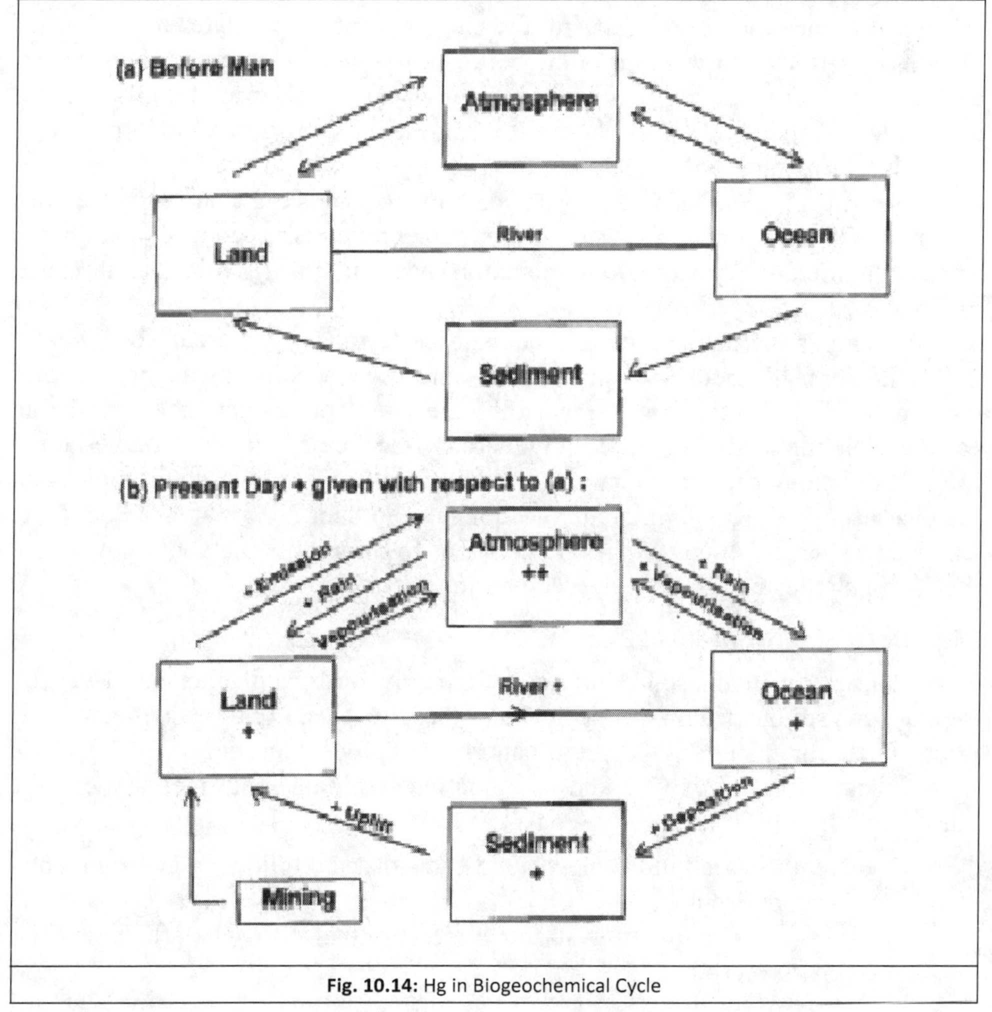

Fig. 10.14: Hg in Biogeochemical Cycle

Hg cycle is shwn in tow stages: pre-man and present day. this concept is adapted from Garrels et al., 1976. The main differences is indicated in teh Hg- mining in the present day cause additional flux-vaporisation-from land to atmosphere. ++ or + singns indicate higher gains in the respectinve sectors the present day cycle relative to pre-man cycle. As a consequence there is also poportional increase in the flxes in the present day cycle between the concerned sectors. Garels et al. (1976) estimate that Hg will be the first metal that is likely to be lost out entiely ot of the solid spheres (litho-and pedospheres) to the fluid (hydro- and atmo-) spheres in foreseabe future.

Mercury in the Environment

Mercury is a silvery, liquid metal at room temperature and is often referred to as one of the "heavy metals." Like water, mercury can evapourate and become airborne. Because it is an element, mercury does not break down into less toxic substances. Once mercury escapes to the environment, it circulates in and out of the atmosphere until it ends up in the bottoms of lakes and oceans. Mercury can be found as the elemental metal or in a wide variety of organic and inorganic compounds. Depending on its chemical form, mercury may travel long distances before it falls to earth with precipitation or dust.

Bacteria and chemical reactions in lakes and wetlands can change the mercury into a much more toxic form known as methylmercury. Fish become contaminated with methylmercury by eating food (plankton and smaller fish), which has absorbed methylmercury.

As long as the fish continue to be exposed to mercury, mercury continually builds up in fish's bodies. Fish that eat other fish become even more highly contaminated. Thus, the largest tend to be the most contaminated. When people eat the contaminated fish, the methylmercury can remain in their bodies for a long time. If they eat fish containing methylmercury faster than their bodies can get discharge it, the methylmercury accumulates in their bodies and can be toxic. Many states have fish consumption advisories to inform people about how many meals of fish they can safely eat over a period of time.

Mercury's Environmental Effects

Fish are the main source of food for many birds and other animals, and mercury can seriously damage the health of these species. Loons, eagles, panthers, otters, mink, kingfishers and ospreys naturally eat large quantities of fish. Because these predators rely on speed and coordination to obtain food, mercury may be particularly hazardous to these animals.

- Recent research in Minnesota indicates that the following environmental effects are occurring:
- Loons are accumulating so much mercury that it may be affecting their ability to reproduce;

- Elevated levels of mercury have been found in mink and otters;
- Walleye reproduction may be impaired by the fish's exposure to mercury.

Similar effects are being documented for other fish and fish-eating species around the United States and Canada. Has there always been mercury contamination, or is this a recent problem? This is a difficult question to answer,in part because of a lack of adequately preserved fish specimens of preindustrial age to compare against contemporary samples.

However, several lines of evidence from recent studies on Wisconsin lakes suggest that increased emissions to the atmosphere, and subsequent higher deposition rates to lakes, likely translate into higher mercury levels in fish.

The Mercury Cycle and Bioaccumulation

There is a constant biogeochemical cycle of mercury. This cycle includes:

- Release of elemental mercury as a gas from the rocks and waters (degassing);
- Long-range transport of the gases in the atmosphere;
- Wet and dry deposition upon land and surface water;
- Absorption onto sediment particles;
- Bioaccumulation (or biomagnification) in terrestrial and aquatic food chains.

Bioaccumulation means an increase in the concentration of a chemical in an organism over time, compared to the chemical's concentration in the environment. Bioaccumulation can be a normal and essential process for the growth of any species, but the accumulation of unnecessary chemicals or toxins, or even the overaccumulation of essential substances can be detrimental. All animals, including humans, daily bioaccumulate many vital nutrients, such as vitamins A, D, and K, trace minerals, essential fats and amino acids, but unfortunately, they can also accumulate many unnecessary substances, such as lead or mercury. What concerns toxicologists is the bioaccumulation of necessary substances to levels in the body that can cause harm.

With substances such as lead or mercury, any accumulation at all can be harmful. Compounds accumulate in living things any time they are taken up and stored faster than they are broken down (metabolized) or excreted.

Understanding the dynamic process of bioaccumulation is important in protecting humans and other organisms from the adverse effects of chemical exposure, and it has become a critical consideration in the regulation of chemicals.

Bioaccumulation varies among individual organisms as well as among species. Large, fat, long-lived individuals or species with low rates of metabolism or excretion of a chemical will bioaccumulate more than small, thin, short-lived organisms. Thus, an old lake trout may bioaccumulate much more than a young bluegill in the same lake.

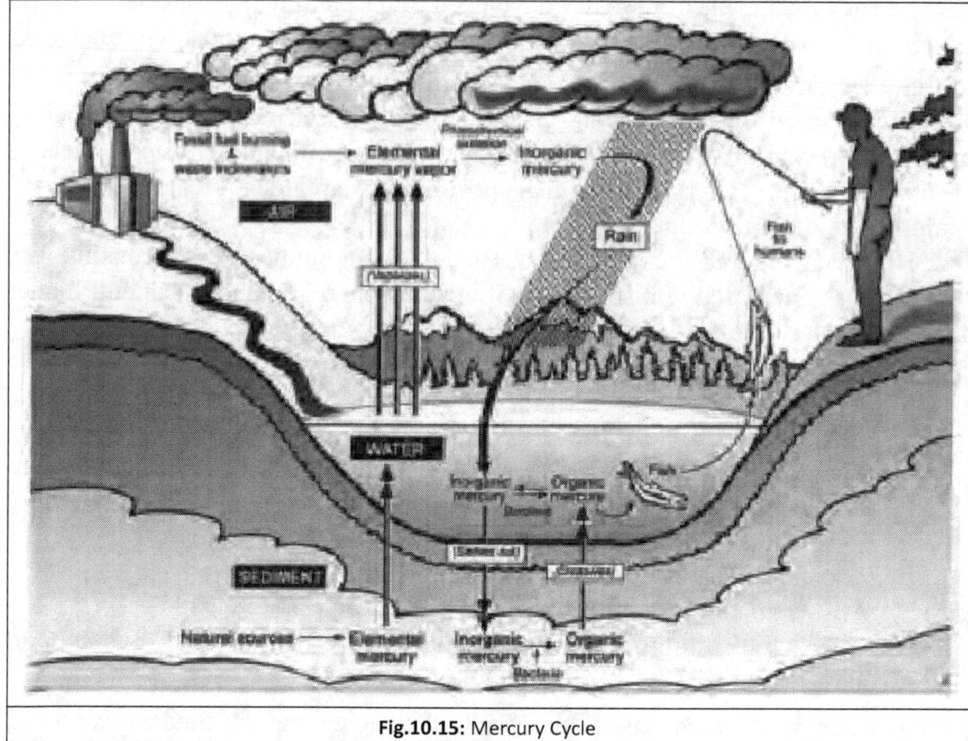

Fig.10.15: Mercury Cycle

Above is a schematic drawing of mercury cycling in an aquatic ecosystem. With the exception of isolated cases of known point sources, the source of most mercury to most aquatic ecosystems is deposition from the atmosphere, primarily associated with rainfall.

In the aquatic environment, mercury can be:

- Dissolved or suspended in the water
- Trapped in the sediments
- Ingested by living things (biota)

Methylmercury is the form of mercury most available and most toxic to biota (including zooplankton, insects, fish, and humans). This form of mercury is easily taken up by biota and bioaccumulates in their tissues. Unlike many other fish contaminants, such as PCBs, dioxin, and DDT, mercury does not concentrate in the fat, but in the muscle tissue. Thus, there is no simple way to remove mercury-contaminated portions from fish that is to be eaten.

LIFE AND BIOGEOCHEMICAL CYCLES

The Sulfur Cycle

Another example of a major biogeochemical cycle of significance to climate and life is the sulfur cycle. Living things require certain safe, low levels of this nutrient. The sulfur cycle can be thought of as beginning with the gas sulfur dioxide (SO_2) or the particles of sulfate (SO_4) compounds in the air. These compounds either fall out or are rained out of the atmosphere. Plants take up some forms of these compounds and incorporate them into their tissues. Then, as with nitrogen, these organic sulfur compounds are returned to the land or water after the plants die or are consumed by animals. Bacteria are important here as well since they can transform the organic sulfur to hydrogen sulfide gas (H_2S). In the oceans, certain phytoplankton can produce a chemical that transforms to SO_2 that resides in the atmosphere. These gases can re-enter the atmosphere, water, and soil, and continue the cycle.

In its reduced oxidation state, the nutrient sulfur plays an important part in the structure and function of proteins. In its fully oxidized state, sulfur exists as sulfate and is the major cause of acidity in both natural and polluted rainwater. This link to acidity makes sulfur important to geochemical, atmospheric, and biological processes such as the natural weathering of rocks, acid precipitation, and rates of denitrification. Sulfur is also one of the main elemental cycles most heavily perturbed by human activity. Estimates suggest that emissions of sulfur to the atmosphere from human activity are at least equal or probably larger in magnitude than those from natural processes. Like nitrogen, sulfur can exist in many forms: as gases or sulfuric acid particles. Sulfuric acid particles contribute to the polluting smog that engulfs some industrial centres and cities where many sulfur containing fuels are burned.

Such particles floating in air (known as sulfate aerosols) can cause respiratory diseases or cool the climate by reflecting some extra sunlight to space. The lifetime of most sulfur compounds in the air is relatively short (e.g. days). Superimposed on these fast cycles of sulfur are the extremely slow sedimentary-cycle processes or erosion, sedimentation, and uplift of rocks containing sulfur. In addition, sulfur compounds from volcanoes are intermittently injected into the atmosphere, and a continual stream of these compounds is produced from industrial activities. These compounds mix with water vapour and form sulfuric acid smog. In addition to contributing to acid rain, the sulfuric acid droplets of smog form a haze layer that reflects solar radiation and can cause a cooling of the earth's surface. While many questions remain concerning specifics, the sulfur cycle in general, and acid rain and smog issues in particular are becoming major physical, biological, and social problems.

BIOGEOCHEMICAL CYCLES

The Earth is a closed system for matter, except for small amounts of cosmic debris that enter the Earth's atmosphere. This means that all the elements needed for the structure and chemical processes of life come from the elements that were present in the Earth's crust when it was formed billions of years ago. This matter, the building blocks of life, continually cycle through Earth's systems, the atmosphere, hydrosphere, biosphere, and lithosphere, on time scales that range from a few days to millions of years. These cycles are called biogeochemical cycles, because they include a variety of biological, geological, and chemical processes. Many elements cycle through ecosystems, organisms, air, water, and soil. Many of these are trace elements. Other elements, including carbon, nitrogen, oxygen, hydrogen, sulfur, and phosphorus are critical components of all biological life. Together, oxygen and carbon account for 80 per cent of the weight of human beings. Because these elements are key components of life, they must be available for biological processes.

Carbon, however, is relatively rare in the Earth's crust, and nitrogen, though abundant in the atmosphere, is in a form that is not useable by living organisms.The biogeochemical cycles transport and store these important elements so that they can be used by living organisms. Each cycle takes many different pathways and has various reservoirs, or storage places, where elements may reside for short or long periods of time. Each of the chemical, biological, and geological processes varies in their rates of cycling. Some molecules may cycle very quickly depending on the pathway. Carbon atoms in deep ocean sediments may take hundreds to millions of years to cyclecompletely through the system. An average water molecule resides in the atmosphere for about ten days, although it may be transported many miles before it falls back to the Earth as rain.

How fast substances cycle depends on its chemical reactivity and whether or not it can be found in a gaseous state. A gaseous phase allows molecules to be transported quickly. Phosphorous has no gaseous phase and is relatively uncreative, so it moves very slowly through its cycle. Phosphorus is stored in large amounts in sediment in the oceans or in the Earth's crust and is recycled back to the surface only over very long periods of time through upwelling of ocean waters or weathering of rocks.

Biogeochemical cycles are subject to disturbance by human activities. Humans accelerate natural biogeochemical cycles when elements are extracted from their reservoirs, or sources, and deposited back into the environment (sinks). Humans have significantly altered the carbon cycle by extracting and combusting billions of tons of hydrocarbons in fossil that were buried deep in the Earth's crust, in addition to clearing vegetation that stores carbon. Global release of carbon through human activities has increased from 1 billion tons per year in 1940 to 6.5 billion tons per year in 2000. About half of this extra carbon is taken up by plants and the oceans, while the other half remains in the atmosphere.

In addition to carbon cycle, humans have altered the nitrogen and phosphorus cycles by adding these elements to croplands as fertilizers, which has contributed to over-fertilization of aquatic ecosystems when excess amounts are carried by runoff into local waterways.

Researchers are trying to understand all of the various pathways and flows of each of the biogeochemical cycles in order to understand how human activities affect these cycles. While many important processes have been understood for more than century, there are many phenomena that scientists are just beginning to investigate. Satellite technology, among other tools, has revealed new information about interactions between the oceans and atmosphere that contribute to knowledge about the carbon cycle, but there remain many unanswered questions.

Human Modifications of Climate Services

Human activities are significantly perturbing all of these biogeochemical cycles as well as other earth system processes, both directly through industrial processes and indirectly through changing distributions and abundance of life. The atmosphere is of particular importance to the perturbations due to its crucial role in mediating all energy that enters and leaves earth. Overall, the atmosphere is the component that controls the dominant energy flow in the earth's climate system, and solar radiation from the sun provides the energy to make the weather machine work.

Embedded in this process are the biogeochemical cycles we have described that operate on a variety of time and space scales and help to regulate flows of energy and materials throughout the earth system Yet, while we understand much about the functioning of separate parts of this system, there is still a great deal to be discovered about the feedbacks and linkages that allow these interconnected parts to function as a whole and, in turn, how they will respond to human modification.

Human Disturbance

Life influences the amount of CO_2 in the atmosphere through photosynthesis, respiration, and oceanic absorption. As ecosystems are altered, the balance of these processes will be altered. Human activities are upsetting this balance and increasing CO_2 in the atmosphere through the burning of fossil fuels and clearing of forests.A significant increase in CO_2 could have dramatic consequences. Mathematical models of the climate suggest that when CO_2 doubles (sometime in the middle of the next century should population, economic and technology trends continue as typically projected), the world will warm up somewhere between 1 and 5 degrees Celsius by 2100 A.D. unless other factors counteract or amplify the CO_2-induced change Even the lower end of that range is a projected warming at the rate of one degree per hundred years, a factor of ten faster than the one degree per thousand that has been the typical average rate of natural sustained global temperature change from the end of the ice ages to warmer interglacial times. Should the higher end of the 1 to 5 degree Celsius range occur, then we could see rates of climate change some fifty times faster than sustained, natural average conditions. Climate largely determines the types of ecosystems that occupy an area. Global climate change at such a rapid rate would force many species to shift their ranges in an attempt to keep up with changing climatic conditions, as occurred during the ice age-interglacial transition ten to fifteen thousand years ago.

Migrations of species such as slow growing trees with large seeds would have to occur much faster than they did in the past to keep up with rapidly shifting climates. Other species could move more easily, raising the likelihood that communities of species could be disassembled. Estimating the rates of global warming in the next century, however, are very controversial because of the uncertainties involved with multiple interacting feedback mechanisms Humanity can control climate in ways other than changing greenhouse gas concentrations. Consider the amount of moisture released to the atmosphere through transpiration in the tropical rainforests.

The dense vegetation in areas such as the Amazon basin typically recycles the precipitation that falls on it many times over, helping to form heavy cloud cover in the region. The clouds, in turn, reflect sunlight and produce more rain, directly influencing regional climate as well as indirectly perturbing global climate through altering large-scale circulation patterns over the tropics. As humanity deforests regions like the Amazon, not only is CO_2 released into the atmosphere, but changes in the hydrologic cycle will almost certainly affect regional climate and possibly even global climatic patterns. In deforested areas of northeastern Brazil, the cutting of the tropical forests has led to desertification, changing both surface reflectivity and the rate of transpiration. This change in ecosystem character can lead to a destabilizing positive feedback, which may cause an even further reduction in precipitation.

In a recent study on the possible climatic impacts of tropical deforestation, researchers suggest that conversion of forest into crop land or pastures would cause significant changes in the local microclimate Expected changes include reduction in soil moisture, larger diurnal fluctuation of surface temperature and humidity deficit, and increased surface runoff during the rainy season and decreased runoff during the dry season. Results from general circulation model simulations of large-scale deforestation and conversion to grassy vegetation in the Amazon basin indicate an increase in surface temperature, decrease in evapotranspiration, and significant reduction in precipitation.

Depending on the scale of the disturbed areas, local climate changes can lead to regional climate changes which, in turn, may cause alterations in the global climate through atmospheric connections between tropical circulation and large-scale circulation patterns outside of the tropics. The effect on the ecological systems through changes in the hydrological cycle, an increase in the dry season, and the disruption of plant-animal interactions may make it difficult for the rainforests to re-establish themselves if they are destroyed. Climate change aside, the implications of this scenario for the conservation of biodiversity are serious.

The provision of fresh water and regulation of its flows through precipitation, evapouration, transpiration, and run off is mediated by all ecosystems. Forests and other vegetation types are critical components of this ecosystem service providing free flood and drought relief among other things. The loss of these services, through landuse change, can exacerbate disasters like spring floods in the Midwest and Southeast resulting from large expanses of land cleared for agriculture as well as the drainage of wetlands and swamps which otherwise might have acted as reservoirs for holding excess water or filtering toxic wastes.

Valuing Carbon

There is already a historic background on the evaluation of carbon that includes climate change policies such as the introduction of carbon taxes that reduce greenhouse gas emissions through increasing prices of carbon-based fuels proportional to the amount of carbon they emit Another mechanism for reducing greenhouse gas emissions is through an international tradable emissions permit system intended to limit emissions of certain pollutants. These policies and others constitute ways of balancing the economic costs of emissions with some assumed benefit of averting the loss of ecosystem services (called "climate damage"). Any comprehensive attempt to evaluate the societal value of climate change should include such things as loss of species diversity, loss of coastline from increasing sea level, environmentally displaced persons, and agricultural losses. first estimated the climate damage at 1% reduction in GNP based on market sector losses for a central estimate of climate change.

This was criticized as too narrow a view of climate as a type of public good since it reflected neither non-market values (e.g. species loss) nor climate "surprise" scenarios conducted a survey of conventional economists, environmental economists, atmospheric scientists, and ecologists.

Their estimates of loss of gross world product (GWP) resulting from a 3 degree Celsius warming by 2090 varied between a loss of 0 and 21% of GNP with a mean of 1.9%). For a 6 degree Celsius warming scenario, the respondents predicted a loss of the world economy ranging from 0.8 to 62% with a mean estimate of 5.5%. A striking difference was noted between respondents from different academic disciplines, with natural scientists' estimates of economic impact 20 to 30 times higher than conventional economists'. Even a two per cent loss of GWP in 1995, however, represents climate damage of hundreds of billions of dollars annually.

While it is impossible to estimate credibly a numerical value on all of the ecosystem services provided through the maintenance of the carbon cycle at present state, it may be useful to look at land use change and loss of biomass, mostly through deforestation, as a source of atmospheric CO_2. In a very simplistic and preliminary evaluation, we can use the rates of net deforestation to calculate a value for carbon. Global loss of above-ground biomass from deforestation in the tropics is approximately 1-3 gigatons/year over the past 10 years.

This amounts to between 2-5 gigatons of carbon in carbon dioxide released into the atmosphere each year from deforestation and forest degradation (this does not include the 6 gigaton carbon emissions from the burning of fossil fuels). Much of the carbon from biosphere emissions is taken up immediately by vegetation, however, leaving approximately 1-2.5 gigatons of net carbon added to the atmosphere each year. We can apply the concept of carbon taxation for emissions to an ecosystem service valuation of retaining the carbon in the forests.

Using a range of carbon taxes from typical macroeconomic models between $1 per ton and $100 per ton of carbon, the net value of carbon lost each year amounts to between $1 and $250 billion/year. However, use of optimizing economic models to estimate climate damage is highly unsatisfying since these studies use very limited and often ad hoc assumptions that both over and underestimate the likely damages to various market and non-market sectors.

Methods of Valuation

The need for alternative methods of evaluation of these climate-related ecosystem services is quite clear when examining preliminary public opinion responses of global warming. In a controversial method called contingent valuation respondents are surveyed to determine how much they would be willing to pay to prevent a given global climate change scenario from happening or accept if so much change were to be allowed. The difficulties with this type of valuing of environmental goods and processes are immense, especially since much of the evaluation is subjective. Public opinion depends, in part, upon people's exposure to the issues and the level of education and information on these issues they have received.

In a Southern California study, the contingent valuation technique was applied to the determine the influence of potential changes in temperature and precipitation resulting from global warming on respondents willingness to pay. Factorial survey methods were used to present a variety of hypothetical climate scenarios to a sample of 600 Southern California residents. Respondents were provided with a baseline microclimate for the region before future climate scenarios were evaluated.

For residents living in coastal communities, the baseline climate over the past 10 years was described as having a summer average high temperature of 75 degrees Fahrenheit, with daily highs ranging between 70 and 80 degrees, and an average of 13 inches per year of rain. One possible future scenario over the next 10 years included a summer average high temperature of 100 degrees Fahrenheit, with daily highs ranging from 80 to 120 degrees (the latter typical of Death Valley, California), and an average of 20 inches per year of rain.

With these and other scenarios, predicted probabilities were determined from the respondents willingness to pay for the abatement of different mean high temperatures. In this scenario, respondents were willing to pay an average $140 to offset a mean high temperature of 100 degrees, while a mean high temperature of 80 degrees was worth approximately $100 This represents a 40% increment in willingness to pay for a 20 degree rise in temperature and other scenario characteristics. residents reached a plateau in their willingness to pay to prevent 120 degree Fahrenheit mean temperatures as compared to 110 degree Fahrenheit mean temperatures.

However, the actual damages to the L.A. basin residents of mean high temperatures of 110 or more degrees Fahrenheit (which would imply occasional extreme heat waves similar in temperature to Death Valley mean highs) would be orders of magnitude more costly, we believe, than a 100 degree Fahrenheit mean high temperature, as such extreme heat would decimate most existing vegetation and threaten the lives of tens of thousands of elderly and other persons vulnerable to heat stroke.

For just such reasons, Berk and Shulman strongly caution against taking the dollar values from the survey literally or using them in cost-benefit analyses as they confound several source of value including stewardship and altruism. In addition, some of the climate increases were well above the range of current scientific estimates of greenhouse warming.

The survey was not done in conjunction with atmospheric scientists and climatologists who could provide more realistic climate scenarios or ecologists, public health officials, or others who could help the respondents realise what such warming might mean for trees, birds, or people. Contingent valuation of the hypothetical good is possible when people believe the survey scenario.

We present this type of evaluation study to highlight how difficult it is to find acceptable methods to place values on the climatic components of ecosystem services. In this survey case, the background of the respondents as well as their (limited) prior knowledge of the impacts of greenhouse warming played a large role in the survey outcomes.

At the same time, however, contingent valuation points out that people are willing to pay to preserve ecosystem services as well as the tremendous need for education to help citizens more realistically value climate and other environmental

The ongoing disturbances of the atmosphere that affect the biogeochemical and physical processes that determine the climate may influence human and natural systems in profound ways. We have attempted to outline a few of the major ecosystem services that are associated with climate and the atmosphere, as well as introduce the challenging task of quantifying, and ultimately monetizing, these services.

Current monetized estimates of climate damage by the middle of the 21st century from typical climate change scenarios range from slight economic benefit to a trillion or more dollars lost annually, with most macroeconomic assessments assuming a one to two per cent annual loss to GWP from climate change. Moreover, the interacting processes and biogeochemical cycles occurring across a wide spectrum of scales lead to synergistic effects that are not usually considered and sometimes not even known when we attempt to disaggregate and value ecosystem services.

The deforestation of the Amazon basin is one example of interacting scales where land use change affecting local and regional climate may also produce a net global residual. Even if the mosaic of regional effects average themselves out globally, there could be residual effects arising from heterogeneous forcing of the climate in areas outside of the tropics (i.e. regional high concentrations of sulfate aerosols or tropospheric ozone).

Ecosystems both mediate and respond to the climate system through a variety of physical, biological, and chemical feedback cycles. The uncertainty of resulting synergisms and potential global effects, as exemplified in the Amazon basin, points to the important challenge of defining and understanding the processes that link species and ecosystems with climate. With increasing knowledge, we can better anticipate ecological responses under changing climate scenarios. Meanwhile, humanity continues to perform this potentially trillion dollar unnatural experiment on "Labouratory Earth".

Air, Climate and Weather

The Earth's atmosphere is a blanket of gases approximately 350km (218 miles) thick. It is a large and complex system that interacts with the Sun, the land, and the oceans in order to produce both the Earth's weather and climate.
The Earth's atmosphere has four distinct layers

- Thermosphere
- Mesosphere
- Stratosphere
- Troposphere
- Thermosphere

The layer of atmosphere most distant from the Earth is the thermosphere, which begins approximately 80km in altitude. It is also the hottest layer; "thermo" being Greek for heat. The temperatures in the thermosphere increase with altitude due to the absorption of intense solar radiation by the limited amount of remaining molecular oxygen. The source of this heat is through bombardment of solar particles carried on the solar wind that do not reach deeper into the atmosphere.

Mesosphere

The mesosphere extends from 50 to 80km in altitude with very sparse atmosphere, accounting for only about 0.1 per cent of the mass of the atmosphere as a whole. Temperatures decline within the mesosphere as altitude rises, containing the coldest temperatures within the Earth's atmosphere. At its upper boundary, the mesopause, average temperatures are near -110°C in the summer and -60°C in the winter.

Stratosphere

The stratosphere extends from approximately 10-12 km to around 50 km above the Earth's surface. The air temperature remains relatively constant up to an altitude of 25 km, then increases gradually having a stabilizing effect on atmospheric conditions. The stratosphere contains nearly 90 per cent of the atmospheric ozone, which plays a major role in regulating temperatures as solar energy is converted to kinetic energy when the ozone molecules absorb ultraviolet radiation, resulting in the heating of the stratosphere.

Troposphere

The troposphere is the layer closest to the Earth's surface, containing more than 80 per cent of total atmospheric mass composed of 78 per cent nitrogen, 21 per cent oxygen, other trace gases, water droplets, dust, and other particles.

The troposphere is where most weather occurs; the circulation of air in intensive vertical movements results in the formation of clouds, while horizontal movements results in wind. Both temperature and water vapour content decrease rapidly as altitude increases within the troposphere. Nearly 99 per cent of atmospheric water vapour is contained within this layer, which plays a major role in regulating air temperature as it absorbs solar energy and thermal radiation from the planet's surface.

Scientists believe that the Earth's climate is changing and is, in fact, heating up. However, there are considerable differences among the views with regard to the rate of change, the impact on our environment, and what can or should be done about it. Current environmental concerns are increa-singly focusing on the particles and gases released into the atmosphere as a result of human activity. Frequently, unusual weather events, such as Hurricane Katrina in 2005, are being linked to global warming. But "weather" and "climate" are two very different things, and it is important to understand the difference, as well as the relation, between the two.

In the troposphere, air rises as it is heated by the sun, falls earthward as it cools, and intermixes with evapourated water from the planet's bodies of water to form clouds and precipitation. The uneven heating of the Earth's surface by the sun, along with Earth's rotation, creates rising (convection), falling (advection), and horizontal air movements (winds). The result of these processes occurring in the form of rain, snow, heat or freezing cold, at a particular place and time, is called weather. The longer term trends in patterns of temperature, rainfall, and other weather indicators over time, usually in blocks of 30 years that can affect the entire Earth, is called climate.

Land and Land Use

Land is the solid part of the Earth's surface, also known as the lithosphere. The surface of the Earth is shaped by a combination of physical processes, including earthquakes and volcanoes, shifts of rocks and sediments, and flows of river and ice. Human activities also shape the land in many ways, including expanding agriculture and population, excavating mineral and forest resources, and changing the flow of rivers with dams and channels.

Land cover is the physical and biological material found on the land surface – which can be either vegetation or man-made structures. Land cover differs from land use, although the way land is used is often either a cause or an effect of its cover type. Land use describes the various ways in which human beings make use of and manage the land and its resources. The way land is used and managed can alter land cover, and can also have an impact on surrounding habitats, the flow of surface waters, and on biogeochemical cycles.

Land cover and land use issues, therefore, can play a significant role in environmental change, including climate change. When land cover is altered, it can result in the destruction of natural carbon sinks that absorb atmospheric carbon dioxide, a heat-trapping gas that contributes to climate change.

Deforestation, is a major contributor to observed increases of carbon dioxide in the atmosphere. There are often competing interests and tradeoffs related to land use that must be taken into consideration. Land use planners and decision makers often struggle with how to meet human needs for food and the utilization of our natural resources while also minimizing impacts and preserving important habitats. These choices are typically made by those who own or control the land; although the choices can also be limited by the physical and biological characteristics of the land, including the climate, soils, and geography.

Land use choices are also limited by institutional factors and, in the United States, can be governed by a mix of local, county, state, and federal laws. Although most of the land. is privately owned, the federal government is the single largest owner of land. Analysis of land use and land cover has been greatly facilitated in the last few decades with the availability of new technologies, including remote sensing. The development of geographic information systems (GIS) has made it possible to combine physical characteristics of the land with information about population, economic data, and other information for further analysis

Water

"Water, water, everywhere, nor any drop to drink." More than 70 per cent of the Earth's surface is covered by water. When the Apollo space missions beamed back pictures of the Earth, public attention was caught by the image of a "Blue Planet," because of the vast expanses of ocean covering the globe. Yet despite this seemingly inexhaustible supply (approximately 370 billion billion gallons) of water on earth, access to clean drinking water remains a major challenge for a large percentage of the world's population. Almost 97 per cent of the Earth's water is in the oceans and is too salty to drink. Less than 1 per cent of the water on the earth is fresh water (the remaining two to three per cent is frozen fresh water contained in glaciers and ice caps).

Water is indispensable for life, not only for drinking water, but also for raising crops and animals for food. Fishing provides a major source of protein for much of the world's population. Throughout history, cities and villages have grown up near sources of water, for drinking, sustenance, and transportation. In this century, population growth has increased in areas where fresh water is relatively scarce.

Water in these areas is a valuable commodity and, in some cases, great technological feats have been undertaken to provide a reliable source of water. In some areas, water quality is a serious concern and waterborne diseases remain a significant source of mortality in many developing countries. The United States has made considerable progress over the last few decades in reducing direct discharges of pollutants into lakes, rivers, and streams. However, indirect sources of pollutants, including both nutrients and toxic substances that are carried by flowing surface waters into larger bodies of water, have proven more difficult to control.

Ecosystems

An ecosystem is generally defined as a community of organisms living in a particular environment and the physical elements in that environment with which they interact. But where does one particular ecosystem end and another begin? An ecosystem can be as small as a field or as large as the ocean, depending on the scale that the researcher is examining. The borders of an ecosystem may be clear, such as a pond; other borders may be less easy to define, such as grassland that gradually changes into brush.

Just as there is an immense diversity of individual species on the planet, so is there a rich diversity of ecosystems, from the icy arctic zones to tropical forests lush with plants and animals. Even the depths of the oceans, once thought to be barren, are now known to be teeming with living microorganisms and other life. There is much that remains to be discovered. Biologists do not know with any certainty how many species there are or even why some areas, such as the tropics, are richer in biological diversity than others.

It is known that human activities, mostly unintended, threaten biodiversity by altering habitats and introducing non-native species. There is general agreement on the importance of protecting biological diversity, especially since humans depend on the services provided by living organisms and ecosystems. There is less agreement on the best approaches to conservation and how to balance preservation of habitats with meeting human needs.

Energy

The development of modern civilization has been dependent on both the availability and the advancement of energy. We have witnessed a progression from animal and steam power, to the internal combustion engine and electricity generation and to the harnessing of alternative sources of energy. Because of our reliance on energy sources, it is also important to understand the impactsof energy use on the environment. All aspects of energy the way it is produced, distributed, and consumed can affect local, regional, and global environments through land use and degradation, air pollution, the acidification of water and soils, and through global climate change via greenhouse gas emissions.

The majority of our current energy stems from fossil fuels such as coal, oil, and natural gas. Coal dominates in our production of electricity, while oil is the world's primary transportation fuel. Natural gas use, which is most commonly as heating fuel, is currently growing faster than other fossil fuels. While cleaner and less carbon intensive than both coal and oil, its use still emits significant amounts of carbon dioxide.

Over the foreseeable future, it is very likely that fossil fuels will remain our largest source of energy. However, fossil fuels are finite resources and there is concern not only about both domestic supply and reliance on foreign supplies, but also with the increasing cost of these fuels. The burning of fossil fuels is also the largest source of carbon dioxide emissions, which contributes to the greenhouse effect. As these fuels continue to decrease in supply and impact the environment, there will be an increasing need to utilize alternative fuel sources.

Nuclear energy provides nearly a fifth of the world's electricity and does not produce harmful by-products, only non-radioactive water vapour. However, controversy remains over the use of nuclear energy due to concerns over safe storage and disposal of radioactive waste, and the potential for accidents and radiation contamination and exposure.

Hydrogen is the most abundant element in the universe and could be an important factor in our energy future. Hydrogen fuel cells can produce power ithout emitting any pollutants; their only byproducts are water and heat. ydrogen can both carry and store energy and can be used in a wide variety f applications, including portable devices that use batteries, transportation vehicles, and a number of stationary power sources. The majority of hydrogen is currently produced by processing fossil fuels (which emits pollutants through the initial process), but it can also be manufactured from renewable energy sources and feedstocks through the electrolysis of water.

With the increased cost and environmental impacts related to fossil fuel use, along with the controversy over nuclear power, research and development in the area of renewable sources of energy continues to flourish. Alternative sources – such as wind, solar, geothermal, and water – have been used in one form or another for many centuries, but will require additional advancement before they can become truly cost-competitive with conventional energy sources.

Wind is an abundant, inexhaustible, and clean source of energy; and, advancements in the design of wind turbines has lowered the cost of wind-generated electricity. However, wind variability, noise pollution caused by the turbines, and the potential to harm birds continue to be areas of concern.

Solar technology obtains power from sunlight. There are three types of solar technology: solar electric, solar hot water, and concentrating solar power. The cost of solar power has declined over the past 30 years, and will continue to decline with increased technological efficiency of technology and manufacturing improvements. However, it is also subject to changes in weather (cloud cover) and pollution, which can block sunlight and decrease overall productivity. Geothermal energy taps into heat underneath the Earth's crust to boil water that is then used to drive electric turbines to heat buildings, homes, or in other non-electrical purposes. Over the past two decades, geothermal energy costs have decreased nearly 50%. However, while abundant, in only a few locations is it close enough to the Earth's surface to provide for large scale power.

Water–or hydroelectric–power has been harnessed for thousands of years, beginning with the water wheel. For more than a century, the technology for sing falling water to create hydroelectricity has evolved to its current use of turbines and generators to convert the energy into electricity, which is then fed into the electric grid. Although a widely accessible energy source in many countries–concerns over land use, impacts on fish populations, and water flow and quality remain.

One effective way of reducing environmental impacts from the current production of energy is to increase energy conservation. By using energy more efficiently, both costs and emissions are lowered; therefore, a significant incentive exists to improve efficiency in manufacturing processes, the construction of new buildings, the production of new appliances, and for many other energy uses.

Several developed countries are already taking action to increase energy efficiency and, on a larger scale, new measures are being explored to assist developing countries with energy conservation as they continue to increase their energy consumption. Under the Kyoto Protocol, developed countries an provide technological assistance through "clean technology" projects with developing nations and can even put the emission reductions toward compliance in their own countries.

As human consumption of energy increases and we come closer to using up our supply of fossil fuels, further research and development will be necessary to produce alternative and/or renewable sources of energy that are readily available, affordable, and less harmful to the environment. While our dependence on energy is not likely to decrease, there will continue to be an increase of new innovations in energy technologies with a larger focus on energy conservation and efficiency.

Energy Information Administratin (EIA): The EIA also provides information on the various types of energy, including renewable, non-renewable, and secondary sources. Congressional Research Service: Energy Issues Policy papers on energy issues can be found on this site, including articles and several reports related to energy policy and environmental protection. Food

Human beings began domesticating plants and animals as long as 10,000 years ago. Success in this endeavor led to drastic changes in how and where human beings lived. It changed how humans interacted with each other and how they interacted with the earth. Agriculture led humans from a nomadic existence to one based in permanent and semi-permanent settlements. Human civilization has its roots in the early domestication of plants and animals.

The unprecedented growth in global population that occurred in the twentieth century was made possible by remarkable advances in agriculture and technology.

The Green Revolution brought high-yield crops and advanced growing techniques to developing countries and improved nutrition and health for most parts of the world. Continuing population growth will require more acres of land to plant and more water for irrigation, increasing pressure on habitats and natural resources.

Advanced agricultural methods have made it possible to grow more food on fewer acres of land, which permits much land to be returned to forest and other uses. However, there are costs associated with high-yield methods, including their heavy reliance on pesticides and fertilizers. Runoff of fertilizers from agricultural lands has affected the water quality of some lakes, rivers, and bays.

Erosion is, and has been since crops were first planted, a serious environmental problem in the absence of preventative measures. Irrigation, required in many parts of the world, can be a significant drain on water supplies in arid areas and, if improperly managed, can lead to buildup of salt deposits, which may severely degrade soil quality.

One of the most critical challenges of the next few decades is to find ways to increase production of food while minimizing environmental degradation. Considerable research in this and other countries has led to advanced methods to preserve soil and prevent erosion, including conservation tilling and computer-controlled application of pesticides, fertilizers, and water. Pesticide use is controversial. Other methods, especially new types of genetic modification, or "genetic engineering" are more controversial. To reduce reliance on pesticides, methods such as biocontrol (using other species to control pests) and biotechnology to create pest-resistant crops are being tested. Biocontrol using wild-type (naturally occurring) species is usually accepted, especially by organic producers. Naturally occurring mutants are accepted as well. More controversial is the use of genetically modified biocontrol agents changed by DNA deletion or insertion of new traits. Concerns about the potential impacts of these methods have been raised.

Environment and Society

Throughout history humans have both affected, and been affected by, the natural world. While a good deal has been lost due to human actions, much of what is valued about the environment has been preserved and protected through human action. While many uncertainties remain, there is a realization that environmental problems are becoming more and more complex, especially as issues arise on a more global level, such as that of atmospheric pollution or global warming. Interactions between human society and the environment are constantly changing. The environment, while highly valued by most, is used and altered by a wide variety of people with many different interests and values. Difficulties remain on how best to ensure the protection of our environment and natural resources. There will always be tradeoffs and, many times, unanticipated or unintended consequences. However, a well-managed environment can provide goods and services that are both essential for our well being as well as for continued economic prosperity.

❐

11

Human Population and Environmental Issues

POPULATION GROWTH, VARIATION AMONG NATIONS

Our global human population, 6 billion at present, will cross the 7 billionmark by 2015. The needs of this huge number of human beings cannot be supported by the Earth's natural resources, without degrading the quality of human life. In the near future, fossil fuel from oil fields will run dry. It will be impossible to meet the demands for food from existing agro systems. Pastures will be overgrazed by domestic animals and industrial growth will create ever-greater problems due to pollution of soil, water and air.

Seas will not have enough fish. Larger ozone holes will develop due to the discharge of industrial chemicals into the atmosphere, which will affect human health. Global warming due to industrial gases will lead to a rise in sea levels and flood all low-lying areas, submerging coastal agriculture as well as towns and cities. Water 'famines' due to the depletion of fresh water, will create unrest and eventually make countries go to war. The control over regional biological diversity, which is vital for producing new medicinal and industrial products, will lead to grave economic conflicts between biotechnologically advanced nations and the biorich countries.

Degradation of ecosystems will lead to extinction of thousands of species, destabilizing natural ecosystems of great value. These are only some of the environmental problems related to an increasing human population and more intensive use of resources that we are likely to face in future. These effects can be averted by creating a mass environmental awareness movement that will bring about a change in people's way of life.

Increase in production per capita of agricultural produce at a global level ceased during the 1980's. In some countries, food shortage has become a permanent feature. Two of every three children in South Africa are underweight. In other regions famines due to drought have become more frequent. Present development strategies have not been able to successfully address these problems related to hunger and malnutrition.

On the other hand, only 15% of the world's population in the developed world is earning 79% of income! Thus the disparity in the extent of per capita resources that are used by people who live in a 'developed' country as against those who live in a 'developing' country is extremely large. Similarly, the disparity between the rich and the poor in India is also growing.

The increasing pressures on resources place great demands on the in-built buffering action of nature that has a certain ability to maintain a balance in our environment. However, current development strategies that essentially lead to short-term gains have led to a breakdown of our Earth's ability to replenish the resources on which we depend.

GLOBAL POPULATION GROWTH

The world population is growing by more than 90 million per year, of which 93% is in developing countries. This will essentially prevent their further economic 'development'. In the past, population growth was a gradual phenomenon and the Earth's ability to replenish resources was capable of adjusting to this increase. In the recent past, the escalation in growth of human numbers has become a major cause of our environmental problems. Present projections show that if our population growth is controlled, it will still grow to 7.27 billion by 2015.

However, if no action is taken it will become a staggering 7.92 billion. Human population growth increased from: 1 to 2 billion, in 123 years. 2 to 3 billion, in 33 years. 3 to 4 billion, in 14 years. 4 to 5 billion, in 13 years. 5 to 6 billion, in 11 years. It is not the census figures alone that need to be stressed, but an appreciation of the impact on natural resources of the rapid escalation in the rate of increase of human population in the recent past. The extent of this depletion is further increased by affluent societies that consume per capita more energy and resources, that less fortunate people.

This is of great relevance for developing a new ethic for a more equitable distribution of resources. In the first half of the 1900s human numbers were growing rapidly in most developing countries such as India and China. In some African countries the growth was also significant. In contrast, in the developed world population growth had slowed down. It was appreciated that the global growth rate was depleting the Earth's resources and was a direct impediment to human development.

Several environmental ill-effects were linked with the increasing population of the developing world. Poverty alleviation programs failed, as whatever was done was never enough as more and more people had to be supported on Earth's limited resources. In rural areas population growth led to increased fragmentation of farm land and unemployment. In the urban sector it led to inadequate housing and an increasing level of air pollution from traffic, water pollution from sewage, and an inability to handle solid waste. By the 1970s most countries in the developing world had realised that if they had to develop their economics and improve the lives of their citizens they would have to curtail population growth.

Though population growth shows a general global decline, there are variations in the rate of decline in different countries. By the 1990s the growth rate was decreasing in most countries such as China and India. The decline in the 90s was greatest in India. However, fertility continues to remain high in sub Saharan African countries. There are cultural, economic, political and demographic reasons that explain the differences in the rate of population control in different countries. It also varies in different parts of certain countries and is linked with community and/or religious thinking. Lack of Government initiatives for Family Welfare Programmes and a limited access to a full range of contraceptive measures are serious impediments to limiting population growth in several countries.

POPULATION EXPLOSION–FAMILY WELFARE PROGRAMMES

In response to our phenomenal population growth, India seriously took up an effective Family Planning Programmes which was renamed the Family Welfare Programmes. Slogans such as 'Hum do hamare do' indicated that each family should not have more than two children. It however has taken several decades to become effective. At the global level by the year 2000,600 million, or 57% of women in the reproductive age group, were using some method of contraception.

However the use of contraceptive measures is higher in developed countries–68%, and lower in developing countries-55%. Female sterilization is the most popular method of contraception used in developing countries at present. This is followed by the use of oral contraceptive pills and, intrauterine devices for women, and the use of condoms for men. India and China have been using permanent sterilization more effectively than many other countries in the developing world. The best decision for the method used by a couple depends on a choice that they make for themselves.

This must be based on good advice from doctors or trained social workers who can suggest the full range of methods available for them to choose from. Informing the public about the various contraceptive measures that are available is of primary importance. This must be done actively by Government Agencies such as Health and Family Welfare, as well as Education and Extension workers.

It is of great importance for policy makers and elected representatives of the people–Ministers, MPs, MLAs at Central and State levels–to understand Human the great and urgent need to support Family Welfare. The media must keep people informed about the need to limit family size and the ill effects of a growing population on the worlds resources. The decision to limit family size depends on a couple's background and education. This is related to Government Policy, the effectiveness of Family Welfare Programs, the educational level, and information levels in mass communication. Free access to Family Welfare information provided through the Health Care System, is in some cases unfortunately counteracted by cultural attitudes.

Frequently misinformation and inadequate information are reasons why a family does not go in for limiting its size. The greatest challenge the world now faces is how to supply its exploding human population with the resources it needs. It is evident that without controlling human numbers, the Earth's resources will be rapidly exhausted. In addition economically advanced countries and rich people in poorer countries use up more resources than they need.

As population expands further, water shortages will become acute. Soil will become unproductive. Rivers, lakes and coastal waters will be increasingly polluted. Water related diseases already kill 12 million people every year in he developing world. By 2025, there will be 48 countries that are starved for water. Air will become increasingly polluted.

Air pollution already kills 3 million people every year. The first 'green revolution' in the '60s produced a large amount of food but has led to several environmental problems. Now, a new green revolution is needed, to provide enough food for our growing population, that will not damage land, kill rivers by building large dams, or spread at the cost of critically important forests, grasslands and wetlands. The world's most populous regions are in coastal areas.

These are critical ecosystems and are being rapidly destroyed. Global climate change is now a threat that can affect the very survival of high population density coastal communities. In the sea, fish populations are suffering from excessive fishing.

Once considered an inexhaustible resource, over fishing has depleted stocks extremely rapidly. It will be impossible to support further growth in coastal populations on existing fish reserves. Human populations will inevitably expand from farm lands into the remaining adjacent forests. Many such encroachments in India have been regularised over the last few decades.

But forest loss has long-term negative effects on water and air quality and the loss of biodiversity is still not generally seen as a major deterrent to human well-being.

The extinction of plant and animal species resulting from shrinking habitats threatens to destroy the Earth's living web of life. Energy use is growing, both due to an increasing population, and a more energy hungry lifestyle that increasingly uses consumer goods that require large amounts of energy for their production, packaging, and transport. Our growing population also adds to the enormous amount of waste. With all these linkages between population growth and the environment, Family Welfare Programs have become critical to human existence.

Planning for the Future

How Governments and people from every community meet challenges such as limiting population size, protecting the natural environment, change their consumer oriented attitudes, reduce habits that create excessive waste, elevates poverty and creates an effective balance between conservation and development will determine the worlds future.

The Urban Challenge

Population increases will continue in urban centers in the near future. The UN has shown that by 2025 there will be 21 "megacities" most of which will be situated in developing countries. Urban centers are already unable to provide adequate housing, services such as water and drainage systems, growing energy needs, or better opportunities for income generation. Methods of Sterilization

India's Family Welfare Programmes has been fairly successful but much still needs to be achieved to stabilize our population. The most effective measure is the one most suited to the couple once they have been offered all the various options that are available. The Family Welfare Programmes advocates a variety of measures to control population.

Permanent methods or sterilisation are done by a minor surgery. Tubectomy in females is done by tying the tubes that carry the ovum to the uterus. Male sterilization or vasectomy, is done by tying the tubes that carry the sperm. Both are very simple procedures, done under local anesthesia, are painless and patients have no post operative problems.

Vasectomy does not cause any loss in the male's sexual ability but only arrests the discharge of sperm. There are several methods of temporary birth control.

Condoms are used by males to prevent sperms from fertilizing the ovum during intercourse. Intrauterine devices (Copper Ts) are small objects which can be placed by a doctor in the uterus so that the ovum cannot be implanted, even if fertilized.

They do not disturb any functions in the woman's life or work. Oral contraceptive tablets (pills) and injectable drugs are available that prevent sperms from fertilizing the ovum. There are also traditional but less reliable methods of contraception such as abstinence of the sexual act during the fertile period of the women's cycle and withdrawal during the sexual act.

Urbanization

In 1975 only 27% of the people in the developing world lived in urban areas. By 2000 this had grown to 40% and by 2030 well informed estimates state that this will grow to 56%. The developed world is already highly urbanized with 75% of its population living in the urban sector.

Urban Environments

Nearly half the world's population now lives in urban areas. The high population density in these areas leads to serious environmental issues. Today, more than 290 million people live in towns and cities in India. There were 23 metros in India in 1991, which grew to 40 by 2001. Urban population growth is both due to migration of people to towns and cities from the rural sector in search of better job options as well as population growth within the city.

As a town grows into a city it not only spreads outwards into the surrounding agricultural land or natural areas such as forests, grasslands and wetlands but also grows skywards with high rise buildings. The town also loses its open spaces and green cover unless these are consciously preserved. This destroys the quality of life in the urban area. Good urban planning is essential for rational landuse planning, for upgrading slum areas, improving water supply and drainage systems, providing adequate sanitation, developing effective waste water treatment plants and an efficient public transport system.

Unplanned and haphazard growth of urban complexes has serious environmental impacts. Increasing solid waste, improper garbage disposal and air and water pollution are frequent side effects of urban expansions. While all these issues appear to be under the preview of local Municipal Corporations, better living conditions can only become a reality if every citizen plays an active role in managing the environment. This includes a variety of "Dos and Don'ts" that should become an integral part of our personal lives.

Apart from undertaking actions that support the environment every urban individual has the ability to influence a city's management. He or she must see that the city's natural green spaces, parks and gardens are maintained, river and water fronts are managed appropriately, roadside tree cover is maintained, hill slopes are afforested and used as open spaces and architectural and heritage sites are protected. Failure to do this leads to increasing urban problems which eventually destroys a city's ability to maintain a healthy and happy lifestyle for its dwellers. All these aspects are closely linked to the population growth in the urban sector. In many cities growth outstrips the planner's ability to respond to this in time for a variety of reasons.

Mega Cities in India	Population (in Milions) in 2001	Projection (in Milions) for 2015
Mumbai	16.5	22.6
Kolkata	13.3	16.7
Delhi	13.0	20.9

Small urban centers too will grow rapidly during the next decades and several rural areas will require reclassification as urban centers. India's urban areas will grow by a projected 297 million residents. In India people move to cities from rural areas in the hope of getting a better income. This is the 'Pull' factor.

Poor opportunities in the rural sector thus stimulates migration to cities. Loss of agricultural land to urbanisation and industry, the inability of governments to sustainably develop the rural sector, and a lack of supporting infrastructure in rural areas, all push people from the agricultural and natural wilderness ecosystems into the urban sector. As our development strategies have focused attention mostly on rapid industrial development and relatively few development options are offered for the agricultural rural sector, a shift of population is inevitable.

As population in urban centers grows, they draw on resources from more and more distant areas. The "Ecological footprint" corresponds to the land area necessary to supply natural resources and disposal of waste of a community. At present the average ecological footprint of an individual at the global level is said to be 2.3 hectares of land per capita.

But it is estimated that the world has only 1.7 hectares of land per individual to manage these needs sustainably. This is thus an unsustainableuse of land. The pull factor of the urban centers is not only due to better job opportunities, but also better education, health care and relatively higher living standards. During the last few decades in India, improvements in the supply of clean water, sanitation, waste management, education and health care has all been urban centric, even though the stated policy has been to support rural development. Thus in reality, development has lagged behind in the rural sector that is rapidly expanding in numbers. For people living in wilderness areas in our forests and mountain regions, development has been most neglected. It is not appropriate to use the development methods used for other rural communities for tribal people who are dependent on collecting natural resources from the forests. A different pattern of development that is based on the sustainable extraction of resources from their own surroundings would satisfy their development aspirations. In general the growing human population in the rural sector will only opt to live where they are if they are given an equally satisfying lifestyle.

The Wilderness–Rural-urban Linkage

The environmental stresses caused by urban individuals covers an 'ecological footprint' that goes far beyond what one expects. The urban sector affects the land at the fringes of the urban area and the areas from which the urban center pulls in agricultural and natural resources. Urban centers occupy 2% of the worlds' land but use 75% of the industrial wood. About 60% of the world's water is used by urban areas of which half irrigates food crops for urban dwellers, and one third goes to industry and the rest is used for household use and drinking water.

The impact that urban dwellers have on the environment is not obvious to them as it happens at distant places which supports the urban ecosystem with resources from agricultural and even more remote wilderness ecosystems.

Urban Poverty and the Environment

The number of poor people living in urban areas is rapidly increasing. A third of the poor people in the world live in urban centers. These people live in hutments in urban slums and suffer from water shortages and unsanitary conditions. In most cases while a slum invariably has unhygenic surroundings, the dwellings themselves are kept relatively clean. It is the 'common' areas used by the community that lacks the infrastructure to maintain a hygienic environment. During the 1990s countries that have experienced an economic crisis have found that poor urban dwellers have lost their jobs due to decreasing demands for goods, while food prices have risen. Well paid and consistent jobs are not as easily available in the urban centers at present as in the past few decades. One billion urban people in the world live in inadequate housing, mostly in slum areas, the majority of which are temporary structures.

However, low income groups that live in high rise buildings can also have high densities and live in poor unhygienic conditions in certain areas of cities. Illegal slums often develop on Government land, along railway tracks, on hill slopes, riverbanks, marshes, etc. that are unsuitable for formal urban development. On the riverbanks floods can render these poor people homeless. Adequate legal housing for the urban poor remains a serious environmental concern.

Urban poverty is even more serious than rural poverty, as unlike the rural sector, the urban poor have no direct access to natural resources such as relatively clean river water, fuelwood and non wood forest products. The urban poor can only depend on cash to buy the goods they need, while in the rural sector they can grow a substantial part of their own food. Living conditions for the urban poor are frequently worse than for rural poor. Both outdoor and indoor air pollution due to high levels of particulate matter and sulphur dioxide from industrial and vehicle emissions lead to high death rates from respiratory diseases. Most efforts are targeted at outdoor air pollution. Indoor air pollution due to the use of fuel wood, waste material, coal, etc. in 'chulas' is a major health issue. This can be reduced by using better designed 'smokeless' chulas, hoods and chimneys to remove indoor smoke. With the growing urban population, a new crisis of unimaginable proportions will develop in the next few years. Crime rates, terrorism, unemployment, and serious environmental health related issues can be expected to escalate. This can only be altered by stabilizing population growth on a war footing.

ENVIRONMENT AND HUMAN HEALTH

Environment related issues that affect our health have been one of the most important triggers that have led to creating an increasing awareness of the need for better environmental management. Changes in our environment induced by human activities in nearly every sphere of life have had an influence on the pattern of our health. The assumption that human progress is through economic growth is not necessarily true.

We expect urbanization and industrialization to bring in prosperity, but on the down side, it leads to diseases related to overcrowding and an inadequate quality of drinking water, resulting in an increase in waterborne diseases such as infective diarrhoea and air borne bacterial diseases such as tuberculosis. High-density city traffic leads to an increase in respiratory diseases like asthma. Agricultural pesticides that enhanced food supplies during the green revolution have affected both the farm worker and all of us who consume the produce.

Modern medicine promised to solve many health problems, especially associated with infectious diseases through antibiotics, but bacteria found ways to develop resistant strains, frequently even changing their behaviour in the process, making it necessary to keep on creating newer antibiotics. Many drugs have been found to have serious side effects. At times the cure is as damaging as the disease process itself. Thus development has created several long-term health problems. While better health care has led to longer life spans, coupled with a lowered infant mortality, it has also led to an unprecedented growth in our population which has negative implications on environmental quality. A better health status of society will bring about a better way of life only if it is coupled with stabilising population.

Environmental Health

Environmental health, as defined by WHO, comprises those aspects of human health, including quality of life, that are determined by physical, chemical, biological, social, and psychosocial factors in the environment. It also refers to the theory and practice of assessing, correcting, controlling, and preventing those factors in the environment that adversely affect the health of present and future generations. Our environment affects health in a variety of ways.

Climate and weather affect human health. Public health depends on sufficient amounts of good quality food, safe drinking water, and adequate shelter. Natural disasters such as storms, hurricanes, and floods still kill many people every year. Unprecedented rainfall trigger epidemics of malaria and water borne diseases. Global climate change has serious health implications. Many countries will have to adapt to uncertain climatic conditions due to global warming. As our climate is changing, we may no longer know what to expect.

There are increasing storms in some countries, drought in others, and a temperature rise throughout the world. The El Niño winds affect weather worldwide. The El Niño event of 1997/98 had serious impacts on health and well-being of millions of people in many countries. It created serious drought, floods, and triggered epidemics. New strategies must be evolved to reduce vulnerability to climate variability and changes.

Economic inequality and environmental changes are closely connected to each other. Poor countries are unable to meet required emission standards to slow down climate change. The depletion of ozone in the stratosphere (middle atmosphere) also has an important impact on global climate and in turn human health, increasing the amount of harmful ultraviolet radiation that reaches the Earth's surface. This results in diseases such as skin cancer.

Bhopal Gas Tragedy

The siting of industry and relatively poor regulatory controls leads to ill health in the urban centers. Accidents such as the Bhopal gas tragedy in 1984 where Union Carbide's plant accidentally released 30 tones of methyl isocyanate, used in the manufacture of pesticides, led to 3,330 deaths and 1.5 lakh injuries to people living in the area. Development strategies that do not incorporate ecological safeguards often lead to ill health. Industrial development without pollution control and traffic congestion affect the level of air pollution in many cities. On the other hand, development strategies that can promote health invariably also protect the environment. Thus environmental health and human health are closely interlinked. An improvement in health is central to sound environmental management. However this is rarely given sufficient importance in planning development strategies.

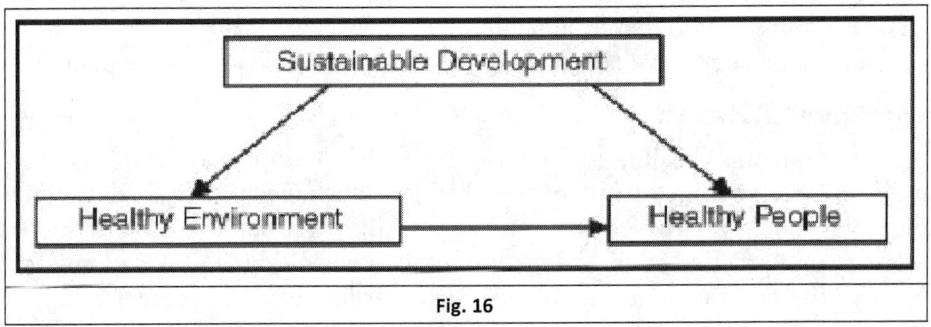

Fig. 16

Examples of the Linkages

- Millions of children die every year due to diarrhoea from contaminated water or food. An estimated 2000 million people are Encyclopaedia of Environmental Science affected by these diseases and more than 3 million children die each year from waterborne diseases across the world. In India, it is estimated that every fifth child under the age of 5 dies due to diarrhoea. This is a result of inadequate environmental management and is mainly due to inadequate purification of drinking water. Wastewater and/or sewage entering water sources without being treated leads to continuous gastrointestinal diseases in the community and even sporadic large epidemics. Large numbers of people in tropical countries die of malaria every year and millions are infected. An inadequate environmental management of stagnant water, which forms breeding sites of Anopheles mosquitoes is the most important factor in the spread of malaria. The resurgence of malaria in India is leading to cerebral malaria that affects the brain and has a high mortality.

- Millions of people, mainly children, have poor health due to parasitic infections, such as amoebiasis and worms. This occurs from eating infected food, or using poor quality water for cooking food. It is estimated that 36% of children in low-income countries and 12% in middle income countries are malnourished. In India, about half the children under the age of four are malnourished and 30% of newborns are significantly underweight.
- Hundreds of millions of people suffer serious respiratory diseases, including lung cancer and tuberculosis, from crowded homes and public places. Motor vehicle exhaust fumes, industrial fumes, tobacco smoke and cooking food on improper 'chulas', contribute to respiratory diseases.
- Millions of people are exposed to hazardous chemicals in their workplace or homes that lead to ill health due to industrial products where controls are not adhered to. Tens of thousands of people in the world die due to traffic accidents due to inadequate management of traffic conditions. Poor management at the accident site, and inability to reach a hospital within an hour causes a large number of deaths, especially from head injuries.
- Basic environmental needs such as clean water, clean air and adequate nutrition which are all related to environmental goods and services do not reach over 1000 million people living in abject poverty.
- Several million people live in inadequate shelters or have no roof over their heads especially in urban settings. This is related to high inequalities in the distribution of wealth and living space. Population growth and the way resources are being exploited and wasted, threatens environmental integrity and directly affects health of nearly every individual.
- Health is an outcome of the interactions between people and their environment. Better health can only come from a more sustainable management of the environment.
- The world must address people's health care needs and sustainable use of natural resources, which are closely linked to each other. Strategies to provide clean pottable water and nutrition to all people is an important part of a healthy living environment. Providing clean energy sources that do not affect health is a key to reducing respiratory diseases.
- Reducing environmental consequences of industrial and other pollutants such as transport emissions can improve the status of health.
- Changing patterns of agriculture away from harmful pesticides, herbicides and insecticides which are injurious to the health of farmers and consumers by using alternatives such as Integrated Pest Management and non-toxic biopesticides can improve health of agricultural communities, as well as food consumers.
- Changing industrial systems into those that do not use or release toxic chemicals that affect the health of workers and people living in the vicinity of industries can improve health and environment. There is a need to

change from using conventional energy from thermal power that pollutes air and nuclear power that can cause serious nuclear disasters to cleaner and safer sources such as solar, wind and ocean power that do not affect human health. Providing clean energy is an important factor that can lead to better health.

- The key factors are to control human population and consume less environmental goods and services which could lead to 'health for all'. Unsustainable use of resources by an ever growing population leads to unhealthy lives. Activities that go on wasting environmental goods and destroying its services by producing large quantities of non degradable wastes, leads to health hazards.
- Poverty is closely related to health and is itself a consequence of improper environmental management. An inequitable sharing of natural resources and environmental goods and services, is linked to poor health. The world's consumption of non-renewable resources is concentrated in the developed countries. Rich countries consume 50 times more per capita than people in less developed countries. This means that developed countries also generate proportionately high quantities of waste material, which has serious health concerns.

Important Strategic Concerns

- The world must address people's health care needs and sustainable use of natural resources, which are closely linked to each other.
- Strategies to provide clean pottable water and nutrition to all people is an important part of a healthy living environment.
- Providing clean energy sources that do not affect health is a key to reducing respiratory diseases.
- Reducing environmental consequences of industrial and other pollutants such as transport emissions can improve the status of health.
- Changing patterns of agriculture away from harmful pesticides, herbicides and insecticides which are injurious to the health of farmers and consumers by using alternatives such as Integrated Pest.
- Management and non-toxic biopesticides can improve health of agricultural communities, as well as food consumers.
- Changing industrial systems into those that do not use or release toxic chemicals that affect the health of workers and people living in the vicinity of industries can improve health and environment.
- There is a need to change from using conventional energy from thermal power that pollutes air and nuclear power that can cause serious nuclear disasters to cleaner and safer sources such as solar, wind and ocean power that do not affect human health. Providing clean energy is an important factor that can lead to better health.

- The key factors are to control human population and consume less environmental goods and services which could lead to 'health for all'. Unsustainable use of resources by an ever growing population leads to unhealthy lives. Activities that go on wasting environmental goods and destroying its services by producing large quantities of non degradable wastes, leads to health hazards.
- Poverty is closely related to health and is itself a consequence of improper environmental management. An inequitable sharing of natural resources and environmental goods and services, is linked to poor health. The world's consumption of non-renewable resources is concentrated in the developed countries. Rich countries consume 50 times more per capita than people in less developed countries.
- This means that developed countries also generate proportionately high quantities of waste material, which has serious health concerns. Important Strategic Concerns

Definition of Health Impact Assessment (HIA) by WHO

Health impact assessment is a combination of procedures, methods and tools by which a policy, programmes or project may be judged as to its potential effects on the health of a population, and the distribution of those effects within the population.

Climate and Health

Human civilizations have adapted mankind to live in a wide variety of climates. From the hot tropics to the cold arctic, in deserts, marshlands and in the high mountains. Both climate and weather have a powerful impact on human life and health issues. Natural disasters created by extremes of weather (heavy rains, floods, hurricanes) which occur over a short period of time, can severely affect health of a community. Poor people are more vulnerable to the health impacts of climate variability than the rich.

Of approximately 80,000 deaths which occur world-wide each year as a result of natural disasters about 95% are in poor countries. In weather-triggered disasters hundreds of people and animals die, homes are destroyed, crops and other resources are lost. Public health infrastructure, such as sewage disposal systems, waste management, hospitals and roads are damaged. The cyclone in Orissa in 1999 caused 10,000 deaths. The total number of people affected was estimated at 10 to 15 million!

Human physiology can adapt to changes in weather, within certain limits. However, marked short-term fluctuations in weather lead to serious health issues. Heat waves cause heat-related illness and death (e.g., heat stroke). The elderly and persons with existing heart or respiratory diseases are more vulnerable. Heat waves in India in 1998 were associated with many deaths.

Climate plays an important role in vector-borne diseases transmitted by insects such as mosquitoes. These disease transmitters are sensitive to direct effects of climate such as temperature, rainfall patterns and wind. Climate affects their distribution and abundance through its effects on host plants and animals. Malaria transmission is particularly sensitive to weather and climate. Unusual weather conditions, for example a heavy downpour, can greatly increase the mosquito population and trigger an epidemic.

In the desert and at highland fringes of malarious areas, malaria transmission is unstable and the human population lacks inherent protective immunity. Thus, when weather conditions (rainfall and temperature) favour transmission, serious epidemics occur in such areas. Fluctuations in malaria over the years have been linked to changes in rainfall associated with the El Niño cycle.

Infectious Diseases

Many infectious diseases have re-emerged with a vengeance. Loss of effective control over diseases such as malaria and tuberculosis, have led to a return of these diseases decades after being kept under stringent control. Other diseases were not known to science earlier and seem to have suddenly hit our health and our lives during the last few decades. AIDS, due to the Human Immunodeficiency Virus (HIV) caused through sexual transmission and Severe Acute Respiratory Syndrome (SARS) are two such examples.

While these cannot be directly related to environmental change, they affect the environment in which we live by forcing a change in lifestyles and behaviour patterns. For example the SARS outbreak prevented people from several countries from traveling to other countries for months, severely affecting national economies, airline companies and the tourism industry. Why have infectious diseases that were related to our environment that were under control suddenly made a comeback? Diseases such as tuberculosis have been effectively treated with anti-tubercular drugs for decades.

These antibiotics are used to kill off the bacteria that causes the disease. However nature's evolutionary processes are capable of permitting bacteria to mutate by creating new genetically modified strains. Those that change in a way so that they are not affected by the routinely used antibiotics begin to spread rapidly. This leads to a reemergence of the disease. In the case of tuberculosis this has led to multi-drug resistant tuberculosis. This is frequently related to HIV which reduces an individual's immunity to bacteria such as mycobacterium tuberculosis that causes tuberculosis. The newer broad-spectrum antibiotics, antiseptics, disinfectants, and vaccines once thought of as the complete answer to infectious diseases have thus failed to eradicate infectious diseases. Experts in fact now feel that these diseases will be the greatest killers in future and not diseases such as malignancy or heart disease.

While antibiotic resistance is a well-known phenomenon there are other reasons for the reemergence of diseases. Overcrowding due to the formation of slums in the urban setting leads to several health hazards, including easier spread of respiratory diseases. Inadequate drinking water quality and poor disposal of human waste due to absence of a closed sewage system and poor garbage management are all urban health issues. This has led to a comeback of diseases such as cholera and an increased incidence of diarrhea and dysentery as well as infectious hepatitis (jaundice). With increasing global warming disease patterns will continue to change. Tropical diseases spread by vectors such as the mosquito will undoubtedly spread malaria further away from the equator. Global warming will also change the distribution of dengue, yellow fever, encephalitis, etc.

Warmer wetter climates could cause serious epidemics of diseases such as cholera. El Nino which causes periodic warming is likely to affect rodent populations. This could bring back diseases such as the plague. Globalisation and infectious disease Globalization is a world-wide process which includes the internationalization of communication, trade and economic organization.

It involves parallel changes such as rapid social, economic and political adjustments. Whilst globalization has the potential to enhance the lives and living standards of certain population groups, for poor and marginalized populations in both the non-formal as well as formal economic sectors of developing countries, globalization enhances economic inequalities.

Tuberculosis

Tuberculosis (TB) kills approximately 2 million people each year. In India the disease has reemerged and is now more difficult to treat. A global epidemic is spreading and becoming more lethal. The spread of HIV/AIDS and the emergence of multidrug-resistant tuberculosis is contributing to the increasing morbidity of this disease. In 1993, the World Health Organization (WHO) declared that tuberculosis had become a global emergency. It is estimated that between 2002 and 2020, approximately 1000 million people will be newly infected, over 150 million people will get sick, and 36 million will die of TB–if its control is not rapidly strengthened. TB is a contagious disease that is spread through air.

Only people who are sick with pulmonary TB are infectious. When infectious people cough, sneeze, talk or spit, they emit the tubercle bacilli into the air. When a healthy person inhales these, he gets infected by the disease. Symptoms include prolonged fever, coughing spells and weight loss. It is estimated that, left untreated, each patient of active tuberculosis will infect on an average between 10 to 15 people every year. But people infected with TB will not necessarily get sick with the disease.

The immune system can cause the TB bacilli, which is protected by a thick waxy coat, to remain dormant for years. When an individual's immune system is weakened, the chances of getting active TB are greater.

- Nearly 1% of the world's population is newly infected with TB each year.
- It is estimated that overall, one third of the world's population is likely to be infected with the tuberculosis bacillus at some point in time.
- Five to ten per cent of people who are infected with TB (but who are not infected with HIV) become sick or infectious at some time during their life. (WHO, 2002). Factors Contributing to the rise in tuberculosis
- TB kills about 2 million people each year (including persons infected with HIV).
- More than 8 million people become sick with TB each year, one person in the world every second!
- About 2 million TB cases per year occur in sub-Saharan Africa. This number is rising rapidly as a result of the HIV/AIDS epidemic.
- Around 3 million TB cases per year occur in South-east Asia.
- Over a quarter of a million TB cases per year occur in Eastern Europe.

Tuberculosis in India

There are 14 million TB patients in India, account for one third of the global cases of TB. Everyday 20,000 Indians contract TB and more than 1,000 die due to this chronic illness. TB attacks working adults in the age group of 15 to 50 years.

HIV is Accelerating the Spread of TB

The link between HIV and TB affects a large number of people, each disease speeding the other's progress. HIV weakens the immune system. Someone who is HIV-positive and infected with TB is many times more likely to become seriously sick with TB rather than someone infected with TB who is HIV-negative. Tuberculosis is a leading cause of death among people who are HIV-positive, accounting for about 11% of AIDS deaths worldwide. Poorly Managed TB programs are Threatening to make TB Incurable

Until 50 years ago, there were no drugs to cure tuberculosis. Now, strains that are resistant to one or more anti-TB drugs have emerged. Drugresistant tuberculosis is caused by inconsistent or partial treatment, when patients do not take all their drugs regularly for the required period, when doctors or health workers prescribe inadequate treatment regimens or where the drug supply is unreliable. From a public health perspective, poorly supervised or incomplete treatment of TB is worse than no treatment at all. When people fail to complete standard treatment regimens, or are given the wrong treatment regimen, they may remain infectious. The bacilli in their lungs may develop resistance to anti-TB drugs. People they infect will have the same drug-resistant strain. While drug-resistant TB is treatable, it requires extensive chemotherapy that is often very expensive and is also more toxic to patients.

Malaria

Malaria is a life-threatening parasitic disease transmitted by mosquitoes. The cause of malaria, a single celled parasite called plasmodium, was discovered in 1880.Later it was found that the parasite is transmitted from person to person through the bite of a female Anopheles mosquito, which requires blood for the growth of her eggs. Today approximately 40% of the world's population, mostly those living in the world's poorest countries, risk getting malaria.

The disease was once more widespread but it was successfully eliminated from many countries with temperate climates during the mid 20th century.

Today malaria has returned and is found throughout the tropical and sub-tropical regions of the world and causes more than 300 million acute illnesses and at least one million deaths annually (WHO). There are several types of human malaria. Falciparum malaria is the most dangerous type of infection and is most common in Africa south of the Sahara, where it accounts for extremely high mortality rates. There are also indications of the spread of P. falciparum malaria in India and it has reappeared in areas where it had been eliminated. The malaria parasite enters the human host when an infected Anopheles mosquito bites an individual. Inside the human host, the parasite undergoes a series of changes as part of its complex life-cycle.

Its various stages allow plasmodia to evade the immune system, infect the liver and red blood cells, and finally develop into a form that is able to infect a mosquito again when it bites an infected person. Inside the mosquito, the parasite matures until it reaches the sexual stage where it can again infect a human host when the mosquito takes her next blood meal, 10 or more days later. Malaria symptoms appear about 9 to 14 days after the mosquito bite, although this varies with different plasmodium species. Malaria produces high fever, headache, vomiting and body ache. If drugs are not available for treatment, or the parasites are resistant to them, the infection can progress rapidly to become life-threatening. Malaria can kill by infecting and destroying red blood cells (anaemia) and by clogging the capillaries that carry blood to the brain (cerebral malaria) or other vital organs.

Malaria parasites are developing unacceptable levels of resistance to drugs. Besides this, many insecticides are no longer useful against mosquitoes transmitting the disease. Good environmental management by clearing pools of stagnant water during the monsoons is effective in reducing the number of mosquitoes. Mosquito nets treated with insecticide reduce malaria transmission and child deaths.

Prevention of malaria in pregnant women, through measures such as Intermittent Preventive Treatment and the use of insecticide-treated nets (ITNs), results in improvement in maternal health, as well as infant health and survival. Prompt access to treatment with effective up-to-date medicines, such as artemisinin-based combination therapies (ACTs), saves lives. If countries can apply these and other measures on a wide scale and monitor them carefully, the burden of malaria on society will be significantly reduced.

Water-related Diseases

Water Supply, Sanitation and Hygiene Development

Among the main problems are a lack of priority given to this sector, lack of financial resources, erratic water supply and sanitation services, poor hygiene related behaviour patterns, and inadequate sanitation in public places such as schools, hotels, hospitals, health centers, etc. One of the most important aspects is a lack of environmental education and awareness that these disease processes are related to poor environment management in various sectors.

Providing access to sufficient quantities of safe water, the provision of facilities for a sanitary disposal of excreta, and introducing sound hygiene related behaviour can reduce the morbidity and mortality caused by these risk factors. Environmental Sanitation and Hygiene Development About 2.4 billion people globally live under highly unsanitary conditions. Poor hygiene and behaviour pattern increase the exposure to risk of incidence and spread of infectious diseases. Water improperly stored in homes is frequently contaminated by inadequate management at the household level. This can be easily reduced through education and awareness of how waterborne diseases are transmitted.

Health and Water Resources Development

An important aspect related to water-related diseases (in particular: water-related vectorborne diseases) is attributable to the way water resources are developed and managed. In many parts of the world the adverse health impacts of dam construction, irrigation development and flood control is related to increased incidence of malaria, Japanese encephalitis, schistosomiasis, lymphatic filariasis and other conditions. Other health issues indirectly associated with water resources development include nutritional status, exposure to agricultural pesticides and their residues.

Water Borne Diseases

Arid areas with rapidly expanding populations are already facing a crisis over water. Conservation of water and better management is an urgent need. The demand and supply balance is a vital part of developing sustainable use of water. This is being termed the 'Blue Revolution' and needs Governments, NGOs and people to work together towards a better water policy at International, National, State, regional and local levels. Locally good watershed management is a key to solving local rural problems. Present patterns of development are water hungry and water wasters. They do not address pollution and overuse. The linkages between managing water resources and health issues are have not been prioritised as a major source of environmental problems that require policy change, administrative capacity building and an increased financial support.

There are 4 Major types of Water Related Diseases:

1. **Water Borne Diseases:** These are caused by dirty water contaminated by human and animal wastes, especially from urban sewage, or by chemical wastes from industry and agriculture. Some of these diseases, such as cholera and typhoid, cause serious epidemics. Diarrhoea, dysentery, polio, meningitis, and hepatitis A and E, are caused due to improper drinking water. Excessive levels of nitrates cause blood disorders when they pollute water sources. Pesticides entering drinking water in rural areas cause cancer, neurological diseases and infertility. Improving sanitation and providing treated drinking water reduces the incidence of these diseases.
2. **Water Based Diseases:** Aquatic organisms that live a part of their life cycle in water and another part as a parasite in man, lead to several diseases. In India, guinea worm affects the feet. Round worms live in the small intestine, especially of children.
3. **Water Related Vector Diseases:** Insects such as mosquitoes that breed in stagnant water spread diseases such as malaria and filariasis. Malaria that was effectively controlled in India, has now come back as the mosquitoes have become resistant to insecticides. In addition, anti-malarial drugs are now unable to kill the parasites as they have become resistant to drugs. Change in climate is leading to the formation of new breeding sites. Other vector born diseases in India include dengue fever and filariasis. Dengue fever carries a high mortality. Filariasis leads to fever and chronic swelling over the legs.
4 **Water Scarcity Diseases:** In areas where water and sanitation is poor, there is a high incidence of diseases such as tuberculosis, leprosy, tetanus, etc. which occur when hands are not adequately washed.

Arsenic in Drinking Water

Arsenic in drinking-water is a serious hazard to human health. It has attracted such attention since its recognition in the 1990s of its wide occurrence in wellwater in Bangladesh. It occurs less frequently in most other countries.

The main source of arsenic in drinking water is arsenic-rich rocks through which the water has filtered. It may also occur because of mining or industrial activity in some areas. WHO has worked with other UN organizations to produce a state-of-the-art review on arsenic in drinking water.

Drinking water that is rich in arsenic leads to arsenic poisoning or arsenicosis. Excessive concentrations are known to occur in some areas. The health effects are generally delayed and the most effective preventive measure is supplying drinking water which is free of arsenic. Arsenic contamination of water is also due to industrial processes such as those involved in mining, metal refining, and timber treatment. Malnutrition may aggravate the effects of arsenic on blood vessels. Water with high concentrations of arsenic if used over 5 to 20 years, results in problems such as colour changes on the skin, hard patches on the palms and soles, skin cancer, cancers of the bladder, kidney and lung, and diseases of the blood vessels of the legs and feet. It may also lead to diabetes, high blood pressure and reproductive disorders. Natural arsenic contamination occurs in Argentina, Bangladesh, Chile, China, India, Mexico, Thailand and the United States. In China (in the Province of Taiwan) exposure to arsenic leads to gangrene, known as 'black foot disease'.

Long term solutions for prevention of arsenicosis is based on providing safe drinking-water:

- Deeper wells are often less likely to be contaminated.
- Testing of water for levels of arsenic and informing users.
- Monitoring by health workers-people need to be checked for early signs of arsenicosis-usually by observing skin problems in areas where arsenic in known to occur.
- Health education regarding harmful effects of arsenicosis and how to avoid them.

Arsenic Poisoning–Bangladesh

More than half the population of Bangladesh is threatened by high levels of arsenic found in drinking water. This could eventually lead to an epidemic of cancers and other fatal diseases. Rezaul Morol, a young Bangladeshi man, nearly died from arsenic poisoning caused by drinking arsenic-laden well-water for several years. The doctor advised Rezaul to stop drinking contaminated water and eat more protein-rich food such as fish. Since then Rezaul feels a lot better and is happy that his skin is healing.

Diarrhoea

Though several types of diarrhoea which give rise to loose motions and dehydration occur all over the world, this is especially frequently observed in developing countries. It causes 4% of all deaths. In another 5% it leads to loss of health. It is caused by gastrointestinal infections which kill around 2.2 million people globally each year. Most of these are children in developing countries.

The use of contaminated water is an important cause of this group of conditions. Cholera and dysentery cause severe, sometimes life threatening and epidemic forms of these diseases.

Effects on Health

Diarrhoea is the frequent passage of loose or liquid stools. It is a symptom of various gastrointestinal infections. Depending on the type of infection, the diarrhoea may be watery (for example in cholera caused by vibrio cholera) or passed with blood and mucous (in dysentery caused by an amoeba, E Histolitica). Depending on the type of infection, it may last a few days, or several weeks. Severe diarrhoea can become life threatening due to loss of excessive fluid and electrolytes such as Sodium and Potassium in watery diarrhoea.

This is particularly fatal in infants and young children. It is also dangerous in malnourished individuals and people with poor immunity. The impact of repeated diarrhoea on nutritional status is linked in a vicious cycle in children. Chemical or non-infectious intestinal conditions can also result in diarrhoea Causes of Diarrhoea

Diarrhoea is caused by several bacterial, viral and parasitic organisms. They are mostly spread by contaminated water. It is more common when there is a shortage of clean water for drinking, cooking and cleaning. Basic hygiene is important in its prevention. Water contaminated with human feces surrounding a rural water source, or from municipal sewage, septic tanks and latrines in urban centers, are important factors in the spread of these diseases. Feces of domestic animals also contain microorganisms that can cause diarrhoea through water. Diarrhoea is spread from one individual to another due to poor personal hygiene. Food is a major cause of diarrhoea when it is prepared or stored in unhygienic conditions. Water can contaminate food such as vegetables during irrigation. Fish and seafood from polluted water is a cause of severe diarrhoea. The infectious agents that cause diarrhoea are present in our environment.

In developed countries where good sanitation is available, most people get enough safe drinking water. Good personal and domestic hygiene prevents this disease which is predominantly seen in the developing world. About 1 billion people do not have access to clean water sources and 2.4 billion have no basic sanitation. In Southeast Asia, diarrhoea is responsible for 8.5% of all deaths. In 1998, diarrhoea was estimated to have killed 2.2 million people, most of whom were under 5 years of age. Interventions: Key measures to reduce the number of cases of diarrhoea include:

- Access to safe drinking water.
- Improved sanitation.
- Good personal and food hygiene.
- Health education about how these infections spread.
- Key measures to treat diarrhoea include:
- Giving more fluids than usual, (oral rehydration) with salt and sugar, to prevent dehydration.
- Continue feeding.
- Consulting a health worker if there are signs of dehydration or other problems.

In rural India, during the last decade public education through posters and other types of communication strategies has decreased infant mortality due to diarrhoea in several States. Posters depicting a child with diarrhoea being given water, salt and sugar solution to reduce death from dehydration has gone a long way in reducing both a serious condition requiring hospitalisation and intravenous fluids, as well as mortality.

Risks Due to Chemicals in Food

Food contaminated by chemicals is a major worldwide public health concern. Contamination may occur through environmental pollution of the air, water and soil. Toxic metals, PCBs and dioxins, or the intentional use of various chemicals, such as pesticides, animal drugs and other agrochemicals have serious consequences on human health. Food additives and contaminants used during food manufacture and processing adversely affects health.

Diseases spread by food: Some foodborne diseases though well recognized, have recently become more common. For example, outbreaks of salmonellosis which have been reported for decades, has increased within the last 25 years. In the Western hemisphere and in Europe, Salmonella serotype Enteritidis (SE) has become a predominant strain. Investigations of SE outbreaks indicate that its emergence is largely related to consumption of poultry or eggs.

While cholera has devastated much of Asia and Africa for years, its reintroduction for the first time in almost a century on the South American continent in 1991 is an example of a well recognised infectious disease re-emerging in a region after decades. While cholera is often waterborne, many foods also transmit infection. In Latin America, ice and raw or underprocessed seafood are important causes for cholera transmission.

Infection with a specific type of Escherichia coli (E. coli) was first described in 1982. Subsequently, it has emerged rapidly as a major cause of bloody diarrhoea and acute renal failure. The infection is sometimes fatal, particularly in children. Outbreaks of infection, generally associated with beef, have been reported in Australia, Canada, Japan, United States, in various European countries, and in southern Africa. Outbreaks have also implicated alfalfa sprouts, unpasteurized fruit juice, lettuce, game meat (meat of wild animals) and cheese curd.

In 1996, an outbreak of Escherichia coli in Japan affected over 6,300 school children and resulted in 2 deaths. Listeria monocytogenes (Lm): The role of food in the transmission of this condition has been recognized recently. In pregnant women, infections with Lm causes abortion and stillbirth. In infants and persons with a poor immune system it may lead to septicemia (blood poisoning) and meningitis. The disease is most often associated with consumption of foods such as soft cheese and processed meat products that are kept refrigerated for a long time, because Lm can grow at low temperatures.

Outbreaks of listeriosis have been reported from many countries, including Australia, Switzerland, France and the United States. Two recent outbreaks of Listeria monocytogenes in France in 2000 and in the USA in 1999 were caused by contaminated pork tongue and hot dogs respectively. Foodborne trematodes (worms) are increasing in South-east Asia and Latin America.

This is related to a combination of intensive aquaculture production in unsanitary conditions, and consumption of raw or lightly processed fresh water fish and fishery products.

Foodborne trematodes can cause acute liver disease, and may lead to liver cancer. It is estimated that 40 million people are affected worldwide. Bovine Spongiform Encephalopathy (BSE), is a fatal, transmittable, neurodegenerative disease of cattle. It was first discovered in the United Kingdom in 1985. The cause of the disease was traced to an agent in sheep, which contaminated recycled bovine carcasses used to make meat and bone meal additives for cattle feed.

Recycling of the BSE agent developed into a common source epidemic of more than 180,000 diseased animals in the UK alone. The agent affects the brain and spinal cord of cattle which produces sponge-like changes visible under a microscope. About 19 countries have reported BSE cases and the disease is no longer confined to the European Community.

A case of BSE has been reported in a cattle herd in Japan. In human populations, exposure to the BSE agent (probably in contaminated bovine-based food products) has been strongly linked to the appearance in 1996 of a new transmissible spongiform encephalopathy of humans called variant Creutzfeldt-Jakob Disease (vCJD). By January 2002,119 people developed vCJD, most from the UK but five cases have been reported from France.

Cancer and Environment

Cancer is caused by the uncontrolled growth and spread of abnormal cells that may affect almost any tissue of the body. Lung, colon, rectal and stomach cancer are among the five most common cancers in the world for both men and women. Among men, lung and stomach cancer are the most common cancers worldwide. For women, the most common cancers are breast and cervical cancer.

In India, oral and pharangeal cancers form the most common type of cancer which are related to tobacco chewing. More than 10 million people are diagnosed with cancer in the world every year. It is estimated that there will be 15 million new cases every year by 2020.Cancer causes 6 million deaths every year–or 12% of deaths worldwide. The causes of several cancers are known. Thus prevention of at least one-third of all cancers is possible.

Cancer is preventable by stopping smoking, providing healthy food and avoiding exposure to cancer-causing agents (carcinogens). Early detection and effective treatment is possible for a further one-third of cases. Most of the common cancers are curable by a combination of surgery, chemotherapy (drugs) or radiotherapy (X-rays). The chance of cure increases if cancer is detected early. Cancer control is based on the prevention and control of cancer by:

- Promotion and strengthening of comprehensive national cancer control programs.
- Building international networks and partnerships for cancer control.
- Promotion of organized, evidence-based interventions for early detection of cervical and breast cancer.
- Development of guidelines on disease and programmes management.

- Advocacy for a rational approach to effective treatments for potentially curable tumours.
- Support for low-cost approaches to respond to global needs for pain relief and palliative care.

Prevention of Cancer

Tobacco smoking is the single largest preventable cause of cancer in the world. It causes 80 to 90% of all lung cancer deaths. Another 30% of all cancer deaths, especially in developing countries include deaths from cancer of the oral cavity, larynx, oesophagus and stomach which are related to tobacco chewing. Preventive measures include bans on tobacco advertising and sponsorship, increased tax on tobacco products, and educational programs which are undertaken to reduce tobacco consumption. Dietary modification is an important approach to cancer control.

Overweight individuals and obesity are known to be associated with cancer of the oesophagus, colon, rectum, breast, uterus and kidney. Fruit and vegetables may have a protective effect against many cancers. Excess consumption of red and preserved meat may be associated with an increased risk of colorectal cancer. Infectious agents are linked with 22% of cancer deaths in developing countries and 6% in industrialized countries.

Viral hepatitis B and C cause cancer of the liver. Human papilloma virus infection causes cancer of the cervix. The bacterium Helicobacter pylori increases the risk of stomach cancer. In some countries the parasitic infection schistosomiasis increases the risk of bladder cancer. Liver fluke increases the risk of cancer of the bile ducts. Preventive measures include vaccination and prevention of infection. Excessive solar ultraviolet radiation increases the risk of all types of cancer of the skin.

Avoiding excessive exposure to the sun, use of sunscreens and protective clothing are effective preventive measures. Asbestos is known to cause lung cancer. Aniline dyes have been linked to bladder cancer. Benzene can lead to leukaemia (blood cancer). The prevention of certain occupational and environmental exposure to several chemicals is an important element in preventing cancer.

HUMAN RIGHTS

Several environmental issues are closely linked to human rights. These include the equitable distribution of environmental resources, the utilisation of resources and Intellectual Property Rights (IPRs), conflicts between people and wildlife especially around PAs, resettlement issues around development projects such as dams and mines, and access to health to prevent environment related diseases.

Equity

One of the primary concerns in environmental issues is how wealth, resources and energy must be distributed in a community. We can think of the global community, regional community issues, national concerns and those related to a family or at the individual level. While economic disparities remain a fact of life, we as citizens of a community must appreciate that a widening gap between the rich and the poor, between men and women, or between the present and future generations must be minimised if social justice is to be achieved.

Today the difference between the economically developed world and the developing countries is unacceptably high. The access to a better lifestyle for men as against women is inherent in many cultures. Last but not the least, we in the present generation cannot greedily use up all our resources leaving future generations increasingly impoverished. Rights to land, water, food, housing are all a part of our environment that we all share. However, while some live unsustainable lifestyles with consumption patterns that the resource base cannot support, many others live well below the poverty line. Even in a developing country such as ours, there are enormous economic inequalities.

This requires an ethic in which an equitable distribution becomes a part of everyone's thinking. The people who live in the countries of the North and the rich from the countries in the South will have to take steps to reduce their resource use and the waste they generate. Both the better off sectors of society and the less fortunate need to develop their own strategies of sustainable living and communities at each level must bring about more equitable patterns of wealth. The right to the use of natural resources that the environment holds is an essential component of human rights. It is related to disparities in the amount of resources available to different sectors of society.

People who live in wilderness communities are referred to as ecosystem people. They collect food, fuelwood, and nonwood products, fish in aquatic ecosystems, or hunt for food in forests and grasslands. When landuse patterns change from natural ecosystems to more intensively used farmland and pastureland the rights of these indigenous people are usually sacrificed. Take the case of subsidies given to the pulp and paper industry for bamboo which makes it several times cheaper for the industry than for a rural individual who uses it to build his home.

This infringes on the human right to collect resources they have traditionally used free of cost. Another issue is the rights of small traditional fishermen who have to contend against mechanised trawlers that impoverish their catch and overharvest fish in the marine environment. These people's right to a livelihood conflicts with the powerful economic interests of large-scale organised fisheries.

There are serious conflicts between the rights of rural communities for even basic resources such as water, and industrial development which requires large amounts of water for sustaining its productivity. The right to land or common property resources of tribal people is infringed upon by large development projects such as dams, mining and Protected Areas. Movements to protect the rights of indigenous peoples are growing worldwide.

Reversing actions that have already been taken decades ago is a complex problem that has no simple solutions. In many cases a just tradeoff is at best achieved through careful and sensitively managed negotiations. This needs a deep appreciation of local environmental concerns as well as a sensitivity to the rights of local people.

Nutrition, Health and Human Rights

There are links between environment, nutrition and health which must be seen from a humanrights perspective. Proper nutrition and health are fundamental human rights. The right to life is a Fundamental Right in our constitution. As a deteriorating environment shortens life spans, this in effect has an impact on our fundamental constitutional right. Nutrition affects and defines the health status of all people, rich and poor.

It is linked to the way we grow, develop, work, play, resist infection and reach our aspirations as individuals, communities and societies. Malnutrition makes people more vulnerable to disease and premature death. Poverty is a major cause as well as a consequence of ill-health. Poverty, hunger, malnutrition and poorly managed environments together affect health and weaken the socioeconomic development of a country.

Nearly 30% of humanity, especially those in developing countries–infants, children, adolescents, adults, and older persons are affected by this problem. A human rights approach is needed to appreciate and support millions of people left behind in the 20th century's health revolution. We must ensure that our environmental values and our vision are linked to human rights and create laws to support those that need a better environment, better health and a better lifestyle.

Health and sustainable human development are equity issues. In our globalized 21st century, equity must begin at the bottom, hand in hand with a healthy environment, improved nutrition, and sustainable lifestyles. Putting first things first, we must also realise that resources allocated to preventing and eliminating disease will be effective only if the underlying causes such as malnutrition and environmental concerns, as well as their consequences, are successfully addressed.

Intellectual Property Rights and Community Biodiversity Registers

Traditional people, especially tribals living in forests, have used local plants and animals for generations. This storehouse of knowledge leads to many new 'discoveries' for modern pharmaceutical products. The revenue generated from such 'finds' goes to the pharmaceutical industry that has done the research and patented the product. This leaves the original tribal user with nothing while the industry could earn billions of rupees.

To protect the rights of indigenous people who have used these products, a possible tool is to create a Community Biodiversity Register of local products and their uses so that its exploitation by the pharmaceutical industry would have to pay a royalty to the local community. This however has still not been generally accepted. Mechanisms have to be worked out so that the local traditional users rights are protected.

Traditional Medicine: Traditional medicine refers to health practices, approaches, knowledge and beliefs that incorporate plant, animal and mineral based medicines, frequently of local or regional origin. It may be linked to spiritual therapies, manual techniques and exercises. These may be used singly or in combination to treat, diagnose and prevent illnesses or maintain well-being. Traditional medicine is often handed down through the generations or may be known to a special caste or tribal group.

Traditional medicine has maintained its popularity in all regions of the developing world and its use is rapidly spreading in industrialized countries. In India, some of our primary health care needs are taken care of entirely by traditional medicine, while in Africa, up to 80% of the population uses it for primary health care. In industrialized countries, adaptations of traditional medicine are termed "Complementary" or "Alternative" Medicine (CAM). While there are advantages to traditional medicine as it is cheap and locally available, there are diseases which it cannot treat effectively.

This is a risk, as patients who use these alternative medicinal practices may rely on an ineffective measure. The consequences could be a serious delay in diagnosis and effective treatment of a treatable condition. There is a need to carefully research the claims of traditional practices to ensure that they are effective. In addition to patient safety issues, there is the risk that a growing herbal market and its great commercial benefit poses a threat to biodiversity through the over harvesting of the raw material for herbal medicines and other natural health care products.

This has been observed in the case of several Himalayan plants. If extraction from the wild is not controlled, this can lead to the extinction of endangered plant species and the destruction of natural habitats of several species. Another related issue is that at present, the requirements for protection provided under international standards for patent law and by most national conventional patent laws are inadequate to protect traditional knowledge and biodiversity.

There are tried and tested scientific methods and products that have their origins in different traditional medicinal methods. Twenty-five per cent of modern medicines are made from plants first used traditionally. Yoga is known to reduce asthma attacks. Traditional Medicine has been found to be effective against several infectious diseases. Over one-third of the population in developing countries lack access to essential allopathic medicines. The provision of safe and effective TM/CAM therapies could become a tool to increase access to health care.

VALUE EDUCATION

Value education in the context of our environment is expected to bring about a new sustainable way of life. Education both through formal and non-formal processes must thus address understanding environmental values, valuing nature and cultures, social justice, human heritage, equitable use of resources, managing common property resources and appreciating the cause of ecological degradation. Essentially, environmental values cannot be taught.

They are inculcated through a complex process of appreciating our environmental assets and experiencing the problems caused due to our destruction of our environment. The problems that are created by technology and economic growth are a result of our improper thinking on what 'development' means. Since we still put a high value only on economic growth, we have no concern for aspects such as sustainability or equitable use of resources. This mindset must change before concepts such as sustainable development can be acted upon.

Unsustainable development is a part of economic growth of the powerful while it makes the poor poorer. Consumerism is one aspect of this process favoured by the rich. As consumption of resources has till recently been an index of development, consumerism has thrived. It is only recently that the world has come to realise that there are other more important environmental values that are essential to bring about a better way of life.

Values in environment education must bring in several new concepts. Why and how can we use less resources and energy? Why do we need to keep our surroundings clean? Why should we use less fertilizers and pesticides in farms?

Why is it important for us to save water and keep our water sources clean? Or separate our garbage into degradable and non-degradable types before disposal? All these issues are linked to the quality of human life and go beyond simple economic growth. They deal with a love and respect for nature. These are the values that will bring about a better humanity, one in which we can live healthy, productive and happy lives in harmony with nature.

Environmental Values

Every human being has a great variety of feelings for different aspects of his or her surroundings. The Western, modern approach values the resources of Nature for their utilitarian importance alone. However true environmental values go beyond valuing a river for its water, a forest for its timber and non-wood forest products, or the sea for its fish.

Environmental values are inherent in feelings that bring about a sensitivity for preserving nature as a whole. This is a more spiritual, Eastern traditional value. There are several writings and sayings in Indian thought that support the concept of the oneness of all creation, of respecting and valuing all the different components of Nature. Our environmental values must translate to pro conservation actions in all our day to day activities.

Most of our actions have adverse environmental impacts unless we consciously avoid them. The sentiment that attempts to reverse these trends is enshrined in our environmental values. Values lead to a process of decision making which leads to action. For value education in relation to the environment, this process is learned through an understanding and appreciation of Nature's oneness and the importance of its conservation.

Humans have an inborn desire to explore Nature. Wanting to unravel its mysteries is a part of human nature. However, modern society and educational processes have invariably suppressed these innate sentiments. Once exposed to the wonders of the wilderness, people tend to bond closely to Nature.

They begin to appreciate its complexity and fragility and this awakens a new desire to want to protect our natural heritage. This feeling for Nature is a part of our Constitution, which strongly emphasises this value. Concepts of what constitutes right and wrong behaviour changes with time. Values are not constant.

It was once considered 'sport' to shoot animals. It was considered a royal, brave and much desirable activity to kill a tiger. In today's context, with wildlife reduced to a tiny fraction of what there was in the past, it is now looked down upon as a crime against biodiversty conservation. Thus the value system has been altered with time. Similarly with the large tracts of forest that existed in the past, cutting a few trees was not a significant criminal act.

Today this constitutes a major concern. We need a strong new environmental value system in which felling trees is considered unwise behaviour. With the small human numbers in the past, throwing away a little household degradable garbage could not have been considered wrong. But with enormous numbers of people throwing away large quantities of non-degradable waste, it is indeed extremely damaging to the environment and our value system must prevent this through a strong environmental value education system.

Appreciating the negative effects of our actions on the environment must become a part of our day to day thinking. Our current value system extols economic and technical progress as being what we need in our developing country.

Environmental Values Based on the Constitution of India

- Article 48A: "The state shall endeavour to protect and improve the environment and to safeguard the forests and wildlife in the country."
- Article 51A (g): The constitution expects that each citizen of the country must "protect and improve the natural environment, including forests, lakes, rivers and wildlife, and to have compassion for all living creatures."

While we do need economic development, our value system must change to one that makes people everywhere support a sustainable form of development so that we do not have to bear the cost of environmental degradation. Environmental problems created by development are due neither to the need for economic development, nor to the technology that produces pollution, but rather to a lack of awareness of the consequences of unlimited and unrestrained anti-environmental behaviour.

Looked at in this way, it deals with concepts of what is appropriate behaviour in relation to our surroundings and to other species on Earth. How we live our lives in fact shapes our environment. This is what environmental values are about. Each action by an individual must be linked to its environmental consequences in his/her mind so that a value is created that leads to strengthening pro-environmental behaviour and preventing anti-environmental actions.

This cannot happen unless new educational processes are created that provide a meaning to what is taught at school and college level. Every small child while growing up asks questions like 'What does this mean?'. They want an explanation for things happening around them that can help them make decisions and through this process develop values. It is this innate curiosity that leads to a personalized set of values in later life.

Providing appropriate 'meanings' for such questions related to our own environment brings in a set of values that most people in society begin to accept as a norm. Thus pro environmental actions begin to move from the domain of individuals to that of a community. At the community level, this occurs only when a critical number of people become environmentally conscious so that they constitute a proenvironment lobby force that makes governments and other people accept good environmental behaviour as an important part of development. What professions require making value judgements that greatly influence our environment? Evidentally nearly every profession can and does influence our environment, but some do so more than others. Policy makers, administrators, landuse planners, media, architects, medical personnel, health care workers, agriculturalists, agricultural experts, irrigation planners, mining experts, foresters, forest planners, industrialists and, most importantly, teachers at school and college level, are all closely related to pro environmental outcomes.

Environmental values have linkages to varied environmental concerns. While we value resources that we use as food, water and other products, there are also environmental services that we must appreciate. These include Nature's mechanisms in cleaning up air by removing carbon dioxide and adding oxygen by plant life, recycling water through the water cycle of nature, maintaining climate regimes, etc. But there are other aesthetic, ethical values that are equally important aspects of our environment that we do not appreciate consciously.

While every species is of importance in the web of life, there are some which man has come to admire for their beauty alone. The tiger's magnificence, the whale and elephant's giant size, the intelligence of our cousins the primates, the graceful flight of a flock of cranes, are parts of nature that we cannot help but admire. The lush splendor of an evergreen forest, the great power of the ocean's waves, and the tranquility of the Himalayan mountains are things that each of us values even if we do not experience it ourselves. We value its being there on Earth for us.

This is called its 'existance value'. The list of wondrous aspects of Nature's intricate connections is indeed awe-inspiring. This is also a part of our environment that we must value for its own sake. This is the oneness of Nature. We must equally look at our environment beyond the wild sphere. There is incredible beauty in some man-modified landscapes, the coloured patterns of farmland or the greens of a tea or coffee plantation in the hills.

Urban gardens and open space are also valuable and thus must be of prime concern to urban planners. These green spaces act as not only the lungs of a city, but also provide much needed psychological support. The mental peace and relaxation provided by such areas needs to be valued, although it is difficult to put a price tag on these values.

Nevertheless, these centers of peace and tranquility give urban dwellers an opportunity to balance their highly man-modified environments with the splash of green of a garden space. Environmental values must also stress on the importance of preserving ancient structures. The characteristic architecture, sculpture, artworks and crafts of ancient cultures is an invaluable environmental asset.

It tells us where we have come from, where we are now, and perhaps where we should go. Architectural heritage goes beyond preserving old buildings, to conserving whole traditional landscapes in rural areas and streetscapes in urban settings. Unless we learn to value these landscapes, they will disappear and our heritage will be lost.

As environmentally conscious individuals we need to develop a sense of values that are linked with a better and more sustainable way of life for all people. There are several positive as well as negative aspects of behaviour that are linked to our environment. The positive feelings that support environment include a value for Nature, cultures, heritage, and equity. We also need to become more sensitive to aspects that have negative impacts on the environment. These include our attitude towards degradation of the environment, loss of species, pollution, poverty, corruption in environmental management, the rights of future generations and animal rights. Several great philosophers have thoughts that have been based on, or embedded, in pro environmental behaviour.

Mahatma Gandhi and Rabindranath Tagore are among the many internationally well-known scholars whose thought have included values that are related to environmental consciousness. We need to appreciate these values to bring about a better way of life on earth for all people and all living creatures.

Valuing Nature

The most fundamental environmental sentiment is to value Nature herself. Appreciating Her magnificence and treasuring life itself leads to positive feelings that are a manifestation of pro environmental consciousness. The one-ness of our lives with the rest of nature and a feeling that we are only a miniscule part of nature's complex web of life becomes apparent, when we begin to appreciate the wonders of nature's diversity. We must appreciate that we belong to a global community that includes another 1.8 million known living forms.

Nothing makes us more conscious of this wonderous aspect of our earth's diversity than a walk through the wilderness, feeling and exploring its beauty and experiencing its infinite variety. The tiny creatures that live complex lives and the towering trees are all a part of this phenomenon we call 'life'. Today, man does not even know if other complex forms of life exist outside our own solar system in distant space. We may be alone in space or may be accompanied by other, completely different, living forms. But for now we only know for sure that the Earth's life forms are unique.

We thus have a great responsibility to protect life in all its glorious forms and must therefore respect the wilderness with all its living creatures, where man's own hand has not created changes that have led to perturbing natural habitats. We need to develop a sense of values that lead us to protect what is left of the wilderness by creating effective National Parks and Wildlife Sanctuaries. However this cannot be done to the detriment of the millions of tribal or indigenous people who live in wilderness ecosystems. There are thus conflicting values that need to be balanced carefully. On the one hand we need to protect natural ecosystems, while on the other, we must protect the rights of local people. Yet apart from valuing the diversity of life itself, we must also learn to value and respect diverse human cultures. Many of the tribal cultures of our country are vanishing because those with more dominant and economically advanced ways of life do not respect their lifestyles, that are in fact closer to nature and frequently more sustainable. We believe that our modern technology-based lifestyles are the sole way for society to progress. Yet this is only a single dimension of life that is based on economic growth.

Deep Ecology

In the 1970s a new thinking on environmental concerns began to emerge, protecting nature and the wilderness for its own sake, which is now referred to as 'Deep Ecology'.

It was fostered by the thinking of Arne Naess, a Norwegian professor of Philosophy and a great believer in Gandhian thinking and Buddhism. It recognises the intrinsic value of all living beings and looks upon mankind as a small segment of a great living community of life forms.

It teaches that the wellbeing and flourishing of human and non-human life on Earth have value in themselves and that these values are independent of the usefulness of the non-human world for human purposes. While currently the environmental movement focuses on issues that are concerned with the management of the natural environment for the 'benefit' of man, Deep Ecology promotes an approach that is expected to bring about a more appropriate cological balance on Earth and is akin to a spiritual approach to Nature. This has great long-term implications not only for humans but for the whole of Nature. For example some environmentalists emphasise the need to preserve wilderness for its aesthetic and utilitarian functions.

Wilderness is being preserved today in PAs because it is scenic and serves the purpose of tourism for nature lovers, and has recreational and economic value. Other environmentalists stress that the goal is for protecting the useful ecological functions of the wilderness, its services and goods that we use. Deep Ecologists on the other hand stress that wilderness preservation is a means to achieve the conservation and protection of biological diversity. Thus it is not enough to protect bits of what is left of the wilderness but to make attempts to restore degraded areas to their former natural ecological state.

In a country such as India, with its enormous population coupled with poverty on the one hand and the need for economic industrial growth on the other, this will be extremely difficult to achieve. Another new approach is that of 'Gaia', the hypothesis that the Earth is itself like one giant form of throbbing life consisting of all the unquantifiable numbers of individuals of its millions of known and unknown species.

Valuing Cultures

Every culture has a right to exist. Tribal people are frequently most closely linked with Nature and we have no right to foist on them our own modern way of life. The dilemma is how to provide them with modern health care and education that gives them an opportunity to achieve a better economic status without disrupting their culture and way of life. This will happen only if we value their culture and respect their way of life.

Social Justice

As the divide widens between those people who have access to resources and wealth, and those who live near or below the poverty line, it is the duty of those who are better off to protect the rights of the poor who do not have the means to fight for their rights. If this is not respected the poor will eventually rebel, anarchy and terrorism will spread and the people who are impoverished will eventually form a desperate seething revolution to better their own lot.

The developing world would face a crisis earlier than the developedcountries unless the rights of poor people that are fundamental to life are protected.

Modern civilization is a homogenous culture, based until recently on a belief that modern science holds the answer to everything. We are now beginning to appreciate that many ancient and even present day sequestrated cultures have a wisdom and knowledge of their own environments that is based on a deep sense of respect for nature.

Tribal cultures have over many generations used indigenous medicines which are proving to be effective against diseases. They have produced unique art forms such as painting, sculpture, and crafts that are beautiful and can enrich living experiences for everyone.

They have their own poetry, songs, dance and drama-all art forms that are unfortunately being rapidly lost as we introduce a different set of modern values to them through television and other mass media.

The world will be culturally impoverished if we allow these indigenous people to loose their traditional knowledge which includes sustainable use of water, land and resources with a low impact on biodiversity.

They will soon lose the beauty within their homes that is based on the things they make from Nature.

The art of the potter will be lost forever to the indestructible plastic pot. The bamboo basket weaver who makes a thing of beauty that is so user friendly and aesthetically appealing, will give place to yet another plastic box. Much that is beautiful and hand-crafted will disappear if we do not value these diverse aspects of human cultures.

Human Heritage

The earth itself is a heritage left to us by our ancestors for not only our own use but for the generations to come. There is much that is beautiful on our Earth-the undisturbed wilderness, a traditional rural landscape, the architecture of a traditional village or town, and the value of a historical monument or place of worship. These are all part of human heritage. Heritage preservation is now a growing environmental concern because much of this heritage has been undervalued during the last several decades and is vanishing at an astonishing pace.

While we admire and value the Ajanta and Ellora Caves, the temples of the 10th to 15th centuries that led to different and diverse styles of architecture and sculpture, the Moghul styles that led to structures such as the Taj Mahal, or the unique environmentally-friendly Colonial buildings, we have done little to actively preserve them. As environmentally conscious individuals we need to lobby for the protection of the wilderness and our glorious architectural heritage.

Equitable use of Resources

An unfair distribution of wealth and resources, based on a world that is essentially only for the rich, will bring about a disaster of unprecedented proportions. Equitable use of resources is now seen as an essential aspect of human well being and must become a shared point of view among all socially and environmentally conscious individuals.

This includes an appreciation of the fact that economically advanced countries and the rich in even poor nations consume resources at much greater levels than the much larger poorer sectors of humanity in the developing world. In spite of the great number of people in the more populous developing countries, the smaller number of people in developed countries use more resources and energy than those in the developing world. This is equally true of the small number of rich people in poor countries whose per capita use of energy and resources, and the generation of waste based on the one time use of disposable products, leads to great pressures on the environment. The poor while polluting the environment have no way to prevent it.

The rich damage the environment through a carelessness that proves only that they have no appreciation for environmental safety. As we begin to appreciate that we need more sustainable lifestyles we also begin to realise that this cannot be brought about without a more equitable use of resources. Common Property Resources

Our environment has a major component that does not belong to individuals. There are several commonly owned resources that all of us use as a community. The water that nature recycles, the air that we all breathe, the forests and grasslands which maintain our climate and soil, are all common property resources. When Government took over the control of community forests in British times, the local people who until then had controlled their use through a set of norms that were based on equitable use, began to overexploit resources on which they now had no personal stake.

Bringing back such traditional management systems is extremely difficult. However, in the recent past managing local forests through village level forest protection committees has shown that if people know that they can benefit from the forests, they will begin to protect them. This essentially means sharing the power to control forests between the Forest Department and local people.

Ecological Degradation

In many situations valuable ecological assets are turned into serious environmental problems. This is because we as a society do not strongly resist forces that bring about ecological degradation. These consist of sectors of society that use a 'get-rich-quick' approach to development. While ecological degradation has frequently been blamed on the needs of fuelwood and fodder of growing numbers of rural people, the rich, urbanized, industrial sector is responsible for greater ecological damage.

Changes in landuse from natural ecosystems to more intensive utilization such as turning forests into monoculture forestry plantations, or tea and coffee estates, or marginal lands into intensive agricultural patterns such as sugarcane fields or changes into urban or industrial land carry an ecological price.

Wetlands, for example, provide usable resources and a variety of services not easily valued in economic terms, and when destroyed to provide additional farmland, in many cases produce lower returns. A natural forest provides valuable non-wood forest products whose economic returns far outweigh that provided by felling the forest for timber. These values must form a part of a new conservation ethic. We cannot permit unsustainable development to run onwards at a pace in which our lives will be overtaken by a development strategy that must eventually fail as Earth's resources are consumed and ecosystems rendered irreparable.

HIV/AIDS

The Human Immunodeficiency Virus (HIV) causes Acquired Immunodeficiency Syndrome (AIDS) through contact with tissue fluids of infected individuals, especially through sexual contact. As it reduces an individual's resistance to disease, it causes infected individuals to suffer from a large number of environment related diseases and reduces the ability of infected individuals to go about their normal lives.

It affects their income generation and/or their ability to utilise natural resources. As more and more people are affected, this disease will also have impacts on our natural resource base, as utilisation patterns change to unsustainable levels. The inability of these patients to have the strength to access natural resources also affects the outcome of the disease process, as their overall health and well being is likely to worsen the course of the disease when their nutritional status suffers. In sub Saharan Africa where the infection has become highly prevalent, it is leading to great suffering and worsening poverty.

The capacity of these patients to work for their usual sources of income generation is lost. An increasing proportion of the poor are affected. It is evident that it is going to be increasingly difficult to manage environments sustainably, as natural resources on which the poor debilitated patients depend continue to be degraded. Incomes lost due to the stigma of HIV/AIDS must be met by the sufferers by overexploiting their resource base. People affected by the disease inevitably try to get whatever they can from their natural resource base as they are not in any position to think of the long-term future.

In Africa, this has led to degradation of the ecosystem and an increase of pressures from other impacts such as overuse of medicinal plants and poaching for wildlife. In South Africa, for example, people have a mistaken belief that turtle eggs can cure HIV/AIDS, thus leading to the eggs being over harvested. As males die of the disease, work on agricultural land has to be taken over by already overworked women and their children, affecting land management and productivity. Providing balanced diets and nutritional support for these poverty stricken patients can be partially addressed by better natural resource management such as afforestation, access to clean water and wholesome food. HIV/AIDS seriously affects the patient's working environment. It creates an incorrect fear in the minds of co-workers. It must be clearly understood that AIDS is not spread by casual contact during work. Patients have a right to continue to work as before along with unaffected individuals. As patients are unable to continue their original hard labour related work, it is essential that alternative sources of work must be created for them. Educators and extention information, in the formal and non-formal educational sectors, must address the issues related to the linkages between natural resource management and this disease, as well as the need to remove the social stigma attached to it.

HIV/AIDS has a serious impact on the socioeconomic fabric of society. By 2002, India had an estimated 3.97 million infected individuals. There is a great need to organise AIDS education on prevention and management of the disease. This needs to be done through the formal educational sector and by using non-formal methods. Education is also important to reduce the stigma and discrimination against these patients. In India, women who are not socially empowered are at a great disadvantage as they are powerless to demand safe sex from their partners. Women also have an added burden of caring for HIV infected husbands. This produces enormous economic stresses on their family. HIV in India is rapidly moving from a primarily urban sector disease to rural communities. Research in Nepal has shown a linkage between rural poverty, deforestation and a shift of population to urban areas resulting in a rising number of AIDS patients. Prior to 1992, it was mainly seen in males who migrated to urban centers. In more recent times, a growing number of women are moving to Indian cities as sex workers. Women engaged in prostitution find it difficult to make partners take protective measures, such as the use of condoms that provide safe sex. A large proportion became victims of the disease. Blood transfusion from an infected person can also lead to HIV/AIDS in the recipient, as well as drug abuse by sharing needles with an infected person. In sexually transmitted AIDS, the use of condoms during intercourse is a key to preventing the disease. Behavioural change, where the number of individuals who have multiple partners, towards strictly single partners, reduces the risk of HIV/AIDS and thus reduces incidence of the disease in society. However, the most important measure to prevent AIDS is the proper use of condoms that form a barrier to the spread of the virus during intercourse.

WOMAN AND CHILD WELFARE

There are several environmental factors that are closely linked to the welfare of women and children. Each year, close to eleven million children worldwide are estimated to have died from the effects of disease and inadequate nutrition. Most of these deaths are in the developing world. In some countries, more than one in five children die before they are 5 years old. Seven out of 10 of childhood deaths in developing countries can be attributed to five main causes, or a combination of them. These are pneumonia, diarrhoea, measles, malaria and malnutrition. Around the world, three out of every four children suffer from at least one of these conditions.

TABLE: The Diagnosis of Common Childhood Disease Conditions

Presenting Complaint	Possible Cause or Associateed Condition
Cough and/or fast breathing	Pneumonia Severe anaemia P. falciparum malaria
Lethargy or unconsciousness	Cerebral malaria Meningitis Severe dehydration Very Severe pneumonia
Measles rash	Pneumonia Diarrhoea Ear infection
"Very sick" young infant	Pneumonia Meningitis Sepsis

Crowded ill-ventilated homes and living in smokey households with open fires can trigger respiratory conditions especially in children. Pneumonia:

Acute respiratory infections (ARI), most frequently pneumonia, is a major cause of death in children under five, killing over two million children annually.

Upto 40% of children seen in health centers suffer from respiratory conditions and many deaths attributed to other causes are, in fact, "hidden" ARI deaths. Children may die very quickly from the infection and thus need treatment urgently. Most patients of pneumonia can be treated with oral antibiotics. Correct management could save over 1 million lives per year globally.

Gastro Intestinal Conditions

Contaminated water and food causes widespread ill health especially in children. Diarrhoea: Diarrhoea is caused by a wide variety of infections. Urgent diagnosis and treatment of diarrhoea is a priority for saving a child's life.

Treating malnutrition that often accompanies diarrhoea can further reduce mortality. Increasing vigilance to detect other diseases that can occur concurrently with diarrhoea, such as measles or malaria, is an important measure.

Two million children die each year in developing countries from diarrhoeal diseases, the second most serious killer of children under five worldwide. In most cases diarrhoea is preventable and children can be saved by early treatment. Correct management of diarrhoea could save the lives of up to 90% of children who currently die by promoting rapid and effective treatment through standardised management, including antibiotics and simple measures such as oral rehydration using clean boiled water with salt and sugar.

In severe cases intravenous fluids must be started. Improved hygiene and management of the home and surroundings is the most important preventative measure, as well as improved nutrition. Increased breastfeeding and measles vaccination have also been observed to have reduced the number of cases of diarrhoea.

Measles

Measles is a rash that appears with fever and bodyache in children and is caused by a virus. It infects over 40 million children and kills over 800,000 children under the age of five. Prevention includes wider immunization coverage, rapid referral of serious cases, prompt recognition of conditions that occur in association with measles, and improved nutrition, including breastfeeding, and vitamin Λ supplementation. Measles is prevented by a vaccine. Young children with measles often develop other diseases such as acute respiratory infections, diarrhoea and malnutrition that are all linked to poor environmental conditions in their surroundings. Children who survive an attack of measles are more vulnerable to other dangerous infections for several months. Effective prevention and treatment could save 700,000 lives per year.

Malaria

This condition is closely linked to pooling and stagnation of water in tropical environments. Malaria is a widespread tropical disease which is caused by a parasite transmitted to humans by mosquitoes. It has proved difficult to control because mosquitoes have become resistant to insecticides used against them and because the parasite has developed resistance in some areas to the cheap and effective drugs that used to provide good protection in the past.

However, alternative newer drug therapies have been developed for use in areas where resistant parasites are found. In India the disease was nearly wiped out a few decades ago but has now re-emerged in many parts of the country. Correct management could save 500,000 lives per year. Approximately 700,000 children die of malaria globally each year, most of them in sub-Saharan Africa. Young children are particularly vulnerable because they have not developed the partial immunity that results from surviving repeated infections. Deaths from malaria can be reduced by several measures, including encouraging parents to seek prompt care, accurate assessment of the condition of the child, prompt treatment with appropriate anti-malarial drugs, recognition and treatment of other co-existing conditions, such as malnutrition and anaemia, and prevention by using mosquito-proof bednets. Because fever may be the only sign of malaria, it can be difficult to distinguish it from other potentially lifethreatening conditions.

Poverty-Environment-Malnutrition

There is a close association between poverty, a degraded environment, and malnutrition. This is further aggravated by a lack of awareness on how children become malnourished. Malnutrition: Although malnutrition is rarely listed as the direct cause of death, it contributes to about half of all childhood deaths. Lack of access to food, poor feeding practices and infection, or a combination of the two, are major factors in mortality.

Infection, particularly frequent or persistent diarrhoea, pneumonia, measles and malaria, undermines nutritional status. Poor feeding practices-inadequate breastfeeding, providing the wrong foods, giving food in insufficient quantities, contribute to malnutrition. Malnourished children are more vulnerable to disease. Promoting breastfeeding, improving feeding practices, and providing micronutrient supplements routinely for children who need them are measures that reduce mortality. The nutritional status and feeding practices of every child under two years of age, and those with a low weight for their age must be intensively managed. Counseling of parents on the correct foods for each age group and helping them to overcome various feeding problems is an essential health care measure. Children between 6 months and 2 years of age are at increased risk of malnutrition when there is a transition between breastfeeding and sharing fully in the family diet.

Changing family habits and the kinds of food offered to children is an important measure. Talking to mothers individually about home care and their child's feeding, with relatively simple changes to better feeding practices, such as helping them to eat rather than leaving them to fend for themselves, can ensure that a child gets enough to eat. A minor increase in breastfeeding could prevent up to 10% of all deaths of children under five: When mothers breastfeed exclusively during at least the first four months and, if possible, six months of life, there is a decrease in episodes of diarrhoea and, to a lesser extent, respiratory infections.

Even small amounts of water-based drinks decreases breastmilk intake and lead to lowered weight gain. This increases the risk of diarrhoea. Continuing to breastfeed up to two years of age, in addition to giving complementary foods, maintains good nutritional status and helps prevent diarrhoea.

Encouraging maximum support to mothers to establish optimal breastfeeding from birth, equipping health workers with counseling skills, and providing individual counseling and support for breastfeeding mothers are measures that reduce malnutrition.

Mothers often give their babies other food and fluids before six months because they doubt their breastmilk supply is adequate. A one-on-one counseling with mothers on breastfeeding techniques and its benefits helps reduce incidence of malnutrition. There are strong connections between the status of the environment and the welfare of women and children in India. Women, especially in lower income group families, both in the rural and urban sector, work longer hours than men.

Their work pattern differs and is more prone to health hazards. The daily collection of water, fuelwood and fodder is an arduous task for rural women. In urban areas, where lower economic group women live in crowded smoke filled shantys in unhygenic slums, they spend long hours indoors, which is a cause of respiratory diseases. In urban centers, a number of women eke out a living by garbage picking. They separate plastics, metal and other recyclable material from the waste produced by the more affluent groups of society. During this process, they can get several infections. Thus they are providing an environmental service of great value, but earn a pittance from this work.

Women are often the last to get enough nutrition as their role in traditional society is to cook the family meal and feed their husband and children. This leads to malnutrition and anemia due to inadequate nutrition. The sorry plight of women includes the fact that the girl child is given less attention and educational facilities as compared to boys in India.

Thus Case Study

Karnataka's GIS scheme, Bhoomi, has revolutionized the way farmers access their land records. Farmers can now get a copy of the records of rights, tenancy and crops from a computerized information kiosk without harassment and bribes. Karnataka has computerized 20 million records of land ownership of 6.7 million farmers in the State. they are unable to compete with men in later life. This social-environmental divide is a major concern that needs to be corrected throughout the country.

ROLE OF INFORMATION TECHNOLOGY IN ENVIRONMENT AND HUMAN HEALTH

The understanding of environmental concerns and issues related to human health has exploded during the last few years due to the sudden growth of Information Technology. The computer age has turned the world around due to the incredible rapidity with which IT spreads knowledge. IT can do several tasks extremely rapidly, accurately and spread the information through the world's networks of millions of computer systems.

A few examples of the use of computer technology that aid environmental studies include software such as using Geographical Information Systems (GIS). GIS is a tool to map landuse patterns and document change by studying digitized toposheets and/or satellite imagery. Once this is done, an expert can ask a variety of questions which the software can answer by producing maps which helps in landuse planning. The Internet with its thousands of web sites has made it extremely simple to get the appropriate environmental information for any study or environmental management planning. This not only assists scientists and students but is a powerful tool to help increase public awareness about environmental issues. Specialised software can analyse data for epidemiological studies, population dynamics and a variety of key environmental concerns.

The relationship between the environment and health has been established due to the growing utilisation of computer technology. This looks at infection rates, morbidity or mortality and the etiology (causative factors) of a disease.

As knowledge expands, computers will become increasingly efficient. They will be faster, have greater memories and even perhaps begin to think for themselves. ❐

12

Nutrient Cycles

The main flow of energy that helps shape conditions on Earth's surface comes from space; and when the energy's work here is done, to space it returns. With respect to energy, then, Earth is an open system. With respect to its chemical endowment, however, Earth is a closed system. That is, the amounts of carbon, hydrogen, oxygen, iron, gold, and other elements in the planet-atmosphere system do not change with time; the chemical arrangement and physical distribution of these elements can and do vary, but essentially nothing enters and nothing leaves the system.

The elements in this closed system that are essential to life are called nutrients. They can be divided into three categories: the 4 main chemical building blocks of living matter (carbon, oxygen, hydrogen, nitrogen); 7 macronutrients, of which smaller but still significant quantities are required for life; and 13 micronutrients, or trace elements, of which tiny quantities perform essential functions. The suitability of any terrestrial or aquatic environment for the support of life depends on the availability of nutrients in appropriate forms and quantities. The processes that govern this availability (or lack of it) are known collectively as nutrient cycles, because of the way the individual basic stocks of physical material move cyclically through the living and non-living parts of the physical world.

These cycles function in support of life not merely by making nutrients continuously available—in other words, by maintaining the fertility of the environment -but also by limiting the accumulation of material in quantities, forms, and places in which it would damage organisms. It is important to recognize in this connection that the same elements and compounds that serve as nutrients for some organisms in some concentrations are often toxic for other organisms, or even for the same organisms at higher concentrations. For example, molecular oxygen (O_2) is toxic to anaerobic organisms and, at high enough concentrations, even to mammals. Ammonia (NH_3) is an important source of nutrient nitrogen for many plants but is toxic to people.

Hydrogen sulfide (H_2S) is a nutrient for certain types of bacteria but is extremely toxic to mammals.

DYNAMICS OF NUTRIENT CYCLING

The steps that make up all nutrient cycles involve two basic processes: physical transport and chemical transformation. They are:

The hydrologic cyclc, including evaporation and precipitation as well as the flow of streams, groundwater, and ice;

- Winds;
- Ocean currents;
- Geologic movement, especially the upward and downward motions at the boundaries of tectonic plates and geologic uplifting on continents.

A fifth agent of physical transport that is important in a few circumstances is the movement of organisms. For example, fish-eating birds that deposit their excrement on land provide a significant pathway by which nitrogen and phosphorus are transferred from the sea to the land. Fish such as salmon, that feed mostly in the ocean but migrate up freshwater rivers to spawn and die, perform a similar function, as do the ocean-caught fish consumed by continent-dwelling human beings.

How the various mechanisms of transport link the principal "compartments" of the physical world. Obviously, the ease (and, hence, the rate) with which a given substance can move through these pathways depends on its physical state (solid, liquid, gas), on other physical properties (solubility, volatility), and on the possibility of chemical transformations that can change those properties. Such chemical transformations take place in living organisms and in the non-living environment. They are considered in more detail below. Each of the compartments indicated often is divided into subcompartments for the purpose of tracing nutrient flows. Thus, for example, it is customary to consider three subcompartments of material in both the soil and the oceans: living organic matter, dead organic matter, and the inorganic medium itself. These subcompartments can be further subdivided—for example, organic matter, into plants, animals, and microorganisms; and seawater, into the well-mixed surface layer and the stratified deeper layers. Finally, within each subdivision of each subcompartment, the magnitudes of the stocks, or pools, of the important elements and compounds are of interest, as are the magnitudes of the flows into and out of the pools (where flows can mean physical movement or chemical transformation or a combination of the two). For any pool that is in steady state—that is, for which inflows balance outflows so the size of the pool does not change with time—an average residence time for a compound can be calculated as the ratio of the size of the pool to the sum of the inflows or outflows. Examination of the residence times for important pools provides quick insight into the dynamics of nutrient cycling, and, indirectly, into the probable vulnerability of specific components of nutrient cycles to disruption by human intervention. For example, a long residence time means the annual flows are small compared to the pool, which means in turn that unless human disruptions cause a very large alteration in inflow or outflow they will affect the size of the pool only very slowly. A short residence time, by contrast, means that significant changes in inflow or outflow can affect the size of the pool very quickly.

CHEMISTRY OF NUTRIENT CYCLES

Two kinds of chemistry play crucial roles in the principal global nutrient cycles: the chemistry of the sedimentary cycle, associated with the very slow transformation of very large pools of material, and the chemistry of life, associated with the more rapid transformation of much smaller pools. Both kinds of chemistry depend very heavily on the very special roles of water as a source of hydrogen (H+) and hydroxyl (OH–) ions for many chemical reactions, as a medium for holding other chemical compounds in solution where they can react, and as a vehicle for physical transport of compounds from place to place.

Sedimentary Cycle

Nutrients are mobilized in the sedimentary cycle when water comes in contact with rock. Rainwater is slightly acidic, having absorbed some atmospheric carbon dioxide to form weak carbonic acid:

$$H_2O \rightarrow CO_2 \rightarrow H_2CO_3$$

$$H_2CO_3 \rightarrow H \rightarrow HCO_3$$

The pH of rainfall in areas free of industrial pollution is around 5.7. (Recall from basic chemistry that pH is the negative of the log to base 10 of the concentration of hydrogen ions in moles per litre. The pH of distilled water is 7.0, corresponding to 10–7 moles of hydrogen ion per litre.) This slight acidity is enough to facilitate greatly the dissolution of many minerals present in exposed rock—a process called weathering.

Kaolinite is a clay; it is left behind in this process as a contribution to the formation of soil, whereas the dissolved silica, potassium ion, and bicarbonate ion are carried away in the runoff. The mobilization and removal of positive ions such as (in this case) potassium is called leaching. Note that the reaction consumes some water and that it reduces the acidity of the solution (that is, it removes hydrogen ions) as it proceeds. These properties are characteristic of weathering reactions. Essentially the same reaction also occurs in the weathering of the aluminosilicates of sodium and calcium ($NaAlSi_3O_8$,$CaAl_2Si_2O_8$). Other silicate minerals also weather in the same general manner to produce kaolinite or other clays.

The weathering of certain other minerals leads to their complete dissolution; no solid trace is left behind to be incorporated into soil. The most important of these minerals are the limestones, or carbonate rocks, which undergo the following reactions:

$$CaCO_3H \rightarrow H\,CO_3 \rightarrow Ca \rightarrow 2H\,CO_3\,-$$

$$CaMg(CO_3)2 \rightarrow 2(H \rightarrow HCO_3\,)\,Mg \rightarrow Ca \rightarrow 4HCO_3$$

Most of the products of the weathering reactions mentioned here eventually reach the sea, either as dissolved ions or as suspended solids. Between the time they are freed from rock and the time they reach the sea, of course, some of these elements have made one or more passages through the living part of the biosphere as nutrients. In the sea, too, the nutrient substances are captured from the water by living organisms, eventually to be returned to the water or to the sediments on the ocean floor.

The reverse of many of the weathering reactions also can occur in seawater in the absence of intervention by living organisms, forming solid precipitates that join the organic debris on the ocean bottom.

Apparently these outputs of the major nutrient elements from the oceans roughly balance the inputs, because considerable physical evidence indicates that the salinity of seawater has been approximately constant for at least the past 200 million years.

The sedimentary cycle is finally closed by the geologic processes that is, by the processes of transformation and transportation that turn seafloor sediments into continental rocks, and by volcanic action.

Living Matter

Ecologist Edward S. Deevey, Jr., has pointed out that living matter can be regarded as a chemical compound with the empirical formula

$$H_{2960}\,O_{1480}\,C_{1480}\,N_{16}P_{1.8}\,S.$$

This enormous (and completely intentional) oversimpli-fication serves both to call attention to the chemical side of life the side concerned -and, more specifically,the relative proportions in which the six most abundant elemental constituents of living material exist.

Actually, there is substantial variation in the ratios of the chemical elements of life in different kinds of organisms, the above formula being a weighted average in which woody plants predominate. The numbers are, in any case, not known with great accuracy. The basic reactions that link the water cycle, the carbon dioxide cycle, and the oxygen cycle through living matter are photosynthesis:

$$6CO_2 \rightarrow 12H_2O \rightarrow \text{solar energy} \rightarrow 6O_2 \rightarrow C_6\,H_{12}O_6 \rightarrow 6H_2O$$

and its opposite, respiration, or oxidative metabolism:

$$6O_2 \rightarrow C_6\,H_{12}\,O_6 \rightarrow 6CO_2 \rightarrow 6H_2O\text{ heat}$$

The compound ($C_6H_{12}O_6$, or $(CH_2O)_6$, is glucose, the simplest carbohydrate produced by photosynthesis. The photosynthesis reaction is written with what appear to be six extra H_2O molecules on each side because twelve molecules of H_2O are actually broken up for each molecule of $C_6H_{12}O_6$ synthesized; radioactive-tracer experiments show that all twelve oxygen atoms from the 12 H_2O end up in the $6O_2$, and the oxygen in the six new water molecules comes from the CO_2. The amount of energy transformed from light into stored chemical energy by the photosynthesis reaction is 112 kilocalories (469 kilojoules) per mole of carbon fixed. The net result of the respiration reaction is to transform the same 112 kilocalories per mole of carbon from chemical energy into heat. Photosynthesis is carried out not only by higher green plants (such as oak trees, grasses, and ferns) but also by various kinds of algae (green, brown, red, blue-green) and by certain bacteria. In the bacteria, substances other than water serve as the sources of hydrogen for the reduction of carbon. For example, the green sulfur bacteria use hydrogen sulfide instead of water, according to the equation

$$12H_2\,S_6 \rightarrow CO_2 \rightarrow 6CO_2 \rightarrow \text{light} \rightarrow C_6\,H_{12}O \rightarrow 6H_2O \rightarrow 12S$$

producing elemental sulfur rather than oxygen. (In fact, these bacteria are obligate anaerobes—they would be poisoned by oxygen.) The reverse reaction (respiration) consumes sulfur and carbohydrate and liberates heat, carbon dioxide, and hydrogen sulfide. Still other photosynthesis reactions use hydrogen, obtained from H_2O or H_2S or other hydrogen donors, to reduce substances other than carbon dioxide. For example, most higher plants can reduce nitrate, instead of CO_2, in photosynthesis:

$$9H_2O \rightarrow 2NO_3 \rightarrow \text{light} \rightarrow 2NH_3\ 6H_2\ O\ 4_1\ O_{22}$$

and nitrogen-fixing photosynthetic organisms can reduce atmospheric nitrogen:

$$6O_2 \rightarrow C_6\,H_{12}\,O_6\ 6CO_2\ 6H_2O \rightarrow \text{heat}$$

These variations notwithstanding, the great bulk of photosynthesis on Earth consists of the reduction of carbon dioxide, with water as the hydrogen source. The photosynthesis and respiration reactions, as we have written them here, actually represent only the end points of vastly more complicated sequences of chemical events. For example, although burning glucose in air is described essentially completely by the respiration reaction given to achieve

$$6O_2 + C_6H_{12}O_6 \rightarrow 6\,CO_2 + 6H_2O + \text{heat},$$

the same result in a living cell (at a temperature far below that of combustion) requires many intermediate steps. The intermediate steps in both respiration and photosynthesis rely heavily on three kinds of organic compounds: the energy carrier ATP (adenosine triphosphate); the electron carrier NADP (nicotinamide adenine dinucleotide phosphate); and various catalysts called enzymes. (Catalysts speed up chemical reactions without themselves being consumed.)

A closer look at these compounds provides some clues to the importance of the elements nitrogen, phosphorus, and sulfur in the chemistry of life. All biochemical reactions of importance (not just photosynthesis and respiration) require enzyme catalysts if they are to proceed at significant rates. All enzymes are proteins. Proteins other than enzymes also serve as hormones; as vehicles to transport oxygen; as important structural components of skin, tendons, muscle, cartilage, and bone; and in numerous other roles. All proteins are made up of building blocks called amino acids, and all amino acids contain the amino group—NH_2. A single protein contains from fifty to some tens of thousands of amino acids, and hence at least that many nitrogen atoms. (Some amino acids contain more than one nitrogen.) The shape and stiffness of many proteins—properties essential to their functions—are governed and maintained by bonds in which sulfur plays a crucial role, so it, too, is essential to the chemistry of life.

ATP and NADP are examples of nucleotides, compounds consisting of a five-carbon sugar to which one or more phosphate groups ($–PO_4$) and a nitrogen base are attached. A nitrogen base is a ring compound containing both nitrogen and carbon in the ring structure. Nucleotides not only perform critical energy-transfer and electron-transfer functions in the chemistry of living cells, but they are also the building blocks of nucleic acids. Each nucleic acid molecule contains hundreds to thousands of nucleotides, the arrangement of which is a chemical means of storing information. Nucleic acids control the synthesis of proteins and a variety of other processes in living cells. The nucleic acid DNA (deoxyribonucleic acid) is the genetic material—the bearer of the coded information that is passed from parent cells and organisms to their offspring to "tell" them what to become. The nucleic acid RNA (ribonucleic acid) plays essential roles in carrying out the instructions coded in DNA molecules.

In addition to its role as a constituent of DNA and RNA, phosphorus, in the form of phosphate, is one of the principal anions (ions with a negative charge) that maintains electric neutrality in the fluids of living organisms. Along with calcium, it is also an important component in shells and bony structures. Sulfur, besides playing its role in proteins, in the form of sulfate (SO_4) is another principal anion in body fluids and cells. The third principal anion is the chloride ion (Cl). The principal cations (positively charged ions) are sodium, potassium, calcium, and magnesium. Magnesium is an essential constituent of many enzymes and of the green pigment chlorophyll. Iron is a key component of many enzymes and of the oxygen-transporting protein hemoglobin.

Oversimplifying considerably, one can say that the quantity of living matter in a given environment is limited by the stock of that requisite of life that is in shortest supply. This rather obvious proposition carries the name Liebig's law of the minimum. Of course, the limiting factor varies from one type of organism to another and from one environment to another.

Generalizations are difficult, but it is fair to say that under many circumstances the quantity of plant material in a given environment is in fact controlled by a limiting nutrient (although the density of a given kind of plant is often controlled by a herbivore). In desert regions the limiting nutrient is generally water. Under many conditions where water is not limiting, as in freshwater lakes, the limiting nutrient turns out to be phosphorus; in the open ocean it may be iron. In some richly productive environments, nitrogen and phosphorus are jointly limiting—producing any more plant material would require supplying more of both nutrients. Obviously, the limiting nutrient in a given biological system represents a leverage point where the effects of natural or human-induced perturbations manifest themselves rapidly.

CYCLES OF THE PRINCIPAL NUTRIENTS

The cycles of many nutrients are tightly linked chemically and biologically. This is so much the case for the cycles of carbon, oxygen, and hydrogen that we consider them together. The cycles of nitrogen, phosphorus, and sulfur, on the other hand, can usefully be examined separately.

Carbon-Oxygen-Hydrogen

It is appropriate to begin our closer examination of the carbon-oxygen-hydrogen cycles with their intersection in the green plant, the basis of almost all life on Earth.

Carbon and Energy in Living Plants

Most living plants not only fix (store) energy by means of photosynthesisbut they also use some of that stored energy to drive their own life-sustaining internal processes by means of respiration.

The total rate at which a plant or a plant community stores solar energy by means of photosynthesis is called gross photosynthesis or gross primary production (GPP). The figure obtained by subtracting from gross photosynthesis the amount of energy used by the plants in respiration (R) is called net photosynthesis or net primary production (NPP). In abbreviated form, then, the relation is GPP-R = NPP. Several interchangeable units are used in the biological literature for the measurement of these quantities: calories or kilocalories per unit of time, watts or kilowatts (different units for measuring the same thing—energy flow), grams of carbon fixed into organic compounds per unit of time, moles of carbon fixed per unit of time, or grams of dry organic matter produced per unit of time.

A wide range of estimates of global net primary production and the associated standing crop of living plants (phytomass) have been published. One recent review indicates that the estimates of net primary production are converging on a value around 160 billion (1.60 x 10) metric tons per year for the globe.

Global phytomass was estimated in the same review as about ten times larger 1840 billion metric tons, dry weight and gross photosynthesis was estimated to be about twice as large as net production (that is, half of gross photosynthesis is consumed in plant respiration and half remains as net production). The estimates for phytomass and net primary production, separated into terrestrial and oceanic components. The roughness of these estimates deserves emphasis. The range of recent estimates of land phytomass 30 per cent larger; the range of published values of ocean phytomass is from twenty times smaller than the value to three times larger. The factor-of-60 range of informed opinion with respect to oceanic phytomass is especially remarkable.

The range in recent published estimates of oceanic net primary production is only ± 10 per cent from the value, and the range in terrestrial net production is from 15 per cent less to 60 per cent more than the value in the table. The fraction of gross primary production consumed in plant respiration is known to vary between 20 per cent and 75 per cent, depending on type of plant and region, but the global average is unlikely to differ much from 50 per cent. Energy and Material Flow in the Carbon Cycle

For the purpose of reconciling estimates of annual production presented in grams of carbon, grams of dry organic matter, and kilocalories, one widely used convention assumes that 10 kilocalories of energy are stored per gram of fixed carbon and that dry organic matter contains 45 per cent fixed carbon by weight. (The actual range of values for different kinds of dry organic matter extends roughly ± 15 per cent from these nominal figures.) Thus, measurements in grams or metric tons (1 MT = 106 g) of carbon can be converted to grams or metric tons of dry organic matter by multiplying by 2.2, and 1 gram of dry organic matter can be taken to represent 4.5 kilocalories of stored energy. Useful conversions following directly from this convention are that I gram of dry organic matter per day corresponds to an average energy flow of about 0.2 watts; 1 gram of carbon per day, to about 0.5 watts; and one metric ton of dry organic matter per year, to about 600 watts. Also, 1 metric ton of dry organic matter contains 37,500 moles of carbon.

Although we shall use the convention just described, the reader may encounter others elsewhere. If the dry organic matter were pure carbohydrate (CH2O)n which stores 112 kilocalories per mole of carbon, then the appropriate conversions would be: 3.7 kilocalories per gram of dry organic matter, 9.3 kilocalories per gram of carbon, 0.4 gram carbon per gram of dry organic matter [Actually, CH2O is a good approximation of the average composition of all living matter because so much of the world's biomass is cellulose tied up in woody plants, and cellulose has nearly this formula. However, most of the plant material produced each year is not cellulose—remember, a big pool does ot necessarily mean a big flow. The annual production of organic matter contains a higher proportion of fixed carbon and higher energy content per gram than does the cellulose-dominated, slow-toturn-over standing crop.] The rounded-off value of 4 kilocalories per gram of dry organic matter is in fairly widespread use in conjunction with the figure of 10 kilocalories per gram of fixed carbon, itself a rough average among several kinds of plant matter.

Taking the range of estimates into account, one can state several conclusions with reasonable assurance:

- The annual NPP of the oceans is only one-third to one-half that of the continents, even though the oceans cover more than 70 per cent of Earth's surface;
- The pool, or standing crop, of living plant material on land is at least several hundred times that in the oceans;
- The average turnover time (life expectancy) of oceanic plants must be in the range of days to weeks, in order for so small a phytomass to be associated with so large an annual production;
- The average turnover time of terrestrial living plants, by contrast, is in the range of a decade or two;
- Oceanic and terrestrial phytomasses are both so poorly known that man-induced or natural changes large enough to be very important in the long term could easily escape notice for some time. In the latter regard, although there are good reasons to suppose that the activities of civilization have diminished the terrestrial phytomass significantly, some experts consider this assertion to be debatable.

Notwithstanding the uncertainties in the data, it is instructive to examine the large regional variations in plant productivity that are concealed in the global totals. Of particular interest are the facts that tropical forests account for more than 40 per cent of terrestrial NPP (although they cover only about 16 per cent of the land area) and that up- welling regions, continental shelves, reefs, and estuaries yield almost 25 per cent of oceanic NPP while accounting for less than 9 per cent of the ocean's area.

The rate at which energy is being stored in net primary production is about 100 million megawatts, roughly two-thirds on land and one-third in the ocean. The rate of gross primary production can be assumed to be about 200 million megawatts. The net primary production is roughly twelve times the rate at which civilization used commercial energy in 1975, where commercial energy refers to energy that is sold (for example, fossil fuels, hydropower, and nuclear power, but excluding food) as opposed to energy that is gathered by the user (for example, most wood, dung).

Very little of the solar energy that reaches Earth's surface is captured by photosynthesis. The net primary production is about 0.1 per cent of the incident solar energy at the surface-0.25 per cent of what strikes the land and 0.05 per cent of what strikes the ocean. Some of the reasons these percentages are so small are obvious. Plants do not by any means cover all of Earth's surface; in many areas of the world the growing season lasts only part of the year; and in some circumstances, as already noted, plant growth is limited by lack of one or more critical nutrients. Under favourable conditions, on the other hand—as in a well watered and fertilized cornfield or a fertile swamp—NPP may reach 2 per cent or so of incident solar energy averaged over the growing season and as much as 6 per cent or 7 per cent of incident solar energy on the most favourable days.

One additional reason for this is that only about one-fourth the solar energy reaching the ground is in the part of the wavelength range that stimulates photosynthesis (blue and red light). The other big reason is the large amount of water that must be evaporated through the leaves of plants to maintain the nutrient-bearing flow of water through them. Evaporating this water typically takes 40 per cent or more of the energy that falls on a plant, and a much higher percentage of what is actually absorbed by the plant. (For production of every ton of wet weight of many crops—of which about one-fourth is dry organic matter around 100 tons of water are evaporated. This means 400 grams of water are evaporated, requiring about 0.6 kilocalories per gram evaporated, for every gram or 4.5 kilocalories of dry organic matter produced. The ratio in such a case is about 50 kilocalories for evaporation per kilocalorie of net primary production.)

OXYGEN BALANCE: METABOLISM BY ANIMALS AND MICROBES

The overall effects of the net production of plants on the cycles of carbon, oxygen, and hydrogen are as follows:

- More oxygen, O_2, is added to the environment than is removed from it;
- Less carbon dioxide, CO_2, and water, H_2O, are added to the environment than are removed from it;
- The CO_2 and H_2O subtracted from the environmental pools show up as additions to the pool of stored organic matter, CH_2O, and to the atmospheric oxygen pool. Quantitatively, for every metricton of dry organic matter produced, there is a net loss from the environment of 0.6 metric tons of water and about 1.5 metric tons of CO_2, and a net gain of about 1.1 metric tons of oxygen.

Why, then, is oxygen not building up continuously in the atmosphere, while water and carbon dioxide are depleted?

Because plant matter is not accumulating, but rather is approximately in steady state. It is being broken down into carbon dioxide and water again by the metabolism of animals and microbes, using up molecular oxygen just as rapidly as new plant material and oxygen are being produced.

Consider the possible fates of plant carbohydrate produced as part of net primary production. Plants either are consumed by herbivores, or they die and add their carbohydrate to the pool of dead organic matter. Of the part consumed by herbivores, some is metabolized and thereby returned to the environmental pool of CO_2 and H_2O, some is added as excrement to the pool of dead organic matter, and some is incorporated into herbivore tissue, either to be eaten by carnivores or to be added at the herbivore's death (by causes other than being devoured) directly to the pool of dead organic matter. The same fates await the carbohydrate that reaches the carnivores. The cycle is closed by the action of decomposers—both animals and microbes (bacteria)—which by metabolizing the fixed carbon in dead organic matter return CO and H O to the environmental pools.

There are some important side pathways in the carbon cycle in which the initial products of decomposition include methane and carbon monoxide (CH_4 and CO) in addition to carbon dioxide. The main reactions are:

$$2CH_2O \;||\; CH_4 \;||\; CO_2$$

$$2CH_2O \;||\; O_2 \;||\; 2CO \;||\; 2H_2O$$

The CH_4 and CO may then be oxidized in the soil or in the atmosphere according to the following reactions:

$$CH_4 \;||\; 3O_2 \;||\; 2CO \;||\; 4H_2O$$

and

$$2CO \;||\; O_2 \;||\; 2CO_2$$

The details concealed by these simple formulas can be quite complex. In steady state, the metabolism of carbohydrate by herbivores, carnivores, and decomposers returns to the environment exactly as much CO_2 and H_2O as the production of the carbohydrate by photosynthesis originally removed. And it consumes exactly as much oxygen as the photosynthesis produced. There is therefore no long-run buildup (or depletion) of oxygen in the environment as a result of this cycle, unless net photosynthesis and metabolism get out of balance.

THE SEQUESTERING OF FIXED CARBON

How did oxygen accumulate in the atmosphere in the first place? The biogeochemical evidence suggests that molecular oxygen in the atmosphere is of biological origin, there having been essentially none before the evolution of photosynthetic organisms, perhaps 2 billion years ago. For oxygen produced by photosynthesis to have accumulated over the long term, part of the carbon fixed by this process must have been withdrawn from the cycle, escaping the oxidation step back to carbon dioxide. Indeed, it is precisely such a break in the carbon cycle that produced the fossil fuels. They originated as plant material that escaped breakdown by decomposers, in many cases by being buried in swamps and bogs under conditions in which both oxygen and anaerobic decomposer organisms were absent. Hundreds of millions of years of burial under considerable pressure and heat converted these plant materials into the hydrocarbon fossil fuels—coal, petroleum, and natural gas. The oxygen released when the carbon in these fuels was first fixed by photosynthesis remained in the atmosphere.

The different ranks of coal—lignite, subbituminous, bituminous, anthracite—represent different ages (listed here from youngest to oldest) in terms of the origin of the plant material they derived from, and somewhat different chemical compositions. Peat is the partially decomposed organic matter that is the precursor of coal. Petroleum, natural gas, oil shale, and tar sands are the results of different histories of biogeochemical transformation, but all originated, like coal, with dead organic matter. Not all of the fixed carbon that has been sequestered out of reach of natural oxidative processes is in deposits as concentrated as the fossil fuels. A much larger amount, in fact, is dispersed in sediments whose hydrocarbon content is on the order of 1 part per 1000 (1 g of hydrocarbon material per kg of sediment) or less.

Since the oxygen reservoir in the atmosphere has been built up by the sequestration of fixed carbon in various forms, could it be jeopardized by any forseeable activities of civilization that would cause the oxidation of these fixed-carbon pools? Deforestation, for example, could lead to the oxidation of part of the standing crop of fixed carbon in living organic matter, and combustion of fossil fuels oxidizes those pools. Yet, examination of the best estimates of the sizes of the fixed-carbon pools shows little cause for concern in the short term. Removal and oxidation of the entire terrestrial biomass would deplete atmospheric oxygen by at most 0.2 per cent. Combustion of all the conventional fossil fuels (coal, petroleum, natural gas—of which about 90 per cent is coal) thought to be recoverable ever by human effort would deplete atmospheric oxygen by 1.8 per cent. This would reduce the oxygen content of the atmosphere at sea level to what it now is at an elevation of 150 meters.

The 1975 global rate of fossil-fuel combustion would amount to a depletion of atmospheric oxygen by about 0.001 per cent per year—far below the threshold of detectability with existing measurement techniques. On the other hand, the amount of reduced carbon thought to be present in oil shales is sufficient, if it were ever totally recovered and oxidized, to consume more than half the atmospheric oxygen pool.

Some observers believe that most of the oil shales are too dispersed an energy resource ever to be economically attractive. But, in the event this view is not correct and civilization commences to use oil shale at a very high rate—say, ten times the rate at which all forms of energy were consumed worldwide in 1975—it would still take many centuries for the resulting depletion of atmospheric oxygen to become serious.

The oxygen situation is made more complicated than indicated in the foregoing paragraphs by the existence of important pools of oxygen elsewhere than in the atmosphere. It is apparent from the information that other oxygen pools must exist, because the number of moles of oxygen in the atmosphere is much smaller than the number of moles of reduced carbon in the sediments and elsewhere.

The oxygen produced when all that carbon was fixed and sequestered was tied up with the oxidation of ferrous oxide (FeO) to ferric oxide (Fe_2O_3) and sulfur to sulfate ($-SO_4$) in sedimentary rocks. It may also have been tied up in the oxidation of carbon of inorganic origin (such as carbon monoxide from volcanoes) to carbon dioxide and carbonate ($-CO_3$). Although all these reactions are slow, they can account for the necessary enormous amounts of oxygen if they are assumed to have operated over geologic time spans. Similarly, alterations in conditions governing the reaction rates could lead to significant changes in atmospheric composition over very long periods. On shorter time scales, however, organic carbon pools and associated reactions are almost certainly more important, and they are unlikely to produce significant changes in atmospheric oxygen in less than centuries.

CARBON DIOXIDE

Although atmospheric oxygen depletion evidently is not a short-term threat, buildup of atmospheric carbon dioxide is. Every mole of oxygen removed from the atmosphere means a mole of carbon dioxide added. Since the number of moles of CO in the atmosphere is 700 times smaller than the number of moles of O, a change of a specified number of moles has 700 times the effect on the CO concentration as on the O, concentration. The combustion of fossil fuel that depletes oxygen by 0.001 per cent per year augments the atmospheric CO by 0.7 per cent per year—not an insignificant increment. Not all of the added CO remains in the atmosphere, however. Some is absorbed by the oceans, and it is conceivable that some is being taken up in an increase in biomass. (Recall from the earlier discussion that the size of the biomass is not known with sufficient accuracy to detect even rather substantial changes.) An increase in the concentration of CO_2 in the atmosphere should increase the rate of net photosynthesis (where other nutrients are not limiting), and this would produce a growing biomass if the rate of oxidative metabolism by consumers and decomposers did not increase correspondingly.

Accrate measurements of atmospheric CO_2 have been made since about 1957. They have been coupled with indirect evidence to produce a model of the rise of atmospheric CO_2 from about 280 molecular parts per million incpreindustrial times to around 325 parts per million in 1975. An additionalckind of evidence that can shed some light on the sources and fate of the CO_2 added to the atmosphere is the abundance of radioactive carbon-14 relative tothe amount of non-radioactive carbon-12 in various pools. Carbon-4 is produced naturally by bombardment of nitrogen by cosmic rays in the stratosphere and also by civilization when nuclear bombs are detonated in the atmosphere; it then mixes throughout the atmosphere and is incorporated, along with other carbon, into plant matter.

Since the haf-life of carbon-14 is only 5770 years, the carbon-14 buried for millions of years in fossil fuels has decayed away entirely. Thus, the CO_2 added to the atmosphere by the combustion of fossil fuels contains no carbon- 14, whereas CO added from pools with faster turnover does contain carbon-14. The reduction combustion of fossil fuels is called the Suess effect. The use of data on the Suess effect, combined with other information needed to help interpret it (such as the way the rate of natural and artificial 14C production has varied over time), has allowed development of the estimate that about half of the CO ever added to the atmosphere by combustion of fossil fuels has stayed there. The same work indicates that the land biomass should have increased between 1 per cent and 3 per cent since preindustrial times, having absorbed some of the industrial CO_2, while the oceans absorbed the rest. This change in land biomass is much too small to measure, and the exact amount is a matter of controversy.

GLOBAL CARBON CYCLE

The pools and flows of the global carbon cycle. We caution the reader once more that substantial uncertainties exist about many of the numbers, as noted in detail earlier, and that further research might therefore alter the picture considerably. The pools of living and dead organic matter on land and in the oceans are assumed to be in equilibrium. Half the CO_2 added to the atmosphere by combustion of fossil fuels is assumed to dissolve in the surface layer of the oceans.

The carbon added to the atmosphere by the respiration of animals and microbes, only a few per cent is in the form of methane (CH_4) and carbon monoxide (CO), while the rest is carbon dioxide. Still, the natural sources of carbon monoxide—incomplete decomposition of organic matter and partial oxidation of methane in the atmosphere—exceed man-made carbon monoxide from fossil-fuel combustion severalfold. Methane is apparently not toxic at any concentrations that occur in natural environments, and carbon monoxide is toxic only at concentrations on the order of 100 times the global atmospheric average or more.

NITROGEN

The nitrogen cycle is chemically the most intricate of the major nutrient cycles, and it is probably the least well understood scientifically. The crucial role of nitrogen in all proteins makes this element essential to life, but most organisms can assimilate and use nitrogen only in specific chemical forms. Only a relatively few species of bacteria and algae can convert gaseous molecular nitrogen (N2) into the more complex compounds that can be used by plants and animals.

The principal usable forms are ammonia (NH3) and nitrates (–NO3), of which the latter is needed in greater quantity by most plants. The conversion of N2 to ammonia and nitrate is called nitrogen fixation, and the species that can accomplish this conversion are called nitrogen-fixing organisms. Thus, although the atmosphere represents an enormous reservoir of molecular nitrogen, the continuation of life on Earth depends absolutely on the activities of these tiny, inconspicuous organisms.

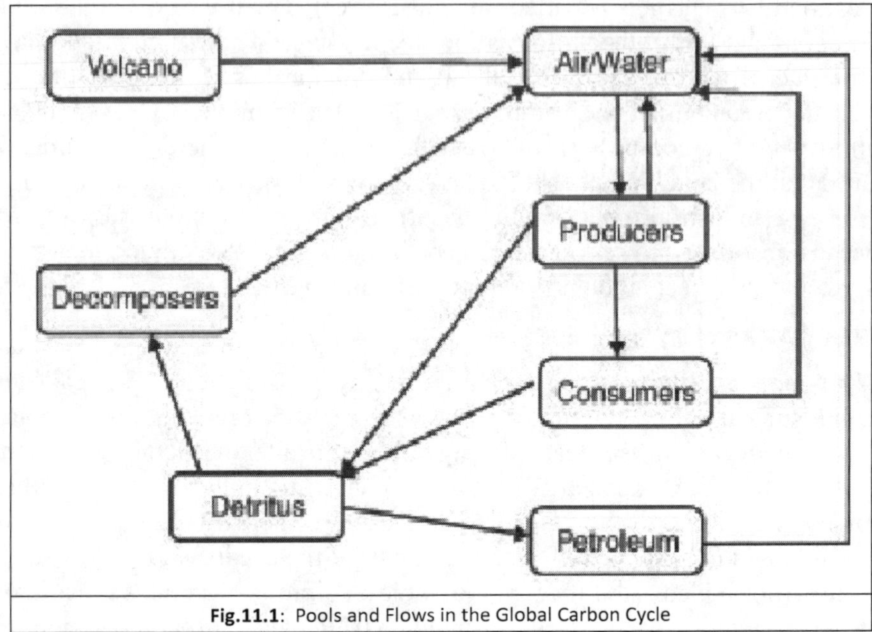

Fig.11.1: Pools and Flows in the Global Carbon Cycle

ORGANISMS AND NITROGEN

The best known of the nitrogen-fixing organisms are the bacteria associated with the special nodules on roots of legumes, which are plants of the pea family. In a symbiotic relationship, the bacteria obtain from the plants the energy they need to carry out the nitrogenfixing reaction;

$$2N_2 + 6H_2O \rightarrow 4NH_3 + 3O_2,$$

and some of the ammonia, in turn, is made available to the plants for the synthesis of amino acids, for example,

$$2NH_3 + 2H_2O + 4CO_2$$

$$2CH_2\ NH_2\ COOH + 3O_2.$$

The capacity of such legumes as alfalfa, beans, peas, and clover to fix nitrogen in large quantities has led to their widespread use in crop-rotation schemes to replenish in the soil the nitrogen depleted by other crops.

Although nitrogen fixation by the legume-bacteria symbiosis is thought to be responsible for most natural nitrogen fixation, the fixation step is also carried out by some bacteria that have looser associations with plants, including the free-living Azotobacter (aerobic) and Clostridium (anaerobic), and by various species of blue-green algae. Blue-green algae are apparently responsible for maintaining the fertility of rice paddies in much of Asia, which are subjected to intensive cropping without the use of nitrogen fertilizers.

The amino acids used by all organisms as the building blocks of protein are synthesized not only from ammonia but also from nitrate produced from ammonia by processes collectively called nitrification. The nitrification reactions yield energy, which supports the life processes of the nitrifying bacteria (Nitrosomonas, Nitrobacter). The actual synthesis of amino acids is carried out by bacteria (which differ widely in how many amino acids they can synthesize), by plants (which can synthesize all the amino acids needed for making proteins), and by animals (which, again, differ in the variety of amino acids they can produce).

Vertebrates—including human beings—cannot synthesize all the amino acids they need and must obtain intact in the food they eat the ones they cannot synthesize. The amino acids an animal must get from its diet are called essential amino acids; those it can synthesize are called nonessential. Fixed nitrogen that has been incorporated into organisms returns to the soil in animal wastes and in dead organisms (microbes, plants, animals) or parts of organisms (for example, leaves). Animal wastes are rich in urea—$(NH2)_2$ CO—which is the principal product of the metabolism of proteins in many organisms.

The proteins of dead organisms are broken down into amino acids and other residues by bacteria and fungi of decay. The latter compounds—urea, amino acids, and other breakdown products of protein—are then converted into ammonia by yet another group of bacteria. This step is called ammonification. (The chemical reaction for the ammonification of an amino acid is just the opposite of aminoacid synthesis.)

So far we have described how nitrogen is fixed by living organisms and how the fixed nitrogen is further processed by them. A logical question at this point is: How is nitrogen unfixed? For without a mechanism for returning nitrogen to its molecular form (N_2), even the vast reservoir of this compound that exists in the atmosphere would have been depleted long ago; much of the nitrogen would be tied up as nitrates in soil, in the oceans, and in sediments.

The mechanism that has prevented this outcome by closing the atmospheric loop of the nitrogen cycle is the action of a relatively few groups of denitrifying bacteria that make their living converting nitrate to nitrous oxide (N_2O) and N_2. (These transformations release energy.) The nitrous oxide is reduced to N 2 by further bacterial action, or in the atmosphere by photochemical reactions. Another set of denitrifying reactions, also carried out by bacteria, transforms nitrate to nitrite and nitrite to ammonia. The chemistry of the main steps in the nitrogen cycle that are mediated by bacteria is summarized. For each reaction, the "energy" column indicates whether energy must be supplied to accomplish it ("in") or whether the reaction makes energy available ("out"). The part of the terrestrial nitrogen cycle directly associated with organisms is represented schematically. The oceanic cycle is substantially similar, relying both on bacteria and algae.

❒

13

Energy and Environment Perspectives

BIOGAS TECHNOLOGY DISSEMINATION

Despite numerous benefits derived from biogas technology is yet to develop a major programmes that will promote the dissemination of biogas plants on a larger scale. This chapter reviews biogas installations and investigates challenges facing the design, construction, and operation of biogas plants. It further captures the current status and functions of biogas plants as well as the impact of these plants on the people who use them. The study was done by surveying fifty (50) biogas installations, and conducting interviews with both plant users and service providers.

From the survey, twenty-nine (58%) installations were institutional, fourteen (28%) were household units, and the remaining seven (14%) were community plants. Fixed-dome and water-jacket floating-drum digesters represented 82% and 8% of installations surveyed, respectively. It was revealed that sanitation was the main motivational reason for people using biogas plants. Of the 50 plants, 22 (44%) were functioning satisfactorily, 10 (20%) were functioning partially, 14 (28%) were not functioning, 2 (4%) were abandoned, and the remaining 2 (4%) were under construction.

Reasons for non-functioning include non-availability of dung, breakdown of balloon gasholders, absence of maintenance services, lack of operational knowledge, and gas leakages and bad odour in toilet chambers of biolatrines. This chapter recommends the development of a national biogas programmes focussing on three major areas–sanitation, energy, and agricultural fertilizer production; it further supports the development of standardized digester models.

The founding of a national body or the establishment of a dedicated unit within an existing organization with the sole aim of coordinating and managing biogas dissemination is proposed. Active harnessing and development of renewable energy (RE) sources began in the mid 1980s. However, after more than 20 years in the recognition of RE potential in contributing substantially to the energy mix of the developing country still lacks the capacity to developing her own RE sources.

In 1997, the Energy Commission (EC) was founded, among other functions, to develop, regulate, and manage RE. In 2006, the EC developed the Strategic National Energy Plant (SNEP)–a policy document that defined the role of various energy sources, setting target for each within a twenty year span. From SNEP, biomass based energy, apart from the direct use as woodfuel (firewood and charcoal), has been exploited to a very limited extent. Woodfuel represent the traditional energy source and accounts for 60 per cent of total energy used.

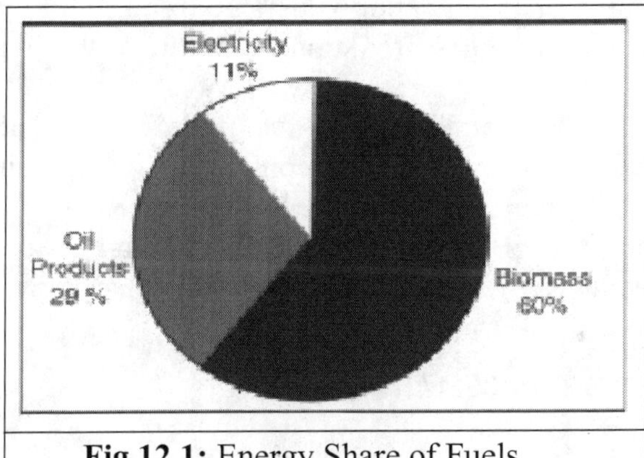

Fig.12.1: Energy Share of Fuels.

Biofuels (biogas, biodiesel, and bioethanol) have not been adequately developed to play a major role in the energy mix. For instance, in SNEP, the potential contribution of biogas technology towards the growth of the energy sector was not captured compared to other renewable energy options such as wind and solar. In order to realise a reduction in the share of woodfuel in the national energy mix from 60% in 2006 to 40% in 2020 as stipulated in SNEP, there is the need to promote research and development in other renewable energy options including biogas technology.

Biogas (anaerobic fermentation) technology is noted for improving sanitation, generating clean energy, and producing rich organic fertilizer. In China, India, and Nepal, household and institutional biodigesters have gained widespread acceptance. Since 2001, China has disseminated over 2 million household digesters annually; in addition, the Chinese government has supported over 200 large and medium livestock farms to own large and advanced biogas units.

From 2001 to 2007, over 18 million households adopted the technology leading to the production of over 7 billion m3 of biogas; moreover, 87 million tonnes of animal waste were treated by 3,556 biogas plants and more than 300 Clean Development Mechanism (CDM) projects involving biogas power generation, with a total capacity of 1 GW and an annual emission reduction of over 20 Mt of CO2, were also developed. In India, over 3 million domestic digesters and 3000 community and institutional plants were constructed by the end of 2002, and since 2005, more than 100,000 biodigesters have been disseminated annually.

Other successful biogas promoting Asian countries include Nepal, Vietnam, and Thailand. In Africa, biogas technology dissemination has been relatively unsuccessful.

Njoroge attributes the non-progressiveness of most biogas programmes to failure of African governments to support biogas technology through a focused energy policy, poor design and construction of digesters, wrong operation and lack of maintenance by users, poor dissemination strategies, lack of project monitoring and followups by promoters, and poor ownership responsibility by users. Despite the relative stagnation of biogas programmes in Africa, the future prospects are encouraging.

Aside energy (cooking and lightning, fuel replacement, shaft power), several biogas plants in recent years have been constructed as environmental pollution abatement system in several countries including Ghana, Kenya, Tanzania, Rwanda, Burundi, and South Africa. Between 4000–5000 digesters is estimated to have been built in Tanzani, while Kenya is said to have disseminated about 2000 digesters as at October 2007.

BIOGAS TECHNOLOGY DISSEMINATION

Before 1990

Interest in biogas technology is began in the late 1960s but it was not until the middle 1980s did biogas technology receive the needed attention from government. Dissemination programmes before the mid 1980s focused on the provision of energy for domestic cooking. Most plants, however, collapsed shortly after duration of project due to immature technologies and poor dissemination strategies. In order to resuscitate the technology, a cooperative agreement between Ghana and China led to the construction of a 10 m3 plant at the Bank of Ghana (BoG) cattle ranch in the Shai Hills and the start of the Appolonia Household Biogas Programmes in 1986.

The Appolonia Household Programmes focused on energy for cooking for cattle-owning households at Appolonia, a rural community in the Greater Accra region. A total of nineteen fixed-dome digesters comprising six 15 m3 and two 30 m3 Deenbandhu digesters, and eight 10 m3 and three 25 m3 Chinese dome digesters were constructed by engineers from the Ministry of Energy (MoE) and the Institute of Industrial Research (IIR). This was followed by the construction of two household demonstration plants at Jisonayilli and Kurugu, all in the Northern Region in 1987, under the aegis of the United Nations Children Fund (UNICEF).

Apart from the Ministry of Energy, biodigesters were also been promoted by Dr. Elias Aklaku, an engineer and a senior lecturer at the Agricultural Engineering Department of KNUST, mostly with support from the German Agency for Technical Cooperation (GTZ).

From 1990 to 1999

In June, 1992, the Ministry of Energy commissioned the first large scale community-based biogas plant in Appolonia. The Appolonia Integrated Rural Energy Project was aimed at providing street lighting and electricity for small load appliances for all the households in the community. Cow dung and human excreta were used to feed the digesters, and the gas produced was used to run a 12.5 kVA generator which provided power for street and home lighting, while the bio-slurry was used for agriculture.

The project experienced several setbacks and did not performed satisfactorily as planned due to multiplicity of factors including feedstock availability problems, distance of kraals (1/2 km) from the community, maintenance problems, and uncooperative attitude of some of the inhabitants.

Problems also arose in the utilization of the digested slurry as farm manure. In the initial stages, the liquid organic fertilizer from the plant was successfully used on farms even though farmers complained of intense labour involved in carrying liquid fertilizer from the plant site to their farms.

Another major problem was the drudgery involved in collecting dung from kraals situated hundreds of meters away from the community. Furthermore, Fulani herdsmen prevented women from collecting dung from the kraals on the basis that women could cause pregnant cows to give premature births. Apart from the Ministry of Energy, the Catholic Secretariat and GTZ have been involved in biogas promotion. The Secretariat financed the construction of biogas plants at the Catholic Mission at Kaleo in the Upper West Region; and in three Catholic hospitals–Holy Family and St.

Dominic hospitals in the Eastern Region, and Battor hospital in the VoltaRegion–between 1994 and 1995. Some biogas projects financed by GTZ has been disseminated to treat slaughterhouse waste at Ejura and KNUST in the Ashanti Region. From 1993 and beyond, direct involvement of the MoE in biogas dissemination slumped mainly due to lack of donor support and unfulfilled expectations of the Appolonia projects. Attempts were made to rekindle the involvement of government in 1996 when the MoE financed a study aimed at assessing biogas resources in Greater Accra, Volta, and the three Northern Regions.

This study was intended to be the first step in planning and developing a national biogas programmes. Ampofo estimated a potential of 88,144 m3 of biogas a day which could generate about 193 GWh of energy annually. Ampofo did not estimate the quantum of liquid fertilizer that could be generated but Bensah and Brew-Hammond have shown that an amount of 360,000 tonnes of liquid organic fertilizer could be produced yearly, which would be capable of fertilizing about 70,000 hectares of irrigated farmland or 140,000 hectares of dry farmland. Intriguingly, after more than a decade since the study was completed and the report submitted to the MoE, the Ministry has shown no interest in developing a national biogas programmes. Furthermore, the Institute for Industrial Research, a parastatal organization.

From 2000 to Date

According to Edjekumhene et al, barriers that have plunged biogas dissemination include unfavourable policies, non-availability of feed materials, poor financing arrangements, problems with social acceptance, absence of market, and lack of information. Following the low involvement of biogas projects by government, a number of private biogas companies have marketed the technology on purely business grounds, and mainly based on the ability of biogas plants to improve sanitation. The two leading companies–Biogas Technology West Africa Limited and Beta Civil Engineering Limited–have been constructing brick-based fixed-dome and concrete-based Puxin digesters, respectively. According to SNV, the potential to realise the dissemination of about 270 thousand domestic biogas plants. This is contained in the Biogas for Better Life (B4BL) initiative for Africa–an ambitious programmes conceived by developing partners of Africa and launched in Nairobi in May 2007.

The major aim is to combat poverty by providing over two million households in Africa with biogas plants, with collateral benefits of improved family health through reduction of indoor pollution and drudgery involved in firewood collection and usage. Other objectives of the initiative include the establishment of more than 800 private biogas companies and 200 biogas appliance manufacturing workshops and the construction of one million biotoilets.

In a study to assess domestic biogas potential from cow dung in the three Northern and Ashanti Regions, KITE estimated a technical potential of 80,000 household biogas installations and a market potential (estimated based on the ability and willingness of users to pay) of about 8,000 (8% of 80,000) plants. According to KITE, households have not shown much interest in using biogas in their kitchen or the digested slurry in their farms; thus, biogas technology has been disseminated mostly in institutions such as schools, hospitals, prisons, and slaughterhouses.

Current State of Biogas Technology

A survey of 50 biogas plants was conducted in order to ascertain the true state of biogas technology. Field visits to biogas installations were conducted between

June, 2008 and February, 2009. The sample size (50 plants) was determined from the population (100 known biogas plants as captured in a survey by KITE) using stratified and convenience sampling techniques. The population was stratified into seven strata, with each stratum representing the number of known plants constructed by each of the seven major biogas service providers.

The sample size was selected from the strata using convenience sampling. The survey technique used included direct observation of the various components of biogas installations, and structured and unstructured open-ended interviews with both users and experts. Major challenges encountered during the field visits included inadequate funding, difficulty in locating biogas plants in cities, towns and villages where they have been disseminated, inability of service providers to provide full records of the location and number of plants constructed, and the absence of a national body that keeps track of developments in the industry.

All the aforementioned challenges affected the smooth gathering of data and information; nonetheless, the depth of data gathered was enough to makegeneralizations on biogas technology dissemination. Figure shows the number of installations built over the last three decades as captured by the survey.

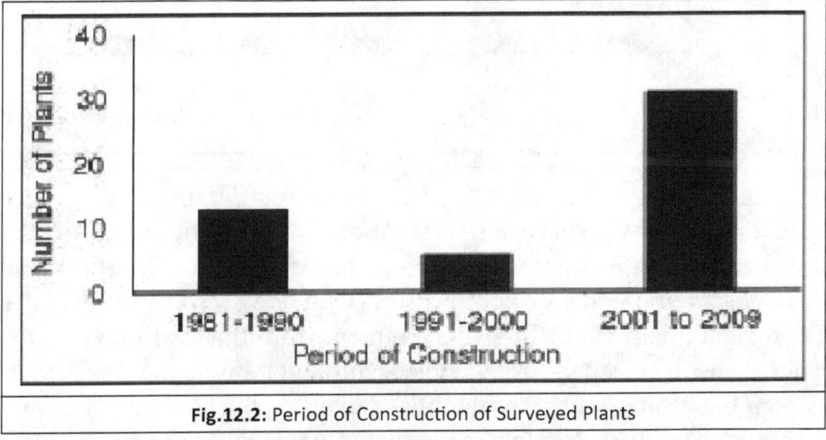

Fig.12.2: Period of Construction of Surveyed Plants

The 1981-1990 period saw the implementation of the domestic biogas programmes at Appolonia and Okushibli–two major cattle-rearing communities in Greater Accra Region. In 1992, the Appolonia Electricity programmes was commissioned; in addition, biogas plants were built in three hospitals at Battor, Nkawkaw, and Akwatia. The relatively large number of plants disseminated from 2001 to 2009 can be ascribed to the founding of BTWAL, the largest biogas company, by John Idan in 2000. Some of the biogas installations disseminated by BTWAL can be found at the Golden Jubilee House (Presidential Palace) in Accra, Central University College–Miotso campus, and Tamale Teaching hospital. Another major factor was the decision by Beta Civil Engineering Limited–a construction company–to disseminate biogas plants; the company has constructed over 40 Puxin digesters and a few in Nigeria since 2006.

Apart from the aforementioned companies, the other biogas service providers have not been very active; for instance, IIR (Institute of Industrial Research) has disseminated a few biotoilets at various locations between 2004 and 2006, in addition to a 200 m3 plant at Ankarful Prison in the Central Region.

Our of 50 installations studied, 22 were in good condition, 10 were functioning even though some defects (including deteriorated gasholders, gas pipelines and appliance) were observed, and 14 were broken-down. This observation is shown in Figure.

The oldest operating plants were found at the St. Dominic Catholic hospital which has been functioning uninterrupted for 15 years despite intermittent problems with the gas delivery systems.

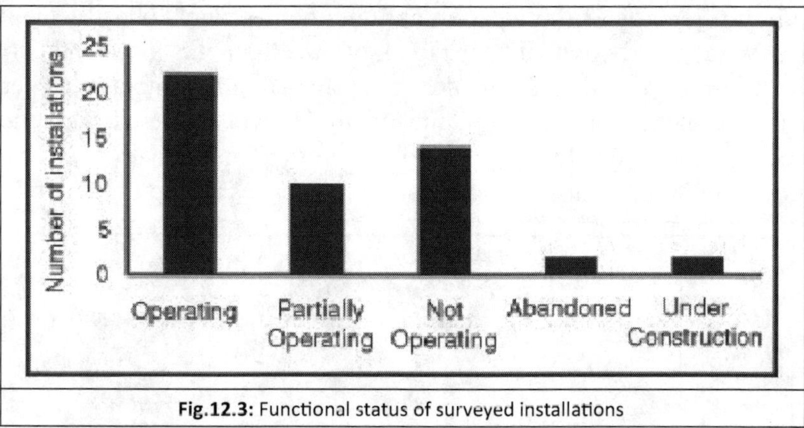

Fig.12.3: Functional status of surveyed installations

More than 50% of surveyed installations were institutional plants while educational and health institutions accounted for 55% of the 29 institutional units. It was also observed that majority of plants (76%) had been constructed mainly in the cities for the treatment of human excrement from flushing toilets. At Obuasi, Anglogold Ashanti Limited has adopted biosanitation technology in treating sewage from her estates and other facilities including the company's hospital.

With the exception of the plants constructed under the Appolonia Electrification Programmes, all community plants surveyed were biolatrines (pit latrines attached to biogas plants).

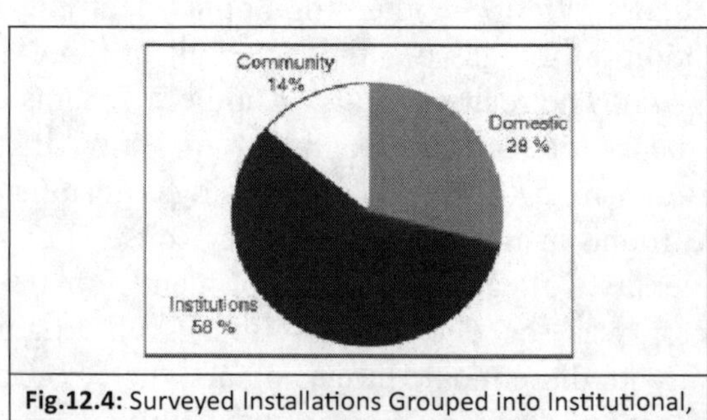

Fig.12.4: Surveyed Installations Grouped into Institutional, Community, and Domestic Plants

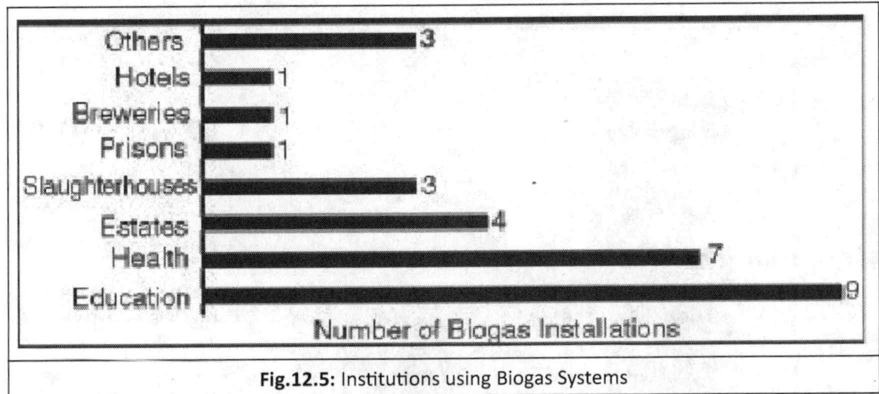

Fig.12.5: Institutions using Biogas Systems

The domestic installations surveyed include six plants built in 1987 at Appolonia and five plants constructed at Okushibli in 1990; most of these installations broke down between 1996 and 2001 due to several challenges including non-availability of cow dung, availability of woodfuel, deterioration of gas stoves, and lack of follow-up services by biogas service providers. After the Appolonia Electrification programmes in 1992, no community installation digesting cow dung has been constructed due to the high cost of biodigesters, reduction of financial support from GTZ, and the absence of specific programmes designed to promote community digesters in cattle-rearing terrains.

MOTIVATION FOR USING BIOGAS PLANTS

As part of the survey, biogas users were asked to state the main motivational factor that influenced them to adopt biogas technology. The response is shown in Figure. It is obvious that most users of biogas units patronize the technology solely on the ability of biodigesters to treat nightsoil, thereby replacing septic tanks.

Biogas service providers have achieved success in promoting the technology as the most efficient and cost effective way of treating sewage despite the high cost of biodigesters. There are a number of sewage rehabilitation projects where old sewage handling systems such as septic tanks and Kumasi Ventilated Improved Pits (KVIPs) are gradually making way for biosanitation interventions.

Apart from improving sanitation, biogas plants are also known for generating energy and organic fertilizer. It appears that the ability of biogas plants to produce organic fertilizer for agriculture has not been given priority by service providers. Most users of biogas systems are content with the performance of the plant as long as the plant treats their sewage without any problem. The low level of interest to disseminate biogas plants focusing on energy and agriculture can be attributed to lack of a concerted programmes targeting the dissemination of biogas plants among livestock farmers, majority of them living in rural communities and with limited/no access to agricultural inputs and modern energy services. These are people who will value the products of the biogas plant and therefore make use of both the gas and the fertilizer.

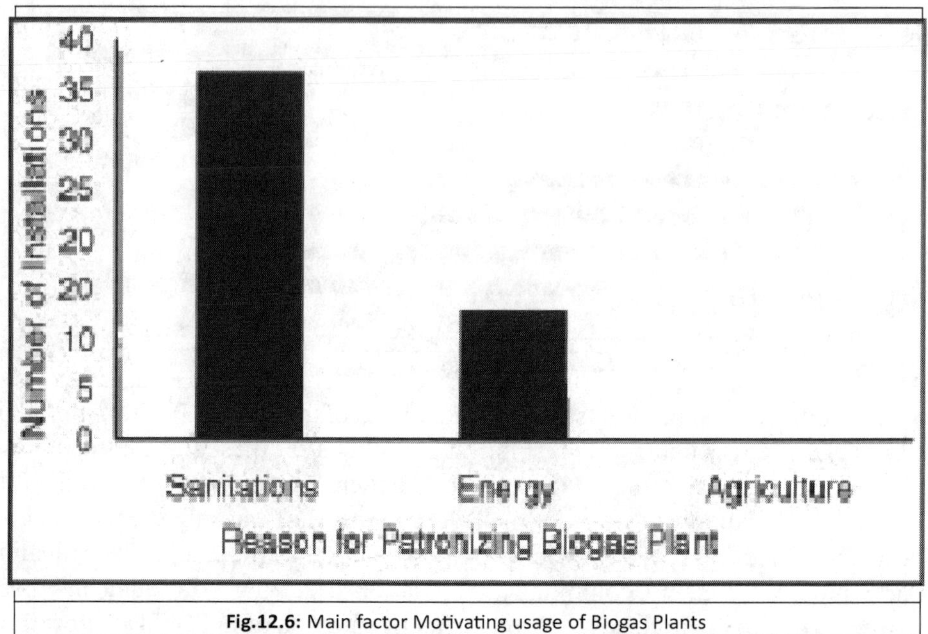

Fig.12.6: Main factor Motivating usage of Biogas Plants

The major problem, however, lies with the inability of rural cattle farmers to afford the full cost of biogas plants which are very expensive. For example, in 2009 the average investment cost of a 10 m3 digester ranged from GH¢ 4,000 to 6,000 ($ 2,800 and 4,200). These figures are far above the financial capability of the rural farmer and it is imperative that special microcredit schemes are developed as a means of promoting biogas plants among poor farmers.

FEED MATERIAL USED IN BIOGAS PLANTS

It has been shown already that the majority of households and institutions patronize biogas plants mainly as waste (nightsoil) treatment systems and as replacement for septic tanks. Most sanitary plants are either designed to treat nightsoil from flushing toilets or nightsoil from KVIP toilets. Apart from nightsoil, slaughterhouse wastes have also been treated using biogas plants.

However, due to the high fibre content of slaughterhouse wastes, most plants constructed to handle the waste have broken down due to the formation of scum, poor operation and maintenance, and inappropriate siting of plants, as bserved Tepa slaughterhouse. Scum is formed when fibrous feed materials accumulate at the top of the slurry, thicken and harden, thus obstructing the flow of gas into the gasholder. The gas is then forced to escape into the atmosphere via the inlet and outlet pipes of the digester. Figure gives an idea about the composition of feeding materials used in biogas plants.

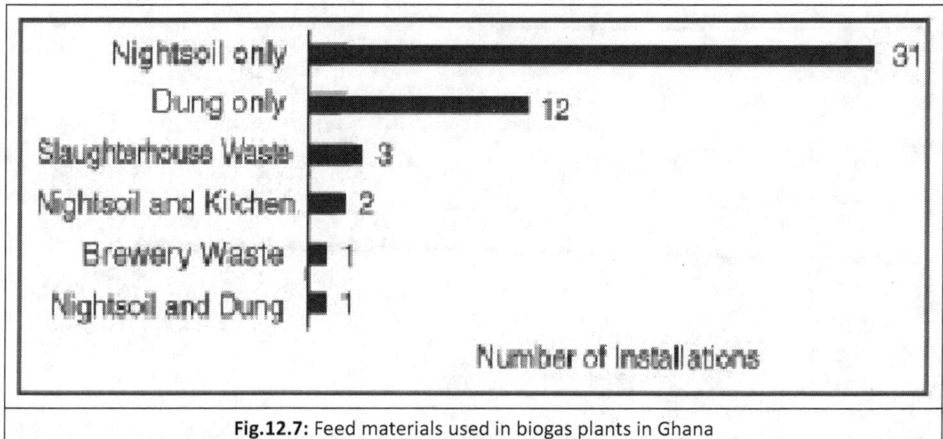

Fig.12.7: Feed materials used in biogas plants in Ghana

TYPES AND SIZES OF PLANTS DISSEMINATED

With the exception of an Upflow Anaerobic Sludge Blanket (UASB) plant constructed at Guinness Ghana Breweries Limited in Kumasi, most of the plants surveyed are either floating-drum or fixed-dome digesters.

All the floating-drum digesters are of the water-jacket type with a spherical digester (BORDA model) while the gasholders are fabricated from mild steel apart from two glass-fibre gasholders at Tepa slaughterhouse and Holy Family hospital, Nkawkaw, and a high density polythene gasholder at GIMPA (Ghana Institute of Management and Professional Administration).

Despite the merits of the floating-drum digester, the fixed-dome digester has found favour among most biogas service providers, accounting for more than 80% of installations sampled. Figure highlights the main types of biodigesters found in the survey. Fixed-dome plants are less expensive compared to floating-drum digesters. Fixed-dome models disseminated by the various companies are the CAMARTEC (Centre for Agricultural Mechanization and Rural Technology) model, the Deenbandhu model, the Chinese dome model, and lately the Puxin digester.

It should, however, be emphasized that most of the biogas companies have also disseminated designs with slight variations from the aforementioned models just to suit the topographic conditions of a particular area.

The sizes of surveyed plants (digester and gasholder as one unit) rangefrom 10 to 100m3.Most household and institutional plants disseminated have volumes of 10m3 and 50 m3 respectively.

Series (tandem) digesters have also been constructed. The most important criterion in sizing the digester is to be able to meet the minimum retention time; this is very important if the plant is to treat human excreta.

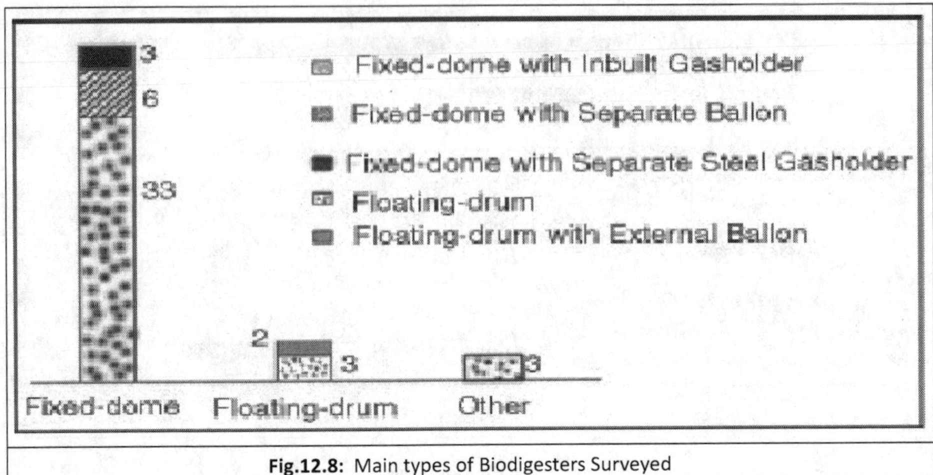

Fig.12.8: Main types of Biodigesters Surveyed

Before the designer settles on the digester volume to use, he needs to consider factors such as daily availability of feed material and the retention time. The choice of the retention time in turn is influenced by the type of feed material, the digestion temperature, the plant type, and financial capabilities of the user, among others. Table gives the retention times used by some major biogas companies.

TABLE: Retention Time used by some Biogas Service Providers Under Mesophilic (25–30°C) Conditions

Biogas Service Provider	**Retention Time [days]**
Biogas Technology Wesh Africa Limited (BTWAL)	30-60
Biosanitation Company Limited (BCL)	20-30
Beta Civil Engineering Limited (BCEL)	15-30
Institute of Industrial Research (IIR)	30-40
Biogas Engineering Limited (BEL)	30-60

If a biogas plant is designed to treat waste such as nightsoil, then the retention time should be long enough to allow for effective treatment of the waste, such that the effluent do not pose a health threat to the environment. The average retention time must be high enough to enable the degradation of the biomass; on the other hand, it must be kept as low as possible, because a high retention time means a high investment cost. Considering the high pathogen levels of human excreta, it is obvious that retention time of 30 days or less may not be adequate to kill enough pathogens under mesophilic digestion. According to Sasse et al, plants less than 30 m3 should not be used for treating blackwater from flushing toilets because of the danger of reducing the retention time. It has also been recommended that a retention time between 70 and 80 days is needed for plants handling nightsoil, in order to ensure the reasonable destruction of pathogens.

As seen in Table, some service providers use a retention time lower than 30 days for designing plants to digest nightsoil, which is inappropriate and should be a matter of concern to environmentalists. However, BCL, BCEL, IIR and BTWAL have been installing systems that have post-treatment facilities ranging from filtration tanks to solar concentrators for further treatment of the effluent. There is still the need to conduct a thorough analysis of effluent samples to determine the safety of the 'treated' effluent.

TANDEM DIGESTERS

Tandem plants have two or more digesters working in series such that the effluent of one digester becomes the feed material for the second digester and so on. They may comprise two or more of the same type of digester or different digesters working together as one unit. The main advantages of tandem digesters (two-stage systems) over one digester (one-stage) systems include higher biomass conversion efficiency, higher methane concentration in the gas, better process reliability and stability, and high quality organic fertilizer. Despite the aforementioned merits, tandem systems are not widely disseminated worldwide because they are expensive and difficult to engineer, construct and operate. Tandem plants have been constructed to treat sewage by BTWAL, BCEL, IIR, and BCL. Most common tandem digesters found consist of two series digesters except the plant at Pope John's Seminary in Koforidua and Hebron Prayer camp at Hebron, where BCEL has constructed systems comprising four 10 m3 Puxin digesters. The main problem with tandem digester designs used by service providers pertains to the inappropriate way of connecting adjacent digesters via a horizontal pipe near the base of the digesters.

This arrangement leads to inefficient digestion since the second digester merely serves as an expansion chamber of the first digester rather than serving as a digester on its own. Moreover, digestion in the first digester will also be inefficient as slurry flows freely into the second digester. Figure shows the layout of tandem digesters as used by biogas service providers as opposed to the recommended layout.

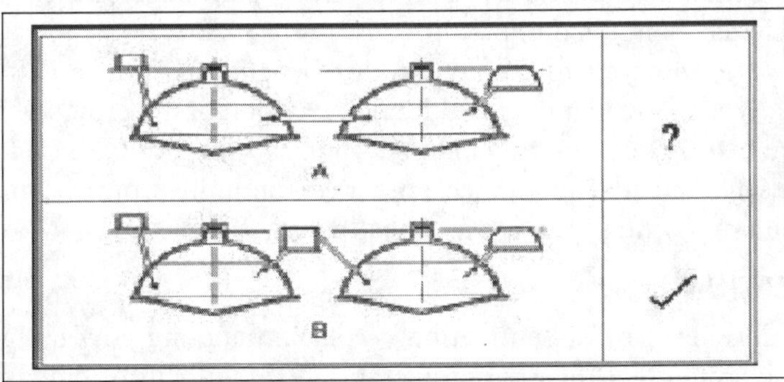

Fig. 12.9: (A) Tandem Digester Layout used by BTWAL, BCEL, and BCL, and (B) Appropriate and recommended Layout.

Tandem plants require a lot of calculations before designing them. The designs with horizontal pipes connecting adjacent digesters can work efficiently only if valves are placed at vital points to control the movement of fermentation slurry from one digester to another, and also the flow of effluent from the last digester. However, such an arrangement will be labour intensive in terms of operation and will also require skilled attendants.

The correct arrangement of tandem plants, is to connect a pipe near the base of the first digester to an expansion chamber which empties into the next digester. This arrangement ensures both plants function as fixed-dome systems, and ensures the effluent from the last digester undergoes thorough digestion.

COMMUNITY BIOSANITATION PROJECTS–BIOLATRINES (BIOTOILETS)

A total of 9 biolatrines were surveyed. Of the 9 biolatrines visited, four were not functioning due to design and operational challenges. The major problem with the design of biolatrines pertains to the connection between the toilet seat and the inlet pipe of the digester. If faeces got stucked to the inlet pipe, aerobic digestion takes place and bad odour is generated. This is believed to be the cause of breakdown of biolatrines at Kaase and Parkoso in the Ashanti Region.

Even though, experience from the Kaase and Parkoso project have led to improved designs, there are still problems with gas leakage through the inlet pipe into the toilet house, as observed in the biolatrines at Abeman, Asokore-Mampong, Kotoku, and Tamale West hospital. Another major challenge concerns the mode of operation of the biolatrine. Even though, biolatrines are designed to handle faeces with little or no water, they rather perform efficiently if water is frequently used to flush the faeces.

Thus, most biolatrines having water reservoirs enable users to flush the toilet with water after use. The use of water regularly avoids the occurrence of aerobic decomposition of faeces entrained in the inlet pipe of the digeste.

From the survey, two biolatrines were found to have been disseminated very close to rubbish dumps. Caretakers attributed low patronage of these biotoilets to unbearable stench from the dump.

It was also observed that communities benefiting from biolatrine projects were not involved in the planning and implementation of the projects. The absence of interaction between community members and service providers on the functions of biolatrines has created ignorance among communities on the importance of biolatrines in improving sanitation and health.

GAS UTILIZATION

Due to low level of dissemination of plants handling cow dung, especially after the Appolonia Electrification project in 1992, the focus of biogas technology shifted from provision of energy (use of gas) to improvement in sanitation (treatment of waste).

This development has created a situation where most plants have been constructed without adequate arrangements for the usage of the gas thereby leading to gas leakage.

Gas leakage is a serious problem. In addition to the discomfort and health implications caused by inhalation of the gas, there is the potential of fire outbreak since mixtures of biogas and air in certain concentrations (6–12% biogas) could be explosive and may cause damage to human life and property. Moreover leakage of biogas, which contains about 60% methane, into the atmosphere offsets environmental gains made by putting up the plant.

Out of the 32 functioning plants studied, 12 of them experienced gasleakage problems. At both the Holy Family and the St. Dominic hospitals, the balloon gasholder was completely deteriorated and gas leaked freely into the atmosphere. Likewise, gas produced by most biotoilets was not used neither was it flared. Gas flaring is regularly done in a few locations, including the biogas plants financed at AngloGold Ashanti in Obuasi, where caretakers have been trained to flare the gas at regular periods in order to avoid high pressures within the digester.

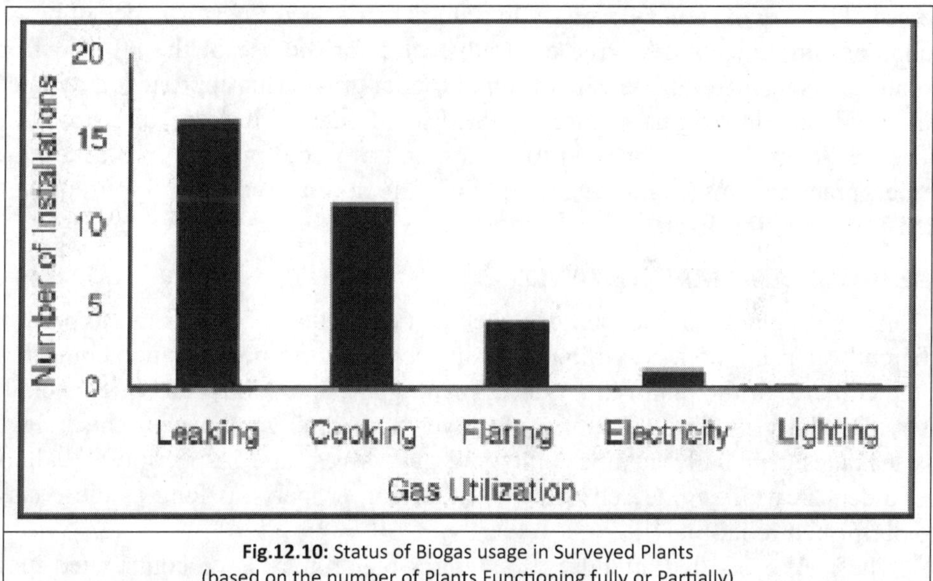

Fig.12.10: Status of Biogas usage in Surveyed Plants (based on the number of Plants Functioning fully or Partially)

EFFLUENT (DIGESTED SLURRY) UTILIZATION

Biogas plants built with the ultimate aim of generating organic fertilizer are rare. The digested slurry is not seen as a resource and most plants discharge their effluent into the surrounding bush or the public drain. Some plants built by BCEL and BTWAL have systems designed to pump the treated effluent for reuse in flushing toilets.

The major concern pertains to the safety of the effluent discharged into the public drain. There is the need to conduct a comprehensive analysis of samples of the effluent in order to determine the efficiency of the digestion process of sanitary plants.

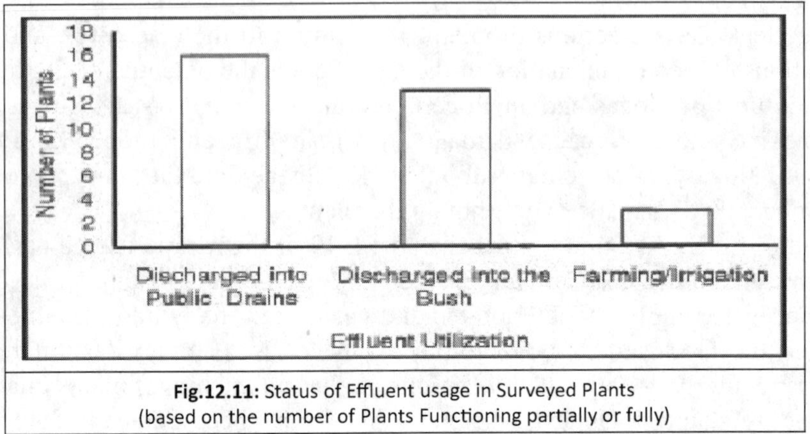

Fig.12.11: Status of Effluent usage in Surveyed Plants (based on the number of Plants Functioning partially or fully)

The effluent from biolatrines is not used because of cultural attitude towards the use of human excreta as manure. In a study to assess the performance of biogas technology in Tanzania, Marree et al advocates for the use of the effluent from biolatrines as fertilizer in the cultivation of fodder grass. This approach, if adopted, will eliminate the stigma regarding the use of human manure in agriculture. Moreover, there is also the need to design and promote programmes that target farmers practicing mixed farming where the effluent can be easily and appropriately used as organic fertilizer.

OPERATION AND MAINTENANCE

A biogas plant may be well designed and constructed but still fail to perform efficiently if it is operated wrongly. Poor operation and maintenance could lead to the collapse of the plant. In a typical plant, good operational activities include, inter alia, daily feeding, cleaning of stoves and other appliances, checking of gas leakages especially at pipe joints and gas valves, and inspection of balloon gasholders. A well engineered plant will function properly as long as these tasks are performed reliably.

The lack of skilled attendants of biogas plants may have contributed more than any factor to the breakdown of most biogas plants. It was observed that users of biogas plants have little or no knowledge of the functions of the biogas plant. In addition, users were not sufficiently educated by the service companies. Of the 50 plants visited, only the installations (3 out of 24 plants were visited) at Anglogold Ashanti Limited were monitored daily by a trained engineer (caretaker); this has ensured continuous operation of the plants since 2001. Biogas plants disseminated in the Catholic hospitals visited have received some level of maintenance; caretakers of the plants had frequently repaired plant components such as the digester, the mixing tank, the expansion tank, and the steel gasholder in floating-drum plants. Unfortunately, balloon gasholders were found deteriorated at St.

Dominic and Holy Family hospitals at Akwatia and Nkawkaw, respectively; moreover, gas appliance were not been used since the gas escapes freely into the atmosphere through holes on the balloon. Other notable maintenance problems were found in biotoilet systems. Service providers do not spend enough time to take users through the rudiments and the functions of various components of the biogas plant.

Broken down plants were not repaired because owners were reluctant to spend additional financial resources on the upkeep of plants; this makes the floating-drum digester especially unfavourable due to the regular maintenance of the steel drum. Follow-up services were almost nonexistent. Service providers attributed this to the unavailability of funds to carry out follow-up programmes.

DEVELOPMENT OF A NATIONAL PROGRAMMES ON BIOGAS TECHNOLOGY

There is the need to develop a national biogas programmes with a three-pronged focus: agriculture (organic fertilizer production), sanitation, and energy. This national initiative must have distinct programmes for both domestic and institutional plants. Promotion of farm plants among livestock farmers or those interested in improving their yields will encourage the utilization of digested slurry in agriculture and reduce the dependence on imported inorganic fertilizers.

The gas generated could be used for cooking, water heating, and lighting. Large-scale dissemination of sanitary plants will improve hygiene in both urban and rural areas.

Replacement of KVIPs with well-engineered biotoilets will ameliorate sanitation problems experienced in public toilets especially in communities where water is scarce. Moreover, the incidence of excretarelated diseases such as dysentery and cholera will be substantially minimized since micro-organisms that cause these diseases are eliminated during the anaerobic digestion process.

Large-scale local industries such as oil palm processing and brewing companies generate large volumes of wastewater that pollute the environment when discharged untreated or partially treated. The use of anaerobic fermentation systems in the treatment of industrial wastewater will reduce pollution and improve the quality of the environment. For the programmes to be successful, government must provide funding to support biogas training and microfinance programmes targeted at social groups with high prospects for adopting biogas technology such as farmers in the three Northern Regions. Fig. Proposed Three-pronged Approach for Future Biogas Dissemination Programmes

FORMATION OF A NATIONAL BIOGAS PROMOTION BODY

From the experience of biogas technology dissemination in developing countries such, Tanzania, China, India, and Nepal, it is obvious that biogas Energy and promotion cannot be left in the hands of private companies alone. The success story of the technology dissemination in China, India, and Nepal can be partly attributed to the direct involvement of the State through special national or parastatal bodies empowered to lead the campaign in biogas dissemination.

There is therefore the need to either set up a national body that is solely responsible for the promotion of biogas technology, or establish a dedicated unit within one of the existing energy organizations for the same purpose.

❐

14

Managing the Environment in a Changing Climate

ENVIRONMENT

Protection of the environment has to be a central part of any sustainable inclusive growth strategy. This aspect of development is especially important in the Eleventh Plan when consciousness of the dangers of environmental degradation has increased greatly. Population growth, urbanization, and anthropogenic development employing energy-intensive technologies have resulted in injecting a heavy load of pollutants into the environment. More recently, the issue assumed special importance because of the accumulation of evidence of global warming and the associated climate change that it is likely to bring.

An important feature of any environmental strategy is that environmental objectives require action in several areas, which typically lie in the purview of different ministries. The Ministry of Environment and Forests (MoEF) has the important role of monitoring the development process and its environmental impact in a perspective of sustainable development and to devise suitable regulatory structures to achieve the desired results. While this role is crucial, environmental objectives can only be achieved if environmental concerns are internalized in policymaking in a large number of sectors. This would require sharing of responsibility at all levels of government and across sectors with respect to monitoring of pollution, enforcement of regulations, and development of programmes for mitigation and abatement. Regulatory enforcement must also be combined with incentives, including market and fiscal mechanisms to encourage both industry and people in their day-to-day working lives to act in a manner responsive to environmental concerns. Sustainable use of natural resources also requires community participation with a responsible role assigned to the communities for conservation.

In this chapter, we outline the main issues in the traditional area of environment and also consider the new challenges posed by climate change. The Tenth Plan was a period of extensive review of environmental processes and law. The first National Environment Policy was put into place in May 2006. Also, the re-engineering of the environmental clearance process and Environmental Impact Assessment (EIA) Notification and review of the Coastal Regulation Zone Notification were undertaken to improve the quality of environmental governance. The Eleventh Plan must build on this experience by integrating environment considerations into policymaking in all sectors of the economy—infrastructure, transport, water supply, sanitation, industry, agriculture, and anti-poverty programmes. It also calls for strengthening the oversight and regulatory framework for environment management so that development decisions do not impinge adversely on sustainability.

Some initiatives needed to integrate environmental concerns into planning and developmental activities across all the sectors are given below:

- Environment is a residual Central subject. Since regulation and enforcement in this area cannot be handled by the Central Government alone and the responsibility of maintaining the environment rests at all levels of government, we need to consider whether environment can be made a concurrent subject in the constitution. This will help the State Governments and the local authorities to enact and notify their own enforcement laws and rules to ensure compliance of relevant environmental norms.
- There is a case for setting up an independent, statutory body on sustainable development with the specific responsibility of guiding government policies and programmes for making them more socially and environmentally sustainable, and to monitor and evaluate their outcomes. This body should comprise eminent environmental experts and citizens with a long and publicly known record in environmental research or action. It should have adequate powers to address environment concerns and not be a purely advisory body.
- The State Pollution Control Boards should be restructured into statutory Environment Protection Authorities with the mandate of developing regulations, standards and upgraded facilities for enforcing compliance.
- At the district level, the scheme of Paryavaran Vahinis, or committees of concerned citizens, should be revived to serve as environmental watchdogs and undertake selective first hand monitoring of the environmental situation in the districts.
- For improving the quality and transparency of the environmental clearance process, prior informed consent of the local self governments of the

respective area, as specifically provided in PESA, may be introduced for proposals requiring environmental clearance.

- Public hearing should also be made mandatory for the activities specified in the EIA notification.

ENVIRONMENTAL IMPACT ASSESSMENT (EIA)

EIA is an important management tool for integrating environmental concerns in the development process and for improved decision making. The MoEF in 1994 issued the first EIA Notification prescribing mandatory environmental clearance for 32 categories of developmental activities. The process involved consideration of environmental consequences before starting projects and included procedures like EIA study, public hearing in certain cases, and consideration of the outcome by expert groups before issuing of final clearance by MoEF.

This process takes time, often causing long delays in commencement of projects. MoEF organized a re-engineering of the process and a fresh notification was issued in September 2006. The specific modifications made in the EIA notification are summarized in Table.

TABLE: Changes in Environment Clearance Notification

Step	Pre 2006 Notification Position	As per 2006 Mptofocatopm
Activities requiring dearance	32 actovotoes including infrastructure in respect of petroleum and chemicals, exploration, hazardous industries, large tourism projects, thermal power plants, construction and industrial estates, manufacturing/fabrication,	39 activities in eight categories including mining/exploration/ power generation, primary processing, material processing, material production, service sector, infrastructure, and Construction.
Process	Clearance based on recommendation Expert Committees set up by the Ministry. Dimensions of the projects in All the Central and State level, provide recommendations.	Clearance by designated EIA of authorities at the Central or State level based on prescribed the Notification. Expert Appraisal Committees for decision-making.
Categories	No categories defined as MoEF the sole authority for granting clearance.	Two categories prescribed of projects was namely, 1. To be dealf in the Central level land 2. Delegated to the State EIA Authority for dearance.
Processing Time excl. EIA/EMP	180 days.	Cat. A: 120 days. Cat B. 30 days.
Public	Requirement of public hearing at the Consultation discretion of government.	Public hearing prescribed for all Category A and some of the Category B projects.

The new process came into effect on 14 September 2006. As State level authorities and committees are in the process of being set up, the impact of the new process is yet to be seen.

AFFORESTATION

Forests play a critical role in environment protection in two ways. They help to absorb carbon dioxide in the atmosphere thus mitigating the build up of GHGs. They also help to cope with water stress by trapping water that would otherwise run off and this is especially so when forests exist in upper ends of catchments. Unfortunately, both benefits are eroded by deforestation largely as a result of human pressure to exploit forest resources. While some exploitation of forests may be appropriate, the sustainability of forest resources is determined by the quantum of withdrawals. The existing information on forest/tree cover based on the satellite data of 2002 indicates a green cover of 23.68%. There has been a net improvement of 0.65% between 2000 and 2002.Presuming that the same growth rate continues till 2007, the Tenth Plan target of 25% may have been achieved by 2007.

This will be variable in the data of 2008, to be used for the State of Forest Report (SFR) 2009. The present extent of forest lands in the country is 77.47 million ha. However, not all the forest lands are under the tree canopy and trees grow outside forests also. SFR 2003 of Forest Survey of India indicates a tree cover of over 67.83 million ha, that is, 20.64% of the land area. This extent aggregates the blocks of more than 1 ha area under tree cover. Outside this, nearly 10 m ha area has been assessed as under tree cover in smaller than 1 ha patches outside forests.

This brings the total green cover up to 77.83 million ha or 23.68% of the total area. Out of this, dense forests have declined from 416809 sq km in 2001 to 390564 sq km, that is, a fall of 6%. Though over the last few years the extent of forest cover has stabilized, the low extent of good forests is a matter of concern. Severe pressure for meeting growing livelihood, industrial, and development needs have been some of the critical contributing factors. A demand–supply gap of almost 64 million cubic meters has been projected for timber in 2006.

The shortage is being met through import since 1985 amounting to nearly `9000 crore (2003–04). Against a sustainable availability of fuel wood of only 17 million tonnes from the forests, the annual fuel wood requirement has been estimated to be above 200 million tonnes. With development of technologies for use of even smaller sizes of wood for structural uses, small wood is also not available for fuel. India may have sufficient food to eat but it does not have sufficient fuel wood to cook it. Forests also meet about one-third of fodder requirement.

The total production of Non Timber Forest Produce (NTFP) has been assessed as worth about `4188 crore annually. Ever increasing herbal markets are only to put further pressures. While the gaps are huge, potential for improvement of productivity exists.

The status of green cover is an indicator of status of productivity of forests. The following strategies have been evolving in the sector for improving the status of green cover.

Participatory Forest Management or Joint Forest Management (JFM) Initiated with the circular of the MoEF on 1 June 1990 on people's involvement in forest conservation and management, the JFM regime has evolved gradually and at present 106479 such committees (22 million participants) are functioning in 28 States covering 22.02 million ha of forests. This participatory regime is seen as a potential strength of forest management for the forest fringes. The challenge now is to effectively empower the local communities with appropriate rights and responsibilities, and ensure that substantial benefits from forest conservation accrue to them. A similar approach has been attempted in wildlife management also. In Project Tiger, India, Ecodevelopment Project was implemented during 1995–2005 under Global Environment Facility (GEF) with focus on village eco-development through optimum use of local resources and involvement of local people in conservation of protected areas. Integration of this approach to protected area management through shared decision-making and full integration of conservation and livelihoods across the landscape are yet to be realised. Also, the recognition of a large range and diversity of community traditions or new initiatives towards conservation is very weak.

Social Forestry

The First Five Year Plan, as far back as 1951, indicated the scope for establishment of village plantations. Subsequently, a series of externally aided social forestry projects during the 1980s and allocation of 25% of District Rural Development Agency funds under National Rural Employment Programmes, Rural Landless Employment Guarantee Programmes, and so on, for social afforestation provided impetus to social forestry. This provision was not pursued since the Eighth Plan. Regeneration of forests and growing of trees being a long-term activity, did not receive due focus in the rural development programmes. Even the Watershed Management Programmes aims more at equitable sharing and caring of watershed benefits, and afforestation remains a latent component. The cost norms of these programmes are not able to support substantial afforestation activities requiring strong wage contribution.

Agro and Farm Forestry

The social forestry programmes also include extension and promotion of agro and farm forestry in the farm sector. The farmers of the States like Punjab, Haryana, Himachal Pradesh, Uttar Pradesh, Karnataka, and Tamil Nadu have adopted agro forestry. However, the forward linkages with the user industry and facilitation by the State agencies have not existed. Thus, the sector suffers from problems of unorganized markets driven by middlemen, depriving the farmers of optimum prices for their produce. Conservative exim regulations, cutting and transit controls act as disincentives for diversification into the species which face controls but are economically important.

Strategy for the Eleventh Plan

The monitorable target of the Eleventh Five Year Plan is to increase the forest cover by 5% of the total geographical area. This would require an additional cover of about 16 million ha. Out of this, 5 million ha could be brought under the tree cover within the recorded forest area while the rest would be added through agro forestry and social forestry. However, the policy target is to have 33% forest and tree cover ultimately which will require additional coverage of about 10–11 million ha. This will have to be done mainly outside the recorded forest area.

The target of 33% forest and tree cover reflects the tree component without accounting for other vibrant non-tree natural biomes like grasslands. The amendment of the target of the 1952 policy, of one-third area under forests, to forest/tree cover resulted in a shift of focus from ecological habitats to tree cover. Further recognition of biodiversity characteristics and ecological services rendered by habitats like grasslands, natural desert ecosystems, alpine, and riparian habitats suggests that several biomes, even if devoid of tree component, can be recognized as 'green cover' and accounted so. Recognizing these facts, the following strategic principles will be adopted for dealing with the green cover.

The policy objective of 33% tree/forest cover should be revisited for its definition on ecological considerations. The green cover should include the existing natural ecosystems within which the tree cover constitutes a sub-sct. Enabling environment for social and participatory regimes should be the aim of Central efforts, as is being done through the JFM mode under the National Afforestation Programmes. Implementation of the Central programmes has met with difficulties on fund flow management at the State level. It is advisable to undertake activities in project mode with earmarked funding, as is done in the externally aided projects.

IMPROVING AIR QUALITY

The Air (Prevention and Control of Pollution) Act, 1981 is the main legislation for regulating air quality, through the Pollution Control Boards (PCBs) in the States. The Central Pollution Control Board (CPCB) has identified 2301 medium and large scale polluting industrial units under 17 highly polluting categories. The requisite pollution control devices are reported to have been provided in 1927 units, while 235 have been closed and the 139 are still defaulting.

CPCB also monitors ambient air quality at 308 stations covering 115 cities/ towns in 28 States and four UTs in the country to:

- Determine the status and trend in ambient air quality on significant parameters like benzene and polyaromatic hydrocarbons (PAH);

- Assess health hazard and the damage to materials;
- Develop preventive and corrective measures;
- Understand the natural cleansing process. Presently, the criteria pollutants monitored by Central/State PCBs and associate agencies include sulphur dioxide (SO_2), NOx, and Respirable Suspended Particulate Matter (RSPM). Other parameters for toxic trace matters PAH are also monitored for select cities.

A programmes for real time air quality monitoring for cities with population of more than 1 million was started during the Tenth Plan. The automatic air quality monitoring systems are operational in Jodhpur, Patna, Pune, and Sholapur while those in Kanpur, Varanasi, Jharia, and Kolkata will be functional soon. As many as 76 cities/towns are found to exceed acceptable limits of these parameters mainly due to vehicular and industrial pollution measured in terms of ambient air quality in residential, industrial and sensitiveareas for SO_2, oxides of nitrogen, Suspended Particulate Matter, RSPM, ammonia, and carbon monoxide.

Figure displays the average air quality levels of select Asian cities between 2000 and 2004. The chart shows that New Delhi, Mumbai, and Kolkata have failed to meet the WHO standards as far as Total Suspended Particulate Matter and RSPM are concerned. Beijing and Shanghai, though better than Indian cities, also could not meet the WHO standards in this respect. The other Asian cities which failed to meet the required standards were Jakarta, Katmandu, Seoul, Dhaka, and Colombo. As far as SO_2, and NO_2 are concerned, New Delhi and Mumbai were much better than the prescribed WHO standards whereas several Asian cities exceeded the safe limits. Action Plans for improvement of air quality have been drawn for 16 identified cities on the advice of the Supreme Court. Based on the basic format prepared by CPCB for this purpose, the State PCBs have been asked to evolve such plans for their respective areas. During the Eleventh Plan period, all Central programmes on outdoor pollution should be reorganized under a National Air Quality Plan, which will cover:

- City based Clean Air Action Plans (CAAPs)
- Pollution Control and Prevention in Industrial Areas programmes.

The ongoing national programmes for monitoring air quality will be taken forward for achieving real time data. This would help in creating early warning systems and thus enforcing the Plan target of conforming air and water quality to WHO standards. The entire Air Quality Monitoring network should be expanded from the current 308 stations to 1000 stations. Real time monitoring of PM 2.5 (Particulate Matter 2.5 micron), ozone, VOCs (Volatile Organic Compounds), PAH, secondary pollutants—sulphates and nitrates—will be organized in about 15 cities per year, to cover the 76 cities which currently exceed the four specified levels of air pollution. Source monitoring of VOC, BTX (benzene, toluene, and xylene), and toxic heavy metals will be initiated to develop control measures.

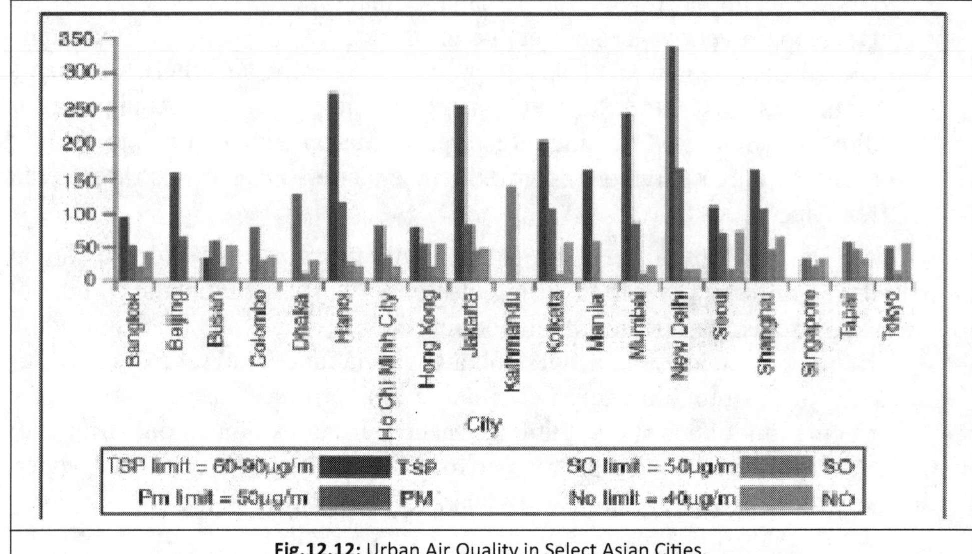

Fig.12.12: Urban Air Quality in Select Asian Cities.

Vehicular pollution is a major source of air pollution in our cities and control of vehicular pollution should have high priority in planning for a clean urban environment. Some of the initiatives needed to control vehicular emissions are listed below. More generally, effective urban transport planning, with appropriate initiatives for the development of public transport is particularly important. CAAP should therefore be integrated with the National Urban Transport Policy and the JNNURM to ensure optimal development of public transport including both bus and rail based mass rapid transit systems. CAAP should also be the basis of all Central and State funding with appropriate provision of penalties for non-compliance. The monitorable target in air quality should be to achieve WHO standards of air quality in all major cities by 2011–12.

Strategy to Control Vehicular Emissions

- There should be uniform fuel quality and emission standards across India. The road map proposed in the government's Auto Fuel Policy must be accelerated and tightened to cover all cities and to implement Euro IV standards.
- The use of diesel in private vehicles must be discouraged. The present distortions in pricing under which diesel is much cheaper than petrol, encourages the growth of diesel vehicles leading to cheap and toxic motorization.
- All Central funding for the transport sector must be linked to the implementation of car restraint measures including parking charges designed to recover the full cost of using parking space (linked to the real estate and construction costs) and rationalization of road and other transportation related taxes (such as passenger tax, sales tax, and so on) to

reduce the tax burden on public transport and increase it for the personal motorized transport.

- The government should impose higher annual taxes on personal transport, which should be used to create a dedicated fund for public transport.

TABLE: Notified Standards for Water Quality

Designated-Best-Use Drinking water source without conventional treatment, but affair distinfection	Class of Water A	Criteria 1. Total Coliform-Most Probable No. (MPN)/100 ml < 50. 2. pH 6.5-8.5 3. DO > 6 mg/litre 4. BOD (6 days 20oC) -<2 mg/litre
Outdoor bathing (Organized)	B	• Total Coliform (MPN)/100 ml-< 500 • pH 6.5-8.5 • DO-> 5 mg/1 • BOD-(days 20oC)-<3 mg/1
Drinking water source after conventional treatment and disnfection	C	• Total Coliform (MPN)/ 100 ml < 5000 • pH 65-9.0 • DO > 4 mg/litre • BOD(5 days 20oC) < 3 mg/1
Propagation of widlife and fisheries	D	• pH 6.5-8.5 • DO > mg/litre • Free Ammonia (as N) <12 mg/1
Irrigation, industrial cooling, controlled waste disposal	E	• pH 6.5-85 • Electrical Conductivity (25oC) micro mhos/ cm < 2250 • Sodium Absoption Ratio Max. 26 • Boron Max 2mg/ Not Meeting A, B, C, D Below E & Criteria

WATER QUALITY

Monitoring Water Pollution

The Water (Prevention and Control of Pollution) Act, 1974 regulates water quality through the State PCBs. The CPCB, under MoEF, has established a nationwide network for water quality monitoring comprising 1019 stations in 27 States and six UTs. The monitoring is done on a monthly or quarterly basis for surface water and on a half-yearly basis for groundwater.

The monitoring network covers 200 rivers, 60 lakes, five tanks, three ponds, three creeks, 13 canals, 17 drains, and 321 wells. Water samples are analysed for 28 parameters including physical parameters, nutrients, major ions, and organic and bacteriological parameters.

National River Conservation Plan (NRCP)

The different standards of water quality required for drinking, bathing, and irrigation are given in Table. Out of the various parameters monitored, Biochemical Oxygen Demand (BOD) and Coliforms are critical. Maintenance of certain level of flow that could ensure a minimum assimilative capacity in water bodies remains a matter of concern, especially in the light of increasing consumptive demands of various sectors such as household, industry, and agriculture.

Based on monitoring done during the last 10 years, 14% of riverine length is highly polluted (BOD more than 6 mg/litre) and 19% moderately polluted (BOD—3 to 6 mg/litre). A total of 86 polluted stretches have been identified and action plans for improving the water quality in them are being prepared.

As small-scale industries do not have adequate resources, space or skilled manpower to treat their wastewater, a scheme of Common Effluent Treatment lants (CETPs) was initiated in June 1990. The CETPs are partially funded by the GoI. Eighty-eight CETPs covering more than 10000 polluting industries have been set up during the Tenth Plan. However, the performance of these units has been generally unsatisfactory largely due to improper operation and maintenance. This calls for strengthening of enforcement.

Charters on Corporate Responsibility on Environmental Protection in respect of 17 categories of highly polluting industries have also been formulated in collaboration with the concerned industries. It is necessary to consider rivers holistically to evolve plans to restore their ecological health and improve the water quality to bathing standards.

Cities and towns located on the banks have been found to be gross polluters and these have been taken up for planning and implementing schemes to ensure that only treated wastewater is discharged into the rivers. Untreated sewage dumped into our rivers is a major cause of river pollution. The total sewage generation in the country is about 33000 MLD (million litres per day).

Against this, the total average treatment capacity is only 6190 MLD and 40% of that capacity is in Delhi. This underlines the urgent need to expand the sewage treatment capacity in the country. The NRCP has covered 160 towns along 34 polluted river stretches in 20 States and has created 2055 MLD of sewage treatment plant (STP) capacity till now, which is approximately 38% of the approved capacity of 5435 MLD to be set up under the Plan.

The average capacity utilization of existing STP capacity is reported to be about 72% against the desirable capacity utilization of 100%. Under-utilization of capacity is due to factors like irregular power supply, absence of connections between domestic sewage drains and STPs, failure of States to provide for aintenance costs of STPs, and so on.

Under the Ganga Action Plan (GAP) Phase-I, sewage treatment capacity of 869 MLD was set up covering 25 Class-I towns in Uttar Pradesh, Bihar, and West Bengal. This amounted to 35% of the total sewage treatment capacity needed. As a result of the implementation of GAP I, the length of the polluted stretch of the river was reduced from 740 km to 437 km (Kannauj to Varanasi).

Phase-II (1993) covered Ganga's tributaries (Yamuna, Damodar, and Gomti) in 96 towns in seven States.

It is estimated that another 20% of the pollution load of Ganga would be covered on completion of the on-going works under NRCP. However, there will still be a gap of nearly 45% of the pollution load to be addressed in the future. Other major river cleaning projects cover Gomti river in Uttar Pradesh; Yamuna River in Delhi; Musi River in Hyderabad and Pamba River in Kerala. The 22 km stretch of the Yamuna in Delhi between Wazirabad and Okhla, is critically polluted.

A sewage treatment capacity of 2960 MLD was required to handle the entire discharge load in Delhi. However, the STP capacity created is about 2500 MLD only, leaving a gap of about 460 MLD. More importantly, the existing STPs are able to treat only about 1600 MLD (about 64% capacity utilization) due to dilapidated condition of trunk sewers.

The impact reported by MoEF reveals that BOD is within the prescribed standards (BOD less than 3 mg/l and Dissolved Oxygen [DO] more than 6 mg/l for Bathing Quality) at most major cities along the Ganges. However, as shown in Table below, the stretch from Kannauj to Allahabad has not yet achieved bathing quality standard. As shown in Table, the stretch of Yamuna river between Delhi and Etawah in Uttar Pradesh is not even of bathing standard. The rivers Narmada, Mahanadi, Brahmini, Baitarni, Subarnrekha, Beas, and Chambal maintain DO levels of 4.0 mg/l or above throughout the year. In the Ganga, Yamuna, Krishna, Sabarmati, Tapi, Sutlej however, the DO level goes as low as 0.3 mg/l.

TABLE: Water Quality Data for River Ganga (Summer Average, March-June)

Station/Location	1986		2006	
	Do (mg/l)	BOD (mg/l)	Do (mg/l)	BOD (mg/l)
Rishikesh	8.10	1.70	8.30	1.00
Haridwar D/S	8.10	1.80	8.10	1.30
Garhmukteshwar	7.80	2.20	7.70	2.10
Kannauj U/S	7.20	5.50	7.35	1.11
Kannauj D/S	na	na	6.45	4.20
Kanpur U/S	7.20	7.20	6.20	6.80
Kanpur D/S	6.70	8.60	3.90	9.80
Allahabad U/S	6.40	11.40	7.10	4.90
Allahabad D/S	6.60	15.50	8.50	3.20
Varanasi U/S	5.60	10.10	8.70	2.10
Varanasi D/S	5.90	10.60	8.65	2.25
Patna U/S	8.40	2.00	7.40	2.05
Patna D/S	8.10	2.20	8.10	2.30
Rajmahal	7.80	1.80	7.20	1.95
Palta			6.96	2.58
Uluberia			6.46	2.64

TABLE: Water Quality Data for River Yamuna
(Summer Average, March-June)

Station/Location	1986		2006	
	Do (mg/l)	BOD (mg/l)	Do (mg/l)	BOD (mg/l)
Tajewala	11.70	1.20	11.20	1.50
Kalanaur	10.40	1.05	7.30	2.50
Sonepat	9.75	3.00	8.40	2.50
Palla	13.95	6.00	8.00	4.80
Nizamuddin Bridge	0.30	25.00	0.00	31.30
Agra Canal	0.35	26.50	0.00	28.30
Majhawali	0.50	22.00	3.05	26.00
Mathura	8.10	4.00	5.40	15.30
Mathura D/S	8.50	2.50	5.77	15.00
Agra U/S	10.65	4.50	8.40	15.00
Agra D/S	1.65	9.00	8.17	14.80

Batteshwar	13.90	11.00	9.68	14.00
Etawah	11.16	7.00	16.90	15.50
Udi	9.71	2.00	13.30	3.00
Auraiya Juhika	8.14	5.00	8.40	1.80

Strategy for the Eleventh Plan

The country's first river action programmes—GAP—completes over 20 years and the NRCP completes 10 years in 2007.

The experience of the first 20 years needs to be used to design an effective and affordable river cleaning programmes for the future:

- A basic objective must be to maintain minimum flows, which are threatened by withdrawal of water to meet the needs of agriculture and industry, and to ensure treatment of sewage and effluents.
- As urban sewage load is the dominant factor causing river pollution, integration of the NRCP and National Lake Conservation Plan (NLCP) with the urban development programmes and agencies is essential. It should be ensured that by 2011–12 no untreated sewage is drained into the rivers.
- For receiving assistance under the NRCP/NLCP, ULBs should be fully involved in developing the programmes along with specific statutory responsibility for operating and maintaining the treatment facilities created.
- The City Development Plans should incorporate environmental management services as the number one priority in JNNURM and Urban Infrastructure Development Scheme for Small and Medium Towns (UIDSSMT). Mechanisms are needed to ensure that the urban areas covered under these programmes provide for full treatment of the sewage generated.
- For the plan period, the River Conservation Programmes should aim at completing the projects in hand and new projects should be considered only in towns which are not covered under JNNURM and UIDSSMT.
- The NRCP should encourage installations on command area basis to facilitate investment in the treatment of maximum quantity of sewage and reuse in the vicinity. In other words, decentralized systems of wastewater treatment need to be encouraged as against centralized, large, end-of-pipe treatment units.
- The wastewater management strategy needs to emphasize the use of state-of-the-art Geographical Information System (GIS)-based decision support systems.
- Water efficiency in flushes and gadgets should be planned in order to reduce wastewater generation. Recycling/reuse of treated sewage in cities

should be promoted. Resident groups should be sensitized towards water conservation, recycling, and reuse.

National Lake Conservation Plan (NLCP)

The NLCP was approved as a 100% Centrally funded scheme during the Ninth Plan with the objective of restoring the polluted and degraded lakes of the country. To begin with, NLCP proposed to cover urban lakes of tourist importance. The scope of work has been expanded during the Tenth Plan to include rural water bodies also. Under the programmes, conservation of 46 lakes in 13 States has been taken up through 31 projects. So far, projects for 10 lakes have been completed and 10 more are likely to be completed.

Improvement in the quality of water in the lakes in the completed projects is more marked than in the river conservation projects.

The scheme now provides assistance to States on a 70: 30 sharing basis and till date the estimated approved cost is about `565 crore. The activities covered under NLCP include interception and treatment of pollution loads entering the lake, lake cleaning interventions such as desilting, deweeding, bioremediation, catchment area treatment, lake front eco-development like bunding, fencing, shore line development, and creation of facilities for public recreation and entertainment.

Dal Lake Conservation Project (DLCP)

The DLCP was approved by the government at a cost of `298.76 crore in September 2005. As an exception, this project is being implemented entirely by the Central Government. The expected date of completion of the project is March 2010. The works envisaged include interception and diversion of pollutants entering the lake, setting up of six STPs, desilting, de-weeding, removal of encroachments in the project area, and so on.

The GoI has released `70 crore for the project till 2006–07. Utilizing this, three out of six STPs have been made operational. The remaining three STPs are expected to be commissioned by December 2007. The progress of implementation is, however, slow due to delay in relocation of families who need to be shifted from the project area of the lake.

While the lakes conservation plan focuses on pollution abatement in lakes, the larger objective of conservation of lakes is the maintenance of the aquatic ecosystems. Keeping this in view, wetlands will be integrated with NLCP and the programme's objectives will cover conservation of life forms apart from mitigating pollution and augmenting catchments.

Water Pollution

Effluent standards, environmental laboratories, and government analysts have been notified under the Environment (Protection) Act. There is a need for much stronger monitoring and enforcement for expanded coverage by both the Central and State Pollution Control Boards. Greater awareness and involvement of local (affected) communities and local governments in compliance and monitoring is critically important.

While compliance will cover treatment and recycling of used water, monitoring should include the extent and quality of the treated water, and water quality of the water bodies providing or receiving the water in treated or untreated form. The Joint Parliamentary Committee on pesticide residues in and safety standards for soft drinks, fruit juice, and other beverages has suggested notifying quality standards for drinking water. These quality standards should be made mandatory.

WASTE MANAGEMENT

Municipal Solid Waste (Management and Handling) Rules, 2000 stipulated that these were to be complied by December 2003. Under these rules, all local bodies were expected to undertake segregation of waste and its collection, storage, transportation, processing, and disposal. While the Ministry of Urban Development reported that Jodhpur in Rajasthan, Sirsa in Haryana, and Namakkal in Tamil Nadu have been making good efforts towards compliance, no other city/town in the country has taken the required steps.

Some local bodies are reported to have taken initiatives to set up waste processing and disposal facilities. These local bodies are generating about 1.2 lakh tonnes of municipal solid waste every day and treatment is not a priority in their plans.

Urban Solid Waste Management

Management of industrial and municipal waste is a serious challenge because of its magnitude and the resources required. The JNNURM covers programmes for urban waste disposal. The scheme of Balanced Use of Fertilizer under the Ministry of Agriculture strengthens the soil-testing programmes in the country and encourages efficient fertilizer use and composting of urban biodegradable waste. The Ministry of New and Renewable Energy (MNRE, previously MNES) has formulated a National Programmes on Energy Recovery from urban and industrial waste. These programmes need convergence and exploitation of synergies.

Strategy for the Eleventh Plan

Collection, treatment and disposal of solid waste are the responsibility of the ULBs. These bodies must be made specifically accountable in this respect. The Supreme Court directions for municipal solid waste collection and management which require segregation, sanitary landfills, and other safe options of treatment should be part of a Master Plan.

The following measures also need to be taken.

- Processes for waste minimization and segregation should bepromoted and pursued. The existing regulations have to be revisited to see the changes needed and the investments required.

- The role of CPCB and MoEF must be clarified. They should becharged with enforcing the rules through the ULBs and not restsatisfied solely from being the key standard setting and pollution monitoring agency.
- Assistance for projects for treatment and disposal of hazardous and biomedical waste should be provided as incentive for compliance.
- Avoidance of waste going to the landfill should be the priority for all ULBs.
- For smaller and medium towns and cities, regional/shared landfills/waste processing infrastructure should be considered. For larger cities, the problem of availability of land should be addressed.
- Programmes like JNNURM and UIDSSMT need to link assistanceto the progress in solid waste management. During the Plan period, all Class I cities should have sanitary landfills in place.

The PCBs are responsible for enforcing Management of Hazardous Waste and Biomedical Wastes (Management and Handling) Rules. It is assessed that about 4.4 million tonnes of hazardous waste are generated annually by over 13000 units. At the instance of the Supreme Court, an inventory of dumpsites is being carried out. MoEF provides assistance for installing Transport, Storage, and Disposal Facilities (TSDF) for management of hazardous waste generating industrial clusters.

So far, six such systems have been supported in PPP mode. The average capacity of these units is 1–1.2 lakh tonnes per annum. Biomedical waste is generally disposed of by incinerators or conventional methods. At least 17 units, mostly incinerators, were set up by industry associations. Some health care facilities have installed their own biomedical waste treatment facilities and others are availing the services of Common Bio-medical Waste Treatment Facilities (CBWTF). There were 157 CBWTFs, including eight under installation, in the country as on March 2007. During Eleventh Plan period, an institutional mechanism would be required to be put in place to ensure that issues related to handling of industrial, hazardous and bio-medical wastes are also dealt under JNNURM and UIDSSMT.

Funding through these two umbrella schemes should be made conditional upon appropriate measures being taken in respect of these issues. In places not covered under JNNURM and UIDSSMT, an effective mechanism should be put into place to ensure strict monitoring and compliance by the concerned local authorities. A framework for management of e-waste also needs to be put into place.

BIODIVERSITY

Habitat conversion or land use changes, land degradation, and pollution result in the decline of ecological goods and services needed for human welfare.

The Biological Diversity Act, 2002 and Rules, 2004 provide for constitution of State Boards (SBBs) and Management Committees (BMCs) for conservation, documentation and sustainable utilization of biodiversity and for building up capacities of these bodies. Chronicling in the form of People's Biodiversity Registers will be the only way we can identify changes happening due to climate change. There is a need to develop appropriate methodology and models for conducting such an exercise. Documentation of traditional knowledge should also include adaptation of such knowledge in line with the present needs of conservation.

Biosphere Reserves

This programmes was started in 1986 with UNESCO (Management and Biosphere) support for integrating social, cultural, and ecological values of ecologically rich landscapes. The primary focus was on monitoring and designing conservation strategies. With a Tenth Plan outlay of `35 crore, the programmes has not been able to deliver an effective outcome for the existing biosphere reserves. There is a need to reinforce the information and strategic back up for the management of protected areas in the core of biosphere reserves. Keeping in view the facts that village ecodevelopment and participatory management of habitats are part of protected areas management schemes, the primary objectives of this programmes will be research, documentation, and monitoring of the dynamics of human ecosystem interface. The programmes should focus on cross-sectoral linkages between biochemical resources and human livelihood.

Mountain Ecosystems

The MoEF has established an autonomous institute, the G. B. Pant Institute of Himalayan Environment and Development (GBPIHED), with the overall mandate of dealing with issues relating to the environment of the Indian Himalayan Region. The results of project studies of the Institute also have implications on issues such as climate change and land degradation. However, the output falls far too short of what is needed to be done for the Himalayan ecosystems.

During the Eleventh Plan period, GBPIHED should reorient its activities to evolve as a resource centre for the Himalayan States and GoI for advice on sustainable development of the Indian Himalayan Region. The focus of research should include socioeconomic development of the mountain habitations. An Indian Alpine Initiative should also be started for tracking the dynamics of alpine biomes in the context of climate change.

Coastal Zone Management

The earlier Coastal Regulation Zone notification prescribed regulation of activities based on uniform principles of vulnerability of coastal areas against human activity. Coastal areas are currently classified in four categories (CZ 1 to 4) with different principles for development activities. Category 1 includes ecologically sensitive areas, category 4 includes islands, and categories 2 and 3 permit construction activities based on vulnerability. The Swaminathan Committee prescribed that the local circumstances and vulnerabilities should be the basis of coastal zone management and regulations.

For this purpose, use of scientific, social, and local information should be made in environmental management plans for coastal areas. Conservation of life forms (and their habitats such as nesting/spawning sites) and integration of their environments with human wellbeing should be essential components of such plans. Participation of civil society and fishing/coastal communities in the State level coastal zone management committees should be ensured for conservation and sustainable harvests.

Botanical Survey of India (BSI) and Zoological Survey of India (ZSI) The Botanical and Zoological Survey of India are today facing major challenges in view of the new regime on genetic resources, provisions of the Biochemical Diversity Act, and fast evolving knowledge and information environment. Use of recent trends in organizing information and modern skills in exploration and documentation will be given priority. These institutions will develop into prime repositories of information on plants and animals, and as referral institutes. Collaboration and linkages with other institutions as part of a network will be encouraged.

ENVIRONMENTAL AWARENESS AND EDUCATION

The Environment Education in School System project initiated in 1999 strengthens environment education in the formal school curriculum through infusion of appropriate education material. Introduction of environmental concepts in Business/Management Education is another focus area. A committee comprising representatives from management institutions, AICTE, UGC, industry, and MoEF is working onthis.

During the Eleventh Plan, the programmes of Environ-mental Education, Training, and Extension may be continued with further linkages with the publicity and awareness mechanisms of State forest departments. This may include a manual on public participation in all activities of MoEF. Public transport like railways, buses, and even airways can be extensively used for environmental awareness through well designed awareness material. Information generated by student activities on local environmental issues may be integrated with the database under the National Environmental Monitoring Programmes (NEMP). Capacity building, such as training of trainers, should also be focused on. The National Museum of Natural History (NMNH), New Delhi, and three regional museums at Mysore, Bhubaneshwar, and Bhopal will be made more effective in natural history education and awareness with the introduction of the state-of-the-art education and interpretation methods. The MoEF has contemplated setting up of a new regional museum at Sawai Madhopur with focus on the life forms of the region, the details of which are being worked out.

National Environmental Monitoring Programmes (NEMP)

The Steering Committee on Environment and Forests for the Eleventh Plan has suggested a unified NEMP for ecology, environmental chemistry, public health, and socio-environmental studies. This programmes would track the status and change in the socially relevant biophysical parameters and their social impacts, wherever possible. NEMP may have sub-programmes on forest cover and ecosystem services, apart from air and water pollution. The programmes may have linkages with educational, scientific, and social organizations working in the relevant fields.

Accordingly, the existing programmes for Environment Information System (ENVIS) will be reshaped to provide information in interactive formats for effective use. Real time sharing of data on environmental parameters collected under NEMP will be implemented. A special portal could also be organized for environmental data from school and college student projects.

ENVIRONMENTAL RESEARCH AND DEVELOPMENT

The MoEF supports nine Centres of Excellence in research which need to be strengthened. During the Eleventh Plan period, environmental policies and programmes will need strong research backup. The identified research riorities will be met by a combination of open, competitive research grant programmes, and dedicated support to special organizations and centres of excellence.

- An 'Environmental Research Grants' programmes should focus on the relevant areas such as clean technologies, preventive strategies, hazardous substances management, and so on. There should be special programmes on Ecosystem Health, Pollution and Health, Ecological Footprint, NTFP regeneration ecology, Invasive species, Fire Ecology, and Forest-Watershed Services.
- Documentation of traditional and community knowledge should be a special area of research.
- Special mechanisms may be set up for co-ordination and management of research amongst agencies like Indian Council of Forestry Research and Education (ICFRE), ICAR, CSIR, DBT, DST, and UGC, as well as multilateral and bilateral donors and private foundations.
- The All India Coordinated Project on Taxonomy (AICOPTAX) needs strengthening in order to bring more taxa of lower organisms and issues like molecular taxonomy.

Taxonomy Capacity Building

The implementation of the Biological Diversity Act, 2002 and Rules, 2004 and National Environmental Policy, 2006 calls for an adequate number of trained taxonomists. Existing programmes on botanical gardens and taxonomic capacity building assume immense significance in this respect. MoEF provided assistance to 72 botanical gardens during the Tenth Plan and under an AICOPTAX, 11 coordinating centres and two centres for training have been supported at various institutions.

AICOPTAX should be augmented for capacity building (including human resources) in Taxonomy. The ZSI and BSI will be the key institutions to organize taxonomic capacity building programmes and train young Indian taxonomists for forest, wildlife and management.

CONSERVATION OF NATURAL RESOURCES AND ECOSYSTEMS

Inland aquatic systems (excluding paddy fields) cover 5.3% of the country's land area but harbour 15% and 20% of India's floral and faunal diversities, respectively. There are wide gaps in knowledge relating to hydrological parameters, ecosystem processes, and aquatic life forms in these systems. Information on the physico-chemical aspects of water quality in rivers and lakes has become available largely because of the support from the National River Conservation Directorate (NRCD) and the CPCB, but that on aspects such as area, depth, hydrology, and ecosystem processes (functions) is practically non-existent.

The existing programmes on conservation accordingly need a revamp. The Scheme on Conservation and Management of Mangroves, Coral Reefs, and Wetlands has been too small to make an impact on the conservation of these ecosystems. Mangroves and coral reefs are mainly found on government lands. Development of appropriate institutions to motivate people to cooperate in the conservation efforts of these systems will be a priority area. More protected areas are needed for the conservation of coral reefs.

Restoration and plantation programmes for mangroves are necessary to make these eco-systems function as effective barriers against invasion of land by sea.

The existing programmes on wetlands, mangroves, and coral reefs will be extended to mountains, grasslands, and alpine ecosystems. Initiatives on conservation need to be more effectively integrated with development and poverty reduction. The principles of Community Reserves under the Wild Life (Protection) Act may be useful in such efforts.

Long-term plans for conservation of such habitats will be important. Wetlands will be part of NLCP and the objectives will cover conservation of life forms apart from mitigating pollution and augmenting catchments.

- An information system will be set up for islands, coral reefs, mangroves, and wetlands by developing a consolidated and easily accessible database of all recorded species.
- Support activities will include participatory research for inventory and valuation of freshwater ecosystems, measures to tackle serious threats to them and ensure conservation, as also the livelihood security of dependent communities.

- With regard to linkage between biodiversity and climate change, it should be recognized that biodiversity is an important tool to find solutions for tackling the impact of climate change on rural populations that depend on biomass for their livelihood.
- There is a need to recognize the potential of biodiversity to provide solutions for adaptation in the form of traditional crop varieties and donors for creating GMOs.
- National agro biodiversity hotspot mapping and identification of important agro biodiversity landscapes will be undertaken for conservation with the participation of farmers and pastoralist groups. Regions with high concentration of genetic diversity on major and important crops may be declared as 'gene sanctuaries' for conservation and enrichment of diversity, involving communities through diversity based income generation and incentives.
- The regulatory mechanism for trials and adoption of GMOs will be required to be based on scientific, socio-economic, and EIA in the long as well as short term. The process of a regulatory regime, therefore, should be consultative and transparent, based on the prescribed scientific, logical, and progressive norms, taking into account factors of human health and environmental health.
- Awareness generation will be important for ensuring effective outreach of the prescribed process for promotion of appropriately and adequately tested GMOs, mechanisms for enforcement of procedures, and processes for safeguarding ecological and health safety will be established.

OTHER PROGRAMMES

Several ongoing programmes of NoEF on specific subjects will continue to be pursued during the Eleventh Plan period. This includes the Environment Planning and Coordination project for Madhya Pradesh and Nagaland, Environmental authorities and tribunals, the Taj Protection Mission, National Natural Resource Management systems, IT, Adaptation and Capacity Building on Climate Change, and so on. Such programmes, with specific mandate derived from the Supreme Court or externally aided projects, will continue till their logical conclusion.

CLIMATE CHANGE

Global Climate Change due to rising levels of GHGs in the atmosphere is one of the most serious environmental concerns of our time. The Intergovernmental Panel on Climate Change (IPCC), established in 1988 by the World Meteorological Organization and the United Nations Environment Programmes, has worked extensively on evaluating past trends and the future prospects of climate change. The synthesis report of the Fourth Assessment Report (AR4) of the Panel was released in November 2007.

The IPCC reports present a grim picture. It is estimated that the Earth's surface temperature has risen by 0.6 ± 0.2°C over the twentieth century.

In the last 50 years, the rise has been 0.13 + 0.07°C per decade and the recent years have been the warmest since 1860, the year from which regular instrumental records are available.

The Panel has concluded that the fact of global warming is unequivocal and there is enough evidence to indicate that this is due to anthropogenic reasons. Although some of these conclusions have been disputed, the assessment of the IPCC represents a broad and growing consensus in the scientific community worldwide.

The current level of atmospheric CO_2 is estimated as 379 parts per million (ppm) compared with the pre-industrial level of only 280 ppm. The annual growth rate of CO_2 concentration has been greater in the last 10 years (1.9 ppm/yr) compared to the last 40 years (1.4 ppm/yr). Halocarbons (chloroflouro-carbons etc.) in the atmosphere, however, are observed to be decreasing due to their phase out under the Montreal Protocol.

Since warming depends upon the total stock of GHG in the atmosphere, continued emissions beyond the earth's absorptive capacity necessarily imply a rise in temperature. If emissions continue to increase as at present, a warming of about 0.2°C per decade is projected by IPCC. If emissions are stabilized at the 2000 level, the warming could be about 0.1°C per decade. The Special Report on Emission Scenarios projects that surface air temperatures could rise by between 1.1 and 6.4°C over the twenty-first century. In most cases, the temperature rise may be 2–4°C. The sea level rise is projected to be 18 to 59 cm.

IMPACT AND IMPLICATIONS OF CLIMATE CHANGE

Climate changes noted in the IPCC Assessment reports include recession of glaciers, thawing of permafrost, lengthening of mid-to high-latitude growing seasons, pole-ward and altitudinal shifts of plant and animal ranges, decline in some plant and animal populations, early flowering of trees, and changes in insect populations and egg-laying in birds. Associations between changes in regional temperatures and the observed changes in physical and biological systems have been documented in many aquatic, terrestrial, and marine environments.

- The water cycle is totally dependent on the temperature regime. Increasing global temperature and resultant faster retreat of most glaciers is expected to affect the snow fed perennial water regimes. Accordingly, the vegetation status of catchments, magnitude and frequency of floods, precipitation, run-off and groundwater recharge may all be affected. These changes may substantially affect hydropower generation, and may also require an increase in irrigation due to higher crop evaporative demand.
- Through the twentieth century, the average rate of global sea level rise has been 1.7 + 0.5 mm/yr. The maximum rise since 1992 has been observed in the Western Pacific and the Eastern Indian Ocean.

- This trend could result in loss of land due to inundation, erosion, floods, and salt-water intrusion, adversely affecting coastal agriculture, tourism, fisheries and aquaculture, human settlements, and health.
- Changing environmental attributes are sure to affect the species spectrum and the profile and composition of forests is also likely to change. The degree to which terrestrial ecosystems will continue to be net sinks for carbon is uncertain. In arid or semi-arid areas, productivity is expected to decrease. Trends of drying up of wetlands and degradation of ecosystems have been observed.
- Estimated crop yield responses to climate change vary widely, depending upon species, cultivation practices and soil properties; pests, and pathogens; the direct effects of carbon dioxide (CO_2) on plants; and interactions between CO_2, air temperature, water stress, mineral nutrition, air quality, and adaptive responses. Even though increased CO_2 concentration can stimulate crop growth and yield, this benefit may not always overcome the adverse effects of excessive heat and drought. The AR4 projects a decline in global rice production by 3.8% by the end of the twenty first century.
- Changes in the severity and frequency of extreme heat and cold,and of floods and droughts, coupled with local air pollution and aero allergens may result in changes in infectious disease occurrence, in local food production and also cause under nutrition, leading to impaired child development. There will also be health consequences of population displacement and economic disruption. For India, the risks of malaria and heat stress related mortality have been projected by IPCC. Increase in flooding and droughts are associated with increased risk of drowning, diarrhoeal and respiratory diseases, and hunger and malnutrition. Adaptive options like health-oriented management of the environment (including air and water quality, food safety, urban and housing design, and surface water management), and appropriate medical interventions are normally related to income levels. Therefore, the poor are more vulnerable.
- Human settlements are also expected to be affected by changes in resource productivity or market demand for goods and services causing structural and functional adaptations leading to migration, which in turn would result in changes in living conditions.

Global warming will affect us seriously. For South Asia, warming has been projected to be above the global mean and in South East Asia, equal to the global mean. The impact projections for India indicate a rise of 0.68°C in the twentieth century with an increasing trend in the annual mean temperature.

Precipitation is likely to increase and extreme rainfall and other climatic events may occur more frequently. Extreme rains in the South West monsoon and fewer rainy days along the East Coast have been projected.

An increase in temperatures of 0.5°C to 1.5°C could produce a decline of between 2.5% in wheat and maize production in India. Existing research based on formal modeling suggests that a scenario of warming between 2–3°C would lead to a permanent loss of up to 3% in the global output. In countries with predominantly agrarian economies, the projected increase in temperature and reduction in precipitation could, in the absence of adaptive responses such as increased irrigation, related agricultural research and related infrastructural interventions, dampen economic growth by reducing agricultural productivity.

These impacts can be moderated by efforts at adaptation but these are expected to be costly and will impose a burden on both producers and consumers. Climate change would therefore result in lower incomes of the vulnerable populations and increase in the absolute number of people at risk of hunger unless these outcomes can be countered through the development of cost effective technologies.

RESPONSES TO CLIMATE CHANGE

The serious consequences of climate change, including especially the consequences for India, lead naturally to the question of what should be our response. Two types of responses need to be considered. The first relates to adaptation, i.e., measures that have to be taken given the very high likelihood that climate change will occur and will have adverse effects. The second relates to mitigation, i.e., steps to be taken that might reduce the extent of climate change.

Recognizing the importance of climate change issues, the Prime Minister established a Council on Climate Change under his chairmanship in June 2007 to co-ordinate national action for assessment, adaptation, and mitigation of climate change. Earlier, the government had set up an expert committee to study the impact of climate change. One of the important decisions taken by the Prime Minister's Council is that a National Action Plan will be prepared compiling action taken by India for addressing the challenge of climate change and the action that needs to be taken in the future.

Adaptation

Since a substantial adverse change in climate appears unavoidable even with the optimal mitigation response, the process of adaptation to climate change must have priority. The most important adaptation measure is development itself. A stronger economy is more able to adapt both in terms of the cost of adaptation and technological capability. Achieving rapid economic growth as targeted in the Eleventh Plan is therefore a key element in adaptation.

As mentioned earlier, the Prime Minister's Council on Climate Change has decided that a National Action Plan on Climate Change compiling action taken by India for addressing the challenge of climate change and the action that it proposes to take, should be prepared as India's National Report on Climate Change. The Action Plan will deal with key vulnerabilities of India to Climate Change and, in particular, the impact on water resources, forests, coastal areas, agriculture, and health.

An action plan for adaptation to climate change would require:

- Action in the area of agricultural research to evolve varieties that can cope with likely climate changes,
- Action to cope with likely increases in water stress,
- Action to be able to cope with a greater frequency in natural disasters.

The institutional network created at the time of preparation of India's initial National Communication to United Nations Framework Convention on Climate Change (UNFCCC) exists and is currently engaged in the preparation of information on India's second National Communication. In the changing temperature and water regimes, improvement of productivity potential and water use efficiency of agricultural crops, specifically in regimes of water shortage and extreme variations of temperature, will be priority areas of research.

The adaptation response needs to be incorporated in the relevant programmes, including those relating to watershed management, coastal zone planning and regulation, forestry management, agricultural technologies and practices, and health. Work on the potential impact, forecasting systems and adaptations in response to Climate Change will be the key concerns for future development. For projections and early warning systems, understanding of the processes which could show discernible changes well before the actual impact starts is essential.

Mitigation

Recognizing the need for international co-operation in devising an effective mitigation strategy, in 1992, the international community established the UNFCCC which established the principle of 'common but differentiated responsibility' between the industrialized countries and the developing countries, taking account of the respective capacities. The countries were required to bring down their GHG emissions to 1990 levels by 2000. In 1997, the Kyoto Protocol under the UNFCCC was adopted, which set legally binding targets for GHG reduction by individual industrialized countries totaling 5.2% below their aggregate 1990 emissions during the 'first commitment period', 2008–12.

Under the provisions of the Kyoto Protocol, non Annex I Parties (India is in this category) are not obliged to reduce emissions of GHG. So far, 176 countries, including India (and the regional economic organization EEC), have ratified the Protocol. The US and some other countries have not ratified the Protocol till now. Most of the industrialized countries thus far have not achieved significant progress in reducing their GHG emissions in the manner required in the first commitment period.

An optimum approach to mitigation from a global perspective is only possible if all countries cooperate to facilitate collective action. This is because mitigation actions are characterized by the classic problem of externality—the benefits of the action of any one country do not accrue only to the country itself but to the entire global community, while the costs of mitigation are fully internalized. Without co-operative action, all countries taken together will do less mitigation than is collectively optimum.

The challenge lies in determining a basis for collective action which is fair, provides equitable entitlement to the global environment space with burden sharing in a manner which recognizes the very different levels of developmnt and also the very different degrees of historical responsibility for causing the problem in the first place. Such differenciation among the countries of the world has been accepted by the UNFCCC and the Kyoto Protocol. This must be effectively operationalized to create an environment of trust and co-operation.

The first commitment period of the Kyoto Protocol is 2008–2012 and as per the provisions of the Kyoto Protocol and discussion at the 13th Conference of Parties at Bali in December 2007, Bali Action Plan was commenced which intends to enable the full, effective, and sustained implementation of the Convention through long-term co-operative action, now, upto and beyond 2012. As part of the Bali Action Plan, a process has also been initiated to determine the GHG reduction commitments of industrialized countries under the Kyoto Protocol, beyond 2012.

The precise nature of the outcomes of this process would be determined in a two year negotiating time table ending in 2009. There is no doubt that the accumulation of GHG, which is the cause of global warming, has occurred verwhelmingly due to the emissions of industrial countries and it is therefore entirely appropriate that the burden of mitigation must fall on them. It is for this reason that the first commitment period in the Kyoto Protocol entailed commitment for reduction only by the industrialized countries.

With a share of just 4% of global emissions, any amount of mitigation by India will not affect climate change. A substantial commitment to reduce emissions by the major industrialized countrics will be required. Nevertheless, as contribution to the global emissions reduction effort, the Eleventh Plan would focus on efforts to ensure that the emissions intensity of India's GHG continues to decline. The correct approach in mitigation is to define the obligations of different parties on the basis of common but differentiated responsibility and differences in respective capabilities.

India's view on burden sharing is based on this approach. Our National Environmental Policy of 2006 requires 'equal per capita entitlements of global environmental resources to all countries'. At the meeting of the G 8 + 5 in Heiligendamm in July 2007, Prime Minister Manmohan Singh indicated that we are determined to see that India's per capita emissions never exceed the per capita emissions of the industrialized countries. This formulation focuses on per capita emissions rather than total emissions as the relevant variable, and also provides industrial countries with an incentive to reduce their level of emissions as quickly as possible.

While optimal mitigation strategies will no doubt evolve in time from the ongoing international negotiations, India is already taking a number of initiatives in clean energy including renewable energy and action to increase energy efficiency and conservation. One of the objectives of the Eleventh Plan is to reduce the energy intensity per unit of GHG by 20% from the period 2007–08 to 2016–17. We will also initiate action to increase our access to cleaner and renewable energy by fully exploiting existing resources (e.g., hydropower and wind power) developing nuclear power, and also supporting research in newer areas such as biofuels from agro-waste, solar energy, etc.

The CDM is a flexible arrangement under the Kyoto Protocol which enables industrialized countries to meet their emission reduction obligations by encouraging adoption of sustainable and environmentally friendly technologies in developing countries. India's CDM potential is a significant component of the global CDM market. Till 14 February 2008, out of a total of 871 projects accorded Host Country Approval, 314 projects in India have been registered by the CDM Executive Board, so far the highest in any country. However, in terms of the corresponding Certified Emission Reduction (CERs), India is second to China.

ALLOCATIONS IN THE ELEVENTH PLAN

The total projected GBS in the Eleventh Plan for the MoEF is `8842 crore (at 2006–07 price) and `10000 crore (at current price).

DISASTER MANAGEMENT

Many regions in India are highly vulnerable to natural and other disasters on account of geological conditions. About 60% of the landmass is susceptible to earthquakes and over 8% is prone to floods. Of the nearly 7500 kilometers long coastline, approximately 5700 kilometers is prone to cyclones. 68% area is susceptible to drought.

All this entails huge economic losses and causes developmental setbacks. Disasters are no longer limited to natural catastrophes. Man-made emergencies often cause bigger disasters in terms of fatalities and economic losses.

With urbanization and concentration of population in metropolitan cities, more and more people are becoming vulnerable to locational disasters. So, the development process needs to be sensitive towards disaster prevention, preparedness and mitigation. Disaster management has therefore emerged as a high priority for the country. Going beyond the historical focus on relief and rehabilitation after the event, there is a need to look ahead and plan for disaster preparedness and mitigation in order to ensure that periodic shocks to our development efforts are minimized.

Disaster risk reduction has not been highlighted in the policies and programmes of various plan schemes. The country's commitment to mainstreaming disaster risk reduction into the process of development planning at all levels so as to achieve sustainable development is yet to be carried forward across sectors through actionable programmes for achieving the desired result.

TENTH FIVE YEAR PLAN STRATEGY AND APPROACH

The Tenth Five Year Plan (2002–07), prepared in the backdrop of the Orissa super cyclone, Gujarat earthquake, and end of the International Decade of Natural Disaster Reduction, recognized disaster management as a development issue for the first time. The Plan devoted to disaster management and made a number of important prescriptions to mainstream disaster risk reduction into the process of development. The Tenth Plan prescriptions on disaster management can broadly be divided into three categories:

- Policy guidelines at the macro level that would inform and guide the preparation and implementation of development plans across sectors;
- Operational guidelines for integrating disaster management practices into development plans and programmes,
- Specific developmental schemes for prevention and mitigation of disasters.

At the macro level, the Plan emphasized that 'while hazards, both natural or otherwise, are inevitable, the disasters that follow need not be so and the society can be prepared to cope with them effectively whenever they occur' and called for a 'multi-pronged strategy for total risk management, comprising prevention, preparedness, response and recovery, on the one hand, and for initiating development efforts aimed towards risk reduction and mitigation, on the other'. It stated that only then could we look forward to 'sustainable development'.

At the operational level, the Plan made a number of very important suggestions, as given below:

- Streamlining institutional arrangements for disaster response by an integrated approach involving civilian and military resources; setting up a modern, permanent national command centre or operations room with redundant communications and data links to all State capitals; establishing a quick response team particularly for search and rescue operations; developing standard operating system for dealing with humanitarian and relief assistance from non-government sources; and formulating a unified legislation for dealing with all types of disasters.
- Building disaster prevention and preparedness in development planning by introducing a rigourous process of vulnerability analysis and risk assessment, maintaining comprehensive database and resource inventories at all levels, developing state-of-the-art infrastructure for mitigation planning, and establishing a Disaster Knowledge Network for the use of disaster managers, decision makers, community, and so on.
- Developing a nation-wide culture of prevention by introducing disaster management in the school curriculum, including relevant aspects of disaster management in professional courses, enhancing the capacity of disaster managers by better training facilities and creating a massive awareness at all levels.

- Encouraging community level initiatives for disaster preparedness by involving people at the grass roots, particularly those who are more vulnerable, for better preparedness and response.
- Developing appropriate zonal regulations, design standards, building codes, and performance specifications for safe construction.
- Inclusion of disaster mitigation analysis in all development schemes in vulnerable areas through which the feasibility of a project is assessed with respect to vulnerability of the area.
- Building disaster mitigation components into all development projects financed under the Plan.

The Tenth Plan felt that planned expenditure on disaster mitigation and prevention measures was required in addition to a Calamity Relief Fund. The Plan, however, stopped short of recommending any specific plan scheme for prevention, mitigation or preparedness for disasters nor did it allocate any amount for such scheme, except making a general recommendation that 'Creation of faculties in disaster management in all 28 states is proposed to be taken up in the Tenth Plan in addition to community mobilization, human resource development, establishment of Control Rooms and forging international cooperation in disaster management. There is also an urgent need for strengthening the disaster management pedagogy by creating disaster management faculties in universities, rural development institutes and other organizations of premier research.'

IMPLEMENTATION OF SCHEMES IN THE TENTH PLAN

A Central Sector Plan Scheme, the National Disaster Mitigation Programmes (NDMP), has been implemented since 1993–94. The scheme mainly provided for training and capacity building of government functionaries and other stakeholders to manage disasters in an effective manner. This scheme was transferred from the Ministry of Agriculture to the Ministry of Home Affairs in June 2002. The allocation for this scheme was raised from `6.30 crore in the Eighth Plan to `16.32 crore in the Ninth Plan and to `30.77 crore in Tenth Plan.

The annual allocations under various components of the scheme during the Tenth Five Year Plan are in Table. Out of the eight components of the above scheme, only two components namely 'Grants-in-Aid' and 'Professional Services' were developmental in nature, while the remaining components were on revenue expenses.

The major activities under the grants-inaid include the setting up of the National Centre for Disaster Management at the Indian Institute of Public Administration, now upgraded as a statutory organization known as National Institute of Disaster Management (NIDM), creation of 29 disaster management faculties/Centres in 28 States, research and consultancy services, and documentation of major disaster events. The faculties did not have sufficient back-up from the States with the result that only limited activities are undertaken by the Centres. Even meagre allocations provided by the Central Government as grants-in-aid were not fully utilized in many of the States.

The allocation of `21.90 crore to the training institutes during the Tenth Five Year Plan was mainly used on NIDM which incurred an expenditure of `11.56 crore. Nearly 71000 persons have been imparted training under the programmes.

TABLE: National Disaster Mitigation Programmes-Year-wise Allocation of Funds in the Tenth Five Year Plan

(in lakh)

S. No.	Head	2002-03	2003-04	2004-05	2005-06	2006-07	Total
1.	Grants-in-aid	262.00	349.00	474.00	505.00	600.00	2190.00
2.	Professional Services	20.00	29.00	54.00	110.00	150.00	363.00
3.	Machinery & Equipments	2.00	20.00	30.00	15.00	10.00	77.00
4.	Foreign Travel	50.00	30.00	25.00	40.00	50.00	225.00
5.	Contribution	20.00	20.00	20.00	17.00	20.00	97.00
6.	Office Expenses	2.50	3.00	2.00	2.00	2.00	11.50
7.	Advertisement & Publicity	20.00	10.00	25.00	1.00	1.00	57.00
8.	Other Charges	0.50	26.00	10.00	10.00	10.00	56.00
	Total	**407.00**	**847.00**	**640.00**	**700.00**	**843.00**	**3077.00**

The National Cyclone Risk Mitigation Project is another Plan Scheme proposed to be implemented with World Bank assistance. The project is presently at the stage of preparation of Detailed Project Report. It covers all 13 cyclone prone coastal States and UTs.

Investments are to be made in upgradation of cyclone forecasting, tracking and warning systems, construction work of cyclone shelters, road linkages, plantation, retro-fittings of vital lifeline installations, strengthening of national and regional training institutions, setting up a techno-legal regime for cyclone mitigation and for strong management and monitoring system.

As regards upgradation of cyclone forecasting, tracking and warning systems, this activity is proposed to be taken up either as part of the project or independently by the IMD. The Ministry of Home Affairs is also implementing Phase II of a Community based Disaster Risk Mitigation (DRM) Programmes in 169 multi hazard districts in 17 States/UTs with the support of UNDP under a multi-donor programmes at a total estimated cost of `153 crore (US$ 34 million).

Under this programmes, disaster management plans are being prepared from village to district; village volunteers are being trained in first-aid, search and rescue, evacuation and relief and shelter management; disaster management teams are being constituted at the district and sub-district levels and mock drills are being conducted at all levels. This 100% externally funded programmes is, however, outside the framework of the Plan and is being implemented on the non-Plan side.

It is the largest community based DRM programmes in the world. Phase I of the programmes was implemented in the three States of Gujarat, Orissa, and Bihar during 2002–06 and has been rated well in a recent evaluation conducted by a team of consultants. While the general feedback on Phase II of the programmes is good, there is a definite need to increase the scope to cover other deficient areas too, in order to draw the full benefit from this.

INSTITUTIONAL AND OTHER INITIATIVES 4IN THE TENTH PLAN

The following significant initiatives on disaster management were taken during the Tenth Five Year Plan period:

- The Disaster Management Act, 2005 has been enacted for establishing requisite institutional mechanisms for drawing up and monitoring the implementation of disaster management plans, ensuring measures by various wings of the government for prevention and mitigating the effects of disasters, and for undertaking a holistic, co-ordinated, and prompt response to any disaster situation.
- The National Disaster Management Authority (NDMA), under the Chairmanship of the Prime Minister, has been set up in terms of the Act. NDMA is an apex body responsible for laying down of policies, plans and guidelines on disaster management so as to ensure timely and effective response to disasters. The Authority has prepared a Draft National Policy on Disaster Management and has taken up preparation of guidelines on prevention, mitigation, response, and recovery in regard to various types of disasters such as earthquake, flood, landslides, industrial disaster, and so on. The guidelines on management of earthquake, chemical disasters, and chemical (industrial) disasters have been finalized and circulated.
- The State Governments are in the process of setting up State and district Disaster Management Authorities. The provisions of the Act relevant to the States/UTs have been brought into force w.e.f. 1 August 2007. While Arunachal Pradesh, Goa, Gujarat, Himachal Pradesh, Kerala, Mizoram, Puducherry, Punjab, and Uttar Pradesh are reported to have constituted SDMAs as per the Act, the other States/UTs are in the process of constituting the same.
- An eight battalion-strong National Disaster Response Force has been set up comprising of 144 specialized response teams on various types of disasters of which 72 teams are for nuclear, biological, and chemical (NBC) disasters.

- The Civil Defence set-up in the country is proposed to be revamped to strengthen local efforts for disaster preparedness and effective response.
- Similarly, the Fire Services are also proposed to be strengthened/ modernized to convert them into a multi-hazard response force.
- The National Institute of Disaster Management has been set up for training, capacity building, research, and documentation on various natural and manmade disasters. A comprehensive Human Resource Plan for Disaster Management has also been developed. Disaster Management has been included in thc curriculum of middle and secondary school education. The subject has also been included in the post-induction and in-service training of civil and police officers. Modules have also been identified to include disaster management aspects sin the course curriculum for engineering, architecture, and medical degrees.
- A committee of experts has finalized the Model building by-laws for town and country planning legislations, land use zonation, and development control legislations. The municipalities and city development authorities all over the country have been advised to make necessary changes in their respective by-laws and regulations in accordance with the model laws.
- The Bureau of Indian Standards has issued building codes for construction of different types of buildings in different seismic zones in the country. The National Building Code has also been revised, taking into consideration the natural hazards and risks of various regions of the country.
- The National Programmes for Capacity Building of Engineers in Earthquake Risk Management to train 10000 engineers and 10000 architects on safe construction techniques and architectural practices is under implementation.
- A Web-enabled centralized inventory of resources has been developed to minimize response time in emergencies. Over 110000 records from 600 districts have been uploaded. Safe construction practices and dos and don'ts for various hazards are being disseminated for creating public awareness.

ELEVENTH PLAN

Vision and Strategy

The Tenth Plan has set into motion the process of shift in focus from response-centric disaster management covering rescue, relief, rehabilitation, and reconstruction to laying greater emphasis on the other elements of disaster management cycle—prevention, mitigation, and preparedness —as a means to avert or soften the impact of future emergencies. The Eleventh Plan aims at consolidating the process by giving impetus to projects and programmes that develop and nurture the culture of safety and the integration of disaster prevention and mitigation into the development process.s

The guidance and direction to achieve this paradigm shift will need to flow from NDMA, and in the true spirit of the Disaster Management Act, to all stakeholders including State Governments and UTs, right up to the PRIs. Communities at large will need to be mobilized to achieve this common objective as they are the first responders. Even the best of isolated efforts will not bear fruit unless they are part of an overall, well-considered approach, and responsibilities of all stakeholders are clearly spelt out and accountability and sustainability factored in.

Projects and programmes undertaken may be such that they address the felt national needs and with well acknowledged prioritization. With scarce resources, it will be imperative to put in place suitable mechanisms which will facilitate systematic project and programmes appraisal. Support will have to be lent to programmes and projects that will lead to sustainable development, with assurance of disaster risk reduction built in.

The question of vulnerability will also have to be considered, not only in the physical sense but in a comprehensive sense; vulnerability could also be social, ecological, organizational, educational, attitudinal, political, cultural, and economical. For an objective screening and scrutiny of development projects in a speedy manner, it is essential that professionalism in project formulation is insisted upon. Well-considered and well-drafted, need-based project proposals with in-built disaster prevention and mitigation features, as brought out above, should usually sail through the decision making process. These projects would need to be based on sufficient studies, reliable assumptions, and have regard for the latest advances in science and technology. To ensure thoroughness in project formulation, all the expected requirements need to be clearly articulated and the project appraisal procedure kept simple and straightforward. One set of standard guidelines may not be adequate in addressing all the complex set of issues and there is clearly a need to evolve a multi-pronged approach to address a host of interlinked issues.

For the assessment of impact of any major development scheme in a given zone of hazard, it is imperative that the proposal takes in its stride what all could go wrong (damage scenarios) and shows how the project design is fashioned to counter that likely damage scenario. Conceptualization of damage scenarios will require quality scientific studies in a multi-disciplinary environment, and generation of new data, besides filling data gaps. The SoI needs to be entrusted with the responsibility of generating large-scale maps which will form the basis for disaster management studies. They will also need to generate a Digital Elevation Map (DEM) of high resolution (1/2m interval along vertical) for the purpose of 3D terrain modelling. There will be a need to carry out prioritization of the areas on which the SoI will carry out the 1:10000 mapping for the whole country and map all towns and cities on 1:2000 scales. Coastal zones and flood prone areas of the country will need to get priority in the preparation of DEM.

NDMA, in consultation with National Informatics Centre (NIC), National Spatial Data Infrastructure, National Remote Sensing Agency, and DoS is actively promoting the use of GIS in all institutions concerned with disaster management. Priority will have to be given to highly vulnerable areas for generating the database for mapping as well as hazard vulnerability and risk assessment related to disasters like cyclone, flood earthquake, chemical and industrial, etc.

Data-sharing among the various data generators is an even more important aspect and needs to be addressed. All such schemes getting Plan funds will be treated as national asset and data would be made available without any pre-condition for use for disaster management authorities and others.

The national emergency communication network, involving the contemporary space and terrestrial-based technologies, in a highly synergistic configuration and with considerable redundancy, needs to be developed and deployed countrywide. With almost a 100% reliability, this network must ensure real-time dissemination of warnings and information direct to the affected community and local authorities.

Research on all aspects of disaster mitigation which has the potential to save lives and property needs to be encouraged. NDMA could provide the necessary policy direction to Science related ministries and departments both at the Central and State levels to foster, promote, and sustain research and development work through needbased disaster mitigation and management projects.

A well sensitized and prepared community forms one of the most important links in the process of better management of disasters. For that reason, 'Extended Disaster Risk Mitigation Project' has been identified for being taken up for preparation of a Project Report during the Plan. This will be supplemented by activities under various other national/State level mitigation projects. As far as Community Based Disaster Management is concerned, a degree of convergence is required to be brought into community mobilization, participation, awareness, and capacity building aspects of all concerned social sector schemes like women and child development, rural development, drinking water, sanitation, and so on.

The concerned ministries may consider inclusion of disaster management content into their plans and make provision of funds in their respective schemes. It is essential that while clearing plans for State Governments, there is an integrated approach, particularly for creation of adequate capacity for relief and rescue operations.

There is need for evolution of a broad classification of anticipated projects that would come up for clearance by the Planning Commission.

One possible classification could be:

- Projects specifically designed in response to the NDMA Strategy, Guidelines and Action Plan,
- Development projects with built-in environmental preservation and disaster mitigation features in tune with the philosophy of mainstreaming mitigation measures into development projects.

Assistance is needed for State level mitigation projects, especially drafting of State, district, and panchayat level disaster management plans and in continuous sustenance, modernization, and upgradation of disaster management capacity. In view of the focus on disaster mitigation of projects, institution building, aimed at training a new breed of disaster managers, and establishment of centres/institutes fully equipped to carry out specialized investigations and post-disaster studies, would need to be encouraged.

The training and skills of the 72 teams for NBC disasters in the eight battalions of the National Disaster Response Force have to be improved and sharpened. Out of the eight battalions of the National Disaster Response Force already sanctioned by the government, two Battalions already have the basic infrastructure. While the existing infrastructure has to be upgraded for these two Battalions, completely new infrastructure has to be created for the remaining six Battalions.

Mainstreaming Disaster Management into Development Planning

Mainstreaming disaster management into the development planning process essentially means looking critically at each activity that is being planned, not only from the perspective of reducing the disaster vulnerability of that activity, but also from the perspective of minimizing that activity's potential contribution to the hazard. Every development plan of a ministry/ department should incorporate elements of impact assessment, risk reduction, and the 'do no harm' approach.

Examples of this approach are urban planning and zoning, upgradation of building codes and their effective enforcement, adoption of disaster resilient housing designs and construction of school and hospitals, flood proofing, response preparedness planning, insurance, establishment of early warning systems for various types of disasters, generating community awareness, creating technical competence and promoting research among engineers, architects, health experts, and scientists.

State governments need to give priority in their plans for schemes regarding hazard identification and risk assessment once they have prepared the project paper, completed preliminary work, and drawn up the details of the scheme. Hazard identification and risk assessment across the country must be bound by uniformly followed procedures, fine-tuned to local conditions.

In the absence of such procedures, any sporadic activity based on some adhoc procedure carries the potential of doing more harm than good. It will be essential that while clearing plans of State Governments, there is an integrated approach particularly for creation of adequate capacity for relief and rescue operations, for example, for funding projects for construction of school buildings. Some school buildings need to be identified which will be used as relief centres and buildings designed so that they withstand the impact of disasters and also have adequate capacity to provide space as relief centres.

Such schools should be equipped with essential services which become important at the time of a disaster.

State governments will need to make comprehensive Management Action Plans for achieving longterm results in a phased manner. To make these plans more meaningful, workshops and training programmes are to be organized at local/regional/national levels for capacity building and awareness generation/ community participation.

Outside the framework of Plan schemes, many innovative measures can be adopted to encourage disaster risk reduction measures by the corporate sector, nongovernment organizations, and individuals. Fiscal measures like rebates on income and property tax for retrofitting unsafe buildings, compulsory risk insurance for bank loan on all types of properties, and so on, shall definitely help to mobilize resources for safe construction and retrofitting of existing constructions in all disaster prone areas. Similarly, many innovative measures may be taken for promoting public–private–community partnership for disaster risk reduction.

Principles for Project Appraisal from the Disaster Management Perspective

To assist the Planning Commission in appraisal of projects, broad and generic guidelines which are not disaster or theme specific have to be adopted. Conceptualization of hazard scenarios and associated vulnerability and risk assessments in a given situation will necessarily have to depend on available maps, Master Plans and building and land use regulations, National Building Code of India, and the various safety Standards and Codes of the Bureau of Indian Standards.

The guidelines will have to cover the following aspects:

- The location of the project is to be carefully considered, especially if this is in a multi-hazard prone area/district recognized by the NDMA. Multi-hazard prone districts are reported in the latest National Building Code of India of the Bureau of Indian Standards.
- The listing has been revised by the Building Materials Technology Promotion Council of India, currently under the consideration of NDMA.
- The project/scheme should be based on a detailed hazard and risk assessment; wherever required, environmental clearance will also be taken. The risk assessment will usually involve the following factors:
  - Assessment of degree of hazard, based on high resolution single/ multiple hazard maps interpreted in the light of all available historic records, publications, site-specific information, and studies. For all major projects/schemes, where such maps are presently unavailable, the project/scheme should be supported by adequate site-specific seismotectonic, geological, geo-physical, and geotechnical studies and analyses. Data gaps, assumptions made, and their implications

should be brought out. Where high resolution multi-hazard maps are not available, multi-hazard assessment is to be made by coalescing the information on single hazards.

- All such site specific risk assessment studies should be referenced to a national high resolution geospatial database so as to facilitate temporal analysis of future assessment impact studies in the area and also enable integration of all other assessment studies carried out in the vicinity. Through such a process it will be possible to evolve a national database of assessment studies which, in turn, will facilitate refinement of National Hazard and Vulnerability Mapping.
- Assessment of vulnerability against a hazard of a given magnitude should be carried out. The vulnerability of an individual or a group of individuals or of any element or an infrastructure like a flyover or a bridge, for a hazard of a given magnitude, will vary from 0 to 1 depending on the degree of mitigation built into the design. For example, an earthquake of magnitude 7 on the Richter scale may render very unsafe school children in a poorly built school (vulnerability = 1) whereas the residents of an earthquake-resistant neighbouring house for the same magnitude of earthquake, may be safe (vulnerability = 0). This distinction is essential because existence of hazard does not automatically mean vulnerability, and vulnerability does not necessarily have to be 1. The question of vulnerability has to be considered not only in a physical sense but in a comprehensive sense. Vulnerability could be physical, social, ecological, organizational, educational, attitudinal, political, cultural, or economical. Vulnerability assessment may also take note of medical care and casualty management that would be possible in the vicinity in case of natural or man-made disaster.
- Assessment of risk against a given hazard will be a function of hazard and vulnerability.
- Identification of elements at risk like population, properties, economic activities, public services, is to be brought out. By overlaying the infrastructure map of an area on the corresponding hazard map of the same scale, elements at risk can be identified.
- Particularly while carrying out hazard identification and risk assessment for industrial estates, issues like release scenario, consequences in terms of heat generation over pressure and toxicities, identification of hazardous chemicals, processes and operations, identification of important receptors, both environmental and physical, classification of units which have potential for creating an off site emergency, and so on, need tobe addressed.

- The reliability of hazard, vulnerability, and risk assessments will depend upon the quality of maps and other investigational data and various uncertainties involved due to inadequacy of data and other factors. It is therefore important that all major stages of project/scheme development, namely, planning, site investigations and designs, are subject to a process of rigourous peer review and accordingly certified.

Where projects specifically identified and designed for disaster management are to be appraised, the following factors would need to be considered:

- The proposed project or scheme,
  - Is to be need based and demand driven;
  - Must fall within the high priority bracket, linked with the development plan of the area;
  - Should have well stipulated goals, clearly identified stakeholders and beneficiaries;
  - Should be fully backed with analyses of risks and quantified benefits in terms of disaster safety;
  - Should clearly reflect implications of not taking up the project in terms of disaster related risks, environmental protection, and economic development.
- Projects/schemes which yield multiplier effect for the greatest good of the largest number will deserve priority. For example, a well drafted practical disaster emergency plan for a school or disaster management plan for a district can inspire other schools and districts to yield the snow-balling (multiplier) effect. Similarly, development of a knowledge based, multi-media disaster mitigation product when translated in different vernacular languages may at once multiply benefits.
- Project merit rating should also depend on the following factors:
  - Breaking new ground in terms of scientific, technological or management innovation, including peoples' participation. For instance, development of an innovative early warning system against a particular type of disaster using simple, readily monitored indicators.
  - Delivering 'Best Practices' for others to emulate and getting inspired by them. For example, best practices of engineered constructions in a given earthquake-affected hilly area.
  - Choice of appropriate technology. For example, partially prefabricated construction technology will be more appropriate than the cast-in-place construction technology in a post-disaster reconstruction programmes in a given situation.
  - mployment generation. For example, a judicious man-machine mix in the construction of flood prevention works may help generate

employment in an area without unduly compromising on efficiency of work.

- Sustainable capacity building. It is the key to empowering a village, district, State, or a region so that each one of them can, as far as possible, manage their own affairs in the event of a disaster. For example, water sampling and testing capacity at local level when post-disaster situations threaten epidemics.
- Pro-active engagement of communities and spreading the culture of safety in communities and other levels.

Disaster education and community leadership development, public awareness development, gender mainstreaming, special focus on the needs of women and children, vocational training of unemployed youth, and concern for physically challenged persons will add weight to the project/scheme.

Since disasters know no district or State boundaries, projects of interest to two or more districts or States may score over those yielding localized benefits. By the same logic, national level mission-mode projects/schemes should get preference over others.

Financial Provisions for Disaster Management

It is important that a portion of the Plan funds is earmarked for efforts that directly or indirectly help in disaster management. Ideally, each project should provide adequately for the disaster mitigation and management expenditure that is identified by the appraisal process described in the previous section as being necessary. In addition, Central and State Governments may, depending upon their own hazard assessments, earmark a suitable and adequate amount for disaster mitigation schemes that are implemented over a definite time period.

The projects to be taken up should include:

- All schemes for generating basic input data for hazard and vulnerability impact analysis.
- Stand-alone disaster management projects such as mitigation projects, awareness programmes, capacity building projects, community based disaster management projects, upgrading early warning systems, failsafe disaster management communication network, micro-zoning, and so on.
- Mainstreaming disaster reduction into already approved projects in sectors of education, housing, infrastructure, urban development, and the like. For example, projects already under implementation such as the SSA, which caters to the construction of school buildings, could be reviewed. The design of the school building under the programmes could include hazard resistant features, at least in multi-hazard prone (earthquake, cyclone, flood), high-risk areas so that these are safe. Similarly, existing infrastructure like bridges and roads will need to be strengthened and upgraded to mitigate disaster at a subsequent stage. ❐

15

Population Change, and the Rural Environment

Rural environment represents the framework of regulations, institutions, and practices in villages defining parameters for the sustainable use of environmental resources while ensuring security of livelihood and a reasonable quality of life. While the scope of environmental infrastructure is often narrowed down to the provision of suitable water supply, sewerage, and sanitation systems, it has within its purview:

- Acquisition, protection, and maintenance of open spaces,
- Clean up and restoration of degraded lands,
- Integration of existing wildlife or habitat resources,
- Sustainable approaches to controlling flooding and drainage,
- Developing river corridors and coastal areas,
- Forest management.

Rejuvenation of natural resources through activation of watersheds, renewal of wastelands along with enhancement of farm productivity, is a component of environmental infrastructure that is attaining increasing importance as expanding anthropogenic activity stresses natural resources beyond their natural regeneration capability. We focus here on natural resources, common properties, and rejuvenation of rural environment. Here we take stock of the rural environment and propose institutional mechanisms to keep the juggernaut of socioeconomic development rolling without impediments. We present a snapshot of the current rural environment demonstrating the phenomena through which irreversible degradation of the environment has resulted.

We examine the veracity of the widely held position that poor social and economic conditions of villagers compel them to overly exploit the environment, leading to a vicious circle of degradation of natural resources perpetuating poverty. Possible policy and technological solutions to pre-empt environmental degradation, restore, and rejuvenate the degraded environment are discussed further on in the chapter.

STATUS OF THE RURAL ENVIRONMENT

The ecosystem within which all rural activities are conducted encompasses the air we breathe, the water bodies surrounding us, and the land we walk on. Unfettered human activity can compromise the ability of the environment to support our kind, a condition usually referred to as environmental degradation and detrimental to our future.

India supports approximately 16 per cent of the world population and 20 per cent of its livestock on 2.5 per cent of its geographical area, making its environment a highly stressed and vulnerable system. The pressure on land has led to soil erosion, water logging, salinity, nutrient depletion, lowering of the groundwater table, and soil pollution—largely a consequence of thoughtless human intervention.

The extent of land degradation, the loss in capacity of our major water reservoirs and the decline in water level in wells in the past few years is larming. Soil erosion from overgrazing, and intensive cultivation and soil degradation from excessive use of agricultural chemicals, have wide-ranging implications. Agricultural activities that cause land degradation include shifting cultivation without adequate fallow periods, absence of soil conservation measures, cultivation of fragile lands, unbalanced fertilizer use, faulty planning or management of irrigation.

Improper agricultural practices are usually observed under constraints of saturation of good lands and population pressure leading to cultivation of 'too shallow' or 'too steep' soils and ploughing of fallow land before it has recovered its fertility. Overgrazing and over-extraction of green fodder lead to forest degradation through decreased vegetative regeneration, compaction of soil, and reduced infiltration and vulnerability to erosion. Annual environmental costs for India in 1995 were estimated at US$ 9.7 billion, of which surface water pollution, land degradation and deforestation contributed 84 per cent. Deforestation with shifting agriculture, over-exploitation for fuel wood and timber collection, and mining activities are also causes of serious concern. Deforestation causes degradation when the land is steeply sloping, or has shallow or easily erode-able soils, and when clearance is not followed by good management.

Over-cutting of vegetation to obtain timber, fuel wood and other products is frequent in semi-arid environments, where fuel wood shortages are severe. Overgrazing causes a decrease in vegetation cover which is a primary cause of erosion. According to the data provided by the National Remote Sensing Agency and Forest Survey of India based on satellite imagery, 80 Mha of 142 Mha of land under cultivation is substantially degraded and about 40 Mha of 75 Mha of land under the forest department has a canopy density cover less than 40 per cent. Perhaps the most widely recognized environmental problem is the pollution of water resources by industrial discharge, household waste, sewage, and agricultural chemicals. Water scarcity induced by mounting population density and growing economic activity in the face of fixed resources, depleting water tables, and silting of reservoirs has led to rapid decline in the quality of life in rural India. Existing irrigated areas are displaying serious water-stressed situations, as both reservoirs and groundwater sources continue to get depleted.

Consequently, the agricultural output from irrigated areas also seems to be more vulnerable to weather shocks than earlier. The problem is compounded by the fact that provision of cheap power encourages farmers to use excessive water. While this problem is widely acknowledged, a holistic policy framework to address the problem effectively is missing. India's biodiversity is gradually narrowing. Maintaining viable populations of species—whether plant or animal—is crucial in biodiversity conservation requiring the protection of important ecosystems, habitats, and the ecological processes of which they are a part. The loss of biodiversity in shrinking forests as well as in threatened marine and wetland ecosystems has strong adverse impacts in store.

ECOLOGICAL FOOTPRINT

Ecological footprint analysis is an accounting tool that estimates the resource consumption and waste assimilation requirement of a defined human population or economy, in terms of a corresponding land area. Such an analysis encompasses many human needs and aggregates them into a single figure, expressed in global hectares per capita per year, which can be used to shape environmental policies. Because of the complexities involved in determining an eco-footprint, very little analysis has been undertaken at the rural scale globally, although country eco-footprints are estimated and published annually by the World Wildlife Fund. The global eco-footprint, based on 2001 data was 13.5 billion global hectares, which exceeds the global bio-capacity by 21 per cent, or 0.4 global hectares per person.

Although India's eco-footprint over the last forty years is almost constant at around 1 global hectare per person (compared to the global average of 1.8), available bio-capacity per person has fallen as its population has almost doubled. The consumption patterns of the large middle class, residing mainly in urban areas will shape its eco-footprint in the years to come and it is here that the rural areas have a role to play in holding the footprint down to reasonable levels by engaging in sustainable farming practices. It must also be noted that the footprint is a very human-centric concept, talking about hectares of land to be set aside for humans without any consideration for other species.

IMPACT ON HUMAN HEALTH

Globally, among the biggest dangers from farming is the continuous exposure to and the unsafe use of chemicals necessary for agriculture. In India, however, the danger to human health from such environment and pollution related causes are not given their due importance as accidents from farm machinery, with a fatality rate of 22 per 100,000 farmers.

Fatality apart, chronic exposure to air and waterborne chemicals can have adverse health effects, which sometimes, can be difficult to measure because of problems in isolating individual chemical effects.

While certain cause and effect relationships are not easy to identify, cumulative effects are likely to be most critical. Cancer risk could be high from nitrate, metals, as well as pesticides; other problems like adverse hormonal functions, liver damage could also take place, as summarized in Table. Moreover, toxic chemicals and pesticides in air, water, and earth enter body tissues and breast milk, through which they are passed on to infants.

TABLE: Impact of Agro-Chemicals on Human Health

Agro-Chemicals	Maximum Contamination Levels	Health Effects	
		Established	Potemtial
Nitrate	10mg/NO3-N	Methaglominemia	Nitrate conversion to nitrite and N-Nitroso compounds affecting thiyroid, endocine functions
Metals (Cd. As. Cr. Zn. Cu) Pesticides (chloninated hydrocabons like DDT, organophosphate like Malathion, Catbamata like Carbaryl)	0.01 to 0.05 mg/L	Impaired kidney functions, skin disorder, tumour Affects nervous system and enzymes reproduction, affects and muscles	Cancer risk cancer risk

For example, the impact of spraying endrin and endosulfan on cashew plantations in the 1970s to combat pests in the Kasargod district of Kerala has been so persistent that in one village alone, 156 cases of health disorders were recorded from 123 households between 1990 and 2001. These included 49 cancer cases and 23 mental retardation cases among several others. Moreover, endosulfan residues, several times the minimum permissible limit, were found in water, milk, vegetable, and fish, putting the village communities at risk even three decades after spraying of the pesticides.

The State of Kerala's Environment

The 'Kerala model' of development has achieved remarkable milestones in the fields of health, literacy, housing, and so on, and earned worldwide praise for the quality of life of its people. The state is now poised for a quantum leap with greater involvement of the common people through what is now called as the 'people's campaign'. It has identified sectors such as tourism, aquaculture, power production, and software production as priority areas for people's participation. Kerala has long been described as 'God's Own Country' in tourism circles.

However, the state of Kerala's forests and rivers is alarming. The total forest area of the state has come down from 11,241.97 ha in 1994 to 11,236.06 ha in 1996: this means 6 ha of forest has vanished in two years. The forest cover in the Periyar catchment area alone is reported to have dwindled by two-thirds over the last hundred years. Deforestation and encroachment by settlers, with large-scale illegal felling of trees by private estate owners and others, have received frequent media attention but hardly enough policy attention. Soil erosion has resulted along with change in climate patterns, as well as movements of wild animals, leading to increasing conflict between man and animal, and disappearance of rare and endangered flora and fauna. Destruction of the forest and tree cover in Nelliampathy hills have resulted in growing incidence of poaching. Ground-clearing for plantations have led to the destruction of rare and endangered orchid species. Kerala's rivers are assailed by pollution, sedimentation, sand mining, and constriction of flows. Garbage and animal wastes from slaughterhouses as well as sewage water are released directly into rivers. A recent survey of twelve major rivers of the state, including the Periyar, the Bharathapuzha, the Pamba, the Chaliyar, and the Valapattanam, highlighted presence of high levels of coliform bacteria. The problem was reported to be the most serious in the Pamba with pilgrims causing most of the damage.

Soil erosion, caused by deforestation and various agricultural, mining, and construction activity, has led to large-scale siltation of the water bodies. The problem is particularly acute in the case of reservoirs. It has reduced their water-holding capacity and affected the utility of the dams. Siltation at the mouths of rivers causes, besides navigation bottlenecks, flooding and damage to the environment. Siltation in the Vembanad Lake seems to have reached a critical stage.

It threatens to turn the lake into a marsh in a matter of years. The damage to the ecology of the lake caused by siltation and water pollution is appalling. Sand mining, though advocated by some as a solution to the siltation problem, is in fact a threat to the ecology of the river systems. Apart from causing bank erosion, flooding, salt water incursion, and pollution, it is also causing drought in the valleys. A combination of these factors has adversely affected the water ecology of the state, exposing fish to diseases, mass mortality, and loss of habitat.

The indigenous fishery for mahaseer, the major game fish of the state, is reportedly dying out. Damming and channelisation activities are virtually killing the rivers of Kerala. The Periyar has been dammed at fourteen sites impacting its water flow such that in several places the river has become virtually invisible. The fate of the Bharathapuzha is even worse, with a larger number of dams cutting its tributaries. The river is dry in the lower reaches in summer, making it an easy prey to sand mining.

CAUSES OF DEGRADATION

The rural environment has been straining to cope with anthropogenic demands on its services, both as a source and a sink, for the last few decades. As seen from some of the parameters reported in On one hand, as human productive capacity has gone up, whether due to the green revolution or rapid industrialization, so has its ability to generate waste.

On the other, there is a growing demand on nature's ability to provide life support as the population keeps growing and livelihood opportunities decline. We could look at this double squeeze on nature in the context of water resources. Water applied to the field in irrigation either seeps through to underground aquifers, or reappears as 'return flow' and finds its way back to the surface (regeneration).

Pesticides, effluents include pollutants and untreated sewage, they find way into water bodies, which in turn leads to declining biodiversity. Excessive pressure on the environment leads to drought-proneness in certain areas owing to declining water table levels and flood-proneness in others owing to silting of reservoirs and loss of forest cover. A much generalized cause of environmental degradation is the failure of the governments to formulate appropriate policies to ensure sustainable land and water use. Such policy failures include price distortions through government controlled prices, subsidies or taxes which give incorrect price signals, faulty delineation of property rights regimes and other legal structures, government projects which directly cause environmental damage, and weak public institutions.

Furthermore, state appropriation of property rights has undermined traditional (often communal) property regimes, as in the case of our forest policy, and has in several cases led to de facto open access and resource degradation. Some specific causes for environmental degradation are discussed next.

Urban Spillovers and Industrial Pollution

Rapid industrialization and unplanned urban growth have resulted in heavy discharges of toxic chemical effluents and untreated municipal wastes to various water sources including streams, rivers, and tanks causing serious damage to water quality and contamination of groundwater. As human waste is flushed away into sewers and released in waterways, solid waste is collected and dumped beyond urban boundaries.

While smoky fuels have been replaced by electricity leading to radical increases in fuel efficiency as well as declines in respiratory hazards posed, it cannot be denied that this has shifted the location of environmental stress away from the city to its fringes without any policy counter-measures to relieve this stress. The strain on the environment is manifested directly in water quality, which in turn affects the soil quality and has serious impact on crop yield as well as human health. Gastro-enteritis and diarrhoea are common occurrences in villages located in periurban fringes owing to the poor quality of drinking water.

To understand the impact of industrial pollution on the rural environment, a detailed study was carried out in a village in the Patancheru region of Andhra Pradesh. The activities of the Patancheru industrial cluster have led to pollution fall fifteen irrigation tanks in the area, with five of them being completely destroyed. Farm lands have been rendered unproductive and the water unfit for drinking. The inhabitants of the hinterland are compelled to use this contaminated water owing to the paucity of water supplied. They suffer from skin infections, joint pains, and fever. Even the fish population in the nearby Manjira river has dwindled as a result of the polluted streams flowing into it.

Rural Anthropogenic Activity

The agricultural sector accounts for over 85 per cent of the total water consumption in the country. Subsidized or free supply of power and water has resulted in over-exploitation and inefficient use. Wastage in irrigation occurs to the extent of 45 per cent by seepage through unlined channels and about 15 per cent due to excess application. Losses in storage are estimated at 15 per cent.

Inefficient use of canal water and groundwater accompanied by neglect of drainage has resulted in waterlogging and salinity. The share of groundwater in the net irrigated area has increased from a third in 1965–6 to half at present to either supplement surface deliveries of water, or to provide irrigation water when limited or no surface water supplies are available. Groundwater authorities attempt to regulate groundwater withdrawals through licensing, but do not define any limits for withdrawals.

Moreover, inadequacy of resources with the monitoring agencies leads to weak enforcement. Over-extraction exceeds recharge in areas where groundwater is used for agricultural purposes. This has led to progressive lowering of the water table affecting the economy of water. Groundwater overdraft, leading to a long-term decline in water levels, has adverse consequences associated with it. These are land subsidence, deterioration of water quality in aquifers, and ingress of saline water in coastal aquifers. Salinity is likely to render more land unfit for biomass production, especially in the semi-arid and arid regions of Uttar Pradesh, Haryana, Punjab, Rajasthan, and Karnataka. Here, rainfall and available water are insufficient to leach salts from the soil.

Application of chemical fertilizers, which has recorded an increase of 80 per cent between 1984 and 1996, has added to the salinity problem. This has led to increased pH level of the water and poor drainage. Excess salts can also lead to oxygen depletion in due course of time.

Unbalanced and excessive use of chemical fertilizers and pesticides in intensive agricultural practices has led to the degradation of soils and water.

In fact, the use of pesticides has risen by 240 per cent between 1971 and 1995. Fertilizers and pesticides cause acidification and eutrophication of water. Leaching of chemicals (particularly nitrates) from fertilizers results in eutrophication precipitating excessive algal growth, which kills fish. While eutrophication is one extreme, several rural regions have a net negative balance of nutrients and suffer from a gradual depletion in the level of organic matter, a trend that is likely to continue.

Application of sodium and potassium based fertilizers leads to deficienciesof macro and micro intensity, reduced biomass production and increased suspended sediments, which, in turn, adversely impact stream flow.

Maintaining the nutrient balance and preventing nutrient deficiencies is a major concern given that the increasing demand for food production will have to be met through higher intensity of cropping.

Acidification, usually by oxidation, also affects the aquatic biota, pH and water quality. It also increases surface run-off. Soil erosion has led to loss of topsoil and terrain deformation. Moreover, siltation, an off-site effect of erosion, is reducing the reservoir storage capacity by 1–2 per cent annually. Exposed soil surfaces are susceptible to erosion during rain storms. Loss of soil particles and organic matter not only leads to oxygen depletion, but affects turbidity by suspended sediments on water, and influences stream channel morphology.

In addition, credit constraints have often led to an underinvestment in suitable agricultural technology and land conservation, which in the long term has resulted in falling yields from declining land fertility. Subsistence pressures lead to an outward migration to marginal areas when no land is available. The major consequences of such area expansion of agricultural production are deforestation, loss of biodiversity, encroachment into wildlife parks, reduction of grazing lands for livestock, and soil erosion with on and offsite effects.

Cropping may expand onto lands which are fragile and incapable of sustaining production even with conservation efforts. This may be the case on steep slopes with shallow soil, as in the hilly regions of the north east, which would explain the high extent of degradation in this region. The impressive growth rates achieved in agricultural production in Punjab due to the green revolution are well known.

However, recent reports reveal disturbing trends about the future of the green revolution. Productivity of rice and wheat which occupy almost 70 per cent of the state's gross cropped area has reached a plateau.Problems like declining water tables, waterlogging, and salinity have reached serious proportions. The water table in central Punjab is depleting at the rate of 0.23 cm per annum.

In some parts, the water table has risen by four to five meters between 1984 and 1994 leading to problems of salinity and water logging. Widespread deficiency of micronutrients has appeared in the soil and there is an increase in weed infestation, pest, and disease outbreak. Burning of combined harvested rice straw is resulting in serious environmental pollution. Household activities such as unhygienic sanitation practices lead to the contamination of waterbodies and the spread of disease. Similarly, the use of biomass fuel for cooking and heating can give rise to indoor air pollution that threatens health, especially that of women and young children, who spend disproportionately more time indoors than men. The need for fuel wood leads to the felling of trees beyond their natural regeneration rate.

Over time, this leads to deforestation accompanied by a loss of vegetative cover. The fertile topsoil which is held together by the vegetation now starts eroding and lack of trees also result in a gradual lowering of the water table.

Given the rate of population growth, the capacity of agriculture to absorb increased working population is limited. Thus, unless there is a substantial increase in off-farm jobs and opportunities, the pressure on natural resources as a means of sustenance will keep increasing. Over fishing and overgrazing beyond the natural regeneration rate of water bodies and grasslands are all manifestations of the extreme dependence of rural population on natural resources for survival. Moreover, with the outward migration of agriculture, there is pressure on grazing lands, leading to a competition between crops and livestock.

Vanishing Common Property Regimes

The breakdown of common property regimes due to policy failures and social and political conflicts, in combination with rapid population growth, has increased the pressure, particularly on forest resources and grazing lands.

This has resulted in expansion of cropping areas and increased stocking oflivestock, mostly in highly fragile or marginal lands. As 'more users' increase the ambiguity in the custody of the common property, it is overexploited and in the long run becomes completely degraded and unfit for use. With the 1980 Forest Conservation Act, tribal populations living in thc forests for centuries were suddenly devoid of all use rights.

With vast tracts of forest areas being declared national parks and sanctuaries, the tribals have no option but to break the law to earn basic subsistence. Unfortunately, government response to such encroachment has been strict policing rather than leading to deterioration of forest land rather than involving the tribals in forest management. Joint Forest Management, the process of involving tribals in maintaining forests, is still limited to small pockets of the country and there is a dire need to expand such programmes.

Government Failure in Managing the Ghasnies of Himachal Pradesh The Sirmour district in Himachal Pradesh lies in the outer western Himalayan Range. Mixed-farming is the main occupation in the mid-hills and people combine private resources, state-owned, and commonly-owned resources in their day-to-day functioning. Until 1974, grazing lands or ghasnies, part of the forest surrounding villages, river beds, irrigation channels (collectively called shamlats) and common paths belonged to the Panchayats and were used for the common purpose of the village.

On account of grass resource seasonality, customary rule imposed on graziers restricted access to grasslands for a period of time. The rule in practice was to prohibit access to the grasslands during the rainy season until it could be collectively harvested in September. Regulatory in nature, these informal restrictions were monitored and enforced by mutual agreement among the user group. The size of the herd allowed to graze the commons in summer was not a binding criterion. However, households were entitled to graze as many animals in summer as they were able to feed in winter period (when extra source of fodder is not available).

This rule aimed at preventing commercial rearing which could take place during the grazing season. Working within the limits of its labour force, a single household was entitled to remove as much as it could from the common pool resource as long as these benefits were not cash or market-oriented. In this way, there were 'social boundaries' on households' usage of the ghasnies. Due to increased concern over the environment in the 1970s, the State Government enacted legislation through which common land ownership by the Panchayat was transferred to the state unless the land was 'subjected to partition between individual co-sharers before the Act commenced'.

This led to powerful farmers securing individual access to and benefits from the commons before nationalization. This change in access to the village commons had several negative outcomes. Large tracts of the newly private pastures were increasingly converted into cultivated lands. Agripastoralists excluded from 'wild privatisation' were forced to graze their cattle in the forest with intensified green lopping on selected tree species such as Ban and Moru, which have now almost disappeared from the region. Forest lands, which are de jure public lands, are now used as open to all.

The remaining parts of the shamlat are vested with the state which has bifurcated its use into grazing lands and allotable pool land for cultivation to land-less labourers. Productivity of these lands are abysmally low. Finally, encroachments and further political regularization have been important factors responsible for both the shrinkage of grazing lands and consequent intensive use of forest for lopping and grazing.

This example illustrates how inappropriate policy can have adverse ecological effects. The de facto partition of common grazing lands has resulted in a reduced commons area, which led to overgrazing and gradual destruction of forest lands. Moreover, control by the state has resulted in numerous possibilities for the individual to escape both legal as well as customary systems of sanctions which makes the rules-in-use unviable. Under these circumstances, collective action can emerge as an efficient alternative to 'privatization' and 'nationalization'.

Poverty and Environment

Natural resources are the most significant source of rural livelihoods in the country. Not surprisingly, some of the poorest regions in the country are also places where agriculture is least developed and land and water resources are poorly managed. Conventional thinking on environment blamed the poor for over-exploitation of natural resources, as poverty and environment were considered linked in a 'downward spiral' in which poor people, forced to overuse environmental resources for their daily survival, are further impoverished by the degradation of these resources. Population growth and economic change (which often bypasses the poor, or reduces their access to natural resources) are also seen to contribute to this process. It was, therefore, believed that poverty needs to be eradicated in developing countries before they can turn their attention to environmental protection. However, the perception of the 'vicious circle' as characterising the environmental degradation and poverty in developing countries is somewhat simplistic and misleading. For example, if one looks at history, there was not much environmental degradation even when poverty levels were much higher.

Now that poverty levels are declining significantly, it does not seem plausible to attribute environmental degradation to poverty. Evidently, other factors such as property rights, access to resources, use of traditional wisdom, and misappropriation of environmental resources in the name of development play a more important role. As the poor are dependent on nature for their livelihood, they are very vulnerable to natural calamities, environmental degradation and ecological disasters (which are often man-made, like the Bhopal Gas Tragedy, and pollution caused in the river Yamuna by industries and upstream rich farmers).

There are cases of destructive development which aggravate poverty and contribute to environmental degradation at the same time. Thus, there is enough empirical evidence now to establish that environmental conservation must go hand in hand with economic development because any economic development which destroys the environment will create more poverty, unemployment, and disease. However, it is disturbing that neither concern for establishing equity nor the desire to preserve the environment underlies the recent interest in sustainable development among the most developed countries.

Rather, the primary concern continues to lie in sustaining their consumption levels and in maintaining the conditions necessary for economic growth. In contrast, the environmental priorities of developing countries are underpinned by grim and undeniable human realities. In many rural areas in India, environmental issues are issues of life and death. The protests of the people in these areas are responses to social injustice and signal the rising political and ethical consciousness in rural India. In fact, if poverty is indeed the major cause of environmental degradation, the poor actually have a good reason to support conservation movements.

Unplanned Development Projects

The environmental impact of unplanned development projects which includes construction of dams, taking over of huge tracts of land for mining, or converting natural forests to forest farms for wood pulp generation, vary in intensity but have some common and inescapable consequences. When implemented without proper cost-benefit analyses these projects result in millions of oustees, violent disturbance of pristine areas, varying degrees of submergence of land including forest land, adverse impact on flora and fauna leading to a reduction in bio-diversity, and severe impact on the fish population in rivers. Stilling of flowing waters leads to temperature stratification, variations in nutrient content and dissolved oxygen, rendering the water inhospitable to aquatic life.

Building dams, in particular, results in drastic changes in the downstream river regime, including reduced flows affecting aquatic life and riparian communities, reduced capacity for self-regeneration, increased pollution levels, reduced recharge of groundwater aquifers, and adverse impact on estuarine conditions. Acquiring land for development projects is not a new phenomenon and resettlement is usually limited to giving cash in exchange. The displaced are then left to go ahead with their own rehabilitation plans.

These people live at the mercy of nature and disease. Their appalling living conditions, housing, health, and nutrition are an insult to the notion of equity. There need to be proper models of analysis that focus on mitigating risks and formulating alternative strategies. There should be greater participation of local people in the decision making process. Only when people have the power to influence the decisions of the management from the very initial stage will development activities stand a long-term chance of success.

The draft Scheduled Tribes (Recognition of Forest Rights) Bill 2005 would be a step in the right direction if it were to be passed as an Act.

MANAGING THE ENVIRONMENT

Policy Changes

Environmental concern in India is inextricably linked with peoples' lives and well-being. Several parameters currently exceed threshold limits of tolerance to the extent that some environmental problems pose real and immediate threats to health and livelihood. Given that the most serious environmental issues in rural areas emanate from improper agricultural practices, it is essential to integrate agricultural and environmental policies.

Specific policies targetting environmental conservation milestones should be built into sectoral and macroeconomic policy targets. Unless the environmental checks and safeguards are institutionalized with the provision of other rural infrastructure such as roads and communication, the socio-economic condition of rural people cannot improve in a sustainable way. While poor environment undermines development, inadequate development results in a dearth of financial resources which can be devoted to environmental protection.

The policy approach to agriculture, particularly in the 1990s, has been to secure increased production through subsidies in power, water, and fertilizer, and by increasing the minimum support price. The creation of new capital assets in irrigation, power, and rural infrastructure has not received priority.

The fertilizer subsidy is, in fact, meant to cover the high cost of the fertilizer industry. Excessive use of subsidized fertilizer has created an imbalance between Nitrogen, Phosphorus, and Potassium in the soil while excessive use of water–given the 'free' availability of power–has resulted in water logging in many areas. Under-pricing of power and irrigation, rather than bringing in improvements in rural income distribution, has led to inefficient utilization, with negative environmental externalities. The subsidies have grown over the years and are now financially unsustainable.

With deteriorating state finances, subsidies have 'crowded-out' public agricultural investment in roads, irrigation, and expenditure on technological upgrading. The equity, efficiency, and sustainability of subsidized power and fertilizers are, therefore, questionable.

Awareness and Attitudes

Although Environmental Impact Assessments (EIAs) is now a prescribed requirement for all projects, in reality they are often inadequate and unreliably conducted studies, making their cost-benefit analysis a suspect. Even the statutory requirement of a public hearing in relation to the environmental clearance of a project is yet to become a well-established procedure. Awareness and advocacy among the local people is important as they are direct stakeholders in the implementation process of these projects, and community capacity-building is necessary to ensure that these people can seek legal recourse, if required.

The answer, however, does not lie in large, centralized, 'topdown', technology-driven projects: local, decentralized, community-based, people-centred alternatives are available. Problems of scarcity of water, depleted aquifers, declining groundwater tables, and drought proneness have been successfully tackled by water harvesting endeavours in Ralegan Siddhi village in Maharashtra, Sukhomajri in Haryana, and Alwar in Rajasthan.

These are not 'small' instances to be dismissed as one-off phenomenon but examples of significant and sustained success achieved in terms of increased water availability and rise in groundwater table. If these examples could be replicated across India, they could act as key drivers in the process of environmental restoration. Larger projects, also, cannot be ignored, as there are circumstances where benefits from micro-level endeavours may not be enough. The possibility of integrating a large project with smaller surface and underground storages within the sub-basin or watershed must be considered.

A negotiated participatory approach should be adopted in ensuring success of any public intervention, involving communities and land users in creating and maintaining environmental infrastructure. This will leverage local experience and knowledge and ensure maximum involvement.

Decentralization in one form or another has been the primary vehicle for bringing democracy at the local level to facilitate local control. This implies either redistribution of authority from the central government to field delegation (de-concentration), the transfer of specific decision-making and management authority to semi-independent units (delegation), the transfer of authority to nongovernmental sectors (privatization), or strengthening and transfer of power to local governments (devolution). However, there are practical difficulties associated with each of these measures. To manage forest resources for example, community biodiversity registers must be developed before issues of ownership of resources and equitable sharing of benefits are sorted out. Moreover, clear and unambiguous guidelines (to pre-empt any possibility of rent seeking) for managing bio-prospecting agreements need to be drawn up keeping in mind the interests of the locals who derive their livelihood from the forest.

Local communities can be involved in forest conservation by employing them as wildlife guards and forest watchers and by using them as trainers to build anti-poaching awareness among their localities. Indigenous skills can be utilized by employing locals for collecting snake venom and capturing and training wild elephants who devastate standing crops.

Water Harvesting in Alwar: Revival of the Tradition of Johad

A Johad is a dug-out pond, created at a place chosen with native wisdom, informed by remembered patterns of water flow during the rains to harness the rainwater run-off with high embankments on three sides. The height of the embankment depends on the volume of run-off from the catchments. The water storage area varies from 2 hectares to a maximum of 100 hectares.

The water collected in a Johad during monsoon penetrates into the sub-soil and recharges the groundwater, improving soil moisture in vast areas mostly downstream. Apart from arresting and storing rainwater, it stops soil erosion, mitigates flood, and ensures water availability in wells for several successive drought years. The groundwater can be drawn from traditional open wells, built and maintained by the villagers themselves.

The water from the Johad is also directly used for irrigation, watering of domestic animals and other household purposes. During the dry season, when the water gradually recedes in the Johad, the land inside the Johad becomes available for cultivation. This land, by receiving good silt and moisture, allows crops to grow without irrigation.

Johad is built using simple technology and local materials. In the Alwar district of Rajasthan it took three years to build the first Johad. In the fourth year,

Tarun Bharat Sangh, a nongovernmental organization (NGO) actively helping villagers, had built fifty Johads. As on date, 9000 such structures exist catering to water needs of more than 1000 villages.

This area which was classified as 'dark zone' in 1995 was reclassified as 'white zone' in 2005. As water availability improved, agriculture became productive and cattle rearing started in earnest, resulting in increased production of milk. Studies have shown that an investment of `100 per capita on Johad raises village domestic production by `400 per capita per annum.

Non-farm Employment

In addition to policy changes to improve agricultural processes, expansion of the non-farm sector is essential for development. If rural non-farm and urban informal sectors grow at an adequately high rate to absorb the surplus labour and surplus food then the scenario is a positive one. If however, they grow at a lower rate, the terms of trade can turn against agriculture leading to depression of agricultural incomes.

Rural enterprises sector can, in this context, become both an engine of growth as well as a major contributor to the reduction of rural poverty. The potential for labour absorption is high in agro-based industries, small-and medium-scale industries, and the rural services sector. Most of these activities are highly labour intensive and provide employment opportunities for the semiand unskilled rural labour and semi-skilled urban labour force.

When non-farm activities provide stable sources of income to small farmers, they can afford to opt for high-value crops with higher associated risks. Income diversification of farm households and the insurance role of non-farm activities would strengthen the viability of a small farm economy. Thus the income of farmers can be magnified by multiple linkages with the non-farm sector. These are production linkages, both backward, via the demand of farmers for inputs such as ploughs, engines and tools, and forward, via the processing needs of agricultural goods through spinning, milling, canning, and so on.

TABLE: Growth of Employment in Rural Enterprises

State/Union Territories	Annual growth rate in non-fam rural employment	Annual growth rate of rural workforce
Andhra Pradesh	5.37	3.05
Arunachal Pradesh	3.65	3.07
Assam	6.62	2.08
Bihar	4.50	1.79
Chhattisgarh	3.24	3.82
Goa	1.75	2.99

Guiarat	3.11	1.27
Haryana	9.68	8.80
Himachal Pradesh	2.73	2.54
Jammu and Kashmir	7.64	7.65
Jharkhand	3.44	0.66
Karnataka	4.78	2.69
Kerala	7.93	4.21
Madhya Pradesh	1.74	1.69
Maharashtra	4.95	3.29
Manipur	4.46	3.24
Meghalaya	6.48	5.05
Mizoram	8.40	4.96
Nagaland	6.05	1.95
Orissa	3.02	2.54
Punjab	7.34	5.19
Rajasthan	4.15	3.44
Sikkim	8.39	6.41
Tamil Nadu	9.96	5.43
Tripura	9.85	5.84
Uttar Pradesh	7.07	4.98
Uttaranchal	7.72	7.06
West Bengal	4.77	1.70
Andaman and Nicobar Islands	-6.16	-3.90
Chandigarh	15.57	12.11
Dadra and Nagar Haveli	8.65	7.56
Daman and Diu	13.64	15.32
Delhi	-0.91	-2.26
Lakshadweep	1.80	3.53
Pondicherry	3.37	3.83
India	5.53	3.33

Consumption linkages are established with higher agricultural income feeding into an increased demand for goods and services produced in nearby villages and towns. Improving rural connectivity also has an impact on creating employment opportunities. It is fairly common to see tea stalls, bicycle, and tractor repair shops and other establishments clustered around bus stands. Such options can ensure to an extent that rural populace does not need to exploit natural resources for subsistence, contributing to a healthier rural environment.

The annual growth of non-farm rural employment for 2005 exceeded the average annual growth rate of workforce at 2 per cent per annum. The average annual growth in employment for the years 1998–2005 was 2.5 per cent compared to the rate of 1.7 per cent between 1990 and 1998.

While this is heartening, it is important to continue with the thrust on rural employment, through the wide spectrum of employment programmes devised by the central government. In addition, there is a need to increase the value added component of non-farm rural enterprises so as to improve their profits and make them sustainable.

Contract farming facilitates the integration of small farms with agro-processing companies, which extend technical expertise and financial support to the farmers to grow high value crops. Historically, contract farming has been a standard practice in the Indian sugar industry and could be extended to other product groups to trigger the development of agro-processing industries and generate employment for surplus rural workforce. Further, commercialization of agriculture could catalyse growth rates for the sector in a way that agricultural development could diversify into dairying, animal husbandry, fisheries, floriculture, horticulture and other areas. This would spur the growth of agro-processing industries in rural areas to meet domestic as well as export demand.

The tremendous potential of the khadi and village industries to generate employment in rural areas to ease the increasing pressure on the land could be exploited by developing forward looking plans and overhauling the marketing and distribution networks. The lacunae are in the formulation and implementation of the market development plans.

The government or NGOs could play a more proactive role rather than just engage in procurement and selling through Khadi and Gram Udyog network. It is essential to make khadi and village industries sustainable and economically viable. In this context, the focus of the policy should shift from providing subsidies to market development. Provision of market infrastructure services can create incentives for production of high quality marketable goods along with the branding of products.

Decentralization

The notion implicit in the decentralization process is that effective management of local resources requires effective participation of individual users at the community level. People's participation in the decision-making process is an important component of decentralization. The objective of the devolution of power is to involve the target groups of state developmental efforts in a very real sense in the formulation and management of developmental projects and programmes.

The 'local people' is a heterogeneous group defining the primary consumer set for local services. This group is the key stakeholder in the success or failure of grassroot programmes as well as civic and administrative systems in place at the local level. Panchayats, as the form of local government, have existed as a traditional village institution for centuries. In 1992, the 73rd Amendment to the Indian constitution identified areas where these grassroots level selfgovernments could be granted administrative power in 27 areas including agriculture, irrigation, and management of drinking water.

This policy shift towards decentralization promises important social change in rural India, providing as it does a three-tier system of Panchayats: at the village level, the district level, and an intermediate level between the two, called the Block Panchayat. The devolution of power to the Panchayat in the management of natural resources, such as water, is a shift from the centralized and state-driven natural resource management regimes of the past towards decentralized and mainly community based management.

Instead of controlling resources from outside, decentralization could enhance participation in the form of devolution of decision-making power to formal institutions, such as, local governments that possess meaningful autonomy and the capacity to mobilize and spend resources. In particular, if there is a traditional institution of decision-making, then it can be easier and more expedient to use this as a participatory channel rather than creating new committees or institutions.

The whole process, in an ideal situation, enables the local people to debate, negotiate, and have control over the resource. If this really happens, then with time, local governance can become an educative process where the participants learn about democratic norms and practices. Because of the dominant role of natural resources in local livelihoods, it is true that people need to have an effective voice in decisions over the natural resources they depend on. The proponents of decentralization argue that the establishment of local (formal) institutions has the capability to improve people's management and use of common property resources, thereby improving the resource base on which poor people are often disproportionately dependent. It is hoped that through these institutions, participation can better target benefits to the poor through the identification of key stakeholders who are most affected, and can imply an on-going information exchange and discussion through consciousness-raising by shared understanding of problems and a vision for the future that leads to commitment and ownership by the community.

The governance structure, it is hoped, will change as a result of decentralization from centralized to localized, with the 'people' at the centre. Ideally, the higher authorities will not manage natural resources, but through a participatory process, the local people will manage them, thus resulting in a change in the pattern from a 'command and control', to a 'responsive and accountable' operative system.

The new people-centred bottom-up paradigm in development thinking has created the overly optimistic view that decentralization will produce just and equitable outcomes for all, and that engaging the people will also act as a check on state power, thus helping to democratize local governance. The new paradigm stresses the involvement of local people in contrast to the top-down paradigm, and tends to dominate management of natural resources at the local level. Chambers argues that the emergent paradigm for humans living on and with the earth brings together decentralization, democracy, and diversity. What is local, and what is different, is valued. In this paradigm, the trends towards centralization, authoritarianism, and homogenization are reversed. Reductionism, linear thinking, and standard solutions give way to an inclusive holism, open systems thinking, and diverse options and actions.

The importance of traditional ways of combating with problems could be important too: here, informal institutions could be involved. For instance, whenever villagers in Karnataka's Bijapur district sense a drought is imminent, they prepare for war with nature. Harbingers travel from place to place and try to bring rain through magic. Rainmaking may not work but the participants at least endeavour to do something in a situation that would otherwise leave them dispirited and helpless. The accompanying sense of accomplishment enables the locals to cope with drought; it is preferable to sitting idle and wringing one's hands.

Besides the above changes in policy orientation, there is also a need to revamp the existing institutional set-up responsible for rural environment. The current policy environment suffers from lack of coordination and communication between different departments and agencies, the activities of which impinge on natural resources. There is a need to incorporate biodiversity, forestry, and land-use considerations in development planning itself.

A multi-level stakeholder approach for the planning process is essential to obtain socially balanced results in which both the economic and ecological objectives are given due weight. At the district level, a coordination committee headed by the district collector, involving the forest department and other relevant agencies including the District Rural Development Agency could be created to oversee the impact of different projects on natural resources.

Similarly, to ensure that all development activities and actions relating to the utilization, production, or management of natural resources are evaluated before implementation, it may be necessary to create an environmental cell in all the departments with expertise in the management of air, water, land, and biodiversity in the face of developmental compulsions.

Isolated planning and implementation of surface water and groundwater development programmes discourage unitary analysis of water and its use as a single resource, resulting in duplication and ambiguity of functions. Central and state groundwater boards have their own separate observation wells for monitoring water tables, but do not share the data among themselves, and doubt each other's databases. At the same time while the Central Pollution Control Board and the state pollution control boards focus on pollution from industrial and domestic sources, agricultural pollution is overlooked. Institutional transformation towards granting operational autonomy in the irrigation sector is called for in conjunction with the establishment of a multidisciplinary environmental unit to encourage the conjunctive use of surface water and groundwater, for irrigation, which would ensure better drainage in canal command area, thereby checking water logging and salinity.

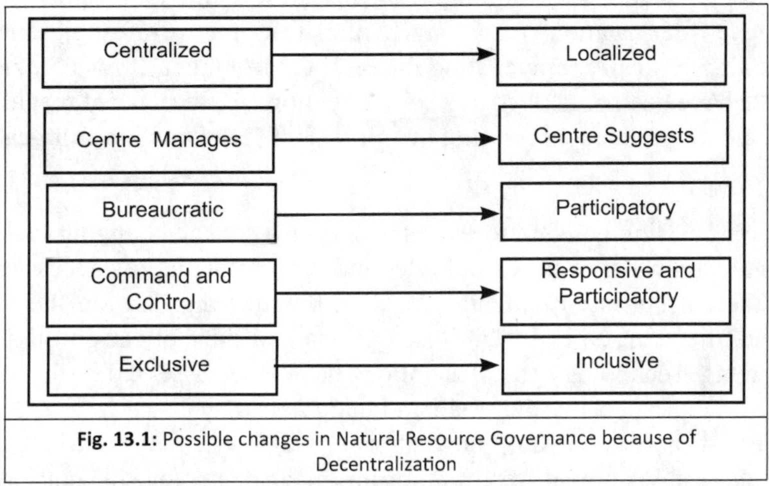

Fig. 13.1: Possible changes in Natural Resource Governance because of Decentralization

Similarly, the utilization of treated wastewater from domestic and industrial sources for irrigation could be promoted. Water management should also be closely synthesized with land management to tackle degeneration of ecosystems regulated by groundwater and surface water.

This should be a crucial aspect of institutional reform along with three other major issues: planning and operation of large surface systems, watershed management, and regulation of groundwater. Indeed, groundwater is one of the country's most valuable natural resources. Despite the urgency to rapidly develop new water supplies, little attention has been paid to or investment made in the preservation of the long term sustainability of groundwater.

Water being a state subject, states are empowered to enact laws or frame policies, but only few have set-up organizations for planning water use and allocation of water for various purposes.

The National Water Policy had recommended the creation of multidisciplinary units in different states for water management and proposed participation of beneficiaries in water management and water price rationalization among others.

However, water management is yet to take off in many states. A major drawback of environmental institutions is that they are slow to incorporate appropriate policies amid a dynamic situation of technological and social change. Environmental policy-making and administrative and implementation mechanisms are subject to the inertia of the historical status quo of special interests.

Improving Managerial Efficiency

While policy and institutions can enable environmental restoration, solutions are often found in technological breakthroughs or suitable managerial interventions. In the following sections, we highlight some of these solutions towards monitoring environmental degradation, managing natural resources, enhancing agricultural growth, and specifically rejuvenating the rural environment. It must be noted that while these have been organized as separate sections for the sake of readability, the solutions are interrelated and cannot be implemented in isolation from each other.

MONITORING

There is a rising concern to examine and monitor the changing health status of soil, land, and water in the rural hinterland. To start with, it is necessary to take stock of the current environmental situation by studying data on the properties of the land, different types of land-use options and their effect on the resilience of ecosystems. Technology transfer options have to be assessed in the context of training farmers, especially the small and marginal.

Projected increase in cropping intensity and consequent nutrient depletion demands the establishment of a monitoring mechanism for soil nutrients. Land capability classification to ensure efficient resource allocation entails completion of land and soil surveys and the development of an inventory of land resources on the basis of land-use classification.

The main objective of collecting land use statistics in India has been to facilitate agricultural accounting. However, the rise in human and livestock population and development of modern agriculture have led to degradation of land, which, if unchecked, may result in irreversible damage. It is important, therefore, that a systematic database be built to assess changes in land use and the type and extent of environmental damage to land resource in each use.

Land use database has to be significantly expanded from revenue and agricultural production assessment to assess and monitor the changes in the ecology and the environment. Data on barren and uncultivated land, already collected by the village Patwari, need to be aggregated as non-agricultural uses have increased substantially and the data would enable environment monitoring and decentralized planning at the district and block level. Soil quality data relating to topsoil erosion, salinity, drainage and waterlogging can be routinely monitored and remediation action initiated wherever needed.

NATURAL RESOURCE MANAGEMENT

Combating pressure on forests would require demand-side management with respect to wood and fodder and supplyside management with respect to enhancing wood yield in forests and increasing the supply of fodder. 'Open forests' representing 26 Mha are already degraded. According to Pachauri and Mehrotra, 15 Mha of this can be managed under the joint forest management (JFM) programmes and the remaining used for commercial timber production.

In fact, nearly 10 Mha is already under the JFM leaving only a remaining 5 Mha to be covered. Of the degraded land earmarked for commercial timber production, 4.2 Mha of scrub (less than 10 per cent cover) can be brought under plantations and 4 Mha currently under cultivable waste and permanent fallow could be used for raising trees. This would imply an increase to about 83.4 Mha under tree cover by 2020.Plantation forestry requires the participation of communities, NGOs, and the private and corporate sectors.

There is a need to increase the area under agro-forestry and farm forestry on agricultural lands by planting trees along farm boundaries and in homesteads in a manner that does not affect crop production adversely. Strategies for promoting agro-forestry should focus on removing legal barriers, bringing about market reforms, fostering farmer–industry linkages, and enhancing the overall profitability of farm forestry.

In addition to managing forest resources, there is a need to manage the requirement of fodder for farm animals and land devoted to pastures. Limiting livestock population through selective breeding and pasture and fodder management could mitigate the pressure on land and forests. Moreover, specific forage resources could be created for livestock so that they do not compete for productive agricultural land. A detailed assessment of fodder requirement at the national and regional levels could be carried out on the basis of which regional allocation for production of fodder can be undertaken.

Encouraging stall feeding to control grazing, providing incentives to farmers to include green fodder in crop rotation, and using crop residue as fodder are alternatives that may be considered. Fodder could also be grown on land under agro-forestry and JFM. Reduced demands on forests for grazing in the long term coupled with wood-demand optimization and strategic forest management will mitigate the pressure on forest resources. This, in turn, will have a positive impact on the country's biodiversity.

AGRICULTURE

Farming systems vary with agro-ecological conditions and no single technical fix will work as a magical cure for improving farm productivity. In some regions, solutions for increasing yields may involve a shortening of fallow periods and extension of cropping periods while in others where soil fertility and/or access to purchased inputs is good, solutions such as annual cropping or multi-cropping without fallow would work.

Again, farming systems based on tree crops, such as tea and coffee, are suitable for some regions only and should be encouraged accordingly. Further, the degree of market integration, choice of crops and cropping systems, use of conservation technologies, and use of purchased inputs and their effects on the farming system, are all important in determining the sustainability of particular farming systems.

REVAMPING AGRICULTURAL EXTENSION

Revival of agricultural dynamism calls for corrective steps to deal with the near collapse of the extension systems in most states and the decline in agricultural research universities. Laboratory-to-land concept should be encouraged and put to practice by providing land-users multidisciplinary technical information and viable land-use options and alternatives identified for various agro-ecological and socio-economic units. Crop combinations and rotations suitable for different agro-ecological regions (as suggested by the Indian Council of Agricultural Research) need to be advocated for better land management.

There is a need to stay abreast with evolving resource conservation technologies and practices and on analysing the conditions and principles of sustainable land use. Efficient use of marginal lands needs to be encouraged and areas of untapped potential developed to ensure optimal utilization. For agricultural diversification to be a major element in the agricultural growth strategy, action on several fronts is called for. Ideally, there should be a shift of land from cereals to non-cereals (increasing both farm incomes and employment) combined with an increase in productivity in cereals to ensure that per capita availability of cereals does not decline.

Improvement in fertilizer application efficiency, integrated with the use of bio fertilizers, to check the degradation of existing resources due to contamination with nitrates could be brought about through on-site farmer training programmes. Success in providing extension services so that breakthroughs in research can be implemented by the farmers entails implementation of extensive irrigation reforms.

BIOTECHNOLOGY

While the possibilities of application of biotechnology to improve the rural environment are limitless, we focus on two specific applications: the improvement of genetic stock of farm animals and forests. Other applications of biotechnology in food and non food-crops like tomatoes and cotton are well known. They have been ascribed more than their fair share of controversy with respect to environmental impact, and these issues are not discussed here.

REJUVENATING THE RURAL ENVIRONMENT

Salinity

Soil salinity has been emerging as a critical problem in both productive and non-productive rural land. While social forestry is a popular means of tackling salinity innon-productive land, salinity management plans in catchments, pastures, and farm areas include a combination of prevention, recovery, containment, adaptation, and pursuing sustainable agricultural practices so that salinity is mitigated over time. The prevention approach involves retaining native vegetation on the land as far as possible to avoid the risk of salinity. Thus, the migration of farmers outwards to marginal areas must be discouraged to retain natural vegetation which can be used for managed grazing of livestock. The recovery approach brings saline land back to full production under its former land use, although this can entail high costs based on currently available solutions. Containment strategies focus on vegetation management so as to reduce the rate of increase in water tables and avoid the spread of salinity. Adaptation refers to changing the land use pattern itself because the soil is saline. This is a good option when there are no cost effective measures for soil recovery and alternate land use such as growing feed for livestock is more profitable than growing crops in the degraded land. Deciding on the right approach to tackle salinity is a measured step combining the interests of ecological balance with socioeconomic needs of the farme. Social forestry has been the most cost effective method to tackle salinity so far. Trees are planted to reduce groundwater recharge, to intercept and use groundwater before it reaches a discharge site, and/or to increase the rate of discharge at a saline site. These approaches work best when they are designed to suit the hydro geological characteristics of the system causing the salinity.

Trees have a higher potential to reduce recharge (compared to that under annual crops and pastures) and to intercept and use groundwater. Strategies that could maximise the reduction in recharge or use of groundwater for a given area of farm forestry include planting on high recharge zones; planting to use groundwater, for example, having tree belts across hill slopes, break-of-slope plantations, and plantations bordering saline water discharge zones.

Trees near annual crops and pastures may be able to reduce the percolation and recharge below the nearby annual plants if they draw water from a zone wider than the tree canopies. This is usually done by planting rows of trees across paddocks, leaving alleys for conventional farming. Also, trees planted on sites where large soil water stores have developed below annual crops and pastures may be able to access this water with their deep roots and prevent it from becoming recharge; this is referred to as phase farming. Trees can also use surface water flowing onto their site if it infiltrates and is stored below ground within reach of their roots.

Rainwater Harvesting

There are many arid zones and drought-prone areas in the country in Rajasthan, Gujarat, Maharashtra, Karnataka, Andhra Pradesh, and Tamil Nadu. Droughts are a recurring feature in these areas, causing untold misery to human beings and livestock, often resulting in large-scale migration. There needs to be a series of area-specific answers by way of local conservation and augmentation to the maximum extent possible to drought proof these areas.

Rainwater harvesting as a method of utilizing rain water for domestic and agricultural use is already widely used throughout India and the world. It is a method which has been used since ancient times and is increasingly being accepted as a practical method of providing potable water. It has wide application also in peri-urban areas where the reliability and quality of piped water is increasingly in question. With most parts of the country subject to a large variation in rainfall leading to spatial and seasonal variation in water supply, rain water harvesting is an ideal tool to take care of local water needs as well as restore the ecological balance of regions through groundwater recharge. Success stories of rain water harvesting abound in villages in different parts of India, a compendium of which can be found at a web site for rain water harvesting maintained by the Centre of Science and Environment.

Common Property Management

Common properties are best managed by social institutional arrangements that enforce a structure of management. Fundamental to understanding the complexities involved with management of a common property resource is the way the government (or the people themselves) form(s) institutions around the resource. The institutions which control decisionmaking regarding natural resources play an important role in providing access for all.

Small group size, well defined boundaries on resources and user groups, and ease of monitoring and enforcement are essential conditions for the success of these institutions in sustaining common property resources and gaining the compliance of generation after generation of appropriators to the rules in use. While informal CPR management arrangements, being socially embedded, have the potential to solve problems of access, they need to work within the framework of formal organizations.

Formal rules could determine the organizational structure and function of a particular informal institution managing a natural resource, and rules and sanctions, while informal arrangements can ensure involvement of locals and sustainability of conservation efforts through cooperative agreements. These organizations are in a better position to analyse the processes involved in conscious inclusion (or exclusion) of certain people in the use of a natural resource.

Of the three solutions that exist to resolve this problem of tragedy of the commons, namely centralized management of the resource, privatization of the resource, and local control of the resource, local control is preferred because it is more inclusive than the other options.

Making local 'user groups' control the common property resource by giving opportunity to people who have a common material interest to participate in management of the resource has been found to be both ecologically and economically efficient.

'Water user organizations', for example, are becoming increasingly popular in India, and are more efficient than the government machinery for managing water resources. However, the application of informal CPR management has, in many ways, been unduly restrictive. It has been focusing on micro-enterprises in management of forest patches at the village level, rather than looking at higher-level forest management and economic processes.

The emphasis has been on the bare bones of participatory development—new rules, one-off participatory planning events, village committees–rather than broader and longer term processes such as the management of trade-offs among diverse objectives, and the need for scaled-up participation. Production of trees, tree products, and fodder has been emphasized upon while broader ecological processes and debates about alternative uses and competing rights and responsibilities have been overlooked.

Information Systems and Accountability

Information Technology and techniques that are freely available, can minimise errors and provide accurate and reliable periodically updated data. Two related issues arc the cost and the system of storing information. Currently, land use data is generated, updated, and maintained by the Patwari at the village level. At each level of aggregation of different types of data, various departments are involved. The information is generated as a by-product of administration and hence, has a low marginal cost. Storage of information is not an issue. However, dissemination of the data should be made as simple as possible so that all departments have access to all the information, which can be viewed holistically before undertaking projects that could impact the environment.

Remediation of Wasteland

Remediation refers to conversion of wasteland which has sustained environmental degradation into land suitable for habitation or cultivation. Approximately 20 per cent of the geographical area of India is now under the wasteland category. Growing demand for fuel, fodder, wood, and food has extensively depleted or eliminated protective plant cover and exposed surface soils to processes of degradation, resulting in partial to complete loss of soil and productivity. As a result the production of vegetation for food and other uses has extended to areas under great ecological stress and with less favourable environment. Several approaches to land remediation have been successfully implemented. For lands that have not sustained too much damage, measures such as building suitable rain water harvesting structures, planting appropriate vegetation and trenching of soil help in improving land productivity. The age old practice of leaving lands fallow for some seasons is also a means of improving productivity of the soil. Jatropha, a valuable crop, owing to its capability in producing bio diesel, is an option which can be chosen to alleviate soil degradation, desertification, and deforestation. The concept of substituting bio-diesel produced from plantations on eroded soils for conventional diesel fuel is gaining acceptance in India.

In recent months, the centre and several state governments have expressed their support for bringing marginal lands, which cannot be used for food production, under cultivation for this purpose.

Management of the Sundarbans

The Sundarbans, home to the endangered Royal Bengal Tiger, cover about 1 million hectares of land and water across Bangladesh and West Bengal, comprising one of the three most extensive mangrove forests in the world. They are the nursery for many coastal and oceanic species of fish. Bio-diversity in the region includes 36 mangrove species, turtles, terrapins, river dolphins, estuarine crocodiles, and several species of birds, many of which are migratory. The Sundarbans are also among the most backward regions in West Bengal. This region is home to about 4 million people, of whom 56 per cent are landless. Literacy rate here is below 35 per cent, and most communities do not have access to electricity or safe drinking water, and agricultural productivity is lower than the state average. Inadequate infrastructure, poor communication facilities, lack of access to clean drinking water and health and education services, and a fragile and limited natural resource base have contributed to a low level of development and high poverty incidence in the region. A diminishing natural resource base is threatening the ecological integrity of the Sundarbans and the livelihood of the inhabitants. 90 per cent of the locals, who depend on agriculture, draw heavily on forest resources to meet their subsistence needs. Thousands of women and children collect wild tiger prawn spawn for sale to shrimp farms, leading to a decline in tiger prawn spawn and concurrent decrease in other fish species. The frequent breaches in the embankments along the riverbanks built to prevent inundation by tidal waters allow saline waters to inundate agricultural lands, destroy crops, and render the soil infertile.

The vulnerability of the region to periodic cyclones forces the local communities to turn to the forest to meet their food and shelter needs. This has resulted in over extraction of wood and non-wood products such as palm leaves, honey, and fuel wood. Low material and technological inputs are constraining agriculture, crab farming, bee keeping, and honey production.

Several initiatives were taken in the past few decades to enhance conservation and development in the area. In 1973, 2585 square kilometres of mangrove forest was declared as a Project Tiger Area. In 1976, three wildlife sanctuaries were created, and in 1984, most of the core of the Project Tiger area was converted into the Sundarbans National Park. The Sundarbans Biosphere Reserve (SBR) was established in 1989 over a larger area.

These initiatives are backed by an enabling policy and regulatory framework, which provides for community involvement in mangrove forest protection and management, and for regulating fisheries. However, effective enforcement of rules and regulations suffers from lack of inter-agency coordination and insufficient enforcement capacity. The extreme poverty in the area led the state government to form the Sundarbans Development Board (SDB) in 1973.

The Sundarbans Affairs Department (SAD) was created in 1994 to give further impetus to socio-economic development in the area, and SDB is now under this department. During 1981–1990, SDB implemented a US$ 9 million integrated rural development project funded jointly by the International Fund for Agriculture and Development and the state government. In 2000, SDB's US$ 4 million development budget focused on civil works and infrastructure development, afforestation, agriculture, fisheries, and small-enterprise development.

The Forest Department (FD) implements central and state-sponsored schemes focusing on mangrove forest rehabilitation, fuel wood and fodder development, and community development. Other state departments including irrigation, public works, public health engineering, and several nongovernmental organizations (NGOs) focus on socio-economic development, mangrove forest restoration, and wildlife conservation.

The United Nations Development Programmes and World Wide Fund for Nature have proposed initiatives to focus on reproductive health and population issues related to Sundarbans conservation; fostering stronger cross-border cooperation, and collaboration between conservation managers in India and Bangladesh; and tiger conservation.

Yet another alternative is the use of mycorrhizal fungi for soil remediation. These fungi are likely to be most beneficial in diverse nutrient deficient ecosystems where the proportion of plants able to form Mycorrizhas is high.

Benefits from use of the fungi include better plant growth by increasing nutrient uptake, efficient recycling of nutrients, and stabilization of the soil in the long term.

Rehabilitation trials are on with Mycorrhizal inoculations for a timefram of eight years at the field station at Gualpahari (Haryana), to yield significant improvement in biomass and nutritional status, while eliminating use of chemical fertilizer completely. Another outlook on reclamation is to use fly ash as a special kind of soil. However, the ash characteristics, particularly, its heavy metal composition, warrants the selection of plant species best suited for reclamation.

WAY FORWARD

The degradation of our rural environment is palpable, albeit gradual. Even though governments acknowledge the gravity of environmental concerns, they have tended to divorce economic growth targets from environmental impact assessment. There is a need to change the policy focus which, so far, has been on a curative (if such a thing is possible) approach to environmental problems and hazards, to a preventive approach. Such an approach would include some of the following measures. For better water management infrastructure there has to be a comprehensive policy integrating water planning with overall economic and social planning. The pursuit of particular goals in water development should also aim at strengthening thc objectives in the management of other resources. This implies developing multipurpose projects with possibilities of adding hydroelectric power production and development of irrigation. Second, the economic value of water must be recognized. Water should be a priced commodity. Third, local people need to be consulted on goals and strategies in water planning. This is important for rural areas and participation must take place from the first stages of decision-making. Fourth, impact assessment should be a must for any water development programmes.

This can bring to attention the threats to water quality as also to human welfare and social stability. This will also help unite the science of environmental analysis with the politics of resource management. In particular, we need to pay attention to the regeneration and revival of old irrigation systems and projects. In the dry land areas of the country, appropriate watershed development is critical.

We need to bring wastelands and degraded lands into productive use, either under crops or under agro-forestry, and to improve credit flows to our farmers through innovative methods. Technological interventions are essential to improve agricultural productivity and to widen the range of products. Institutional structures governing rural activities also need to be reformed and strengthened. It is necessary to evolve a new approach to agricultural policy based on a careful assessment of current constraints and possibilities. Both land and water are crucial constraints for expanding production in agriculture. The extent of forest cover has declined alarmingly. Thus, there is little possibility of increase in the cultivated area in the country, and indeed perhaps an eventual decline as urban demand and environmental imperatives lead to conversion of some agricultural land. There is, therefore, no alternative but to focus on raising the productivity of our land in a sustainable manner.

For this there is a need to develop and disseminate agricultural technologies. Over the years India has developed an extensive system of agricultural research centres and extension services. However, their efficacy is in question. Strengthening the agricultural research and development system and a significant improvement in the sophistication of the technology dissemination methodologies are essential to achieving rapid and sustained growth in agricultural productivity. An increased focus on subsistence crops and technologies in rainfed/dry land areas would be appropriate in this regard. Every effort needs to be made to bring presently uncultivated land into productive use, whether in agriculture or in forestry. For this, it will be essential to evolve a comprehensive land-use policy which will lay out the contours of the ownership and institutional framework that will encourage the productive utilization of such land.

Potential policy options for agricultural watersheds would include controlling sources, and thereby minimizing nutrient and waste input, while processing and moving surplus manure to nutrient-poor areas. Regulation of animal stock densities has to be ensured. On-farm nutrient budgets could be promoted by public institutions and community involvement. Finally, educating the people is important, and has to be done with the involvement of various stakeholders.

❐

16

Water Vapor: Distribution and Trends

It long has been recognized that the first thing to do in cooking a rabbit is to catch the rabbit. Similarly the distribution of water vapour requires that first there shall be water vapour that may be distributed. And this raises questions about that water vapour. Where does it come from? How much is there of it? Where does it go to? How does it get there? Well, no matter what was the origin of the first water vapour about the earth, nearly all of it that at any time is now in the atmosphere got there as a result of evaporation, mainly from water surfaces, snow and ice sheets, damp soil and growing vegetation.

EVAPORATION

In picturing to ourselves the mechanism of evaporation, it is convenient in respect both to gaseous and liquid water (ice presently will be mentioned separately) to think of the ultimate particles or molecules as moving about among themselves along exceedingly short paths (except only when the vapour and all its accompanying gases are together extremely rare) in every which direction and with widely varying speeds with reference, in the case of each particle, to some point in the direction of its path and which we regard as fixed in position. This may sound a little finical, but it isn't, for really we must have something to tie to, as we say. If we measure the velocity of anything it must be in terms of its rate of approach towards, or departure from, something else that for our particular purpose we can consider as at rest, whatever the motion of the thing "at rest" may be in respect to a third object, or of that in turn to still another, and so on indefinitely. For instance, one may consider his face as being at rest while his hands are in motion with reference to it as he eats in a dining car; or the car at rest and himself, face and all, in motion as he comes in or goes out; or the ground at rest while the car is in motion over it; or the sun at rest while the earth has several kinds of motions; or the stars at rest while—but there is no end to this and we would better stop while stopping is good.

The point is this: If we talk about velocity we always must think of, or measure, that velocity in terms of rate of movement directly towards, or directly away from, something else, which for our then particular purpose we may regard as being fixed in position. If this little digression has helped us to think straight about "velocity," so that we know what we mean, we then may return to our consideration of the linear movements (they also spin) of water molecules. In the case of an open vessel of water, for example, the sides and bottom evidently may be regarded as at rest with respect to the moving molecules, and so too may the surface of the water, if it is not agitated. The same also is true of reservoirs, ponds, lakes and other bodies of water.

As just explained, we think of the molecules of water, and also of those of every other sort of liquid or gas, as being continuously in a state of utterly confused migration both as to direction and speed. Furthermore, the greater this average speed the higher the temperature. This is the right way to look at it, but if we wish to do so, and it often is convenient, we also may say, the higher the temperature the greater the average speed of the molecules. Now myriads of molecules at and next to any free water surface, for instance, are moving outwards, and many of them with velocities so great as to carry them into the space beyond and away from that restraining force, whatever it is, that tends to keep them within, and a part of, the body of the liquid.

Outside the liquid the space per molecule is greater, in fact greater without limit, except as restricted by enclosing walls, as in a closed vessel, or by the force of gravity. Hence its free path, the distance it travels from collision to collision, also is vastly greater outside than within. All restraining force of cohesion then is gone and the swarm of free particles is a gas, or, in this case, to be more exact, a vapour, at least at ordinary temperatures—a gas which, at its current temperature, can be compressed to a liquid.

Since great velocity of molecules means high temperature, and small velocity low temperature, and since the greater the velocity of a molecule in a liquid the better its chance to get away from cohesion restraint and become a gas molecule, it follows that evaporation, the changing of a liquid to a gas, decreases the average velocity of the liquid molecules—the faster getting away while the slower remain behind—and hence lowers the temperature of the remaining liquid, or at least keeps it from getting as hot, or hotter as fast, as it otherwise would.

Not only do the molecules in liquid water get away in great numbers and become gas molecules, but also many such gas molecules, in the course of their randomly directed flights, plunge against every water surface with such velocity that they merge with and become a part of the liquid. This is condensation, the reverse of evaporation, more fully explained later. Both processes occur simultaneously, while the circumstances of temperature and pressure determine which shall be the faster of the two. By "evaporation" we generally mean net evaporation, or excess of evaporation over condensation; while by "condensation" we commonly mean net condensation—excess of condensation over evaporation.

Perhaps a duly inquisitive reader may note that nothing at all definite has yet been said about the nature of the forces that keep a liquid just what it is—a liquid—and prevent it from exploding violently into a gas.

To be sure a liquid can be made to explode by heating it to a high temperature under pressure and then suddenly removing that pressure. But under ordinary circumstances a group of liquid molecules does not explode into a group of gas molecules of the same kind, and the problem as a whole still challenges the guidance of clever theories and the test of ingenious experiments.

Although, and confessedly, we soon run out of satisfactory answers to reasonable questions about evaporation we know that it does occur. We know, too, referring now to water, in which we are here particularly interested, that the higher the temperature of the water the faster it evaporates; that the lower the vapour pressure just over the water surface, and therefore also the greater the wind, since it in turn, by remov- ing the accumulated vapour, lowers that pressure, the faster the evaporation; that the greater the extent of the surface the greater likewise, but not in the same proportion, the total amount of water evaporated in a given time; that the saltier the water the slower it evaporates; and that this and that other condition also has its own effects, but generally negligibly small in comparison to those just specifically mentioned.

Although the evaporation of water has been the subject of many studies, as might be inferred from the above list of facts about it the subject indeed of hundreds of investigations no simple and inclusive relation has been found between the evaporation from an open body of water and the different factors that control it. Several empirical equations have been devised to express these relations, but they are far from being satisfactory.

Indeed it seems that the true relation probably is much too complicated to admit of expression in the terms of a simple equation. The best we can do, perhaps, especially in connection with climatological studies, is to obtain evaporation measurements from water pans of the same size and like exposure at favourable localities. Such data do not tell the whole evaporation story, but they have the great merit of being both easily obtained and intercomparable, and they do give a fair idea of the relative evaporativities of the places at which they were obtained.

Ice in any form, from snowflake to glacier, evaporates too sublimes if one insists on using the proper technical term for changing from a solid to a gas without going through the intermediate stage of being a liquid. But it isn't easy to form a satisfactory conception of just how ice, or any other solid, does evaporate. It is agreeable enough to think of the molecules of a gas as darting Water Vapor: in every direction among themselves like a swarm of bees on the wing, only more so and altogether blindly; and we can, without shock, liken the molecules in a drop of water to an excited school of minnows, but the molecules of a solid have no such freedom. They are as fixed in position as barnacles on a rock.

Even the atoms, that together make the molecules, keep to their respective places, as we know from X-ray photographs of crystals. Yet they have some sort of motion, oscillatory, presumably, like a pendulum, or the prongs of a tuning fork, for a crystal has temperature just as well as a liquid or a gas, and temperature is a manifestation of the velocities of molecules with reference to each other.

By virtue of this vibration, surface molecules are shaken off more or less rapidly, as determined by the temperature and the kind of substance involved. Many crystals, such as gems of every sort, do not evaporate at all at ordinary temperatures, while others, such as the snow crystal and the crystals of carbon dioxide evaporate freely. The sources, then, of the water vapour in the atmosphere are sublimation from every snow field and ice sheet; evaporation from water surfaces, soil and vegetation; and volcanic steaming from within the earth negligible in comparison to the least of the others.

HUMIDITY

Humidity is just another word for moisture or dampness; but "the humidity" means the degree of wetness of the air, expressed either as "absolute humidity," that is, the quantity of water vapour present per unit volume, or "relative humidity," the ratio of the amount of vapour actually present in any given volume to the amount necessary to saturate that volume at the same temperature. The mass or quantity of vapour necessary to saturate a given volume increases rapidly with increase of temperature, but is not materially affected by the presence or absence of other gases.

Relation of Humidity to Height

Although water vapour gets into, or becomes a part of, the atmosphere at the surface of the earth, it does not stay at that level, but spreads out by diffusion, is mixed with the other gases of the air by convection and turbulence, and drifted about by every wind from the lightest to the strongest. Of course, if the atmosphere consisted of water vapour and nothing else the density of that vapour would be greatest at the surface of the earth (except wherever, as rarely might happen, the temperature decreased very rapidly with increase of height) because the pressure on it would be greater there than at any higher level by the weight of the vapour up to that level.

But as water vapour is only a decidedly minor member in the atmosphericgas association it has but little control of its own fortune, and so is made to behave in ways that often are not its own. One of the most important of these is its distribution with height. It is not always, nor nearly always, densest at the surface, nor does its density decrease with height according to that beautiful law and order that normally applies to the atmosphere as a whole and that would apply, as stated, to the water vapour if it existed entirely alone. In the early morning, for example, after a warm calm night, especially when the ground is wet, or where considerable bodies of water are close by, the absolute humidity, and often the relative, too, is greatest close to the surface, and decreases with height for 50 to 100 feet or more.

After sunrise, however, vertical convection begins, especially when the sky is clear, and frequently an appreciable wind also sets in, and each of these tends to so redistribute the vapour as to make it the same fraction or percentage of the total air at every level from top to bottom of the agitated or mixed up layer, perhaps (and often) 1,000 feet deep. This evidently decreases both the absolute and the relative humidity near the surface and increases each of them in the upper portion, owing to decrease of temperature, to and beyond the saturation point, and thus causes the formation of cloud. When cloud is produced there obviously is left just that much less moisture that can be carried to still greater levels, and so it happens that as the temperature decreases with increase of altitude so too does the amount of water vapour necessary to give saturation. However, the actual amount of vapour over a given place at a particular time seldom if ever decreases nearly so regularly with gain of height.

A column of saturated air, for instance, may rise to a considerable altitude and there spread out, forming a cloud layer. This layer in turn may evaporate and the humid but now clear stratum drift on with air of less humidity in either sense, relative or absolute, both below and above it.

Furthermore, since layers of air of various origins are thus formed at different levels it follows that, in respect to humidity, the atmosphere at any particular time is quite certain to be more or less stratified, no matter how smooth the curve may be that expresses the average relation of the quantity of water vapour to height.

As just implied, we should expect the curve of the average vertical distribution of absolute humidity to be smooth, that is, to show no tendency towards a maximum or minimum at any height, or, in other words, to show that water vapour does not tend either to accumulate at, or to avoid, any particular level. But this natural assumption is in error. There is a tendency for vapour to accumulate at about the limit of daily convection, the level at which rising air spreads out horizontally. On the average this is about a mile and three quarters high during summer in middle latitudes, and much less than one mile in winter, say six to seven tenths of a mile.

Saturation often occurs also at and near the top of the troposphere, as we know from the frequent occurrence there of the cirrus cloud. The amount of the water vapour per unit volume at the very low temperatures, -60° to -70°F, that prevail at this height, 6 to 7 miles above the surface, is very small, even when the air, or space, is fully saturated. That is why the high cirrus clouds are so thin, gauzy and diaphanous in structure and appearance.

As soon as the limit of the troposphere is passed there are no more clouds above to diminish the intensity of the sunshine by day or dim the brilliancy of the stars at night—no clouds here to smoor the sun or blink the stars—except very rarely when either the so-called nacreous, or the so-called noctilucent cloud appears, clouds of wholly unknown origin. The habitual absence of clouds from this upper region is, at first, rather surprising. As stated above, we know from the cirrus clouds that saturation often occurs at the base of the stratophere, and so we should expect the vapour molecules here to wander on up into the stratosphere—to diffuse, as we say, just as odours diffuse from place to place—and thus to bring about saturation at these upper levels also.

We then should except clouds to form at these great heights whenever, for any reason, a distinct drop in temperature should occur. Now, drops (and of course rises also) in the temperature of this region do occur over any given place, drops sometimes of 15°F to 20°F and yet the expected clouds do not appear. Why not? Because, in the first place, there is, at best, very little vertical convection in the stratosphere to carry up bodily the vapour from below, while diffusion, the only other way by which it can ascend, is a slow process. Second, before diffusion can carry up much vapour, or any of it to a considerable distance, the air of this region is sharply cooled and the excess vapour, if any, precipitated in the form of ice needles or spicules that fall down gravity far more rapidly than the vapour from which they were condensed originally rose.

Now, although this cooling at the base of the stratosphere occurs irregularly, it generally does so at intervals of less than a week. Hence distinct clouds do not form, even occasionally, in the stratosphere, if we ignore the rare nacreous and noctilucent clouds, because it is kept exceedingly dry by the frequent formation of an invisible haze at its base. Contradictorily, then, clouds do not form in the stratosphere because clouds do form in the stratosphere—never well up and perceptible, because continually and at frequent intervals, hence low down and imperceptible.

Daily run of Humidity

There are several things that affect the humidity of the atmosphere at a particular place. Some, if not all, of the more important of these are, temperature, convection, nature of the immediate surroundings, wind velocity and wind direction. In general, increase of temperature carries with it an increase in the absolute humidity, through increase of evaporation; and a decrease in the relative humidity, owning to the consequent increase in the vapour capacity per unit volume, a capacity that increases rapidly with increase of temperature. At any rate this is the initial effect. After a sufficient time for much more water vapour to be provided by evaporation the relative humidity may come up to, or even exceed, its former value.

The effect of convection on the distribution of humidity may seem obvious since, as applied to the atmosphere, it com- monly is lower air being pushed up and staying up, with a slight but more widely spread descent of the upper air; or a local descent of the upper air and slight ascent of the neighbouring lower air; or, ideally, a mere interchange of position between the upper and the lower air, the one coming down and staying down while the other goes up and stays up. If, then, in this latter, or ideal, case the lower air at first is more humid than the upper, presently, after convection sets in, the upper certainly will be more humid, one might suppose, than the lower, because the lower air (the more humid) has become the upper, and the upper air (the less humid) the lower. This delightfully simple relation seems, as stated, perfectly obvious, but it is invariably true only for a definition of humidity different from either of those already used.

For these it may or may not be true. If the relative humidity is greater in the lower air than in the upper, then after convectional interchange it will be even more pronouncedly greater in the then upper air owing partly to the increase of its relative humidity, due to cooling by ascent, and partly to the decrease of the relative humidity of the falling air, owing to its gain in temperature. Clearly, though, the lower air may be so dry, with reference to an upper layer, that it has the smaller relative humidity not only before, but also after, they have interchanged places.

Evidently, then, the relative humidity of the lower air may be either greater or less than that of the upper both before and after the convectional turn-over; it also may be less before and greater after, but as explained above, never greater before and less after. Neither does the simple relation, greater before interchange greater after, less before less after, neccessarily hold for absolute humidity (the mass of water vapour per unit volume), since it decreases with ascent and consequent increase of the volume occupied by a given mass of the vapour and similarly increases with descent.

We thus are brought to the alternative of either giving up the simple and seemingly self evident relation mentioned above or of finding some sort of humidity (so defining it) that is not affected by change of temperature or change of volume. Such a definition may, at first, seem impossible, but so long as water is not removed by precipitation these conditions are fully met by "specific humidity," the mass of water vapour per unit mass of the humid air. This particular humidity (as per definition) is not so extensively used as either of the others, relative and absolute, but it is in use, and occasionally, as in the case under discussion, is, by far, the most convenient.

So much, then, for the effect of convection on the humidity of the air—not exactly what one might have supposed, but for that reason all the more interesting. And there are other things that affect humidity. It was stated, a page or two back, that it varies also with the nature of the immediate surroundings. The extreme cases of a small island in midocean and a sand dune in a desert will make this effect of environment practically self evident.

On the desert there usually is very little water available for evaporation, so that the air once dry, for whatever reason, normally remains dry for a comparatively long time. In the case of a small island, however, the situation is radically different. Here there is water in every direction that always keeps the lower air, at least, quite humid.

It should be noted in passing, however, that while high humidity is essential to precipitation, that result does not necessarily follow. Catalina Island, for example, is as arid as any portion of the adjacent mainland. High humidity is a factor essential to precipitation but it is not of itself alone entirely adequate. There must coexist also an active process, such as vertical convection, that tends to push the humidity to supersaturation, and would so push it if condensation did not occur. Another of the modifiers of humidity is wind velocity, and it acts in different ways.

It expedites evaporation—the wind, as we know from abundant experience, rapidly dries up wet roads, dries them by increasing the rate of evaporation of the water on them. This increase of evaporation gets moisture into the air at a correspondingly increased rate and thereby affects the humidity. Furthermore, winds are accompanied by turbulence, a mixing of the surface air with all the air for some distance above. In this way the humid air over damp soil and bodies of water is mixed with the commonly drier air overhead, and thus the humidity at all levels in the agitated layer more or less altered.

Also wind direction is an important factor in determining the humidity of nearly all places except, perhaps, those of middeserts and tropical oceans, where its effect is at a minimum and even uncertain as to sign. Evidently the air is more humid when it has just come from over a warm body of water than it generally is when from a dry land area. The sea breeze therefore is more humid, as a rule, than the land breeze, and the summer monsoon winds richer in water vapour than the counter winds of winter.

From the foregoing controls of humidity it is evident that at low levels over the ocean it is quite likely to increase during the day, especially when the wind is light, owing to the addition there by evaporation of abundant fresh vapour, and the slight extent to which it is carried off by convection and mingled with the upper air. Obviously, though, this can go on indefinitely, as on the average it does, only because the accumulating vapour is more or less regularly removed from the place in question.

This necessary removal is effected locally and temporarily by the horizontal drift of winds and vertical mixing by turbulence, and eventually completed somewhere else by condensation and precipitation. Over the ocean, too, as there is very little change of temperature between day and night, and but feeble convection, there are no marked humidity maxima and minima at approximately uniform hours, as there are over most land areas. On land, as just implied, the humidity, both relative and absolute, commonly undergoes distinct diurnal fluctuations in the warmer season, not in the colder, with maximum relative humidity of the lower air at the time of minimum temperature and minimum at the time of maximum temperature; and maximum absolute humidity about 8 to 9 o'clock a.m. and again at 9 to 10 o'clock p.m. with, of course more or less pronounced minima at some intervening time, both day and night, as determined largely by wind and convection. During winter the diurnal changes of humidity are comparatively slight since evaporation then is slow, the water being cold, and diffusion incident to thermal convection practically absent. Since change of temperature so greatly affects humidity, and since temperature depends largely on sunshine, it follows that the daily course of humidity is greatly affected by the amount of cloudiness—the greater the cloudiness the less the sunshine, and the less the sunshine the lower the temperature and the less the evaporation.

The Yearly Run of Humidity

Since water vapour is gotten into the air by evaporation, a process that varies directly with the temperature of the evaporating water, and distributed or diffused by horizontal winds and vertical convection, it follows that the annual variation of humidity is very much like unto its diurnal variation. Hence, throughout the first mile or so above the surface of the earth, the absolute humidity of the air is greatest, as a rule, and its relative humidity least, during the summer; and the relative humidity greatest, and absolute least, in winter.

From this in turn it follows that clouds commonly are lower in winter than in summer, both over land and over sea. Of course, too, the annual run of humidity, like the daily, also depends on locality.

Over tropical seas there is but little variation, while in certain other localities the change is all the way from frequent saturation at one season to distinct aridity at another.

Horizontal, or World, Distribution of Humidity

From the fact that all sorts of humidity, the familiar absolute, the more familiar relative and the specific, vary so greatly with temperature, nature of the surroundings and the direction and speed of the wind, it follows that each kind varies also with latitude and longitude as well as with the time of day and from season to season.

However, it is not necessary here to consider specific humidity, the weight of water vapour per unit weight of humid air, since it generally varies in the same manner as the absolute humidity, or weight of water vapour per unit volume. The variation of this latter, or absolute humidity, with location is amazingly great. On the most stiflingly hot and humid days in a tropical jungle, for instance, the humidity may amount to even 5 per cent of the air by volume, or 3 per cent, roughly, by weight, while in the coldest places of the world when the temperature is lowest it may be even less by weight than one part in 300,000.

Thus the greatest amount of water vapour that sometimes is in the air where the temperature is quite high is at least 10,000 times greater than that which is or can be in it at temperatures occasionally reached in eastern Siberia. Usually, too, the absolute humidity is quite small over all desert regions, but probably never nearly so small as it necessarily is wherever and whenever extremely low temperatures prevail.

In general, and owing to the very rapid rate at which the amount of water vapour necessary to produce saturation of a given volume increases with emperature, absolute humidity grows rapidly and continuously with decrease of latitude from either pole to the equator, except across the high pressure belts of 25° to 35°, roughly, both north and south—horse latitudes of the oceans, and desert zones of the continents.

Quite commonly the absolute humidity is less along these belts than on either side of them, just as they are also zones of minimum cloudiness and rainfall. The dryness of these zones, in terms of amount of water vapour lacking per unit volume to produce saturation, is more pronounced over the land or desert areas, where there is but little surface and plant water available for evaporation, than along the horse latitudes of the oceans. But strange as it may seem, even these sea areas are arid regions in the sense of having relatively low humidity and scanty precipitation. This is because the air here usually is slowly descending and for that reason commonly rather dry except near the surface.

In addition to these world-wide variations of absolute humidity with location there also are many others of lesser extent. For instance, it decreases in general with distance from the ocean, but not in any regular sort of way. Where the rainbearing winds are nearly all from the same general direction, as they are, for example, over the central United States, they carry, on the average, less and less water vapour, hence exhibit a progressive decrease in absolute humidity, with increase of path over land after leaving the ocean.

This is owing to the fact that the amount of water vapour taken up into the air by reevaporation over any particular area is less than the amount removed from it by rainfall on that same area—less by the contribution of this area to the stream flow, or runoff. But the winds from the different oceans do not all carry equal amounts of vapour, nor do they, as a rule, penetrate equally far into the land, hence the place of lowest average absolute humidity generally is considerably removed from the most inland point. Furthermore, the decrease of absolute humidity with progress of the wind from the ocean often is very irregular.

Thus a coastal range of mountains, where the prevailing winds are on-shore, commonly has considerable to heavy precipitation on its windward slopes, while to the leeward the humidity usually be so low as to produce semiaridity for hundreds of miles beyond, unless another mountain range is encountered. And even if such mountains do lie across the winds, and precipitation in quantity does occur on them, the average absolute humidity is not thereby increased, but more or less decreased in proportion to the contribution of the given region to the general stream flow. The most important things to remember in considering what the distribution of absolute humidity is, and why, are the facts that its possible value, and generally, therefore, its actual value, except in arid regions, rapidly decreases with decrease of temperature; that it is decreased by precipitation; and that it is increased by evaporation. Change of pressure, and change of temperature, by altering the volume occupied by a given mass of air, also change the current value of the absolute humidity.

These same things, mainly, evaporation, precipitation, change of pressure, and change of temperature, also are the factors that control the relative humidity, that is, the current fraction of saturation, or ratio of the amount of water vapour actually present per given volume to the amount necessary to produce saturation at the same temperature of the same volume. It should be especially noted that while the actual temperature is essential to the determination of at least the maximum possible value of the absolute humidity, it is of no importance in fixing the value of the relative humidity.

Since the vapour capacity of a given space or volume—the maximum amount of water vapour that can exist in that space in the presence, to be exact, of a flat surface of pure water-depends on temperature only and rapidly increases with increase of temperature, it is evident that, with a given amount of water vapour, short of that necessary to produce saturation, present per unit volume, any decrease of temperature will increase the relative humidity.

This is owing partly to decrease of the volume, the pressure remaining the same, occupied by a given quantity of the vapour and partly to the decrease of the vapour capacity per unit volume, both incident to the decrease of temperature.

Conversely an increase of the temperature, other things remaining equal, decreases the relative humidity, and so too does decrease of pressure, temperature remaining constant, owing to the corresponding increase of volume.

From these fundamental principles it is practically obvious how the relative humidity must vary with change of place. Thus, tropical winds on their course to higher latitudes normally become colder, and therefore gain in relative humidity even to saturation and the production of fog and cloud. Similarly the relative humidity of winds from higher latitudes generally decreases, owing to increase of temperature, except where evaporation is particularly rapid. Again, in the summer, relative humidity commonly decreases with progress inland over a continent, owing partly to increase of temperature and partly to an actual decrease of the humidity in the air.

In the winter the relative humidity generally is greater over continents than over oceans, owing to the comparatively low temperature then of the land surface. Naturally, too, the relative humidity is especially low over arid regions, and this condition extends in a measure to the oceans, where the relative humidity averages about 75 per cent along the horse latitudes, and 80% over the rest of the ocean areas.

CONDENSATION

Condensation, or the changing of water from the gaseous state to the liquid state, is exactly the reverse of evaporation. And just as the latter, that is, evaporation, makes the space into which the vapour passes humid—makes the air humid (part water) we commonly say—so obviously its reverse, condensation, always reduces the absolute humidity, or amount of water vapour present per unit volume.

These two processes, evaporation and condensation, are the fundamental controls of humidity, every form of which, so long as the temperature and the ressure are kept constant, is increased by the one and decreased by the other.

It, therefore, is desirable to know something of the conditions under which this important phenomenon, condensation, occurs, and to formulate some idea as to how it occurs.

It is now well established by evidence fascinating and convincing, but too long to include here, that each pure gas, hydrogen, oxygen, water vapour or what not, consists of equal discrete particles so infinitely small that if 500,000 people were each picking them out at the rate of 100 per minute, day and night, it would take them 1000 years to accumulate particles enough to make a volume of the air we breathe as large as the head of a pin. And yet we know to closer than one part in a hundred the weight or mass of each of these particles, and how any kind, oxygen for instance, differs from any and every other kind. We know, too, that all these particles are in swift motion, on the average, among themselves with, seemingly, all the irregularities of a swarm of bees, and more besides. Yet in spite of this apparent utter confusion as to direction, and wide range of velocity, certain surprising regularities or laws do obtain.

Thus the average velocity of the molecules varies directly as the square root of the absolute temperature, inversely, as the square root of its mass. In a mixture of gases, such as constitute the atmosphere, the average energy of each molecule is the same as that of any other—whatever the lighter ones lack in mass they make up for in speed. And that speed is astonishingly great, in fact at ordinary temperatures it is around 1500 feet per second, or say that of an ordinary rifle bullet, while the velocity of the water molecule in the is approximately 1900 feet per second.

But regardless of their great velocities the molecules of a gas, or anyhow much the greater portion of them, seem to come to rest when they impinge onto a solid or liquid surface. That is, at such times they at least momentarily condense where they strike. If the temperature of the object struck is high they are quickly knocked off again, but if its temperature is low, if, in other words, its molecules vibrate slowly, the chance that the arrested molecules will promptly leave it is correspondingly less. Hence, at a certain temperature which we call the dew-point more moisture collects on a water surface, and also on many other objects, than leaves it.

This is condensation. When this condensation occurs on foreign particles in the atmosphere, however minute they may be, the result is a fog or cloud. When every thing of this dust-like nature is taken out of the air, by filtering it through a tube stuffed with raw cotton, or otherwise, fog does not form immediately the dew-point is passed, nor indeed under any degree of supersaturation that possibly can occur naturally and in the open. We know, therefore, that if the atmosphere were pure air, in the sense of containing nothing that could be mechanically filtered out of it, never a fog nor cloud would form, nor rain nor snow, only evaporation from the surface, and condensation again somewhere on the surface—evaporation chiefly in the warmer regions and condensation in the colder.

The atmosphere, however, is not pure, but filled with countless myriads of condensation nuclei of many kinds—kernels as it were about which water shells can and do form, immediately the state of saturation is passed. Though but an infinitesimal part of the ordinary outdoor air they nevertheless are absolutely essential, indeed weather vitamins without which there could be no rain and no snow.

Clouds Yield Rain

The above may describe the making of a cloud droplet, but a cloud droplet is a long way from being a rain drop. Indeed it takes about a million of the former to make one of the latter, and besides the droplets have but little tendency to spontaneously and rapidly unite, as is evident from the fact that clouds often float along for hours and even days without yielding a drop of rain.

On the other hand, rain often does occur and so we wonder why clouds sometimes produce abundant rain and at other times none at all. They look quite alike on the two occasions, as like as peas in a pod, and are alike except that when rain is falling a lot of relatively large drops are present while none are present when no rain is falling.

But that is the point. Why are the clouds not always the same, always consisting entirely of minute droplets or always containing also rain drops? Well, as stated, there is no important difference between these clouds except in respect to the detail of the size of some of the drops, but there is a marked difference in the movement of the air in which they occur, and in this rests the secret we are trying to find. Clouds from which no precipitation, rain or snow, is falling, are floating along in air that has very little vertical movement, whereas there always is appreciable, and at times even great, velocity upward of the air in which rain drops or snowflakes are forming.

Now, water drops fall through still air with a velocity that varies with their size, from about one tenth of an inch per second in the case of the cloud particles, to some 25 feet per second for the largest possible rain drop—one fifth of an inch in diameter, approximately. Clearly, then, whenever the movement of the air has an upward component, clouds will form as soon as expansion has brought the temperature to the dew-point, and will so continue to form in the ever fresh air at the same level as long as that particular stream continues to rise. Yet, despite this constant formation of cloud not a droplet can fall until by growth from condensation, or coalescence, or both together, it has become large enough to overcome by its weight the upward drag of the rising air. And, further, since more and more cloud is formed, and since well up in the cloud the velocity of ascent grows less and less with increase of height, and finally ceases altogether, it follows that after a time there must be a falling of drops of appreciable size; in short, a rain, and at such rate as to equal the fresh supply of cloud, or new condensation.

Other Forms of Precipitation

Rain is not the only form of precipitation: Snow.—When the condensation occurs at temperatures below the freezing point the usual result is a snow crystal, and when, in the course of their fall, many such crystals get entangled together into a single mass the result is a snowflake. Sleet.—Sometimes rain falls into a freezingly cold layer of air only a thousand or so feet thick and reaches the earth as frozen pellets, or sleet of the variety that rattles when it hits a windowpane.

- **Hail:** On the other hand, in a vigourous thunderstorm rain drops frequently are carried to such great heights that they freeze, and often grow to considerable size before they overcome the lift of the rising air and fall to the ground as hail stones.
- **Graupel:** Still another form of precipitation is the rather unfamiliar graupel or soft hail—white, granular bits of ice roughly spherical in shape and an eighth of an inch, more or less, in diameter. It may be formed in subcooled clouds, that is, clouds of water droplets at temperatures below the freezing point, much as rime, or granular snow tufts, are formed on the windward side of exposed objects in a drifting, subcooled fog. Certain particles here and there in the cloud may, and undoubtedly do, fall faster than most others, and grow by capture as they fall. If, now, the cloud is subcooled the droplets are certain to freeze immediately they come into contact with the falling larger mass and adhere to its underside, thus giving it, as it turns every which way in the course of its fall, an approximately spherical form.
- **Fog Drips:** The drippage from trees and other objects when a dense fog drifts through them. Dew and frost are intentionally omitted from the above list of precipitation form, because they do not fall from the clouds as do rain and snow.
- **Qiantity of Precipitation:** What is the quantity of all this precipitation? Well, for the whole world there is lots of it. There is enough humidity in the atmosphere all the time to make a layer of water at least an inch deep over the entire earth, while the average rate of the total precipitation, and of course also the rate of evaporation, is approximately 16,000,000 tons per second, enough to cover the whole earth in the course of a year with a layer of water three feet deep.

Atmosphere Usually is not Saturated

Evidently this precipitation furnishes all the water used in plant growth, and all that runs off as stream-flow except, to be exact and if geologists are right, a portion of that negligibly small amount that comes from gysers and certain hot springs—a portion furnished, some hold, by the primordial rocks themselves. But there is another important thing that results from the occurrence of precipitation that is not so obvious. It is the fact that in this way the atmosphere is kept from being everywhere and at all times intolerably humid, indeed saturated, or very nearly so. Every rain and every snowfall takes that much water out of the air—dries it to that considerable extent and thus keeps the humidity generally far below the saturation point, and therefore normally within the bounds, not of endurance only, but of actual comfort.

DISTRIBUTION AND CHANGES OF ATMOSPHERIC PRESSURE

Although atmospheric pressure, or weight of the air per unit horizontal area directly above the place where the pressure is measured, is not a weather element of any particular importance so far as our feelings are concerned, such as temperature and humidity are, nevertheless it is the element that more than any other locates the weather of today and tells us what tomorrow's will be, and therefore is of great importance. Latterly this importance has appreciably decreased, however, because other methods have come into prominence for finding the positions, magnitudes, speeds and directions of travel of the contrasting masses of air whose conflicts and whose passage keep the weather in ceaseless ebb and flow from fair to foul and foul to fair. Still these other methods have not yet been sufficiently developed to justify ignoring the distribution of atmospheric pressure, and how it is increasing here and decreasing yonder, as aids to a foreknowledge of the weather of the morrow.

Neither is it at all likely that so great a development and use of these other means, while entirely possible, will be effected at an early date, if indeed they ever are so developed. Therefore a knowledge of the distribution of atmospheric pressure, the near equivalent of the distribution of the air itself, and of its changes, still are, and for a long while must remain, matters both of great practical importance and of keen theoretical interest.

Barometer Corrections

Since the pressure of the air at any particular time and place is, as stated, the weight of all the atmosphere above a horizontal unit area then and there it follows that a knowledge of the values of this pressure over a given region, however extensive, tells us, when the proper corrections have been made, just what the distribution of the air itself is over that same region. But what are these corrections? That depends on the kind of instrument used to measure the pressure and on what we want to find, pressure or quantity of air, for we must not forget that the same quantity or mass of air, as also of other things, varies in weight, or pressure, with position not with every change of position but with most changes.

But we can think easier and straighter if we consider one thing at a time, and know exactly what that one thing is. Let us first, then, compare the masses of air over different places. We can do this by means of a mercurial barometer which is only a device for balancing the pressure exerted by a column of mercury at its bottom, tending to make it flow out, against the pressure there of the atmosphere that keeps it back in. Clearly the pressure exerted by the mercury at the bottom of the column, when corrected for the capillary action of the tube, is proportional to the product of its density by the height of the column by the force of gravity, or pull oil a unit mass by the earth, at that place.

Similarly the pressure there of the atmosphere is proportional to the weight, or mass times the force of gravity, of all the air, if still, above that level (the level of the bottom of the mercury column) per horizontal square inch, or other specified area. When the air is in motion, however, as it always is, a further correction of the reading of the barometer is indicated, since the weight (downward push) of a given mass varies with the direction and speed of its motion. But since the value of this correction (never known accurately) at most is negligibly small it will not be discussed here just mentioned for the sake of completeness.

Similarly the effect of the variation of gravity with height also is negligible. In this case then we may consider sauce for the goose as sauce for the gander too, in the sense that gravity applies in equal (almost) measure and exactly alike for both the mercury and the air. Hence the mass of the air above any given horizontal area, such as a square foot, say, at the level of the mercury in the barometer basin, is proportional to the height of the mercury column, or reading, as we say, of the barometer, corrected for whatever slight difference in that height there may be due to the pull, or capillary action, between the mercury and the wall (glass) of the barometer tube, and corrected also for the density of the mercury corrected to a common or standard density, commonly that which it has at the melting point of ice, 32° Fahrenheit, or 0° Centigrade.

In this way the readings of all the mercurial barometers are made strictly intercomparable, and the ratios of their several readings to each other exactly the same (to within an unknown but always negligible amount) as that of the masses of air above them, so that a geographic map on which these values, if simultaneous, are written, or otherwise represented, is also a map of the distribution of the atmosphere over that same region at the time the readings were made. To obtain the instantaneous distribution of the air the barometer readings necessarily must all be taken at that particular time, whatever it is, 8 A.M. 75th meridian time, or what not.

If, however, aneroid barometers are used instead of mercurial for obtaining the mass distribution of the atmosphere, then their readings, which indicate pressures directly, must be corrected for the local values of gravity, because the atmospheric pressure, being the weight per unit horizontal area of a vertical column of air, varies directly with the force of gravity which, in turn, increases with latitude. This quantity, or mass, distribution of the atmosphere is of considerable importance, especially through its influence on the rate of transmission, in or out, of radiation. A thin atmosphere and a thick one are to the earth much like a sheet and a blanket, respectively, to one sleeping in the open.

Although this mass distribution of the atmosphere is very important as a factor that helps to make the weather, especially as to temperature, what it is, still very little use ever is made of it. The distribution of pressure, or push, in the air is on the contrary frequently, carefully and extensively plotted because it is the unequal distribution of push on the atmosphere (or pressure, for they arc equal to each other) that makes the winds to blow and thereby, in large measure, but not wholly, change the weather of one place by bringing to it the air and the weather of another. Clearly, then, a knowledge of this distribution must be, and is, a great help to the fore- caster when judging of the weather of the morrow and the day after. That is why the barometer is so extensively used in making meteorological observations and used almost exclusively for measuring the atmospheric pressure.

Now the pressure exerted by a column of mercury, the column in this case that is balanced by the pressure of the air, is proportional to the height of the column (corrected for capillary drag), to the density of the mercury (which in turn varies with the temperature) and to the local force of gravity. Hence, to compare the pressures at different places, one with another, it is necessary to correct the readings of the barometers to what they would be if all the instruments had the same temperature, say the temperature of melting ice, and were actuated by the same force of gravity its value at sea level at latitude 45°N or some other agreed-upon value.

The readings of aneroid barometers require no corrections for this purpose since, as previously stated, these instruments measure pressure directly. To sum up: If we are measuring the mass distribution of the air the readings of the mercurial barometer must not be altered for gravity differences, while those of the aneroid must be so changed.

On the other hand, when it is the distribution of pressure we are finding, the element we nearly always seek, it is the readings of the mercurial barometer that must be corrected to what they would be under some common force of gravity and not those of the aneroid barometer—they are true pressure values, if the instruments are correct, as they stand, and therefore must not be changed.

The next and final step is to reduce all these pressure values, or correctedreadings of the barometer, to what they would be if taken at some common level. This is done because the travel or flow of the air, due to difference of pressure, is, of course, along the route of least resistance, which route, as a rule, is over a nearly level surface, that is, a surface everywhere normal to the direction of the force of gravity, such as the mean surface of the ocean. Actually the direction of least resistance to a sample of air through the surrounding atmosphere, the direction indeed of no resistance at all, save alone that incident to viscosity, is along what the physicist calls an isentropic surface. However, this exact and refined treatment of the subject will not here be followed to its conclusion, though to do so affords both surprises and mental satisfaction.

But let us have as little boggy ground as possible under our mental feet. Why is the level surface usually the one of least resistance to a sample of moving air? Because ordinarily the change of temperature of the atmosphere with change of height is such that work would have to be done on the sample to carry it to a greater height or push it to a lower level, but no work (save that duc to viscosity which is the same in all directions) to move it horizontally.

Normally this distribution of temperature is such that a sample of airtaken upward would cool, incident to its expansion under a decreased pressure, to a lower temperature than that of the then adjacent air. Being at the same pressure as this adjacent air but cooler, it also would be denser, so that an actual lift would be required to carry it above its initial level. On the other hand, if taken downwards the compression due to the increased mass of air above it would warm it to a higher temperature than that of the other air at its new position.

But being now warmer than the adjacent air, and having the same pressure, it also must be lighter. Hence, normally, only an actual downward push can force a sample of air at any height to a lower plane. In short, air normally can be taken out of its own level only at the cost of work, work against gravity if pushed up, and work against the buoyance of the denser air beneath if forced down.

Clearly then no work against gravity, nor any against buoyancy, is required when, and only when, air is moved over a level surface, or, to be exact, over an isentropic surface—a surface which, as already explained, usually is very nearly level. Obviously any direction over this surface is a direction of least resistance.

Therefore if we would know the magnitude of the cause of the winds, and from that magnitude infer their strength, we must find the values of the atmospheric pressure at the same level at various places over the region in question.

If we had a great number of these pressure readings somewhat evenly distributed over an area of considerable extent, say the size of the United States, or larger, we then could draw on a map of that region a line connecting all the places that, at a particular instant, had the same pressure. This line, according to present practice, we would call all "isobar," or line of equal pressure. By drawing a number of such isobars, corresponding to small but constant differences in pressure, we construct a map or diagram from which the horizontal pressure gradient, or horizontal push per mile, say, can be found. Note that no two isobars (lines of equal pressure) can cross each other. Şuch crossing would signify two different pressures at the same time and place, a manifest impossibility.

The direction of the flow of the air under these unequal pressures alone would be horizontal, as previously explained, and along the course of the shortest distance between the lines of equal pressure, that is, normal to these lines. A sample of air would not flow off slopingly to this normal to the isobars, though that too would be going from a greater to a lesser pressure, because every component of pressure directed to the right along or parallel to an isobar is exactly balanced by a component of pressure in the opposite direction. The only unbalance is strictly normal to the direction of the isobars, and therefore this is the direction of flow, or would be, if there were nothing to disturb or prevent such flow. As a matter of fact this flow is modified by the rotation of the earth, and so profoundly modified as to change it from the direction of the push to the right angles thereto, except in the lower air where the drag of the surface makes the change in direction less than a right angle. But this fascinating paradox must be passed by. Its explanation would be out of place here; besides only the language of mathematics properly fits it anyway.

Since the movement of the air under differences of pressure is horizontal, or very nearly so, as just explained, it is clear that our isobars must be drawn for a common level if we would know from them what the values of the pressure gradients are that are making the winds of the moment to blow, fixing the intensity of a passing storm and largely determining whither and how fast it shall go. But what should be this common level, and how can we find the value of the pressure there?

These questions are old; their answers would be new. For the mariner they are simple enough: sea level is the place, and read your barometer where you are (on that level) is the way. But for the landlubber it is a different and awkward matter, since in general no two of his observing stations are at the same level. For the purpose of weather forecasting one can chart on his map the departures of the barometer readings of the several stations from their respective normal values for the season, or day, in question, and then connect by a smoothly curved line the places of equal, like departures.

Such a map can be, and has been, used with fair success, but it never became a general favourite, though it is the only pressure map that deals with known accurate magnitudes. One thing that renders such a map less helpful to the forecaster than at first one might expect it to be is the fact that the normal, or average, pressure for the given month, say, is one thing for weather of one kind and another for weather of a different kind. In short, the general average or normal for all days is not applicable alike to all sorts of weather conditions—the normal itself, or the observed departure from it, needs to be corrected for each special reading, but by how much we do not know. The pressure-departure map therefore is not entirely satisfactory and can not be made so.

Another promising procedure, and one that has been seriously attempted, is to use the pressures that exist at some definite height above sea level, half a mile, one mile, or what not. Wherever this level is above the surface, surely at that level there is air. Surely, too, instruments can be carried to that level and the actual pressure there measured with considerable accuracy. However, this direct manner of obtaining the desired pressure values is now too expensive to be practicable.

Therefore the method used has been to compute the pressure at a given height from the pressure and temperature at the surface with a general formula and then correct the value found according to the circumstances, especially as to season and direction of the wind, as empirically deduced from data obtained under various conditions by kite, balloon and airplane. But at best more or less doubt as to accuracy always attaches to the values thus obtained. Hence weather maps that ostensibly show the distribution of atmospheric pressure over a horizontal surface at some definite height above sea level have not yet come into general use.

There remains, in this connection, one other obviously possible procedure, the one that has been, and now is, all but universally and solely used. This is, to reduce each pressure reading made on land to what it presumably would be at sea level immediately beneath the point of observation if all below that point, and round about, were not rock and soil, but air in that state and condition appropriate to the location, season and kind and distribution of the weather. The calculation, only approximate of course, is based on the known height of the station above sea level, and the observed temperature and pressure of the air, and then empirically more or less corrected. The chief source of error is the local surface temperature. Sometimes this is abnormally high and sometimes abnormally low. In the first case the reduction to sea level is made on the assumption that the hypothetical air beneath the place of observation is correspondingly warm and light, and in the second that it is correspondingly cold and heavy.

The one yields a pressure too small and the other a pressure too great. Occasionally this error is so big that even the sense of the pressure distribution is reversed—actual cyclones and anticyclones, as indicated by the direction of the winds, changing by reduction to anticyclones and cyclones, respectively, on the map. Pity then the forecaster. But don't pity him overly much. He doesn't deserve it. He just winks the other eye, having in mind any poor devil trying to learn the game from such charts, mentally alters the map in accordance with his background of experience and all data available, some of which are never charted, and goes ahead serenely, if not always confidently. At his own game, and with loaded dice, of course he beats the innocent novice—until he, too, gets wise.

The one thing we definitely know about all reductions of the barometer to sea level, or any other level, is this: They invariably are in error. Generally, however, they are usable, and we do not know of any practical way to improve them. We know too that a "level" surface is not the correct surface on which to draw isobars; but then the surface that is the proper one, the isentropic surface, does not "stay put," as we say, but perpetually so warps and waves as to make pressure reduction to it quite out of the question.

As a rule, however, this correct surface is, as previously stated, nearly level. Hence, and because reduction of pressure to a common height usually is only slightly erroneous, it follows that the sea-level map meets the needs of the weather forecaster fairly well, and better indeed than any other that as yet it has been practical to construct. In this connection let us confess a little inconsistency and confusion of practice in our sea-level reductions. We reduce the pressure, or barometer, readings to sea level, but we leave the temperatures unchanged. Hence our weather maps show sea-level pressures and "mile-high" temperatures (themselves often exaggerated by surface conditions) side by side. This for the forecaster. When the climatologist sums the weather data all up he commonly reduces both temperature and pressure to sea level. Consistency, thou mayest be a jewel, but art not always prized.

GENERAL DISTRIBUTION OF PRESSURE

Since the atmosphere is so exceedingly fluid, moving horizontally under the slightest push, one might suppose that its sea-level pressure would everywhere be the same and therefore its mass distribution substantially equal the world over. However, no such uniformity obtains. In fact, this pressure varies from place to place at any time and from time to time at any place, seasonally, diurnally, semidiurnally, and irregularly in different manners. And there are two fundamental causes of all the larger and more important of these changes of pressure; namely, differences in temperature and the rotation of the earth.

Because, on the average, the atmosphere for miles above the surface of the earth (not all the way out, however) is warmer in equatorial regions than at middle to high latitudes, it there is most expanded and from there has most overflown towards either pole. Hence for this reason alone, if there were no other, the sea-level pressure of the air would be continuously lower over the tropical portion of the earth than over the colder parts on either side.

But there also is another important factor contributory to the final result, namely, the rotation of the earth, in consequence of which the circulation, just entioned, of the air between the equator and the poles seldom moves with the same speed and in the same direction as the surface of the earth itself, though this is the state, a dead calm, towards which surface friction and turbulence always and everywhere tend to bring the air. The normal resultant of this tendency, rarely anywhere satisfied, is a system of more or less continuous winds from easterly points between the latitudes 30°N and 30°S, roughly, and a system of westerly winds in each hemisphere between the latitudes 30° and 60°, or thereabouts. The winds beyond latitude 60°, especially those of the southern hemisphere, again are more or less from the east.

But what has this to do with the pressure of the atmosphere? Much. The bulge of the earth is just such that a frictionless object on its sea-level surface if once at rest would remain at rest in perfect equilibrium, and such would have been the case if the rotation of the earth had been either greater or less than it now is. In the first case the bulge would have been correspondingly more pronounced and in the second less pronounced. If therefore our frictionless object were given a velocity eastward it would find the earth insufficiently bulged, so that to it equatorwards would be down hill. Similarly, if it were given a velocity westward its down hill then would be in the direction of the adjacent pole. Hence the winds of the tropical regions, being from the east, tend to drift off polewards, those south of the equator towards the south pole and those north of it towards the north pole. Therefore in this mechanical way, as well as by virtue of the expansion and consequent overflow of the relatively warm air, a belt of low sea-level pressure is established and maintained along, and for some distance on either side of, the equator.

Also, the poleward push of the tropical winds is roughly balanced by the equatorward urge of the mid-latitude winds- winds prevailingly from the west—from which there results two belts of high pressure, one in the northern hemisphere with its axis roughly along latitude 30°N and the other similarly situated in the southern hemisphere. Beyond each of these belts of high pressure in turn, and for exactly similar reasons, there is a belt of low pressure with its axis at about latitude 60°.

The one in the southern hemisphere is by far the better developed of the two. All along it the average pressure is decidedly lower than it is anywhere else on the globe. The current pressure there, however, undergoes frequent and great changes owing to the passage of storms as presently to be explained. Beyond these circumpolar belts the sea-level pressure again increases, as is evident from the fact that there is a more or less continuous outflow of surface air from each of the polar regions.

In respect, then, to the major features of the distribution of the atmosphere over the earth there are seven distinct great regions: three where it exists in maller amount than the average for the world as a whole, namely, the equatorial belt and the two 60° latitude circumpolar belts; and four, the two high pressure belts of latitudes 30°N, 30°S and the two polar caps, where it occurs in amounts larger than the average.

Centers of Action

Not only are there seven permanent divisions, as just listed, in the distribution of the atmosphere and its sea-level pressure over the earth, but three of these divisions in turn have important semipermanent subdivisions, which, owing to their connection with the direction and strength of winds, and course, intensity and frequency of cyclonic storms, commonly are called centers of action.

Several of these are the peaks of high pressure that normally exist near the eastern ends of the ocean portions or segments of the high pressure belts. These belts are far better developed, and undergo fewer and smaller changes over the oceans than on land, but even here they are not of equal intensity from end to end. Near the eastern limit of each of these links the pressure normally is much greater than it is elsewhere, a circumstance quantitatively related to the strength of the surrounding winds. As this pressure maximum shifts in position and intensity so do the winds of that general region also Water Vapor: Distribution and Trends 197 vary in place and strength. Evidently there must be, and there are, five of these highpressure centers of action, one each in the North Pacific, South Pacific, North Atlantic, South Atlantic and Indian Oceans, respectively.

Besides these there are two recognized low-pressure centers of action, namely, the Icelandic low, southwest of Iceland at about latitude 60°N and the Aleutian low, just south of the eastern Aleutian islands and near latitude 55°N. The former of these, the Icelandic low, though shifting in position, and often being replaced for short intervals by high pressure, or an anticyclone, nevertheless statistically, or on the average, is present month after month and year after year as determined by the distribution of temperature over and round about that region.

The Aleutian low, on the other hand, is only a winter phenomenon when the land areas round about are much colder than the water where the prevailing low pressure occurs. In the summer time when the land areas are comparatively warm, low pressure is not a normal feature of the ocean in the general region of the Aleutian Islands or the Gulf of Alaska. Both these "lows" also frequently are called centers of action since many migratory cyclones either start from them, or, at least, come by way of them from regions beyond. There are no definitely known counterparts to these cyclonic, or low-pressure, centers of action in the southern hemisphere. Perhaps, however, they do exist, one over the Ross Sea where the surface temperature obviously is higher than that of the mostly surrounding great glaciers and snow fields, and possibly another, but much feebler one, over the Weddell Sea.

Pressure Changes

As every one knows, where the air is being heated it expands and where it is being cooled it contracts. Therefore the atmosphere overflows from the summer hemisphere onto the one where winter prevails, and thereby causes a difference between their respective quantities of air and values of their sea-level pressures. The change is small from winter to summer, only about one part in three hundred of the whole. However, in quantity of air this comes to, roughly, 10,000,000,000,000 tons which are shifted across the equator every six months from the warming to the cooling half of the world.

Not only is there a world-wide shift of the atmosphere with the seasons, but also there are like-caused, similar shifts of the air with the seasons between the oceans and the continents of each hemisphere. In winter the average temperature on the land is much lower than that of the adjacent oceans, and in summer much higher. Hence their air pressures also vary from season to season, being greatest over the continents during the colder season and least during the warmest, and over the oceans greatest in summer and least in winter. This difference between land and sea pressures varies widely from place to place, but on the average is greater than that between the summer and winter pressure of either hemisphere (northern or southern) as a whole.

The equatorial belt of low pressure varies a little in position owing to the annual shift of the axis of maximum temperature, or thermal equator—nearly always and everywhere a little north of the geographic equator. Evidently the shifting of this belt necessarily entails some change, though small, in the air pressure of the region covered by it. The high pressure belts also shift north and south, and through several degrees, thereby causing other annual swings in the pressures of many places. The Icelandic low, as previously mentioned, is much stronger in winter than in summer, while the Aleutian low exists only through the winter. Whether there is much change in the low-pressure belt around the border of Antarctica is not entirely certain, owing to lack of adequate observations through the year in that part of the world. Presumably, however, there are some annual changes in the depth and position of this belt also as there are in the intensities and locations of all the others.

In addition to these annual swings in position and intensity of the belts of high and low pressure, there also are marked occasional variations in the locations and strengths of their centers of action, the peaks on the ridges of high pressure, and the depths in the circumpolar valleys of low pressure. When marked shifts of these centers will occur can not yet be accurately foretold, nor indeed do we know in detail any specific chain of circumstances that produces such a result. We do know, though, that they often cause, so long as they last, a great difference in the weather of neighbouring lands, and at times even of places far away.

For this reason the centers of action, both the cyclonic and the anticyclonic, deserve and have received much study. We do not yet fully understand them, at least not so well that we can know in advance when and where they will occur nor how decided they will be, but that is no reason for giving over our study of them; rather is it a challenge to make our study all the more persistent and thorough. We should here emulate the Frenchman who, with the courage of his convictions, said that though we do not understand woman, and may never understand her, we are not for that reason going to give her up. An even greater puzzle among the many pressure changes of the atmosphere are the occasional unheralded surges towards greater pressure or lesser pressure of roughly equal amount simultaneously, or nearly so, over an area of continental size. It may last for days and then disappear as mysteriously as it came. So far as the weather is concerned these surges appear to be of little importance, but until we do understand their cause and know their effects they will remain a taunt of ignorance that leaves us mentally ill at ease. Possibly, too, when we know them better we shall find them more important than we ever before dreamed they could be.

The most important of the pressure changes, due, we must remember, to changes in the amount of air overhead, are those that in middle latitudes occur on the average at least once a week, and often amount to fully three per cent of the whole. These changes are incident to the passage of those systems of wind, and their accompanying phenomena, which we call cyclones in which the pressure is relatively low, and anticyclones in which it is comparatively high. In most tropical regions, but not all, nor anywhere close to the equator, cyclones, often very intense, occur, but never an anticyclone. The anticyclone is the harbinger of fair weather, and the cyclone, wherever it occurs, the bringer of cloud and commonly also of rain or snow according to the current temperature. The forecaster locates on his map these systems of wind and weather, and from his understanding of the subject and intimate familiarity with such maps, quickly estimates what the weather will be a short time ahead, usually 12 to 36 hours, at each place over his territory, large or small. In some cases he makes a general forecast for at least a week ahead, and in others, as for the aviator, a very detailed forecast for only the next two or three hours.

There still are a few more irregularly-occurring changes in the pressure of the atmosphere that should be mentioned. The precipitous drop in the pressure, amounting in some cases to probably one tenth of the whole, that occurs with the central passage of the destructive tornado. This extreme drop in the pressure is owing to the centrifugal force incident to the violent rotation of the air about the axis of the storm. The irregular fluctuations of atmospheric pressure that accompany the earlier stages of a passing thunderstorm, due, in large measure, to irregularities in the downward velocity of the air in the front portions of such storms. The minute fluctuations in pressure of 5 to 10 minutes period that sometimes occur when a shallow (a few hundred feet deep) surface layer of cold air is overrun by a wind of warmer air which throws the former, or cold layer, into waves or billows, much as winds make waves on the ocean. (4) Finally, if we think to have done with all these irregularities in the atmospheric pressure at one full swoop by the use of an automatic registering apparatus of exceeding delicacy, such as has been constructed and tried, we shall find that we have but multiplied our difficulties beyond endurance, for minute changes are continuously occurring that appear to be wholly without order or regularity of any kind.

But even all the above changes, annual and irregular, of the readings of the barometer do not exhaust its known variations. Indeed only a few years after the invention of the barometer in 1643 an article was published in the Philosophical Transactions of the Royal Society of London on the twice daily symmetrical swing of the atmospheric pressure, which in tropical regions amounts, from maximum to minimum value, to roughly one part in 150 of the total pressure.

The amount of this pressure change decreases with increase of latitude, slowly at first and then so rapidly that it becomes uncertain beyond latitude 60°. Its maxima occur at about 10 o'clock, morning and evening, and its minima at 4 o'clock, both A.M. and P.M. There appears also to be a variation of the barometer with an 8-hour period. Furthermore there is a daily variation of the pressure from maximum, in valleys and other low places just before sunrise when the air is coldest and most contracted, to a minimum when the air is most heated and expanded.

On mountain tops, and other places of considerable elevation, the pressure is minimum and maximum when down below it is maximum and minimum, respectively. Still further variations in the surface pressure of the atmosphere are caused by the tidal actions of the sun and the moon. Those due to the action of the sun occur twice daily, but are very small, having a total variation even in tropical regions where it is most pronounced of only about one part in 40,000 of the whole pressure. It therefore can be detected only by averaging a long series of observations hour by hour so as to eliminate irregular variations and then subjecting the result to rigid analysis. As stated, there also are similar, but about two and a half times larger, variations in the pressure incident to the air tides caused by the moon. Even these no novice at the game would ever detect. Lastly—not the last that could be mentioned, but the last that here will be mentioned—there are the annual solar tides due to the swing of the sun north and south of the equator, wholly lost, perhaps, in the much greater thermally induced swings of the atmosphere in the opposite directions; and the lunar-monthly tides caused by the drifting of the moon from one hemisphere to the other across the equator.

As just stated, not all the changes in atmospheric pressure, all the comings and goings of the air but enough of them, no doubt, to make it clear that the ways of the air are devious and strange—full of puzzling problems that it is an intellectual delight to solve, and in every sense profitable to understand.

WIND

"Never mind the weather so the wind don't blow," runs the old song, turgid with the wisdom of experience, that our granddads used to sing in their boyhood days. But the winds do blow—blow comfort when gentle, annoyance when strong, and destruction when violent. Naturally, then, they were of vital importance to our ancestors of the tree tops, and important even to our great-great-great-to-the-nth-generation grandfathers who went to the other extreme and lived in caves, for whatever the shelter they found within the earth, surely the game that hunger compelled them to chase had to be captured on the outside in the realm of Boreas and all his blustery kin.

Nor have the artificial caves we now make of all sorts of materials, from stone and mortar to stubble and straw, and place on the surface of the earth, improved the situation one whit, for they, too, even more than natural caves, leave us subject to stress and storm. There remains the same compulsion to leave them for the open betimes, nor are they always proof against every wind that blows.

And so it has come about that ages ago man asked: "What is a wind?" "Where does it come from?" "Where does it go to?" "What caused it?" They asked a lot of other questions too, but these will do for the present. To the first of these questions, "What is wind?" a Greek philosopher replied: "Air in motion." No better definition of wind than this ever was made nor ever can be try it. The next questions proved to be more difficult. In fact until yesterday, relatively speaking, they were either given up wholly as unanswerable". The wind bloweth where it listeth, and thou hearest the sound thereof, but canst not tell whence it cometh, and whither it goeth"—or else answered in an entirely fanciful and irrational manner, however pleasing the concept and beautiful the poetry that adorned it.

According to one such concept a demigod, Aeolus, had charge of the winds, which he kept confined in a great cave, and from which this or that one was let out as whim, fancy or need might suggest. A similar poetic concept was that of the personified north wind, Boreas, also a demigod, winged and wild with disheveled hair and beard, who likewise dwelt in a cave. Why he was supposed to dwell in a cave is not stated, but of course if, as believed, he was real and possessed of human qualities he had to live somewhere, and being superhuman that somewhere could not be like unto the home of a mere mortal, but rather some strange and mysterious place.

Furthermore a cave, and there are many such, that has two openings at widely different levels gives out, on hot days, and from the lower entrance, a strong and continuous cold blast; and surely in the presence of such a weird phenomenon a personificationalist, such as a polytheistic poet, who knew nothing of its true and simple cause, might well attribute it to the blowing by mouth, as we ourselves would blow, of some giant, superhuman being. Perhaps this is why Boreas occasionally was represented, not as winding a conch shell, as by the more aesthetic Greeks, mindful of the noise of the north wind, but with distended cheeks and rounded lips that urge the whistling blast. Similarly, and presumably for much the same reasons, it was common through the middle ages the period of great faith and piety to picture winds as coming from the puckered mouths of wingless cherubs. The well-nigh universal ignorance of winds and their causes is further emphasized by the world-wide belief that by this means or that, and especially by some form or other of imitative magic, one could obtain any kind of wind he might desire, from calm to tempest, and of whatever direction.

Nor were, neither are, these superstitions confined to recognized savages. Ulysses is said to have received the winds in a leathern bag from Aeolus, and quite analogously, for centuries, in certain civilized countries, winds could be purchased, and were purchased, especially by sailors. Perhaps they still may be obtained at a few places, not in leathern bags, as Ulysses got his, for that would be inconvenient, but in knotted strings or handkerchiefs one knot to be untied for a breeze, two for a blow, and three for a storm.

But our knowledge of winds has improved of late. We now can talk of the how and the why of them, and discuss their whence and their whither in the exact language of the physicist. We know that air has weight and inertia, and that like mass, or matter, of any other kind a force is required to put it in motion, or to change its speed; and also that when it is not somehow restrained a force acting on it gives it motion, or alters the motion it already has. From this we correctly infer that every wind is caused by a push on the moving air.

Furthermore, we know that this push is owing to the unequal distribution of the atmosphere itself over the earth, and that this vastly important irregularity is owing, in turn, mainly to the irregular distribution of temperature.

This, however, is not the whole story, for there is mechanical reaction as well as action. That is, while pressure causes the air to flow the flowing air or wind causes a pressure. That a wind produces a pressure against any obstacle in its path every one will admit. He knows it from the bending, and sometimes breaking, of trees in the blast, and from the push he has felt on his own body when out in a gale.

We also know the magnitude of this push in terms, say, of pounds per square foot of flat surface directly facing the wind, when we know the velocity of the wind and the density of the air. But there also is another and very different force, or reaction, that comes into existence immediately the wind begins to blow, a reaction long unsuspected but of exceeding importance. It comes from the rotation of the earth, which causes an object moving horizontally over the surface continuously to deflect, or at least steadily to push, to the right (assuming one to be facing forward in the direction of travel) in the northern hemisphere, and to the left in the southern hemisphere.

To get some idea as to how this may be, imagine a cannon to be located at the north pole and to be aimed and fired exactly at New York City; let the speed of the projectile be one mile in three seconds, a fair cannon-ball velocity; and let it keep on its course a little above the surface of the earth, and wholly unaffected by it or the winds, until it gets as far south as the original target, or farther if this should be missed.

And it will be missed, for during the time the projectile was on its way from the pole, the earth was continuously rotating from west to east at the rate of one revolution, counted by the fixed stars, in a little less than 24 hours, so that at the end of the projectile's flight of three hours, roughly, necessary for it to get so far south, not New York City, the original target, but some place in California, perhaps San Francisco, will be struck. The projectile therefore in this case has gone or deflected far to the right of its initial course, as viewed from the surface of the earth. If, now, we would keep it directed all the time along the meridian through New York so that the object aimed at ould also be hit, and, not missed a few thousand miles, we must arrange a tube, or groove of some kind, to carry the projectile and force it to keep up with the surface of the rotating earth.

In this case the projectile and the western side of the tube or groove would press against each other—a force would be exerted on the projectile to make it keep up with the eastward movement of the earth, and of course an equal westward push would be exerted by the projectile against the groove. This is the deflective "force" of our projectile to the right (in the northern hemisphere) of its path. In reality the projectile merely keeps straight ahead on its course equatorward while the earth at the same time speedily turns under it. Thus the target runs away from the projectile and a wide miss instead of a hit is scored.

Obviously an exactly similar result, a miss, would occur no matter what the target shot at with our super-cannon from the north pole, or south pole either, only in the latter case the deflection would be to the left. Evidently, then, a wind across either pole and directed towards the equator would deflect-to the right if from the north pole, to the left if from the south pole—unless prevented from doing so by a suitable pressure against it. All this is obvious enough for a projectile, or a wind, from either pole just mentioned, but what happens at the countless places away from these poles? Exactly the same sort of thing, only with less force, or smaller rate of deviation, because every point on the earth's surface, except those along the equator, is a pole about which the earth is turning with an angular velocity equal to that at the north or south pole times the sine of the latitude of the place.

At the north or south pole it rotates once, measured by the fixed stars, in a little less than 24 hours; about any point 30° from the equator it turns once in just twice that time; and similarly for all other places. Nonsense, we say, and it does seem so. But carefully suspend a good-sized metal ball by a wire from a point some distance above, as Foucault did under the dome of the Pantheon in Paris in 1851, and set it swinging in any direction and watch what happens. If the experiment has been carefully made the pendulum will slowly change the direction of its swing.

We can easily agree that if the pendulum were at the north pole and swinging towards New York City the earth would so turn under it that presently it would be swinging towards Chicago, and a little later towards San Francisco, and so on around until, when the earth had made one complete rotation, it again would be swinging towards New York City.

We can also agree that the pendulum would behave in a similar manner at any place about which as a pole the earth was actually turning and only at such a place. Well, that concession has our logic hemmed in with no possible line of escape, for the Foucault pendulum behaves in just that way wherever it is mounted, save along the equator. Always and everywhere, except on the equator, the direction of its swing slowly changes; that is, in reality, the earth slowly turns around under it, at such rate that the time required to make the complete turn is equal to the time of rotation at the north or south pole divided by the sine of the latitude of the place of observation.

Every place then on the surface of the earth away from the equator is a pole of rotation, and everywhere the earth is rotating with a known angular velocity under the wind at that place. Hence every wind tends to its direction—tends to deflect to the right in the northern hemisphere and to the left in the southern hemisphere, and does thus deflect except in so far as by force it is prevented from doing so. It would seem too that at any particular place this force must be, and it is, the same for one pound of air as for any other pound of air going horizontally with the same speed. That is, this force is proportional to the weight of the object exerting it. It also, though that is not so obvious, is directly proportional to the horizontal velocity of the object moving.

This looks like complication enough, but there still are other forces that must be considered. If we fasten a marble or like object to the end of a string and then, holding the other end fast, whirl it around in a circle we shall find that there is a tension on that string—a pull from each end against the other that increases with the mass of the object whirled, the length of the string and the number of turns per second. Also, if in any way the string is shortened, by allowing it to wrap around one's finger, for instance, or otherwise, the rate of the turning will increase and the pull grow stronger.

Similarly, if water in a large vessel, such as a basin or a bath tub, has some rotary movement and then is allowed to flow out rapidly, the spin in the water will increase as it approaches the outlet until, in many cases, a hollow funnel of air extends from the surface even into the upper end of the drain pipe. Now precisely similar things, and for exactly the same reasons, happen also to the atmosphere. It flows hither and yonder over a rotating earth; rises here under local heating and sinks there owing to local cooling, and there is no rest for it, only ceaseless movement and turmoil. Also, making confusion more confounded, the mountains and valleys, hills, hollows, cliffs, trees and every other roughness on the surface of the earth, stir the winds into myriads of mingling eddies, surges and whirls whose only regularity is their utter irregularity, like a swarm of bees on the wing, only worse, for all the bees are of one size while the eddies are of all sizes.

But really the situation is not that bad provided we are willing to consider one question at a time, and willing also to leave some of the details for the amusement of our successors. As already stated, winds anywhere are the combined result of earth rotation and differences in pressure at the same level. Thus continuous heating of the air in the equatorial regions, such as always obtains, together with its ceaseless cooling in the higher latitudes causes it perpetually to overflow from the former to the latter and to underrun in the opposite direction. This looks simple and obvious, and would be if the earth were not rotating and if the heating and the cooling were continuous and uniformly distributed over their respective areas.

But neither the heating nor the cooling is uniform, and the earth is not still but rotating. However, since the tropical regions are all the time warmer than those near the poles we may assume, as a first rough approximation, that the heating in the former and the cooling in the latter are uniform. This, if the earth were not rotating, would lead to an approximately uniform and continuous overflow from the warmer to the cooler regions and a like return undercurrent from the colder to the warmer zone.

But, as stated, the earth is rotating and that leads to endless complications in the movements of the atmosphere, or winds. Everywhere friction between the surface of the earth and the atmosphere tends to bring the latter to rest with respect to the former—to stop all winds and everywhere produce a dead calm. Therefore air pushed to higher latitudes, where the distance around the earth is less, soon outruns the surface and appears as a west wind. Similarly, air on its way equatorward sooner or later is outrun by the earth and therefore become an east wind. Really this difference in distance around the earth along parallels of unequal latitude is only a part of the cause of east winds in equatorial regions and west winds in middle latitudes. The effect of the additional part, conservation of areas, is even greater than that which would result from the incomplete but obvious cause indicated.

From this it would seem that there should be a belt of winds from the east (east winds) along, and to a considerable distance on each side of, the equator, and beyond this belt another in either hemisphere of west winds. But such aroundthe-world winds would be their own undoing. They come into being owing, as we have seen, to the movement of warm and of cold air to higher and lower latitudes respectively. But such winds, if continuous around the earth would create along and near the interface between them an unbroken belt of high pressure parallel to the equator that would be an impenetrable barrier against any further interchange between the warmer air of the lower latitudes and the colder air of the higher latitudes, whereupon surface friction soon would bring all the air to a dead calm.

As a matter of fact these east winds and west winds, and the belts of high pressure between them, are more or less broken up and irregular. They are well developed and fairly continuous over the Atlantic, Pacific and Indian Oceans, but are only fragmentary and fitful over the continents. In great measure the tropical air makes its first movement towards the polar regions in the form of vast currents to the immediate westward of the ocean links of the high pressure belts; and, in more or less equal measure, the polar air moves towards the equator around the eastern ends of these same links of the high pressure belts.

Trade Winds

The tropical east winds that are over the oceans are called trade winds, in the sense of track winds, or more or less steady winds, along the same course. There are two trade winds on the Atlantic Ocean, namely, the North East Trade and the South East Trade. The first of these comes out of the north around the east end of that link of the northern hemisphere belt of high pressure that crosses the Atlantic Ocean. This wind as it flows southward soon is so outrun by the surface of the rotating earth as to become first a northeast wind and then more and more nearly an east wind.

Its velocity varies from around 6 miles per hour in October to about 12 miles per hour in April. Some of this air ascends, on being decidedly warmed, while some of it crosses the ocean and then turns first northwestward, then northward past now the western end of the same section of the high pressure belt by whose eastern end it had begun in flowing southward. Soon, as it moves northward, it outruns the surface of the earth and turns northeastward to eastward, thus merging with and becoming a part of the prevailing westerlies (winds from the west) of the middle latitudes.

The other trade wind of the Atlantic Ocean comes from the middle latitudes of the southern hemisphere, flowing first northward towards the equator past the eastern end of the Atlantic segment of the southern high-pressure belt, then, as the surface of the earth outruns it, turning northwestward and then nearly westward. Its velocity is roughly 12 to 15 miles per hour through the year.

A portion of this South Atlantic trade wind also ascends, owing to surface heating, in tropical regions and flows off at high altitudes towards the adjacent pole, while another portion rounds the western end of the Atlantic segment of the southern high-pressure belt, after which it becomes a portion of the westerlies of the southern hemisphere. Exactly similar trade winds, only feebler as a rule, occur only the Pacific Ocean. Over the Indian Ocean trade winds occur only south of the equator. North of the equator monsoon winds, to be briefly discussed presently, predominate.

Countertrades

So much then for the "trades" which are the steadiest and most persistent of all winds. The next in order of persistence, year in and year out, and which in conjunction with the trades might be of decided advantage in transoceanic aerial travel are the counter trades, or return over-flow branches of the trades, resulting from their ascents in tropical regions. This ascending air flows off towards the higher latitudes of the same hemisphere in which their ascent occurred and after passing latitude 20°, or thereabouts, turn more or less eastward, so that the trade wind beneath and its overhead companion counter trade have very different directions. In this way the counter trades merge with and become portions of the normal or prevailing westerlies.

Westerlies

As already explained, owing to the rotation of the earth, any wind that has come from the general neighbourhood of the equator and attained to latitudes beyond 20° begins to flow eastward, slightly at first and then, with increase of latitude, more and more nearly directly eastward. These winds are the westerlies (winds from westerly points) of middle latitudes. Though this is their average or prevailing direction they are by no means either as regular or persistent as the trades and the countertrades.

Polar Easterlies

Beyond latitude 60° in each hemisphere the surface wind again tends to blow from easterly points, just as it does near the equator. This is because of the flow of the cold air here away from near the pole to places where the distance around the earth along a parallel of latitude is greater, and therefore the eastward speed of the surface correspondingly greater—so much so indeed that the cold air is more or less run away from and thereby rendered an easterly wind. This polar east wind is far better developed over and around Antarctica, where the outflow of the cold air is more pronounced than it is over and around the Arctic Basin where the outflow is comparatively sluggish.

The winds above described, that is, the five trade winds, the five corresponding counter trades, the two westerlies (one in each hemisphere) and the two polar easterlies may be, and often are, called planetary winds, since they have their origin in two general or planetary conditions, namely, (1) rotation, with, in most places, a high surface speed, about an axis that is not far from normal to the direction to the sun, or, as astronomers say, to the plane of the ecliptic; and resulting therefrom, (2) a persistent great difference between the temperatures of the tropical and the polar airs that maintain an equally persistent interzonal circulation.

Monsoons

Another great class of winds, noted for their long duration and constancy of direction, are the monsoons. This word, of Arabic origin, means a season, and is here applied to winds of season-long duration. They occur in many parts of the world but are strongest, most persistent and best known in India and over the adjacent ocean north of the equator. Here, owing to the great contrast in temperature between the continent and the ocean, there is a persistent under or surface flow of air from the land to the water during the colder season, which flow, under the influence of the rotation of the earth, becomes a wind from the northeast—the northeast monsoon.

During the warmer season, on the contrary, the wind is prevailingly from the ocean onto the relatively warm land, and, as before, affected by the rotation of the earth which, in this case, causes the wind to come from the southwest. This is the southwest monsoon that brings the rains to India. The depth of the summer monsoon of India is about 3 miles, and that of the winter monsoon roughly 1.2 miles. Various other places over the earth have greater or less seasonal changes in the directions of their prevailing winds, owing to changes in the temperatures of the adjacent regions, but at no other place are these changes as pronounced, persistent and important as they are in India.

Land Breeze and Sea Breeze

Closely analogous to the monsoon, but vastly less in depth and extent and having only a diurnal period instead of an annual one, are the land breeze and the sea breeze, caused by the difference in temperature between the surface of the water and the surface of the land from the shore back some miles. They are most pronounced where the land is low and arid, skies prevailingly clear and general storm movements only occasional and commonly slight.

The sea breeze, an on-shore daytime wind, is caused by the greater atmospheric pressure, at the time of its occurrence, over the water than over the land, incident to the lower temperature then of the surface of the evaporating water than that of the surface of the immovable soil. At night, especially when there are no clouds to send back radiation to the earth and keep up the temperature of its surface, the temperature contrast is reversed, the water surface then being the warmer of the two. Hence the direction of the horizontal pressure gradient, or push on the air, also is reversed and the surface wind turned back from land to sea. These winds are shallow, roughly 300 to 1500 feet deep, and seldom extend more than 25 miles inland. In most cases they are so graciously refresh- ing as to be as welcome as the family physician or parish priest.

Valley Breeze

Another daytime wind is the so-called valley breeze, a wind, commonly gentle, that on hot and otherwise calm days blows up mountain valleys. It seems to be owing essentially to the heating and expansion of the air over the nearby valley or plain, which heating increases the pressure in the air at a given height, more over the plain than over the adjacent side of the mountain (the air column to the level in question being shorter above the mountain than over the plain) and therefore causes a flow of the air towards and up the mountain sides and valleys.

KATABATIC OR DRAINAGE WINDS

A somewhat analogous class of winds, oppositely directed and ranging from imperceptible to severe, are the mountain breeze, canyon wind, glacier wind, and fall-wind, or, to use their recognized formal name, katabatic winds, of every sort. Another generic name for air movements of this kind is "drainage winds." In all these cases the wind is directed down grade along the surface. In the main it is a night phenomenon and owing to the cooling and consequent contraction and gain in density of the under air, incident partly to its own radiation, but mainly through contact with the relatively cold soil or other substance beneath—cold usually as a result of rapid loss of heat by radiation. In many respects this air drainage is quite analogous to water drainage.

It flows to lower levels, and from uplands runs into gullies, from gullies into drains, from drains into ravines and from ravines into canyons or valleys, and thence, it may be, out onto a plain of lower level. It may flow away gently, or ripplingly like a brook where the grade is moderate, and even plunge over a precipice in the form of an aerial cataract—a cataract, however, that in some cases never reaches the foot of the fall, because in the course of its descent the increasing pressure upon it makes it warmer and therefore lighter than the surrounding air which thereupon quickly arrests its rush to lower levels, and even causes a partial rebound. Drainage winds teach us many a useful lesson.

The mountain camper, for instance, soon learns to pitch his tent above his vening fire, and not below it where the down-flowing air would pester him with the smoke. The Swiss valley dweller sets his house on a knoll so as to be above the nightly flood of cold air that surges around him. The wise orchardist selects for his trees a place part way up a slope, where the air by descent will have appreciably warmed, and yet above the lower levels where the cold air accumulates since it can drain away but slowly.

That is, he plants his orchard along the "thermal belt" where frosts, both early and late, are least likely to occur. The prudent aviator avoids landing at the mouth of a canyon when drainage winds can be strong, because they always are shallow and at times even violent. The cautious boatman when near a steep elevated coast of high latitude keeps his weather eye open for the williwaw, that abrupt, smashing avalanche of frigid air that seems to have sloshed over the mountain brim from the icy bowl beyond. The fisherman of Trieste, Fiume or Novorossiisk, to mention only the most conspicuous of such places, has a wholesome regard for the sudden and chilling Bora of hurricane force that, rushing from the snow-covered uplands pounds the sea into waves and the waves into a cloud of spray. The polar explorer, unless daring the winds to do their worst, as Sir Douglas Mawson seems to have done the year he spent at the "home of the blizzard," keeps away from steep slopes facing the sea.

Of course there are many other examples of gravity winds, as they also are called, or fall winds, another of their various names, but the above are sufficient to emphasize their interest, importance and world-wide distribution—always shallow and ranging from the gentle, balmy drift to a furious, frigid blast of hurricane force.

Tropical Cyclone

A radically different wind from any of those already mentioned, important, often violent and covering a somewhat circular or elliptical area, 50 to 1,000 miles across, is the tropical cyclone—the hurricane of the Atlantic Ocean, Gulf of Mexico, Caribbean Sea and eastern Pacific; baguio of the Philippines; typhoon of the China Sea; and the cyclone of the Indian Ocean. It is characterized by strong winds, up to 100 to 150 miles per hour, directed spirally inward and around a common centre, the place of lowest atmospheric pressure, clockwise in the southern hemisphere, counterclockwise in the northern hemisphere. It is further characterized by dense clouds, much thunder and lightning and torrential rain.

It occurs over all tropical seas and oceans except the South Atlantic, but most frequently over the western North Atlantic, including the Gulf of Mexico and Caribbean Sea, Bay of Bengal and western Pacific, including the China Sea. It starts usually, if not always, in a doldrum or calm region, 5° to 20° away from the equator, and more particularly in that portion which presumably is a very sluggish atmospheric eddy—a section, 100 miles broad, perhaps, of air which though nearly quiet probably has a slight eddy motion owing to the presence on either side of it of oppositely directed "trade" winds.

Here the warm air becomes excessively humid, and for that reason, as well as because of its high temperature, comparatively light. Hence convection here becomes pronounced, clouds form and thunderstorms develop.

This ascent of light air necessarily implies the corresponding inflow of surface air, and so it must happen from time to time that, owing to the rotation of the earth, and to the opposition in the directions of the winds on either side of the place of ascent, this incoming air is set into rotation, gently at first, but more and more vigourously as the centre is approached, like disturbed water as it runs into a drain-pipe at the bottom of a basin.

The ascent of the air causes it to cool by expansion, its cooling leads to condensation of some of the vapour present, condensation liberates much heat and this heat in turn continues and enhances the convection—not indefinitely, of course, but to a very considerable extent. Thus once begun this rotating disturbance is likely to perpetuate itself so long as a great quantity of water vapour may be had for condensation. Hence, as the storm is captured by the major neighbouring wind, commonly the trade wind on its poleward side, and often carried thousands of miles across a tropical ocean before reaching land, or being turned off to higher and cooler latitudes, its duration generally is at least several days, often a fortnight, and in extreme cases three weeks or more from its inception in some doldrum region to its last feeble efforts far away where water vapour is scarce and energy from its condensation adequate to the continuation of the storm therefore not available.

Commonly it travels a long way from east to west in a trade wind stream, then slowly turns around the western end of the adjacent link of a high pressure belt and swings off to the east again, but at much higher latitudes and with properties largely different from, and more complicated than, those it formerly possessed. In many cases it is driven onto land where, for a time, it brings heavy rains, and then, as a rule, fades away.

The rate at which these storms use energy—their horsepower—is amazing. Thus from the exceptionally large amount of excellent data available about a particular typhoon the Japanese savant, Horiguti, computed that the latent heat of condensation was freed at a rate equivalent to 1,200,000,000,000 horsepower, and that the rate of its mechanical dissipation of energy was equal to 27,000,000,000 horsepower.

Yet some people talk seriously about destroying the typhoon, or hurricane, by dropping bombs onto it from aeroplanes. How the aviator is to reach his vantage point over such a storm is, of course, his problem. But there is no occasion to worry about the details, for saying "boo," to a blizzard would be in about the same proportion and quite its effective in stopping it.

Extratropical Cyclone

The great contrast in temperature maintained between the polar and equatorial portions of the earth keeps the atmosphere, as previously explained, on a continuous round of circulation between them, because the equality of temperature this circulation tends to produce never is even approximately attained. And because this interzonal circulation is over a rotating earth it follows, as may be demonstrated mathematically, that in mid-latitudes it must occur, and it does so occur, largely in the form of great swirls between broad currents of warm air moving poleward and cold air flowing equatorward. Furthermore, since the warm air moving to higher latitudes outruns the earth it therefore must be to the eastward of its companion cold current flowing equatorward and thereby tending to lag behind the rotating earth.

Hence the swirl thus produced between these two currents, the place where the air tends to run ahead on the eastern side and to lag behind on the western, is a region of atmospheric deficiency, or of low atmospheric pressure, with winds directed round about it counterclockwise in the northern hemisphere, clockwise in the southern, and, in each case, turned spirally inward. The simultaneous directions of the wind all around the centre of this low pressure, or place where the pressure is least, seem, on first considerations, to indicate that the air is spinning spirally around and towards that centre— around and around St. Louis, for instance. But St. Louis is fixed in position, while the centre of the storm is not, but moves on eastward 20 to 30 miles an hour, as a rule, and so one gives up his first inference that St. Louis is the hub of the winds and jumps to the next "obvious" conclusion, namely, that they spiral around and around the traveling centre like the tire of a rolling wheel about the axle.

Well, that is wrong, too, for the tire and the axle do not keep pace with each other. The storm is not a spinning mass of always the same air sliding over the earth, but a traveling swirl of constant type but of ever renewed material. Well then, if both these inferences are wrong, what course does the air follow in such a disturbance? We have learned something about that from the drift of clouds and of free balloons, and from the state and condition of the air at successive times and places, and know now that this course is not by any means so simple as we once thought it was. But we must not let this fascinating flight of an individual air particle develop into a red herring dragged across the course of our subject—wind.

In the area of an extratropical cyclone, roughly elliptical and ranging from 200 to 2000 miles across, there always and everywhere is wind, usually of 10 to 30 miles an hour, but by no means confined within these limits. The cyclone is of frequent occurrence, two or three to a dozen a month, over most of the earth beyond latitudes 30°N, S and most of the winds of these regions are associated with them. As previously explained, the extratropical cyclone is a product of the interzonal exchange of air over the surface of a rotating earth. It therefore consists of two distinct parts, a warm and humid eastern portion where the air is of tropical origin, and a relatively cool and dry western section where the air has come from a higher latitudes. Wherever the boundary, 100 to 1000 miles long, between these two great counter currents of air passes, as it moves over the earth with the storm, there and then the direction of the wind quickly changes, from southwest, likely, to northwest, and its quality from relatively smooth-flowing to gusty. This traveling interface between the two wind systems, the tropical and the polar, is known as the wind-shift line or the polar front. It brings showers of the so-called clearing-up type, many of which in the summer are accompanied by thunder and lightning and squally winds.

A great deal more could be said about the winds of the extratropical cyclone, a disturbance that can be started in three or four different ways, but that would involve tedious details which, except for the specialists, would tend to obscure the major facts here stated and general outlines just drawn. Anticyclone

The anticyclone, a system of winds over a region where there is more air per unit area, and therefore higher atmospheric pressure, than there is round about, is entirely an extratropical phenomenon, roughly elliptical in shape and commonly 200 to 2000 miles across, which, like the cyclone of the same latitudes, may be originated in more ways than one. Owing to the relatively high pressure within its area the lower air flows outward from an anticyclone, but of course under the influence of the rotation of the earth. Its direction therefore is spirally outward from the centre of highest pressure, and around it, clockwise in the northern hemisphere, counterclockwise in the southern. But as in the case of the extratropical cyclone the direction of the wind in an anticyclone does not mark its course. Near the centre of the anticyclone the winds always are light, while close to its border they are much stronger, sometimes approaching or even reaching gale force. There is, however, a maximum possible velocity which the wind of an anticyclone can have, while no such limit applies to the winds of the cyclone. Another interesting contrast between these major disturbances is the fact that in the cyclone, especially the true cyclone of the tropics and subtropics, the swiftest winds are near the centre of the storm-not at the centre for there the winds often are still, but within 5 to 15 miles of it, while the swiftest winds of the anticyclone are near its border. On the average the velocity of the wind in an anticyclone is only 5 to 15 miles per hour, and distinctly less than that of the cyclone of the same region and season.

One of the most important and striking weather phenomena of the anticyclone is confined to the winter months and is most frequent and severe in the central United States. This is the cold wave that rushing from high latitudes sometimes carries freezing temperatures even to the border of the subtropics. It brings also the dread "norther" of Oklahoma and Texas, and the true blizzard of the Great Plains whose howling, frigid blasts, filled with powdered, blinding snow, give no quarter to man nor beast that ventures forth or is found afield.

Tornado

Child of the cyclone and the anticyclone, of the south wind and the north wind, where brought into closest propinquity, and born among the clouds, is that meteorological enfant terrible, the tornado, the smallest yet most vicious and destructive of all storms. It may be, and generally is, only 1000 feet, or less, in diameter, yet the velocity of its mad, dancing—dervish winds as they wildly spin around and up the central vortex may be 100, 200, even 500 miles an hour. So exceedingly violent is its action that few things can withstand the fury of its blasts. Like one of its parents, the cyclone, its gyration always is counterclockwise in the northern hemisphere and clockwise in the southern, and like its other parent, the anticyclone, it never comes near to the equator. The area covered by this storm is rather small since the length of its path commonly is less than 25 miles. Also its occurrence is, at most, only occasional, hence despite the terror and destruction it works wherever it does hit, the chance of being hit by it at all is so slight that relatively few people, even where it is most likely to occur, show any dread or concern about it.

Waterspout

A usually much smaller and less violent whirling storm than the tornado is the well known waterspout. This is not a rotating column of sea-water, 1000 feet high, as many seem to think, but just a column of cloud, formed in the humid air by its cooling due to the expansion under decrease of pressure incident to rapid rotation. It may be a tornado at sea, in which case it has the same restrictions as to sense, or direction, of rotation, and the same inability to approach the equator. On the other hand it may start from the surface, owing to the local ascent of air that is very humid and therefore relatively light, just as dust whirls are begun by the concentrated ascent of highly heated desert air. Waterspouts of this origin, like their desert brothers, the dust devils, as they often are called, occur at practically all latitudes, even on the equator, and wherever they do occur rotate in either sense, clockwise or counterclockwise. Also in whichever sense a waterspout or a dust whirl starts rotating, that way it continues to rotate as long as it lasts.

The duration of the waterspout commonly is very brief, 15 minutes, or thereabouts. It therefore is an excellent thing with which to prove the efficacy of magic; a little whistling, for instance, or a threat by the pointing of a sword, and it soon vanishes—as it was going to do anyway. Or, if one is more naval-minded, he can fire a gun at it—cannon to popgun, it makes no difference. If a gun is not convenient point a stick at it, and if you haven't got a stick your finger will do quite as well, for whatever you do or don't do the spout soon will be only a thing to remember and talk about.

THUNDERSTORM WINDS

Another special class of winds are those that normally accompany the thunderstorm. Commonly this storm is preceded a few minutes by a gentle wind directed towards its nearest portion. This is abruptly followed, just before the rain begins, by a much stronger and distinctly turbulent wind in exactly, or nearly, the opposite direction. The first of these is the indraft of surface air, on its way to great heights where the vapour it carries, or is a part of it, is to condense to cloud and rain and the rain in turn to play its important role in the production of thunder and lightning. The second, or outflowing turbulent wind, has come down from above, owing to increase of density due to cooling, and is swift because it has retained the faster velocity winds usually have at considerable heights.

Eddies

Sometimes one is surprised when, on climbing a mountain with a strong wind at his back, he finds, on going down the opposite side, a wind in his face. He is puzzled to know how the wind can blow up the mountain on both sides at the same time. The mountain does the trick. As the wind crosses the crest it tends to go on at the same level, but it also catches up some of the air beneath it and drags that along, just as the air in the tube of an atomizer is pulled out, and with it some of the liquid beneath, by the blast across its top.

This depletion of the air between the side of the mountain and the sheet of crossing wind, or start towards the production of a vacuum, necessarily leads to an inflow of an equal amount of air from outside, and the most available source is down the mountain. Hence a wind crossing a mountain range induces a return eddy current up the lee slope that is a great puzzle to the uninitiated, and great worry to the student of physics who tries to explain it with all the mathematical and other details satisfactory to an exacting professor.

Foehn

Sometimes the wind that crosses a mountain is so pulled down on the lee side that it strikes the valley or plain beyond, now gently and again even violently. As this crossing air descends to lower levels its temperature increases owing, as we know, and in proportion to, the gain in pressure on it, or amount of air above it. If, now, there is no cloud or rain on the mountain this gain in temperature by the descending air on the lee side is only equal to its cooling in the course of an equal ascent on the windward side, and nothing strikingly out of the ordinary happens.

When, however, abundant rain does fall from the ascending air the cooling is much less, owing to the latent heat of condensation freed, or rendered sensible, than it is when the air is dry, while the amount of warming is substantially unaltered, the air on the lee side of the mountain being clear, or nearly so, in each case. On such occasions the descending air reaches the valley or plain abnormally warm and dry; indeed so great are these effects that now and again they even change fields of deep snow into a dusty plain in a few hours—yes, hours, not days.

TIMES OF GREATEST AND LEAST WINDS

Nearly everywhere over land, but no much so over the ocean, there are diurnal changes of the wind velocity. Over level regions the velocity of the wind is least a little before sun-up, when the surface air is coldest and convection least; that is, when surface friction has acted longest on the same identical air and brought it nearest to rest. Similarly the winds are greatest, on the average, when convection has effectively brought upper, and swiftermoving, air most abundantly down to the surface, commonly 1 to 2 o'clock P.M. On mountain tops the hours of maximum and minimum wind velocity are just the reverse of those of the plain.

Also the winds are stronger in the winter, on the average than in the summer. his is because the push, or horizontal pressure gradient, that makes the winds, is greater, as a rule, in winter than in summer, and this in turn is because the change in temperature with latitude, which largely determines the corresponding change in pressure, is greater in winter than in summer. Wind Velocity and Height Another interesting and exceedingly important thing about wind is the fact that generally its velocity changes a great deal with change of height.

Many studies have been made of this phenomenon, though some of the best of them are pretty well removed from the view of the average individual by, to him, an opaque wall of mathematics. This seclusion was not intentional, but unavoidable owing to the nature of the subject. It is easy to observe what happens, but the details of why it happens are, in many cases, quite difficult to explain.

Whenever there is an appreciable wind the roughnesses on the surface of the earth throw the lower air into a turmoil of rising, falling and mingling eddies. Hence, as the surface is the place where the friction occurs, the velocity of the wind must at first rapidly increase with increase of height, and also progressively change in direction to the extent of 30 or 40 degrees, owing to the rotation of the earth. In many cases this increase of velocity continues, but at a slower rate, up to the level of the highest clouds.

After this the velocity falls off until at twice that level, roughly, it again is no more than at the surface. At still greater heights the velocity once more increases. Often, in the forenoon especially, the increase of wind velocity with height continues up to the 1,500 foot level, or thereabouts, and then decreases until at twice that height the velocity has fallen to surface values. Beyond this level it again increases at a fairly uniform rate, and all the rest of the way behaves as above explained. Occasionally, on the other hand, the air is practically calm from the surface up to great heights—5 or 6 miles at least, and perhaps higher.

❐

17

Chemical Processes for Environmental Engineering

STUDY

Environmental chemistry is the study of chemical processes occurring in the environment which are impacted by humankind's activities. These impacts may be felt on a local scale, through the presence of urban air pollutants or toxic substances arising from a chemical waste site, or on a global scale, through depletion of stratospheric ozone or global warming. The focus in our courses and research activities is upon developing a fundamental understanding of the nature of these chemical processes, so that humankind's activities can be accurately evaluated.

The field of environmental chemistry is both very broad and highly interdisciplinary. Within the Department of Chemistry we have a core group of faculty whose research interests are in atmospheric and aquatic chemistry, photochemistry, and the chemistry and transport of long-lived pollutants.

We interact with other chemists in the Department, with numerous other researchers at the University who have related interests, and with nearby government agencies. Indeed, the setting for the study of environmental chemistry is ideal. Our graduate Programmes consists of graduate courses which stress the fundamental photochemical, kinetic, analytical and transport aspects of environmental phenomena, regular seminars, and close interactions between the different research groups. We emphasize that students are able to put expertise in their own research field into a global context.

The field of environmental chemistry is rapidly expanding, and excellent employment opportunities exist in the academic, government, industrial and public policy sectors. The term environment refers to the natural elements that make up the Earth and surround living organisms, enabling life. The study of the environment and its proper conservation, has gained enormous importance in last three decades. The term environmental pollution has become one of the most important issues of global concern.

Accessible Components of Environment

- **Atmosphere:** This is a blanket of gaseous layer around the Earth.
- **Hydrosphere:** This includes all sources of water like oceans, rivers, lakes, glaciers, ground water, etc. and comprises about 96% of Earth's surface.
- **Lithosphere:** This comprises of soil, minerals, organic matter etc. and refers to Earth's solid crust containing the outer mineral cover.
- **Biosphere:** This is the domain of living organisms in contact withatmo-sphere, hydrosphere as well as lithosphere.

Man has begun to effect significant changes in his natural environment mainly after the advent of the Industrial revolution. The change in industrial and agricultural practices in the last 300 years and the direction of overall economic, technological and institutional development, has led to a tremendous increase in the use of energy and its impact on the environment. Due to these climatic changes, health hazards, effects on food and water supplies, changing habitats and eco-systems are effecting plants and animals life. Effects on the environment such as ozone depletion, greenhouse effect, acid rains, etc. have surfaced in recent years and are often the hot topics of discussion. The branch of chemistry, which deals with the study of various chemical phenomena occurring in the environment, is called environmental chemistry.

CONCEPT AND SCOPE

CONCEPT

Environmental chemistry" is the scientific study of the chemical and biochemical phenomena that occur in natural places. It should not be confused with green chemistry, which seeks to reduce potential pollution at its source. It can be defined as the study of the sources, reactions, transport, effects, and fates of chemical species in the air, soil, and water environments; and the effect of human activity on these. Environmental chemistry is an interdisciplinary science that includes atmospheric, aquatic and soil chemistry, as well as heavily relying on analytical chemistry and being related to environmental and other areas of science.

Environmental chemistry is that branch of one,which deals with the study of chemical and biochemical phenomena that occur in natural places like air, soil, water Environmental chemistry involves first understanding how the uncontaminated environment works, which chemicals in what concentrations are present naturally, and with what effects. Without this it would be impossible to accurately study the effects humans have on the environment through the release of chemicals.

Environmental chemists draw on a range of concepts from chemistry and various environmental sciences to assist in their study of what is happening to a chemical species in the environment. Important general concepts from chemistry include understanding chemical reactions and equations, solutions, units, sampling, and analytical techniques.

SCOPE

In chemistry the term mole represents a pile or mass of atoms, molecules, ions or electrons. Just as common man measures quantity in terms of kilograms or dozens, a chemical scientist deals with a 'mole' of atoms, molecules, ions or electrons. A mole is the amount of a substance that contains the same number of units of atoms or molecules or ions, as there are atoms in 12 g of the carbon 12 C6

The 12 g of carbon contains a certain number of atoms of carbon. This number is estimated to be $6.023 \times 10_{23}$ and is called Avogadro's number. Thus 12 g of the carbon 12 C is one mole of carbon. 6 Since the number of atoms in 12 g of carbon is called Avogadro's number, a mole may be redefined as the amount of a substance, which contains Avogadro number of molecules.

Avogadro's number is the number of elementary units (atoms, molecules, or ions) present in one mole of a substance, and it is equal to $6.023 \times 10_{23}$.

The concept of a mole being equal to a number $6.023 \times 10_{23}$ is akin to the concept of a dozen being equal to a number 12.

A mole of atoms consists of Avogadro's number of atoms, and has a mass equal to one gram atomic mass of that element.

Therefore, the following illustrations hold true:

- One mole of oxygen atoms contains $6.023 \times 10_{23}$ atoms of oxygen and has a mass of 16 g.
- One mole of sodium atoms contains $6.023 \times 10_{23}$ atoms of sodium, and has a mass of 23g.

Similarly one mole of molecule of a substance has Avogadro's number, $6.023 \times 10_{23}$, of molecules and has a mass equal to one gram molecular mass of that substance.

- One mole of oxygen molecules contains $6.023 \times 10_{23}$ molecules of oxygen and has a mass of 32 g.
- One mole of sodium (mono atomic) contains $6.032 \times 10_{23}$ molecules (or atoms, as it is mono atomic) and has a mass of 23 g.
- One mole of sulphuric acid contains $6.023 \times 10_{23}$ molecules of sulphuric acid and has a mass of 98 g.
- One mole of carbon dioxide contains $6.023 \times 10_{23}$ molecules of carbon dioxide, and has a mass of 44 g.

- One mole of an ionic substance like sodium chloride or calcium chloride contains $6.023 \times 10_{23}$ formula units.
- One mole of sodium chloride contains $6.023 \times 10_{23}$ formula units i.e., $6.023 \times 10_{23}$ sodium ions and $6.023 \times 10_{23}$ chloride ions. It has a mass of 58.5 g (i.e., 23 + 35.5) g.
- One mole of calcium chloride contains $6.023 \times 10_{23}$ formula units i.e., $6.023 \times 10_{23}$ calcium ions and $[2\ (6.023 \times 10_{23})]$ chloride ions. It has a mass of 111 g.
- One mole of aluminium chloride contains 6.023 × 1023 formula units of it, and contains $6.023 \times 10_{23}$ aluminium ions and $(3 \times 6.023 \times 10_{23})$ chloride ions. It has a mass of 133.5 (27 + {3 × 35.5}).

Normally, mass of 1 mole of a substance is called its molar mass. It is expressed as g/mol. For example molar mass of water is equal to 18 g/mol.)

COMPONENTS OF ENVIRONMENT

Components of environment can be Analysed at various levels i.e. At the level of activities, at the level of processes, and at the level of orientations.

Some of these issues and examples are discussed below.

As discussed the following activities have been suggested as components of environmental quality:

- Fire protection.
- Comfort of home.
- Electric service.
- Privacy in your home.
- Relation with fellow workers.
- Postal service.
- Garbage-collection.
- Mechanical helpers in your family.
- Telephone services.
- Public-water system.
- Relation with Neighbours.
- Gas service.
- Freedom to live where you want.
- Sewage disposal service.
- Availability of food around your living place.
- Convenience for getting to important places.
- Noise level in the home setting.
- Beauty of your home.
- Security of your home.
- Topography of land around.
- Product available to the community.
- Medical care in your locality.

- Police protection.
- Quality of water used by the household.
- Natural outdoor recreation.
- Variety of wild life in community.
- Cleanliness of air around.
- Overall weather.
- Public information media.
- Crowding in your residential Neighbourhood.
- Transportation over long distance.
- Level of traffic congestion.
- Job-opportunities.
- Quality of water for recreation.
- Product quality and variety.
- Relations among group in the community.
- Freedom to move from class to class.
- Freedom to move form on job to another.
- Public services-gas, sewage etc.
- Unspoiled natures.
- Man made environment.
- Physical condition of environment at school.
- Isolation of your community.
- Physical condition of environment where you work.
- Amount of open space around.
- Access to parks.
- Control of dogs and other pets.
- Cost of living.

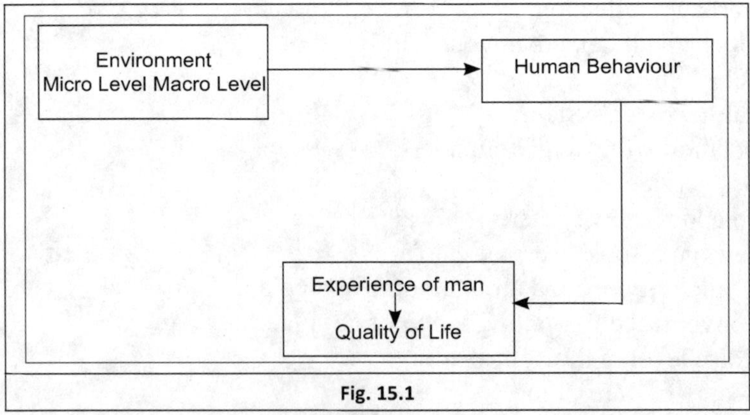

Fig. 15.1

Understanding environment involves studying the complex relationships between the people and the typical physical settings in which they conduct their daily lives. Environmental interest in studying human behaviour in the familiar, everyday physical environment where people live and work as well as its relevance to the environment design and social planning has made it especially responsive to the demands of today's world.

Environmental studies are an area of social sciences where the focus of investigation is the interrelationship between the physical environment and human Behaviour and experience of man. Fig below presents this.

The other perspective of components of environment relates to environmental processes, its multidisciplinary and applied aspects.

ENVIRONMENTAL PROCESSES

Environment is new subject and a complex field of study. Therefore it is important to consider some of the characteristics that describe the what, how and why of the ways environmental processes work. In this chapter we attempt to explain the adaptation focus, physiological processes, the holistic view, interdisciplinary involvement and applied orientation of the subject matter.

Adaptation Focus

The process of adaptation. It is of interest to study how people adapt to the complex demands of the physical environment. For example, how people live in crowded places or overcrowded setting. Robert White defines adaptation as encompassing all the processes with their environment.

This includes most simple ways of dealing with minor environment and irritations to the most complex efforts to cope with major environment changes. These challenges in living system in interaction with the environment are the adaptation processes.

The holistic view of the organism and environment is considered along with the active role of living organism in relation to their environment.

PHYSIOLOGICAL PROCESSES

The adaptation focus of the environment emphasizes the process that mediates the effect of physical setting on human activity e.g. the effect of classroom noise on student's grades.

OVERALL VIEW

Historically, the analysis of the environment was at very small and molecular levels but now the emphasis has shifted from micro to macro level. It is how environment and Behaviour must be seen as interrelated parts of an indivisible whole.

The positive and adaptive ways in which people cope with environmental challenges suggest active roles. This view looks into how people have varied and creative ways to cope with their environments.

The adaptation process model (shown above) shows how environment effect on Behaviour is mediated by a number of processes. It shows that the direction of effect in the environment-Behaviour relationship is reciprocal i.e. people may act on the environmental conditions while the environment in turn also acts on human Behaviour. The negative effects of situation such as over- crowding may be reversed through effective coping processes.

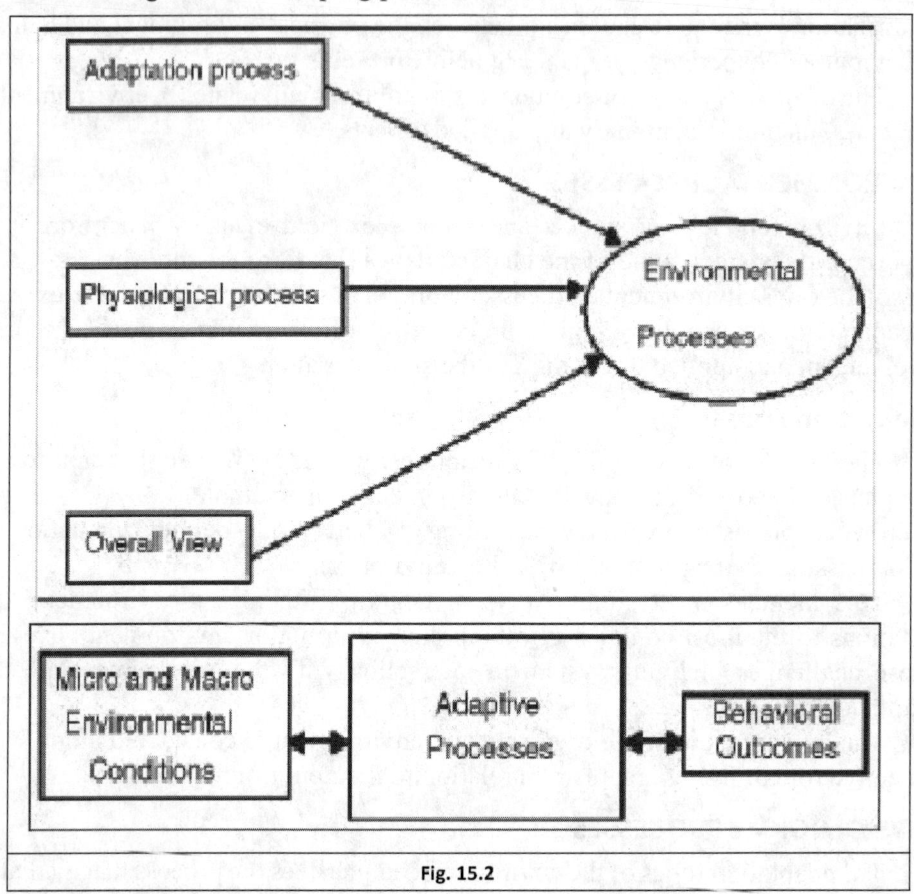

Fig. 15.2

MULTIDISCIPLINARY ORIENTATION

The branches of study here have included not only environment but also other fields such as Sociology, Psychology, Geography, Anthropology, Medicine, Architecture and Planning etc. The study of human Behaviour in physical settings requires the works of researchers in many social sciences as well as of architects and planners responsible for the design of human settings. The areas of study include.

APPLIED ORIENTATION

The environmental study has orientation towards both the resolution of practical problem and the formulation of new theories. Lewins action research is a useful model for the intermingling of practical and theoretical needs of the environment.

VALUATION TECHNIQUES

Various techniques, which attempt to means environmental effects, are below.

Preventive Expenditure Techniques

How much people are prepared to spend to prevent degradation of environment can be used as a measure of the value of the environmental effects. It is, however, usually thought that this technique underestimates the value of the environmental effects for various reasons.

If the full scopes of environmental effect including the long-team consequences are beyond the common people's comprehension, the expenditure they would undertake to prevent degradation of environment would all very much short of the required levels. Secondly, it may always not be possible to fully offset all environmental effects by preventive expenditure especially in case of massive investment projects (such as large dams) which cause large environmental changes.

Replacement Cost Techniques

In this technique, how much people are prepared to spend to restore the environment to its previous state after degradation has occurred is taken as a measure of the value of the environmental effects. This technique also suffers from shortcomings similar to those mentioned in the context of the preventive expenditure technique. It is normally beyond the competence of common people to judge the full scope of damages done to the environment and it is not usually possible to restore the environment to its previous state after it has been damaged.

Property Value Techniques

Real estate values depend on various factors, including environmental factors. A residential accommodation at the lake or with a view to maintain commands a premium in the market. Air pollution depresses price of property in the area. People are willing to pay more for property similar in other respects in areas with clean air.

The difference in market price can be used as a measure of the value to society of the cleaner environment. The measurement of this value in practice may however be quite difficult. In order to separate the effect of cleaner environment on properly value from other factors (such as location, quality of neighbourhood, etc) that affect a site's value, one would require to collect data as the value of various sites throughout the area and analyse the data with the help of an appropriates statistical method.

There have been several studies exploring the relationship between air pollution and property values, most of these studies show a significant negative relationship between properly value and air pollution. These studies have, however, been criticized for underestimating the environmental effects of air pollution. People often do not know or understand the severity of pollution problems at different sites.

It requires special expertise to accurately estimate the pollution level at a site. Even if people are told about the presence of air pollutants in the area, they cannot make any use of such information because they lack knowledge about the effects of pollutants on human health or property. In such conditions it is not logical to expect that difference in properly value would correctly estimate the value of environmental effects of air pollution.

Human Health - Effect Technique

One method to estimate the value of environmental effects as to study the effect of pollutant exposures on human health.

It involves putting a monetary value on illness and death caused by exposure to pollutants. This idea of putting money value on human life does not appeal to everyone. Many regard human life to be of infinite value. But, is society's behaviour always consistent with such thinking? IS society prepared to pay whatever is necessary to prevent a death? If human life were considered of infinite value, we would never allow those activities which pose a risk to human beings, given that virtually no economic activity is completely risk-free, and they all would be stopped.

For example, injurious and deaths caused by road accidents could be reduced significantly if laws pertaining to road safely are made very strict. Several studies have shown that appropriate changes in design of vehicles and lowering of speed limits can reduce the number and severity of traffic accidents significantly. But we know that society does not take all these required steps. This ofcourse, does not mean that human life is cheap. Most people consider it invaluable, if not of infinite value.

One way to determine the effects of exposure to pollutants on human health is to study the incidence of diseases among people. Of course, air pollution is not the role cause of diseases. There are many other explanatory variables such as age, sex, hygienic conditions, climate, living standard and dietary habits. In order to separate the effect of air pollution from other variables, a large body of data on the incidence of various diseases among different sections of the population in various places needs to be collected and analysed with the help of appropriate statistical methods. An alternative method of determining the effects of exposure of pollutants is to conduct Labouratory experiments on plants or animals. Plants animals are exposed to different levels and different kinds of pollutants to study their effects. Rats are often used as study objects for various reasons. The results of such studies, however, cannot be generalized too much. These studies cannot tell much about the effects of pollutants on human health as the human physiology is much different from that of rat or any other animal.

Once the relationship between pollutant exposures and human health effects is determined, the next step is to estimate the value of human health effects in monetary terms. The human health effect technique calculates the value of the health effects by looking at lost economic production and increased health-care costs. The opportunity costs of an individual's premature death can be calculated by estimating the discounted present value of his future earning if he had not died prematurely. This provides a measure of lost economic production and thus a measure of loss to society. In case of illness, the cost of care is added to the losses of earnings. This technique however does not value the psychic costs of ill health.

Many people find such procedure of calculating the worth of human life highly unsatisfactorily because according to this criterion a large section of people would command a very low economic value, especially the old people, disabled and those unemployed or in poorly paid jobs.

Travel Cost Method

This method estimates the time and money that people are willing to spend on 'eco-trips', i.e., visits to places of environmental value such as mountains, forests, lakes and parks. This method, however, severely underestimates the worth of the environmental assets and services because of its highly narrow focus. It only looks at the entertainment, recreational aspects of environmental assets. Environment constitutes the basic life support system; its total value consists of several components as discussed above. This approach ignores most of them.

Contingent Valuation

In this method, the value of environmental assets and services is assessed through public opinion surveys. People are directly asked their valuation of the environment. They are asked how much thing is willing to pay to conserve the environments.

There are problems, however, with approach. People often lack knowledge in that area; therefore their responses do not accurately reflect their concerns. Secondly, there is a general inherent problem with Survey method i.e. the answer to a question very much depends on the way it is asked. If people realise that their answer will have effect on the decisions they may reply in strategic manner.

For example, if how much they are willing to pay is a hypothetical question (i.e., no payments are involved), people will answer it in one way. On the other hand, if there is a probability that they may actually have to pay they will answer it in another way. In the later case, there would be a systematic underestimation of the value of the environment. If the understand that they do not have to pay, the polluter will pay, their responses are likely to overestimate the value of the environment.

If they are asked how much compensation they should be given for the environmental damages, they would certainly exaggerate the worth of the environment. For these reasons, it is quite difficult to interpret the survey responses and come up with a figure which genuinely represents people's valuation of the worth of the environment. Investigators should try to frame questions in such a manner which induces individuals to respond not as interested parties but as disinterested observers, as people concerned with the public interest.

For example The study of university housing environment is one research area where theoretical and practical objectives have been successfully combined. Thus one may understand the components of environment at various levels i.e. At the level of activities, at the level of processes and at the level of orientations.

ENVIRONMENT AND ECOLOGY

A Systems Perspective

In the earlier sections it has been explained that there are many components of the concept and these concerns have always been there ever since the pre-historic time. The fact is that environment and ecology do not stand in isolation. Man. Society and culture are affected and in turn affect these concerns.

These could be viewed as the systems analysis. A systems thinking suggest that all subsystems of the main system interact with each other and if one sub system doesn't make impact it affects the total system. The systems analysis could further be studied as a closed system analysis and an open system analysis. Fig below presents a closed system analysis.

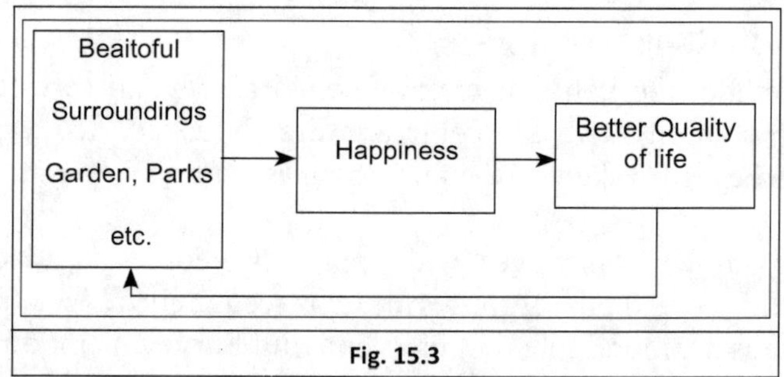

Fig. 15.3

Similarly, there are ongoing impacts of environment on the quality of life of people, when time moves on. In fact, systems approach, views interactions and interdependence among subsystems are important as individual components. If one of the subsystems malfunctions, it affects the performance of the entire system.

Boulding (1956) has pointed out different levels of systems ranging from static structures to cybernetic structures and human systems. He believes that human systems and social systems are most complicated ones and we have yet to discover the rudiments of theoretical models for them.

Keeping these analogies and components analysis in mind one could further analyse ecology and environment within a systems perspective.

MULTIDISCIPLINARY PERSPECTIVE

Taking these as components of environment and ecology this web based course attempts to present a multidisciplinary system approach. The concepts are drawn from all subjects relevant to explain the environment scenario. Appropriate

Technology, Psychology, Sociology, Economics, Management, and Philosophy, Values and Ethics are some subjects from where concepts and concerns are drawn. The course also presents methodological approaches, training, ideologies, globalization, ecological behaviour and knowledge management relating them to environment and ecology.

The 21st century is a century of shorter distances, closer places but one that is plagued by a multitude of problems, man made as well as natural disasters, catalyzed by man made actions.

The growing populace with its increased technology usc is one that is seemingly empowered to control its actions and yet susceptible to the minutest happenings as were its forefathers.

Globalism - refers to an appreciation of the world as a single place or space and the ways in which people have internalized a basic view of the world as a common domain in which activities or events (natural or man-made)in one part might have profound effects in another, distant one. It is an awareness of large scale knock-on effects.

Today Ecology has moved from the realm of biology to become a stand alone discipline that encompasses all facets of human activity be they economic social or developmental. The word comes from the greek work for oikos, houseand logi -logy. Ecology and environment encompass a number of disciplines.
Some of the ecological systems are:

- Community ecology
- Ecosystem ecology
- Biological ecology and diversity
- Geographical ecology
- Oceanographic and marine ecology
- Ecosystems of the world
- Evolutionary ecology
- Political ecology

ENVIRONMENT: A MULTIDISCIPLINARY PERSPECTIVE

Environment as a concept is this multidisciplinary involving a number of disciplines as presented earlier in fig which is repeated here.

Social Science	Technology
Economic	FactorsPolicy related factors
Social factors	Psychological factors

Environmental analysis also includes the understanding of multiple issues related to disciplines.
Some of the examples are:

- Air pollution
- Water pollution

- Industrial pollution
- Social environment
- Psychological environment
- Biological environment
- Ozone depletion
- Global climate change
- Rise in sea level and other natural changes
- Antarctica: a source for resource development
- Marine resources
- Offshore oil and gas
- Manmade habitats etc

This calls for:

- Studies of sustainable development
- Environmental politics
- Environmental audit
- Suitable environmental policies and legislation Therefore, in this section we have visualized an overall perspective of ecology and environment from multidisciplinary angle. The next section presents some of the methodological considerations to Analyse these issues.

ENVIRONMENT ANDECOLOGY ANALYSIS

Scientific Social Research

It would be difficult to comprehend the nature and content of research without an appreciation of the method we designate as scientific. Research, simply put is an endeavour to discover intellectual and practical answer to problems through the application of scientific methods to knowable universe. The scientific methods increase the likelihood of answering questions related to WHAT, HOW And Why in an enquiry. Einstein and Infield state the essence of science in the following words-"Science is the attempt of the human mind to find a connection between the world of ideas and world of phenomena. All the essential ideas in science were born in dramatic conflict between reality and our attempt at understanding the same." This does not mean, however, that science will always come out with an answer... Thompson observes "The belief that science has explained everything is hopeless misunderstanding." There may be many paths leading to truth, however scientific is entails more understandable and reliable. Meaning of Research therefore is 'A careful, critical enquiry or examination in seeking facts or principles in order to as certain something.

Scientific Social Research

It would be difficult to comprehend the nature and content of research without appreciation of the methods we designate as scientific. According to Scientific Social Research have the following properties: Reliance on empirical evidence Use of relevant concepts Commitment to objectivity Ethical neutrality Predictions based on probability Testing of conclusions through replication.

Research Process

There are various measures to study the effects of environment. Earlier strategies were not that sophisticated and reliable, thus mere have been fascinating development in this field.

- Co relational studies.
- Experimental studies in Labouratory settings.
- Small scale fields settings.
- Diverse research strategies.
- Case studies.
- Evaluation studies.
- Econometric studies.
- Time series studies.

Correlation studies based census data: It is based on study relationship between density and social pathology as reflected by cause data or over statistical records e.g. mere may be comparison between mortality, mental illness and serious crime in densely populated region and sparsely populated suburbs.

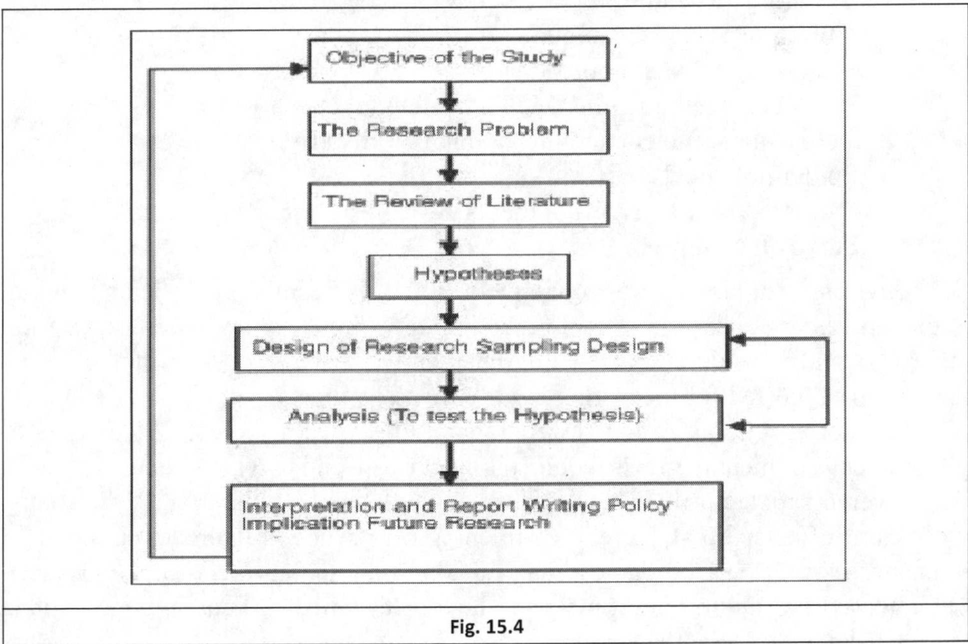

Fig. 15.4

- Experimental studies: A more recent research strategy 1960's involves experiment in lab settings. Initial studies were dared on animals in controlled settings the god of this is study is to establish a casual like between crowding and social effects on humans. Later were also done as humans.
- Study of environment in small-scale setting: Most recent strategy such as - dormitories stores etc include - survey -questionnaires etc.

Diverse Research Strategies

Each strategy has its strength and weakness. In some animals studies it is the size of group that is more important man me space available to it also effect of density have inserted with other variables such as species type situation etc.

Approach for Environmental Studies

There are number of models and approaches to study environment and ecology. Measuring Environmental Values, impact assessment, people's reactions, environmental protection and crisis etc. there are many Economics based approaches issues and approaches.

They are:

- Market values
- Indirect methods
- Contingent valuation

Some of the indirect methods are:

- Preventive expenditure
- Replacement cost technique
- Human capital technique
- Travel cost method Other Methods include
- Technical coefficient and the value of Human Life
- Epidemiological studies
- Property value and Air pollutants
- Survey techniques

Environmental issues are too vital to be left out of economic decision making. Social cost benefit analysis cannot be complete without taking into consideration environmental considerations. In this chapter we discuss various methods that have been suggested for measuring environmental values.

It is not easy to asses the environmental effects of a project. What types of physical environmental effects would a project cause and to what extent, can rarely be answered satisfactorily. There is always an element of uncertainty associated with there effects. Most large investment projects now require environmental impact assessment (EIAs) that estimates the environmental effects of the projects. Often these EIAs churn out estimates of the effects, which are not only incomplete but also, devoid of scientist rigour.

Even if the physical effects are assessed with reasonable degree of accuracy and certainty, the next problem one encounters is how to assign economic values to these effects. It is a widely held view that a market economy does not value all aspects of the environment appropriately. Markets and prices may not exist for many of the environmental effects that require valuation.

Even if market prices exist, they may be distorted for various reasons. The environmental effects of big projects are very large. In such cases, the project may actually alter prices. For these various reasons indirect methods are used for valuation of the environmental effects. There are a number of techniques developed by economists who attempt to value the environmental effects. Before discussing these various techniques it needs to be noted that total economic value (TEV) of environmental assets consists of use and no-use values. The total economic value has four components:

TEV = Direct use values + Indirect use values + Option value + Existence value

Let us consider the case of tropical forest. Direct values are the resources and services provided directly by the forest such as habitat and food for indigenous people, timber and minor non-timber products. Indirect use values of environmental assets are essentially their ecological functions, which indirectly support economic activity and human welfare. Tropical forests, for example, protect adjacent eco-systems, provide watershed protection, and are store of CO_2. An option value is the value placed on an option that allows for use of the environment in the future.

The amount individuals would be willing to pay to conserve a tropical forest, or at least some of its uses, for future would be its option value. Existence value of an environmental asset in the value placed on it independent of its any actual or possible future use. It relates to valuation of the resources as a unique asset in itself, an object of intrinsic value. Existence value of tropical forests arises because people simply want them to exist, even if they are never likely to ever see or use then in any way.

Environmental Impact Assessment: includes Environmental Inventory; Environmental Assessment; Environment Impact Statement; Historical Perspective; Environmental Assessment Process; Goals of EIA Process need to emphasize methodological issues related to following aspects:

- Air Environment.
- Water Environment.
- Noise Environment.
- Socio-Economic Environment.
- Cultural Environment.
- Biological Environment.
- People's Participation.
- Case Studies are also important in this regard. Some examples.
- Air Pollution Monitoring.

Ganga Action Plan

- **The study of Ganga:** The Holy River; Ganga; From Gomukh to Bay of Bengal; Ganga; Its importance, Ganga; Overuse, Abuse and Pollution; What is River Pollution? What are the Principal Sources of Pollution? How to measure the extent of pollution? Concern for the River; Formulation of the Ganga Action Plan; what does the Action Plan seek to do?
- **River Front Facilities:** Low Cost Sanitation; Construction of Electric Crematoria, Pollution from Industries; Scientific Content in the Context of Ganga Action Plan;Public Involvement in the Ganga Action Plan; Monitoring of Water Quality; Recycling Sewage; Where we went wrong. What do we need to do? Can people's participation help in this project?
- **Industrial Pollution:** Some Case Studies: Pollution Control Aspects of Fertilizer Industries; Ammonia Plant Effluents; Removal of Carbon; Oil Removal; Hydrogen Sulphide Removal; Nitrogen-Wash Tail Gas; Ammonium Sulphate Plant; Phosphoric Acid Plant; Solid Effluent; Liquid Effluent; Gaseous Effluent; Pollution Control in Petroleum Refineries and Petrochemical Units; Characteristics of Liquid Effluent; Refineries; Petrochemical Industries; Refinery Liquid Waste-Treatment Methods; Implant Measures; Steam Stripping; Oxidation; Removal of Phenols; Biological Treatment; Disposal of Sludge's; Treatment of Liquid Effluents from Petrochemical Industries.
- **Environmental Audit:** Case studies Indian Scenario Advantage of Audit; Environmental Audit. How to measure the extent of pollution? Concern for the River; Formulation of the Ganga Action Plan; what does the Action Plan seek to do? River Front Facilities; Low Cost Sanitation; Construction of Electric Crematoria, Pollution from Industries; Scientific Content in the Context of Ganga Action Plan;Public Involvement in the Ganga Action Plan; Monitoring of Water Quality; Recycling Sewage; Where we went wrong. What do we need to do? Can people's participation help in this project?
- **Industrial Pollution:** Some Case Studies: Pollution Control Aspects of Fertilizer Industries; Ammonia Plant Effluents; Removal of Carbon; Oil Removal; Hydrogen Sulphide Removal; Nitrogen-Wash Tail Gas; Ammonium Sulphate Plant; Phosphoric Acid Plant; Solid Effluent; Liquid Effluent; Gaseous Effluent; Pollution Control in Petroleum Refineries and Petrochemical Units; Characteristics of Liquid Effluent; Refineries; Petrochemical Industries; Refinery
- Liquid Waste-Treatment Methods; Implant Measures; Steam Stripping; Oxidation; Removal of Phenols; Biological Treatment; Disposal of Sludge's; Treatment of Liquid Effluents from Petrochemical Industries.

CYCLES OF MATTER IN ENVIRONMENT

Environmental problems caused by human interference in matter cycles. In the past decades humans have been the cause of extensive environmental pollution. Pollution is not a one-sided term; all kinds of matter has been applied by humans, resulting in a disturbance in natural processes. Disturbances of natural processes are clearly shown, when you take a look at human interference in matter cycles. This has caused various environmental problems, which are important issues today.

HYDROLOGICAL CYCLE H_2O

There are three ways in which humans cause environmental problems by interfering with the hydrological cycle. Firstly, we remove large quantities of freshwater from rivers, lakes and groundwater supplies. In areas where great amounts of people use the water from groundwater supplies, groundwater may be fully depleted. This effect is enhanced by vegetation removal to create new farmland or underground mines, or to build roads. Secondly, because of vegetation removal, rainwater washes away and it no longer infiltrates, so that the groundwater supplies are not restored. In many cases groundwater supplies will than be filled up with salt water from rivers and lakes. When groundwater is salinated the entire area is influenced. Other effects also occur because of vegetation removal; the risk of flooding is enhanced, and soil erosion and land sliding increase. Finally, people change the quality of water, by adding nutrients and contaminants.

This causes the ecological processes that usually purify the water to be disturbed:

- **Precipitation:** Transported through the circulating atmosphere the clouds move themselves inland, as a result of gravity, and lose their water as it falls back unto the ground. This phenomenon is called rain or precipitation.
- **Infiltration:** Rainwater infiltrates into the ground and sinks to the saturated zone, where it becomes groundwater. Groundwater slowly moves from places of high elevation and pressure to places with low elevation and pressure. It moves from the area of infiltration through an aquifer and out to a discharge area, which can be either a sea or an ocean.
- **Transpiration:** Plants and other forms of vegetation take up water from the soil and excrete it again as water vapour. About 10% of the precipitation that falls on the ground vapourizes again through transpiration of plants, the rest evaporates from seas and oceans.
- **Surface Run-off:** The rainwater that does not infiltrate into the soil will directly reach the surface water, as it will run-off to rivers and lakes. After that it will be transported back to the seas and oceans. This water is called surface run-off.
- **Evaporation:** Due to the influence of sunlight the water in oceans and lakes will warm up. As a result of that it will evaporate and rise up into the atmosphere. There it will form clouds that will eventually cause rainwater

to fall back on earth. The evaporation of oceans is the most important kind of evaporation.

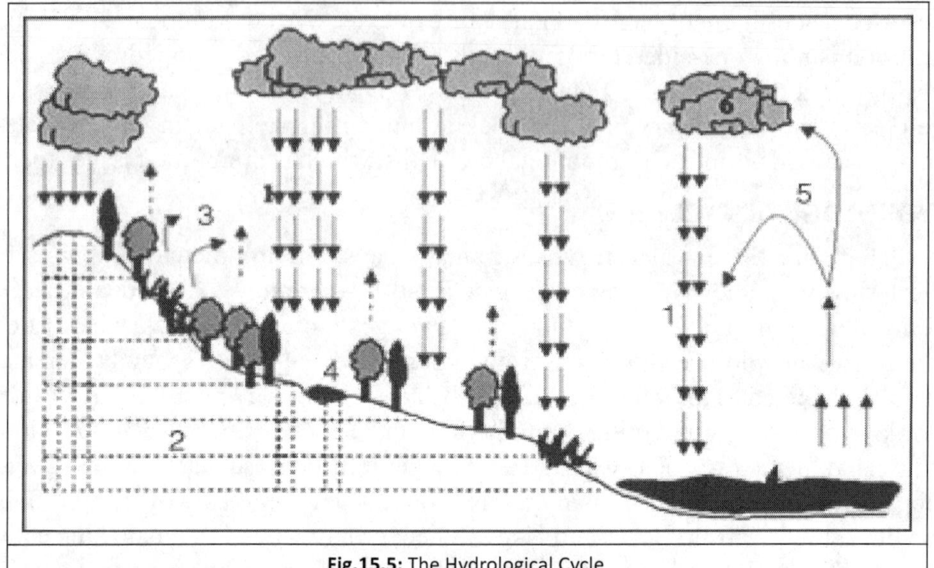

Fig.15.5: The Hydrological Cycle

- Condensation: In contact with the atmosphere the water vapour will transform back to liquid, so that it will be visible in the air. These accumulations of water in the air are what we call clouds.

Carbon Cycle

The geological component of the carbon cycle is where it interacts with the rock cycle in the processes of weathering and dissolution, precipitation ofminerals, burial and subduction, and volcanism. In the atmosphere, carbonic acid forms by a reaction with atmospheric carbon dioxide (CO_2) and water. As this weakly acidic water reaches the earth as rain, it reacts with minerals at the earth's surface, slowly dissolving them into their component ions through the process of chemical weathering. These component ions are carried in surface waters like streams and rivers eventually to the ocean, where they precipitate out as minerals like calcite ($CaCO_3$). Through continued deposition and burial, this calcite sediment forms the rock called limestone. This cycle continues as seafloor spreading pushes the seafloor under continental margins in the process of subduction. As seafloor carbon is pushed deeper into the earth by tectonic forces, it heats up, eventually melts, and can rise back up to the surface, where it is released as CO_2 and returned to the atmosphere. This return to the atmosphere can occur violently through volcanic eruptions, or more gradually in seeps, vents, and CO_2-rich hotsprings. Tectonic uplift can also expose previously buried limestone. One example of this occurs in the Himalayas where some of the world's highest peaks are formed of material that was once at the bottom of the ocean. Weathering, subduction, and volcanism control atmospheric carbon dioxide concentrations over time periods of hundreds of millions of years.

Biological Carbon Cycle

Biology plays an important role in the movement of carbon between land, ocean, and atmosphere through the processes of photosynthesis and respiration. Virtually all multicellular life on Earth depends on the production of sugars from sunlight and carbon dioxide (photosynthesis) and the metabolic breakdown (respiration) of those sugars to produce the energy needed for movement, growth, and reproduction. Plants take in carbon dioxide (CO_2) from the atmosphere during photosynthesis, and release CO_2 back into the atmosphere during respiration through the following chemical reactions:

$$\text{Respiration: } C_6H_{12}O_6 \text{ (organic matter)} + 6O_2 \rightarrow 6CO_2 + 6\,H_2O + \text{energy}$$

$$\text{Photosynthesis: Energy (sunlight)} + 6CO_2 + H_2O \rightarrow C_6H_{12}O_6 + 6O_2$$

Through photosynthesis, green plants use solar energy to turn atmospheric carbon dioxide into carbohydrates (sugars). Plants and animals use these carbohydrates (and other products derived from them) through a process called respiration, the reverse of photosynthesis. Respiration releases the energy contained in sugars for use in metabolism and changes carbohydrate "fuel" back into carbon dioxide, which is in turn released to back to the atmosphere. The amount of carbon taken up by photosynthesis and released back to the atmosphere by respiration each year is about 1,000 times greater than the amount of carbon that moves through the geological cycle on an annual basis.

On land, the major exchange of carbon with the atmosphere results from photosynthesis and respiration. During daytime in the growing season, leaves absorb sunlight and take up carbon dioxide from the atmosphere. At the same time plants, animals, and soil microbes consume the carbon in organic matter and return carbon dioxide to the atmosphere. Photosynthesis stops at night when the sun cannot provide the driving energy for the reaction, though respiration continues. This kind of imbalance between these two processes is reflected in seasonal changes in the atmospheric CO_2 concentrations. During winter in the northern hemisphere, photosynthesis ceases when many plants lose their leaves, but respiration continues. This condition leads to an increase in atmospheric CO_2 concentrations during the northern hemisphere winter.

With the onset of spring, however, photosynthesis resumes and atmospheric CO_2 concentrations are reduced. This cycle is reflected in the monthly means of atmospheric carbon dioxide concentrations shown in Figure.

In the oceans, phytoplankton (microscopic marine plants that form the base of the marine food chain) use carbon to make shells of calcium carbonate ($CaCO_3$). The shells settle to the bottom of the ocean when phytoplankton die and are buried in the sediments.

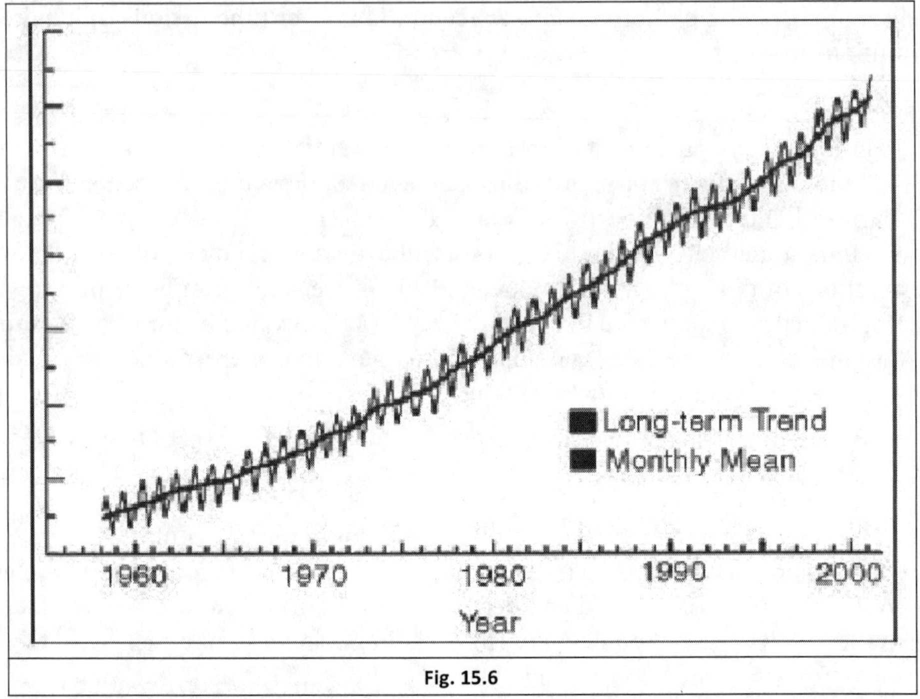

Fig. 15.6

The shells of phytoplankton and other creatures can become compressed over time as they are buried and are often eventually transformed into limestone. Additionally, under certain geological conditions, organic matter can be buried and over time form deposits of the carbon-containing fuels coal and oil. It is the non-calcium containing organic matter that is transformed into fossil fuel. Both limestone formation and fossil fuel formation are biologically controlled processes and represent long-term sinks for atmospheric CO_2.

ALTERATION OF CARBON CYCLE

Recently, scientists have studied both short- and long-term measurements of atmospheric CO_2 levels. Charles Keeling, an oceanographer at the Scripps Institute of Oceanography, is responsible for creating the longest continuous record of atmospheric CO_2 concentrations, taken at the Mauna Loa observatory in Hawaii. His data (now widely known as the "Keeling curve," shown in Figure) revealed that human activities are significantly altering the natural carbon cycle.

Since the onset of the industrial revolution about 150 years ago, human activities such as the burning of fossil fuels and deforestation have accelerated, and both have contributed to a long-term rise in atmospheric CO_2. Burning oil and coal releases carbon into the atmosphere far more rapidly than it is being removed, and this imbalance causes atmospheric carbon dioxide concentrations to increase.

In addition, by clearing forests, we reduce the ability of photosynthesis to remove CO_2 from the atmosphere, also resulting in a net increase. Because of these human activities, atmospheric carbon dioxide concentrations are higher today than they have been over the last half-million years or longer. Because CO_2 increases the atmosphere's ability to hold heat, it has been called a "greenhouse gas." Scientists believe that the increase in CO_2 is already causing important changes in the global climate.

Many attribute the observed 0.6 degree C increase in global average temperature over the past century mainly to increases in atmospheric CO_2. Without substantive changes in global patterns of fossil fuel consumption and deforestation, warming trends are likely to continuc.

The best scientific estimate is that global mean temperature will increase between 1.4 and 5.8 degrees C over the next century as a result of increases in atmospheric CO_2 and other greenhouse gases.

This kind of increase in global temperature would cause significant rise in average sea-level (0.09-0.88 meters), exposing low-lying coastal cities or cities located by tidal rivers such as New Orleans, Portland, Washington, and Philadelphia to increasingly frequent and severe floods. Glacial retreat and species range shifts are also likely to result from global warming, and it remains to be seen whether relatively immobile species such as trees can shift their ranges fast enough to keep pace with warming.

Even without the changes in climate, however, increased concentrations of CO_2could have an important impact on patterns of plant growth worldwide. Because some species of plants respond more Favourably to increases in CO_2 than others, scientists believe we may see pronounced shifts in plant species as a result of increasing atmospheric CO_2 concentrations, even without any change in temperature.

For example, under elevated CO_2 conditions, shrubs are thought to respond more Favourably than certain grass species due to their slightly different photosynthetic pathway. Because of this competitive inequality, some scientists have hypothesized that grasslands will be invaded by CO_2-responsive grass species or shrubby species as CO_2 increases.

In an attempt to understand whether recently observed changes in the global carbon cycle are a new phenomenon, or have instead occurred throughout geologic history, scientists have devoted considerable effort to developing methods for understanding Earth's past atmosphere and climate.

These techniques include analysis of gas bubbles trapped in ice, tree rings, and ocean and lake floor sediments for clues about past climates and atmospheres. Together, these techniques suggest that over the past 20 million years, the Earth's climate has oscillated between relatively warm and relatively cold conditions called interglacial and glacial periods. During interglacial periods, atmospheric CO_2 concentrations were relatively high, and during glacial periods, CO_2 concentrations were relatively low. We are currently in an interglacial warm period, and human activities are pushing CO_2 concentrations higher than they have been for hundreds of thousands of years (Figure).

Understanding and mitigating negative impacts of atmospheric CO_2 enrichment constitute two of the most central challenges that environmental scientists and policy makers currently face. In order to address this issue, the scientific community has formed the Intergovernmental Panel on Climate Change (IPCC), an international, interdisciplinary consortium comprised of thousands of climate experts colLabourating to produce consensus reports on climate change science.

Many nations have agreed to conditions specified by the Kyoto Protocol, a multilateral treaty aimed at averting the negative impacts associated with human-induced climate change. The country, which is currently responsible for approximately one quarter of global CO_2 emissions, has so far declined to participate in the Kyoto Protocol.

NITROGEN CYCLE

The nitrogen cycle involves the uptake of nitrogen form the atmosphere by a process called fixation which is carried out by microbes or industrial processes. Decomposition of biological waste by microbes can return nitrogen to the atmosphere. Nitrogen is mainly used by humans as a fertilizer in farmlands, but its excessive usage can lead to serious problems (such as eutrophication).

Nitrogen (N) is an essential component of DNA, RNA, and proteins, the building blocks of life. All organisms require nitrogen to live and grow. Although the majority of the air we breathe is N_2, most of the nitrogen in the atmosphere is unavailable for use by organisms. This is because the strong triple bond between the N atoms in N_2 molecules makes it relatively inert.

In fact, in order for plants and animals to be able to use nitrogen, N_2 gas must first be converted to more a chemically available form such as ammonium (NH_{4+}), nitrate (NO_{3-}), or organic nitrogen (e.g. urea–(NH_3)2CO).

The inert nature of N_2 means that biologically available nitrogen is often in short supply in natural ecosystems, limiting plant growth and biomass accumulation. Nitrogen is an incredibly versatile element, existing in both inorganic and organic forms as well as many different oxidation states. The movement of nitrogen between the atmosphere, biosphere, and geosphere in different forms is described by the nitrogen cycle (Figure below), one of the major biogeochemical cycles.

Similar to the carbon cycle, the nitrogen cycle consists of various storage pools of nitrogen and processes by which the pools exchange nitrogen (arrows). Five main processes cycle nitrogen through the biosphere, atmosphere, and geosphere: nitrogen fixation, nitrogen uptake (organismal growth), nitrogen mineralization (decay), nitrification, and denitrification. Microorganisms, particularly bacteria, play major roles in all of the principal nitrogen transformations.

As microbially mediated processes, these nitrogen transformations tend to occur faster than geological processes like plate motion, a very slow, purely physical process that is a part of the carbon cycle.

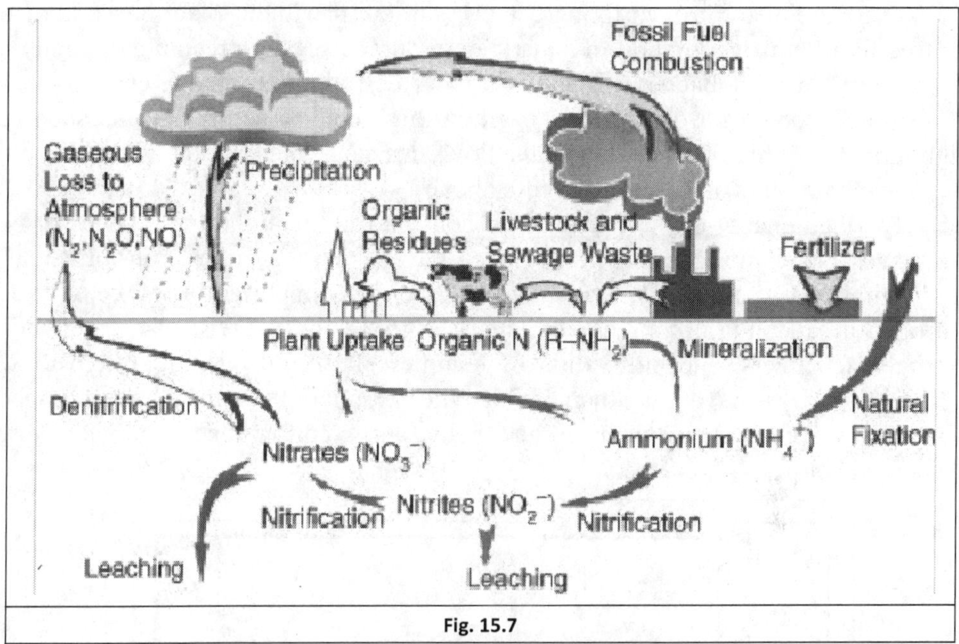

Fig. 15.7

Instead, rates are affected by environmental factors that influence microbial activity, such as temperature, moisture, and resource availability.

Fig. 15.8

Nitrogen Fixation

$N_2 \rightarrow NH_{4+}$ Nitrogen fixation is the process wherein N_2 is converted to ammonium, essential because it is the only way that organisms can attain nitrogen directly from the atmosphere. Certain bacteria, for example those among the genus Rhizobium, are the only organisms that fix nitrogen through metabolic processes.

Nitrogen fixing bacteria often form symbiotic relationships with host plants. This symbiosis is well-known to occur in the legume family of plants (e.g. beans, peas, and clover). In this relationship, nitrogen fixing bacteria inhabit legume root nodules (Figure) and receive carbohydrates and a Favourable environment from their host plant in exchange for some of the nitrogen they fix.

There are also nitrogen fixing bacteria that exist without plant hosts, known as free-living nitrogen fixers. In aquatic environments, blue-green algae (really a bacteria called cyanobacteria) is an important free-living nitrogen fixer.

In addition to nitrogen fixing bacteria, high-energy natural events such as lightning, forest fires, and even hot lava flows can cause the fixation of smaller, but significant amounts of nitrogen (Figure above).

The high energy of these natural phenomena can break the triple bonds of N_2 molecules, thereby making individual N atoms available for chemical transformation.

Within the last century, humans have become as important a source of fixed nitrogen as all natural sources combined. Burning fossil fuels, using synthetic nitrogen fertilizers, and cultivation of legumes all fix nitrogen. Through these activities, humans have more than doubled the amount of fixed nitrogen is pumped into the biosphere every year (Figure below), the consequences of which are discussed below.

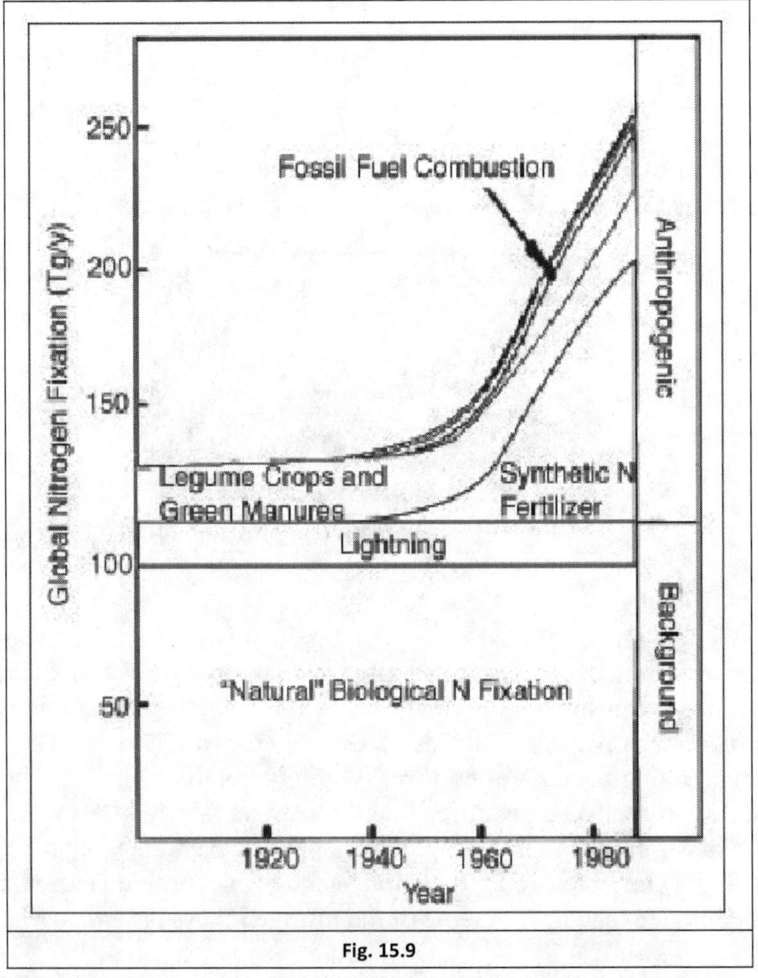

Fig. 15.9

Nitrogen Uptake

NH_{4+} → Organic N The ammonia produced by nitrogen fixing bacteria is usually quickly incorporated into protein and other organic nitrogen compounds, either by a host plant, the bacteria itself, or another soil organism. When organisms nearer the top of the food chain (like us!) eat, we are using nitrogen that has been fixed initially by nitrogen fixing bacteria.

Nitrogen Mineralization

Organic N →NH_{4+} After nitrogen is incorporated into organic matter, it is often converted back into inorganic nitrogen by a process called nitrogen mineralization, otherwise known as decay. When organisms die, decomposers (such as bacteria and fungi) consume the organic matter and lead to the process of decomposition. During this process, a significant amount of the nitrogen contained within the dead organism is converted to ammonium. Once in the form of ammonium, nitrogen is available for use by plants or for further transformation into nitrate (NO_{3-}) through the process called nitrification.

Nitrification

$$NH_{4+} \rightarrow NO_{3-}$$

Some of the ammonium produced by decomposition is converted to nitrate via a process called nitrification. The bacteria that carry out this reaction gain energy from it. Nitrification requires the presence of oxygen, so nitrification can happen only in oxygen-rich environments like circulating or flowing waters and the very surface layers of soils and sediments.

The process of nitrification has some important consequences. Ammonium ions are positively charged and therefore stick (are sorbed) to negatively charged clay particles and soil organic matter. The positive charge prevents ammonium nitrogen from being washed out of the soil (or leached) by rainfall. In contrast, the negatively charged nitrate ion is not held by soil particles and so can be washed down the soil profile, leading to decreased soil fertility and nitrate enrichment of downstream surface and groundwater's.

Denitrification

NO_{3-}→N_2+ N_2O Through denitrification, oxidized forms of nitrogen such as nitrate and nitrite (NO_{2-}) are converted to dinitrogen (N_2) and, to a lesser extent, nitrous oxide gas. Denitrification is an anaerobic process that is carried out by denitrifying bacteria, which convert nitrate to dinitrogen in the following sequence:

$$NO_{3-} \rightarrow NO_{2-} \rightarrow NO \rightarrow N_2O \rightarrow N_2.$$

Nitric oxide and nitrous oxide are both environmentally important gases. Nitric oxide (NO) contributes to smog, and nitrous oxide (N_2O) is an importantgreenhouse gas, thereby contributing to global climate change.

Once converted to dinitrogen, nitrogen is unlikely to be reconverted to a biologically available form because it is a gas and is rapidly lost to the atmosphere.

Denitrification is the only nitrogen transformation that removes nitrogen from ecosystems (essentially irreversibly), and it roughly balances the amount of nitrogen fixed by the nitrogen fixers described above.

ALTERATION OF N CYCLE AND ITS CONSEQUENCES

Early in the 20th century, a German scientist named Fritz Haber figured out how to short circuit the nitrogen cycle by fixing nitrogen chemically at high temperatures and pressures, creating fertilizers that could be added directly to soil. This technology has spread rapidly over the past century, and, along with the advent of new crop varieties, the use of synthetic nitrogen fertilizers has led to an enormous boom in agricultural productivity.

This agricultural productivity has helped us to feed a rapidly growing world population, but the increase in nitrogen fixation has had some negative consequences as well. While the consequences are perhaps not as obvious as an increase in global temperatures or a hole in the ozone layer, they are just as serious and potentially harmful for humans and other organisms.

Not all of the nitrogen fertilizer applied to agricultural fields stays to nourish crops. Some is washed off of agricultural fields by rain or irrigation water, where it leaches into surface or ground water and can accumulate. In groundwater that is used as a drinking water source, excess nitrogen can lead to cancer in humans and respiratory distress in infants. The U.S. Environmental Protection Agency has established a standard for nitrogen in drinking water of 10 mg per litre nitrate-N. Unfortunately, many systems (particularly in agricultural areas) already exceed this level.

By comparison, nitrate levels in waters that have not been altered by human activity are rarely greater than 1 mg/L. In surface waters, added nitrogen can lead to nutrient over-enrichment, particularly in coastal waters receiving the inflow from polluted rivers. This nutrient over-enrichment, also called eutrophication, has been blamed for increased frequencies of coastal fish-kill events, increased frequencies of harmful algal blooms, and species shifts within coastal ecosystems.

Reactive nitrogen (like NO_{3-} and NH_{4+}) present in surface waters and soils, can also enter the atmosphere as the smog-component nitric oxide (NO) and thegreenhouse gas nitrous oxide (N_2O). Eventually, this atmospheric nitrogen can be blown into nitrogen-sensitive terrestrial environments, causing long-term changes.

For example, nitrogen oxides comprise a significant portion of the acidity in acid rain which has been blamed for forest death and decline in parts of Europe and the Northeast United States. Increases in atmospheric nitrogen deposition have also been blamed for more subtle shifts in dominant species and ecosystem function in some forest and grassland ecosystems.

For example, on nitrogen-poor serpentine soils of northern Californian grasslands, plant assemblages have historically been limited to native species that can survive without a lot of nitrogen. There is now some evidence that elevated levels of atmospheric N input from nearby industrial and agricultural development have paved the way for invasion by non-native plants. As noted earlier, NO is also a major factor in the formation of smog, which is known to cause respiratory illnesses like asthma in both children and adults.

Currently, much research is devoted to understanding the effects of nitrogen enrichment in the air, groundwater, and surface water. Scientists are also exploring alternative agricultural practices that will sustain high productivity while decreasing the negative impacts caused by fertilizer use. These studies not only help us quantify how humans have altered the natural world, but increase our understanding of the processes involved in the nitrogen cycle as a whole.

Nitrogen is a part of vital organic compounds in microrganisms, such as amino acids, proteins and DNA. The gaseous form of nitrogen (N_2), makes up 8% of the troposphere. One might think this means we always have plenty of nitrogen available, but unfortunately it does not work that way. Nitrogen in the gaseous form cannot be absorbed and used as a nutrient by plants and animals; it must first be converted by nitrifying bacteria, so that it can enter food chains as a part of the nitrogen cycle. During the conversion of nitrogen cyano bacteria will first convert nitrogen into ammonia and ammonium, during the nitrogen fixation process. Plants can use ammonia as a nitrogen source.

Nitrogen fixation is carried out according to the following reaction:

$$N_2 + 3H_2 \rightarrow 2\ NH_3$$

After ammonium fixation, the ammonia and ammonium that is formed will be transferred further, during the nitrification process. Aerobic bacteria use oxygen to convert these compounds. Nitrosomonas bacteria first convert nitrogen gas to nitrite (NO_{2-}) and subsequently nitrobacter convert nitrite to nitrate (NO_{3-}), a plant nutrient.

Nitrification is carried out according to the following reactions:

$$2\ NH_3 + 3O_2 \rightarrow 2\ NO_2 + 2\ H_+ + 2\ H_2O$$

$$2\ NO_{2-} + O_2 \rightarrow 2\ NO_{3-}$$

Plants absorb ammonium and nitrate during the assimilation process, after which they are converted into nitrogen-containing organic molecules, such as amino acids and DNA. Animals cannot absorb nitrates directly. They receive their nutrient supplies by consuming plants or plant-consuming animals.

When nitrogen nutrients have served their purpose in plants and animals, specialized decomposing bacteria will start a process called ammonification, to convert them back into ammonia and water-soluble ammonium salts. After the nutrients are converted back into ammonia, anaerobic bacteria will convert them back into nitrogen gas, during a process called denitrification.

Denitrification is carried out according to the following reaction:

$$NO_{3-} + CH_2O + H_+ \rightarrow ½\, N_2O + CO_2 + 1½\, H_2O$$

Finally, nitrogen is released into the atmosphere again. The whole process starts over after release. A schematic representation of the nitrogen cycle is shown here:

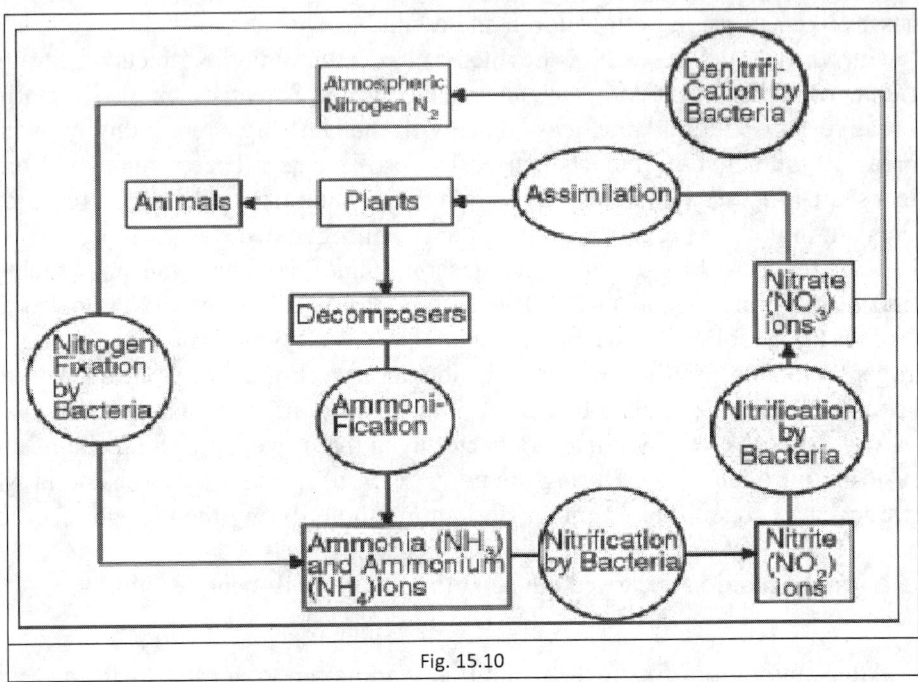

Fig. 15.10

Nitrogen as a Limiting Factor

Although the nitrogen conversion processes often occurs and large quantities of plant nutrients are produced, nitrogen is often a limiting factor for plant growth. Water flowing across the soil causes this error. Nitrogen nutrients are water-soluble and as a result they are easily drained away, so that they are no longer available for plants.

Annamox Reaction

In 1999 researchers at the Gist-Brocades in Delft, The Netherlands, discovered a new reaction to be added to the nitrogen cycle; the so-called annamox reaction. This is now found to occur in the Black Sea, as well. The reaction implies conversion of nitrite and ammonium to pure nitrogen gas (N_2), which than escapes to the atmosphere. The reaction mechanism is triggered by a newly discovered bacterium, called Brocadia anammoxidans. This appears to be a compartmentalized bacterium; within the cell membrane two compartments can be found which are also surrounded by a membrane, a very rare phenomenon. Intermediate products of the reaction included hydroxylamine, and toxic hydrazine compounds. The bacterial membranes were found to consists of badly permeable membranes, which are thought to function as a barrier for hydrazines produced within the cell. This discovery has major consequences, as it alters the entire contribution of oceans to the nitrogen balance.

PHOSPHORUS CYCLE

The phosphorus cycle involves the uptake of phosphorus by organisms. Phosphorus in the environment is mainly found in rocks, and natural weathering processes can make it available to biological systems. After decomposition of biological waste, it can accumulate in large amounts in soils and sediments. Phosphorus is used by humans as a fertilizer in farmlands and in detergents. Overuse of phosphorus can lead to eutrophication.

WATER CYCLE

The water cycle is the process by which water travels in a sequence from the air (condensation) to the earth (precipitation) and returns to the atmosphere (evaporation). It is also referred to as the hydrologic cycle.

PHOSPHORUS CYCLE

Phosphorus is an essential nutrient for plants and animals in the form of ions PO_{43-} and HPO_{42-}. It is a part of DNA-molecules, of molecules that store energy (ATP and ADP) and of fats of cell membranes. Phosphorus is also a building block of certain parts of the human and animal body, such as the bones and teeth.

Phosphorus can be found on earth in water, soil and sediments. Unlike the compounds of other matter cycles phosphorus cannot be found in air in the gaseous state. This is because phosphorus is usually liquid at normal temperatures and pressures. It is mainly cycling through water, soil and sediments. In the atmosphere phosphorus can mainly be found as very small dust particles. Phosphorus moves slowly from deposits on land and in sediments, to living organisms, and than much more slowly back into the soil and water sediment. The phosphorus cycle is the slowest one of the matter cycles that are described here.

Phosphorus is most commonly found in rock formations and ocean sediments as phosphate salts. Phosphate salts that are released from rocks through weathering usually dissolve in soil water and will be absorbed by plants. Because the quantities of phosphorus in soil are generally small, it is often the limiting factor for plant growth. That is why humans often apply phosphate fertilizers on farmland. Phosphates are also limiting factors for plant-growth in marine ecosystems, because they are not very water-soluble. Animals absorb phosphates by eating plants or plant-eating animals.

Phosphorus cycles through plants and animals much faster than it does through rocks and sediments. When animals and plants die, phosphates will return to the soils or oceans again during decay. After that, phosphorus will end up in sediments or rock formations again, remaining there for millions of years. Eventually, phosphorus is released again through weathering and the cycle starts over.

A schematic representation of the phosphorus cycle:

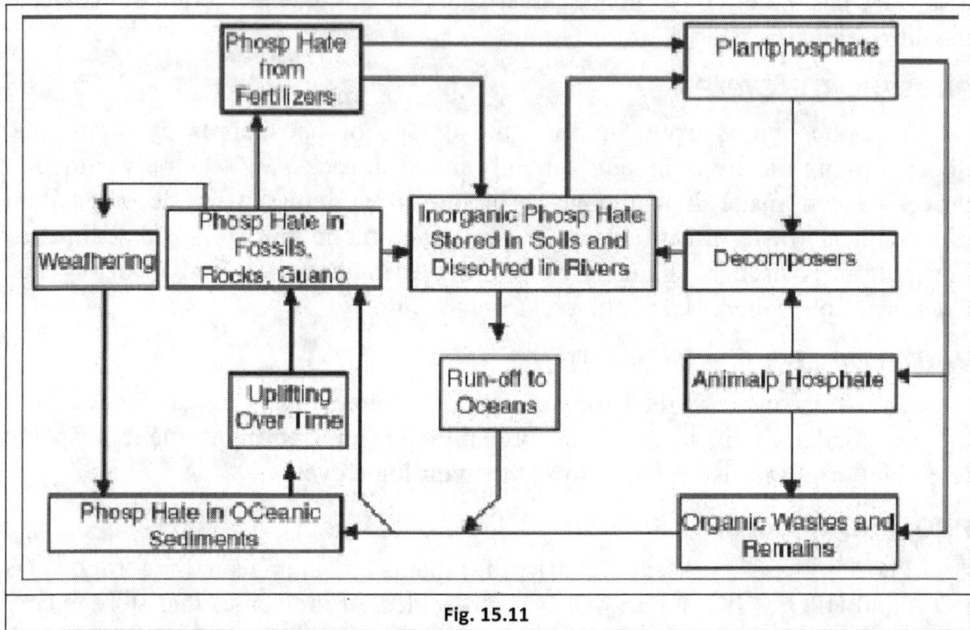

Fig. 15.11

Humans also cause environmental problems by interfering with the phosphorus cycle. We remove phosphates on certain locations, to apply it on farmland as fertilizers. On these locations the phosphate supply is moving, causing phosphate levels of surrounding land-soils and groundwater to become too high. On farmland where phosphates are applied as fertilizers, plants do not absorb all phosphate. The phosphates end up in water and stream towards lakes and reservoirs, where they cause a phenomenon called eutrophication. Eutrophication means that the water is so rich in nutrients that it causes certain water plants, such as green algae, to grow extensively.

As a result the oxygen supply in the water depletes, causing fish and other oxygen-dependent organisms to die and bacteria that are not oxygen dependent (anaerobic) to take over.

When people practise extensive wood chopping in forest areas, such as the tropical rain forest, phosphate supplies deplete, because most of the phosphate can be found in the ground under the trees in such areas. When the trees are removed the rain washes phosphates away, causing the ground to become unproductive.

Phosphate supplies are restored very slowly, because most phosphate in soils and water comes from weathering rocks and oceanic sediments.

SULPHUR CYCLE

Sulphur is one of the components that make up proteins and vitamins. Proteins consist of amino acids that contain sulphur atoms. Sulphur is important for the functioning of proteins and enzymes in plants, and in animals that depend upon plants for sulphur. Plants absorb sulphur when it is dissolved in water. Animals consume these plants, so that they take up enough sulphur to maintain their health. Most of the earth's sulphur is tied up in rocks and salts or buried deep in the ocean in oceanic sediments. Sulphur can also be found in the atmosphere. It enters the atmosphere through both natural and human sources. Natural recourses can be for instance volcanic eruptions, bacterial processes, evaporation from water, or decaying organisms. When sulphur enters the atmosphere through human activity, this is mainly a consequence of industrial processes where sulphur dioxide (SO_2) and hydrogen sulphide (H_2S) gases are emitted on a wide scale.

A schematic representation of the sulphur cycle:

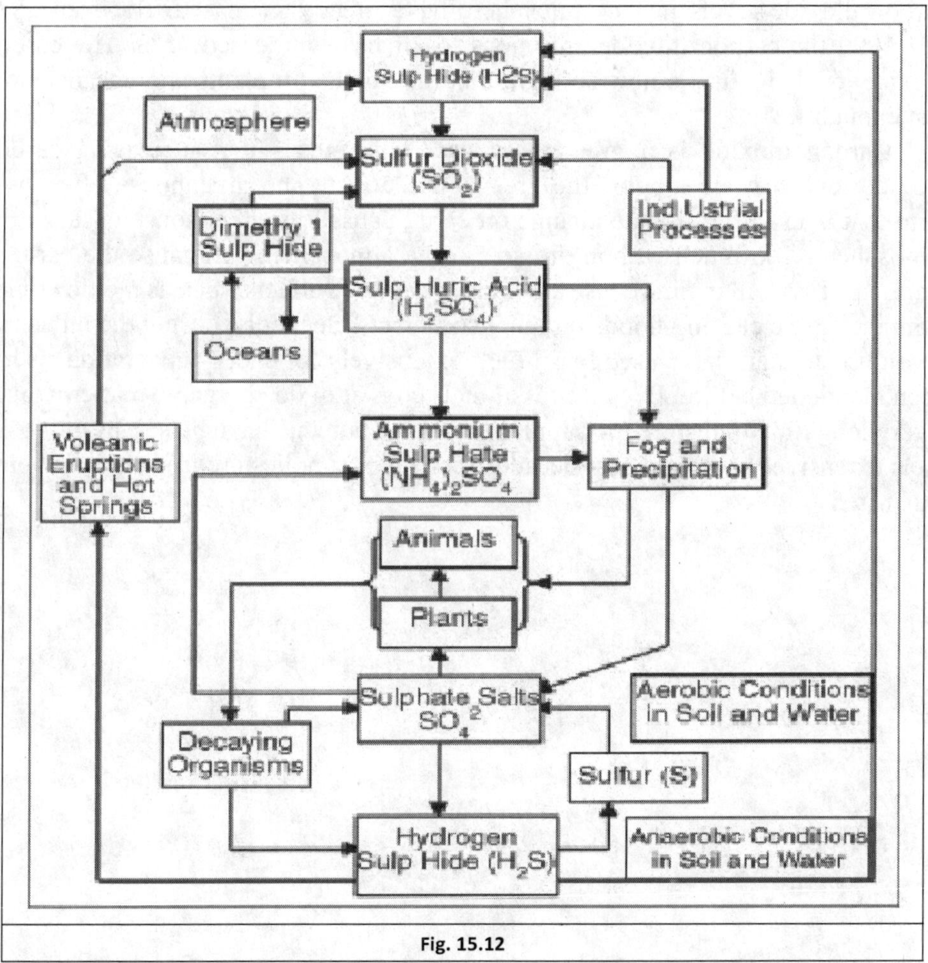

Fig. 15.12

When sulphur dioxide enters the atmosphere it will react with oxygen to produce sulphur trioxide gas (SO_3), or with other chemicals in the atmosphere, to produce sulphur salts. Sulphur dioxide may also react with water to produce sulphuric acid ($H_2 SO_4$). Sulphuric acid may also be produced from demethylsulphide, which is emitted to the atmosphere by plankton species.

All these particles will settle back onto earth, or react with rain and fall back onto earth as acid deposition. The particles will than be absorbed by plants again and are released back into the atmosphere, so that the sulphur cycle will start over again.

Humans cause environmental problems by influencing the carbon cycle in two ways. Firstly, the removal of forests has caused depletion in plants and trees that absorb carbon dioxide. Humans cause great carbon emissions to the atmosphere, which take place during industrial processes, such as coal and oil combustion. These processes serve the generation of energy. As a result of these effects the carbon dioxide levels in the atmosphere have greatly increased. Between 1870 and 1990 the carbon dioxide emissions to air have increased 25%. The carbon dioxide levels in air are now so high, that the uptake by plants and oceans is not fast enough.

Carbon dioxide is a greenhouse gas. This means that increasing carbon dioxide levels in air support the greenhouse effect. The greenhouse effect is a term that is used for climate change on earth, caused by greenhouse gases in the atmosphere. Too much carbon dioxide in the atmosphere can cause the earth to warm up. This will in time cause all kinds of negative effects, such as melting of ice from the Arctic causing floods on other continents. Secondly, during the industrial revolution humans have used fossil fuels extensively for energy generation, as has been mentioned before. Because fossil fuels consist of dead organic matter it takes a very long time to restore the supplies. Because humans have been burning fossil fuels extensively over the past decades fossil fucl supplies are in danger of being exhausted.

❒

18

Forms of Condensation and Precipitation

The meteorologist borrowed the term precipitation from the chemist and then generally gave it a meaning different from that used by its original proprietor. The chemist means by it the process of separating a substance from its solution in such form that it settles or tends to settle uner the force of gravity. The substance itself he calls a precipitate. The meteorologist sometimes makes this distinction too, but more commonly he means by precipitation not a process but a product, such as rain or snow.

The chemist gets nearly all his precipitates from solutions, and commonly, too, as a result of some combination into a new substance. The precipitation (precipitate) of the meteorologist, on the contrary, is not a new substance it all, but the same old substance in a different state, that is, liquid water, such as rain or solid water, as snow, for instance, condensed from gaseous water.

Some of the chemist's precipitates settle quickly, while others fall slowly, and a few take forever-and-a-day to come down. So it is also with the meteorologist's precipitations. Large hailstones fall with a speed that is dangerous, drizzle drops just dawdle along, and dew doesn't fall at all—it is just said to fall.

There is a tendency on the part of many meteorologists to confine the term precipitation to only that water, liquid or solid, which, after dropping out of the clouds, actually reaches the earth. In its broader usage, however, it includes, as explained, all natural condensates from gaseous water to liquid or solid water. It is here used in that more comprehensive sense.

FOG

One of the most interesting, and in some respects an exceedingly important, form of precipitation, is fog, a great swarm-like assemblage in the surface air of hundreds of thousands of droplets per cubic inch so minute that it would take 7,000,000,000 of them to make a teaspoonful of water. It is a surface cloud whose myriad myriad droplets have formed on condensation nuclei, such as ultra-microscopic particles of seasalt, incident to the cooling of the air below its dewpoint. In many cases this necessary cooling occurs on still clear nights in river and creek valleys, as the result, chiefly, of the loss of heat by the surface through radiation. When the sky is overcast with low clouds their radiation to the earth is nearly equal to that of the earth to them, and the cooling of the surface therefore, and, through it, the adjacent air, insufficient to induce condensation.

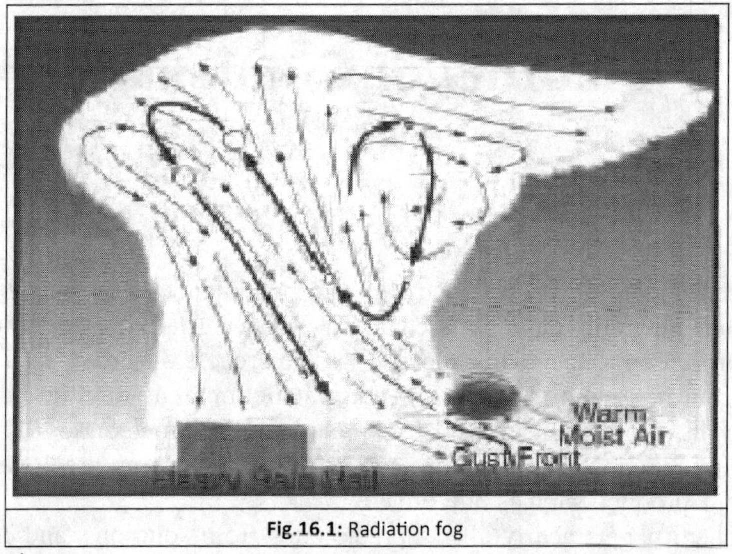

Fig.16.1: Radiation fog

Similarly this type of fog, radiation fog as it commonly is called, can not and does not form, even on clear nights, when there is a considerable wind, as the cooling then also is slight. The loss of beat by radiation is not, of course, affected by wind, except indirectly through its effect on the surface temperature; but the fall in temperature and the depth of the chilled air, and hence the formation of fog, are affected by it. Thus, whenever a dead calm prevails the cooling surface of the earth cools in turn only a very shallow layer of air, practically just that which is in direct contact with the chilled objects on the surface, since it itself is a very poor conductor of heat. This chilling of the mere surface air, no matter how pronounced, does not produce fog, a sort of cloud above the surface, but only dew or frost condensed from the air that is chilled by the cold objects with which it is in direct contact.

If there is a little wind, just enough to mix the air through a layer 10 to 50feet deep, the cooling, though less than when there is no wind at all, often is enough to push the temperature of this mixed air below its dewpoint and thus to fill it with fog. That is, the considerably cooled surface air, by mixture with a rather small amount of air above, cools it in turn below its point of saturation and thus induces throughout its volume condensation on the ever present, innumerable, hygroscopic nuclei, in droplets too small to fall to the ground against the diffusing force of the existing slight turbulence caused by the slow wind movement.

When the winds are strong, however, the consequent pronounced turbulence brings, one after another, all portions of the air through a considerable depth into contact with the surface and thus the loss of heat, though actually greater than it is on still nights, owing to higher surface temperatures, now is shared by a far larger mass of air and the fall in temperature, or actual cooling, thereby much reduced, and the occurrence of fog prevented no matter how bright the stars. In short, even when the skies are clear, no wind at all implies no fog; very light wind, say 2 to 6 or possibly 8 miles per hour, fog, if other conditions are favourable; strong wind, again no fog.

Fog is formed also by the drifting of warm humid air onto a cold surface off Newfoundland where the air from the Gulf has come over the much colder Labrador Current. It forms also, though to a much less extent, where the air is distinctly colder than the water. This is an exceedingly common occurrence on calm cold mornings over ponds, lakes and streams, from which one often sees wisps of fog rising at numberless places like so much steam from hot water.

That is why the usual name for this phe- nomenon is steaming. It is only the condensation of the moisture in the humid surface air as it is buoyed up and mingled with the colder air above. Ordinarily the evaporation is so slow, and the air above so dry, that this steam-fog is only a shallow, gauzy affair of but little consequence. Occasionally, a in high latitudes, the air is so very cold that the "steam" immediately changes to ice needles, or frost smoke, as it usually is called.

Fig.16.2: Advection Fog

Not only may fogs be produced by the drifting of warm humid air over cold surfaces, and by the importation of cold air over warm water, but also, however and wherever formed, they themselves may be transported by gently moving air to other places, even in many cases to localities at which fog seldom is generated in situ. Such advection, or drift, fogs are exceptionally frequent just inland from mid-latitude seacoasts along which the prevailing wind is on-shore.

In certain cases this moving fog is largely caught by trees, notably by the towering redwoods of California, as it filters through them, and thereby condensed into a fog-drip that in every particular is the equivalent of just so much gentle rain. And this rain is so frequent too in various places as to maintain luxurious vegetation where otherwise only the spiny, stunted growths of an arid region could survive.

CLOUD

In general, a cloud, a thing that every one knows when he sees it, though few know much about it, is essentially just a fog in the sky, though on the average its droplets are roughly 35 times as bulky as those of true fog—so bulky indeed that if they were merged together a teaspoon could hold only a paltry 200,000,000 of them without running over. But not all clouds consist either wholly or even in part of these droplets. Some contain also rain drops, each of which is equal in mass to many thousands of the average cloud particle, while many others contain no droplets at all, but only snow in some one or more of its many forms, from discrete, tiny columns or needles to matted tufts an inch or more across.

Clouds differ from each other not only in respect to composition, some consisting of water droplets and others of ice particles, as just stated, but also in height above the surface, appearance and mode of formation. And these conditions are the criteria according to which we classify clouds and name them—a delightful intellectual diversion for which our forefathers felt no need and never practiced.

When they had occasion to speak of a cloud they generally just called it a cloud, and let it go at that, without specifying what sort of a cloud it was. In fact it was not until 1801 that clouds were christened at all in any European language, and until 1803 that they were given names that proved to be generally acceptable. A second edition of this international cloud atlas, only slightly modified and extended, was published in 1910.

In general this atlas was quite satisfactory and became widely known and extensively quoted in books and periodicals. In 1922 another International Cloud Commission got busy and at the end of ten years produced a handsome work that has much to recommend it, especially its many excellent cloud pictures.

But shade of Luke Howard! How some of the good old definitions have been changed, and how unwieldy for practical use, are both it and the abridged form that appeared in 1930! Many of us expect another International Cloud Atlas to appear some day in which the revisions largely will be reversions. Also there is urgently needed a working cloud atlas, cheap and above all convenient to use.

Fig.16.3: Cirrus, hook type.

All the more commonly recognized clouds are listed below, though the definitions and comments are less formal than those given in international atlases.

Cirrus

The highest, generally, of all clouds, with the exception of two rare forms, to be mentioned later, the natures of which are unknown, is called "cirrus," and was so named by Luke Howard, because of its frequent appearance in the form of swirls and ringlets. It is a detached, white cloud, fibrous in structure and too thin to cast a distinct shadow or blur the outline of the sun or moon.

Fig.16.4: Tufted Cirrus.

Owing to the great heights at which it usually occurs, 6 to 7 miles above sea level in mid-latitudes and half as high again near the equator, its temperature commonly is very low, -50° to -80°F roughly. And because of this very low temperat-ure the citrus cloud nearly always, though perhaps not quite invariably, consists of ice particles.Although the cirrus usually is higher than any other cloud, it does not always have that topmost position. Indeed it can and does occur at any level within the convectional portion of the atmosphere-the troposphere—from its highest reaches clear down to the surface of the earth. It occurs now and then beneath a cloud sheet of a different type and even between two such sheets or layers.

Furthermore, it assumes as many different forms as Proteus, to many of which distinctive names, seldom used, have been given. When it slowly disappears it is a harbinger of fair weather, but when it grows denser it frequently is the forerunner of a general rain—a sort of a pilot or scout for the ranks of heavier and lower clouds following in its wake.

Fig.16.5: Cirro-stratus.

Cirro-Stratus

The next highest cloud also is white and commonly too thin to blur the outline of the sun. It may appear as only a formless milky haze or as a more or less continuous, fibrous sheet or layer.

It therefore has been given the appropraite name "Cirro-Stratus," that is, a stratum of cirrus cloud. It shows the well-known large rings or halos around the sun and the moon, from which we know definitely that it consists of minute ice crystals, as they alone so refract the light of the sun or moon as to produce these circles. When this cloud tends to thicken there generally are still denser clouds on the way bringing rain or snow in 6 to 24 hours.

Cirro-Cumulus

Sometimes the cirrus, or cirrostratus, seems to curdle, as the housewife says of milk when it coagulates into lumps. Such a cloud canopy, consisting of small lumpy clouds, often with more or less fibrous edges, but practically without shadow, and irregularly distributed, frequently is called a "curdled sky."

Usually, however, the cloud lumps are arranged in orderly rows, or waves, that remind one of the colour patterns on a mackerel's back. This is the "mackerel sky."

Fig.16.6: Cirro-cumulus.

Not infrequently, too, the larger waves are crossed, diagonally, as a rule, by many much smaller waves, or ripples. Technically all such fine lumpy clouds are called cirro-cumuli, or heaped-up cirrus. They obviously are caused by numerous small convections in a thin unstable layer of air, and evidently contain but little cloud material since they are practically without shadow.

Alto-stratus

The next cloud type in the order of decreasing altitude is the alto-stratus, a heavy, whitish to bluish veil of clouds, sometimes fibrous in texture and again structureless, which we call alto-stratus. Commonly it is thin enough to reveal the approximate position of the sun, as if seen through ground glass, but too thick to show either coronas or halos.

It may be formed in any one of several ways, chiefly by the spreading out by wind of the upper portions of cumulus clouds; by the cooling in place of a layer of moist air; and by the flow of a layer of warm humid air over colder air, the cloud forming along the interface. It is not disturbed by numerous local convections as we know from the fact that they would make of it an alto-cumulus, a cloud it does not resemble; neither is it a product of a wide-spread active convection, since that would cause a more or less steady rain or snowfall, which is not a feature of the alto-stratus. It is recognized that the above definition of alto-stratus is in keeping with that given by the older international cloud committee, appointed in 1891, and not with that of the committee of 1922.

Fig.16.7: Alto-stratus

According to this later definition the altostratus always is fibrous, at least in parts. If so, then the fairly common "ground-glass" cloud of the same level and density that shows no such structure is without recognition. In this case many of us prefer our old friend to the new acquaintance.

Alto-cumulus

Somewhat lower, on the average, than the alto-stratus is the alto-cumulus, a canopy of closely packed but isolated or nearly isolated cloud masses, seldom more than a few diameters of the sun or moon across and nearly always arranged in more or less conspicuous rows, often in two directions. The smaller of these cloud masses are white; the larger appreciably shaded. When these clouds are slowly evaporating, and are 20° to 30° from the sun, they often afford wonderful colour displays—especially the delicate pinks and greens that characterize the finest mother-of-pearl. These exquisite patches of colour are but fragments of coronas produced by the diffraction of sunlight by water droplets. This iridescence in addition to charming us with its beauty, tells us also that the alto-cumulus consists of water droplets and not of ice crystals. Droplets produce coronas, iridescence, and rainbows; the crystals give the many forms of the halo.

The alto-cumulus, like the cirro-cumulus, appears to be caused by local convections in an unstable layer of air, but one of greater moisture content than that in which the cirrocumulus forms. When the quantity of the humidity present is unusually great an isolated cloudlet here and there grows larger and much taller than its neighbours. In this case the cloud sheet may be called alto-cumulus castellatus, and is so called when are especially discriminating.

Fig.16.8: Alto-cumulus.

Such a cloud not infrequently is a forerunner of rain. In practice, it is the whole cloud sheet that is called altocumulus, and not one of its individual parts to which, ordinarily, no name at all is given, though actually it is a cumulus.

Strato-cumulus

There is yet another cloud sheet of irregular density, a sheet of largish, dark grey balls or rolls of cloud that frequently are united by thinner cloud. These masses are larger, denser and commonly at a lower level than those of the alto-cumulus. Also in in many cases, especially just after the passage of a general storm, this cloud has the form of a continuous canopy in irregular folds and layers. Another type of this cloud is low and fluffy in appearance, and often, presumably, produced by the convection due to turbulence.

The tops of the unit masses of this cloud often are raised and rounded, showing that they are produced each by a local convection, while their bases have an approximately common level. In many cases, however, the tops of the separate masses spread out, like smoke under a ceiling, until they nearly or quite merge together. And they spread out because they too, like smoke in a room, have reached a ceiling—the limit to which the distribution of the temperature in the atmosphere permits them to rise. Just how and why this is it would be interesting and instructive to follow up in detail, but to do so soon would have us repeating in part the important story of the structure of the atmosphere and repetition to such an extent is not in order.

Nimbus

When a general, or cyclonic, storm is passing the clouds associated with it often run the changes from cirrus, as the most advanced scouts, to cirro-stratus, and then an altostratus that grows denser, darker, and lower, covering the entire heavens and without distinctive form, until finally rain or snow falls from it steadily. In this final state, and even before that, so long as rain or snow seemed imminent, this cloud sheet formerly was called nimbus, and that is what some of us still want to call it who see no need whatever for giving it the new name "nimbostratus" adopted by the latest cloud committee. The two names refer to one and the same cloud, and therefore the new and longer name can only take more time to write, more space to print and cause more cloud confusion, a confusion that already was so great as to shake ones confidence in many, if not most, cloud observations.

The nimbus cloud normally results from the gradual ascent of humid air over an obstacle in its path, rising land in some cases but more commonly cooler air in front of the storm centre. Some rain or snow can and does fall from nearly, if not quite, every type of cloud, and even, on rare occasions, out of a clear and cloudless sky but that does not justify calling anything nimbus except the cloud just described, from which rain or snow normally falls in appreciable amounts.

Fig.16.9: Strato-cumulus.

Cumulus

When the weather is calm and warm and the sky generally clear, surface heating by sunshine starts rising many a column of air here, there and yonder, and when that air is moderately humid a cloud forms at the top and in the upper portion of each well-developed pillar—a cloud rounded above, with bulging sides, like a great pack of wool or a giant head of cauliflower, and flat at the base. This is the cumulus cloud, brilliantly white where the sun shines on it, and blue-grey to dark over the shaded portion, except along the edge not far from the sun where there is a silver lining—a wondrous capital of marvelous carving supported by an invisible column. Some of these clouds tower higher than ther neighbours, but the bases of all, big and little, at any particular time and place, have substantially the same height, the height at which the current surface air on rising will have cooled to its dewpoint, or temperature at which saturation is attained and condensation begins.

Fig. 16.10: Cumulus.

During its growth the air within a cumulus cloud is in a state of vigourous agitation, as is evident from the way its sides and top continuously change in size and shape, and confirmed by the jolts and tossings the aviator experiences when in a large cloud of this type.

From the origin of the cumulus cloud, namely, vigourous local convection, incident to surface heating, of moderately humid air, one correctly infers that in middle latitudes it is essentially a summer phenomenon; that it is most numerous in tropical regions and rarest near the poles; that at times it forms a chain of balloons, as it were, in the sky that delimits the course of a river, or a seashore, below; that over land it is most numerous by day and over the ocean at night; that it frequents islands far more than the open ocean; and that, among many other things, it is a fair weather phenomenon, since it can not occur when the sky is generally overcast, as adequate surface heating does not then occur, nor at the time of any considerable wind, since that would prevent the establishment of large columns of ascending air.

Cumulus-nimbus

When a cumulus cloud grows to great size and height by the continuous feeding into it from below of humid air, or has been formed along a squall line by the juxtaposition of warm and cold air, the time comes when for one reason or another vast numbers of the droplets become so large and heavy that they fall out as so much rain which commonly reaches the ground, though in arid regions it sometimes evaporates wholly in mid air. As soon as this rain phase of the cumulus cloud begins we heretofore have been ready to change its name to cumulo-nimbus.

As the rain begins thunder and lightning also usually occur. In many cases also the cumulo-nimbus spreads out at the top into a more or less well-defined anvil shape. Finally, the topmost portion of the cloud frequently, but by no means always, nor at first, is combed out by the wind into a sheet of cirrus. And yet this is the one phenomenon by which the 1922 Cloud Commission distinguish between the cumulus and the cumulonimbus. They say: "Masses of cumulus however heavy they may be, and however great their vertical development, should never be classed as cumulo-nimbus unless the whole or a part of their tops is transformed into a cirrus mass."

Again: "None of these cumulus clouds should show ice crystal clouds (hybrid cirrus) at their tops; this would mean that they had reached the cumulo-nimbus stage." And in the formal definition of cumulo-nimbus there is this phrase: " the upper parts having a fibrous texture." Finally, their brief definition of cumulo-nimbus is: "Cumulus clouds of great vertical development, with the tops composed of ice crystal clouds."

Well, if we must accept this definition of cumulo-nimbus, then this is the definition of cumulo-nimbus we must accept, being both official and international. But, for all that, it leaves some of us a little confused in our minds, as we say-those of us who have seen many a heaped-up cloud from which abundant rain was falling and in which there was frequent and vivid lightning, but of which no portion was cirrus nor showed any signs of cirrus.

Fig.16.11: Cumulo-nimbus.

We called such a cloud cumulonimbus, and we very much dislike now to call it just cumulus. Indeed, rain is the really important differentiating feature between cumulus clouds, and not the trifling circumstance of the presence or absence in their uppermost portions of a little cirrus. And our confusion is greater, not less, when we recall other heaped-up clouds, cumuli we then called them, and still would like to call them, which developed abundant cirrus, but never furnished a drop of rain or a flash of lightning.

This is another case where some of us much prefer our old and faithful friend to the new and untried acquaintance.

STRATUS

Sometimes the sky is covered with a comparatively low layer of grey formless cloud from which little or no rain falls, only a drizzle at most. This cloud is called "stratus," and the height of its base varies from just free of the surface to about 3000 feet. It may be formed in several ways, such as the flow of warm humid air over colder air beneath, the cloud forming at their interface; by the drift of fog over relatively warm land, where the lower portion of the fog is burned off, leaving the upper portion as a stratus cloud; and by the relatively shallow turbulence convection induced by surface friction and surface obstacles. Stratus is the lowest and most nearly uniform of all clouds. Indeed it is so devoid of contrasts that it scarcely can be shown in a photograph at all except where it is pierced by a hill or mountain, and even in that case a lively imagination is a great help—not to supply the details, for of these there are none, but to awaken that lonesome, dreary feeling one has on a day of drizzle. It would seem that, in all conscience, Nature had completely exhausted her sartorial ingenuity in the making from vapour alone of all the above sorts and kinds of cloud robes. But here too her ability has no bounds. There were craps left over, as always, when the robes were fashioned.

BILLOW CLOUD

Although the billow cloud is only a special form of the alto-cumulus, it is so strikingly different from the average cloud as to arrest attention and to call for a distinctive name. It occurs in long, narrow and evenly spaced strips of white cloud. Even the individual strips themselves often show considerable detail of structure. It generally is supposed that this cloud occurs on the crests of waves or billows formed by the flowing of one current of air over another.

However, while the name "billow cloud" generally is restricted to a particular form of the alto-cumulus, the same sort of structure though of much finer grain, occurs as a frequent type of the cirro-cumulus, and likewise a far coarser one as occasional variety of the strato-cumulus.

As just implied, two sheets of air of different density flowing horizontally the one over the other develop waves or billows at their interface which billows of course are coolest in their highest portions or crests and warmest in the troughs. Hence when the humidity is just right such billows are cloud-crested and clear-troughed. But according to the theory of these air waves the distance from the crest of any one to that of the next is far greater than that from cloud to cloud in the delicate cirrocumuli ripples, and greater too than the distance between the rows of a wave-like alto-cumulus or billow cloud.

Fig.16.12: Billow clouds.

Hence, while actual air waves, induced in the manner explained, probably are the cause of certain groups of cloud strips of coarse grain, or long wave length, it seems probable that the beautiful alto-cumulus billows are not produced in this manner, and certain that the citrus ripples have some other origin. The probable origin of many, perhaps all, of the finer structured, or shorter wave-length billow clouds, is convection through unstable air beneath into air above of different velocity.

Fig.16.13: Lenticular cloud

When the two superincumbent layers of air are still with reference to each other the ascending masses are roughly as broad as long, horizontally, as indicated by the roundish patches of cloud that often cap them, and usually occur in more or less distinct rows and patterns. When, however, they are flowing the one past the other the clouds are more or less drawn out in the direction of the motion of the upper air with reference to the lower, and if this motion is in the direction of an alignment of the clouds the individual small cumuli in each row parallel to the wind are so intermingled, or so it seems, as to become a uniform cloud wave or billow. This obviously is not the whole story of the genesis of the billow cloud, and frankly it may not be more than half right as far as it goes. At any rate this cloud still challenges us for much more observation, experiment and study than yet has been given to it, as do also cirro-cumuli waves, especially those that are crossed by numerous ripples.

Lenticular Cloud

Sometimes an isolated cloud, whether solitary or one of a group, has the general shape of a double convex lens, that is, thick in the middle and gradually thinning to nothing at the edge. This is the lenticular cloud, whatever else it may be, such as alto-cumulus, alto-stratus or what not.

It may be formed by the arrest of an ascending small cumulus cloud at the base of a layer of stable air—a layer in which the decrease of temperature with increase of height is so slow that ascending air, unless in great volume, can not deeply penetrate it. Here the cloud, arrested by the impenetrable ceiling, spreads out much in the shape of a lens, circular or distorted as the circumstances, especially the direction and strength of the winds, may determine. Another abundant source of lenticular clouds is any large mountain peak, ridge or range. Here, in most cases, the crossing wind is thrown into great standing billows whose crests occasionally are capped with lenticular clouds as fixed in position as a waterfall, and made of as rapidly changing material—condensing on the windward side and evaporating to the leeward.Sometimes this lenticular cloud of obstruction or interference origin, and therefore cresting a stationary air wave, is low down, and again at considerable height, as determined largely by the amount and distribution of the humidity. This cloud also deserves more study than it yet has received.

Crest Cloud

When quite humid wind blows up and over a mountain a cloud forms, such as the famous "TableCloth" Cloth" on Table Mountain near Cape Town, that rests on the ridge and hangs down each side like a giant cloth, fringed on the leeward slope where it is pulled down in evaporating shreds by the descending air. This is the crest cloud. It is interesting and of some importance, and if baptized in Latin might eventually become a member of the aristocratic cloud family.

Fig. 16.14: Lenticular Cloud, Interference Type

Riffle Cloud

Frequently, when the crest cloud is moderatcly heavy and the wind fairly strong one or more long cumulus billows hang fixed in the free air to the leeward of the mountain ridge and at about its height.

Fig.16.15: Crest cloud.

These mark the upper portions of the great air waves caused by the interference of the mountain with the passing wind. Analogously produced waves, called riffles, occur in running water where it passes over an obstruction in the bed of the stream; hence the name "riffle cloud."

Banner Cloud

Any high mountain peak that steaks up like Cleopatra's Needle, and there are such—Mount Assiniboine in Canada, for instance—now then develops a cloud, in an otherwise clear atmosphere, that flutters out like a huge banner made fast to the peak's upper leeward side. This occurs, of course, only in air that is humid well-nigh to the point of saturation. The cooling necessary to pass that critical cloud-producing temperature obviously may be owing in part to contact of the air with the cold walls of the peak, and certainly in part to the decrease of pressure in the eddy region on the sheltered side. It is not a familiar cloud nor one of much importance, but for all that it is both beautiful and interesting.

Boa Cloud

Fig. 16.16: Banner cloud.

Sometimes a great river of cloud, such as may flow along a broad valley, is obstructed by an isolated mountain, a volcanic cone, perhaps, that stands in the midst of the stream and divides it right and left as would an island. Thus the mountain becomes wrapt around by the flowing cloud, as by a giant feather boa. The old Romans called it, when it twined about Mount Etna, the Serpe, and rightly regarded it as a harbinger of rain.

Scarf Cloud

If one will watch carefully just above the top of a large growing cumulus he sometimes will have the pleasure of seeing a little cloud of silken sheen and bowed upward, over and free from the topmost head like a portion of a halo, emerge from the sky where before there was nothing to be seen. In a little while it probably will cover the head of the cumulus, and a bit later rest on its shoulders like a great silken scarf. It seems to form in a thin but quite humid layer of air that is cooled past the saturation point, and cloud thus formed, by the slight lift given to it by the ascending cumulus which it later adorns. Some say that it is a harbinger of showers. Perhaps so; at any rate the towering cumulus that generates it certainly is.

MAMMATO-CUMULUS

If you ever saw the sky covered with huge grapes, or everywhere festooned, or a great mother cloud, as it were, plethoric with a thousand mammae, then, whether you knew it or not, you were looking at the interesting mammato-cumulus cloud, a frequent forerunner of the dread tornado. Not everything about the genesis of this cloud is known, but it appears to be the result of an upsidedown convection of some sort, perhaps induced by the flow of a thin cloudy layer of air over other air beneath that is sufficiently warm near its upper limit to permit of short-range interchanges between the two.

Tornado Cloud

The tornado cloud is a violently rotating, hollow, tapering cloud column 100 to 1000 feet in diameter, as a rule, extending from the surface of the earth, where it is smallest, up into the heavy storm clouds overhead. The cooling that produces this cloud is caused by the expansion of the air incident to the decrease of pressure within the column due to its rapid rotation. The cloud column of the waterspout is produced in the same way. This is a thin cirro-stratus-like cloud of rare occurrence and unknown origin, richly iridescentand about three times as high as ordinary cirrus, that is, around 16 miles.

Noctilucent Cloud

It was stated in the description of the cirrus cloud that usually it is the highest of all save two that are rarely seen. The higher of these, in turn, is the noctilucent cloud, a thin cloud-like patch of something, roughly 50 miles above the surface. Not all the clouds of this great height are lucent, but the first ones observed were, and the name "noctilucent" then given to them may as well be applied alike to all until we know something—now we know nothing-about the substance of which they are composed and how they are formed.

Fig. 16.17: Tornado Cloud

❐

19

Effect of Environment

ATMOSPHERE

The atmospheric chemistry studies the chemical composition of the natural atmosphere, the way gases, liquids, and solids in the atmosphere interact with each other and with the earth's surface.

For convenience of study, atmospheric scientists divide the atmosphere as if it consists of 4 layers. The division is mainly due to temperature variations as the altitude increases.

The 4 layers according to the variation of temperature are:

1. Ionosphere
2. Mesosphere
3. Stratosphere
4. Troposphere

In the given figure as shown below, 100 km is the thermosphere and ionosphere where the temperature increases from 200 K at 100 km to 500 K at 300 km. The temperature goes even higher as the altitude increases. activity as the altitude decrease. In the outer space, most particles consist of single atoms, H, He, and O etc. At lower altitude (200–100 km), diatomic molecules N_2, O_2, NO etc are present. The ionosphere is full of electrically charged ions.

The UV rays ionizes these gases. The major reactions are:

In the ionosphere:

$$O + h\nu \rightarrow O_+ + e$$

$$N + h\nu \rightarrow N_+ + e$$

In the neutral thermosphere:

$$N + O_2 \rightarrow NO + O$$

$$N + NO \rightarrow N_2 + O$$

$$O + O \rightarrow O_2$$

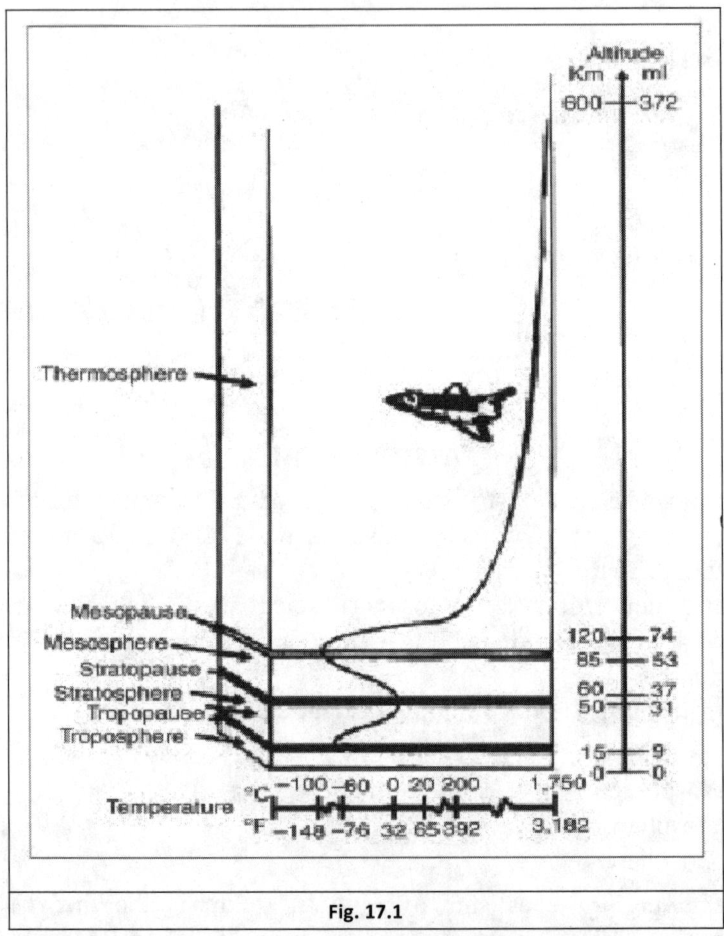

Fig. 17.1

Beyond the neutral thermosphere is the ionosphere and exosphere. These layers are of course interesting for space explorations and environmental concerns and space sciences. The atmosphere in the outer space is more like a plasma than a gas. Below the thermosphere is the mesosphere (100–50 km) in which the temperature decreases as the altitude increase. In this region, OH, H, NO, HO_2, O_2, and O_3 are common, and the most prominent chemical reactions are:

$$H_2O + h\,v \rightarrow OH + H$$

$$H_2O_2 + O \rightarrow OH + OH.$$

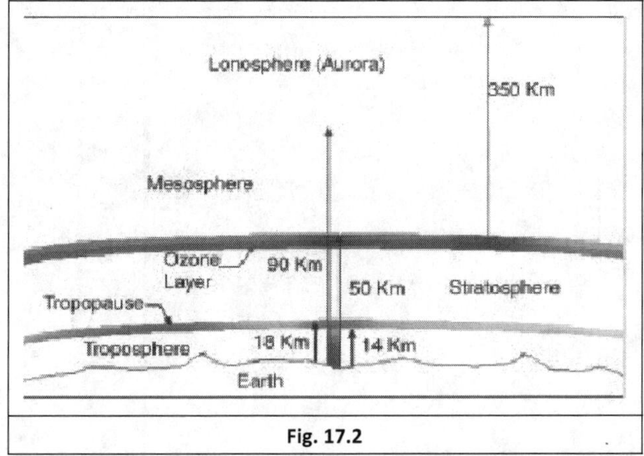

Fig. 17.2

Below the mesosphere is the stratosphere, in which the temperature increases as the altitude increase from 10 km to 50 km. In this region, the following reactions are common:

$$NO_2 \rightarrow NO + O$$

$$N_2O \rightarrow N_2 + O$$

$$H_2 + O \rightarrow OH + H$$

$$CH_4 + O \rightarrow OH + CH_3$$

Air flow is horizontal in the stratoshpere. A thin ozone layer in the uppe stratosphere has a high concentration of ozone. This layer is primarily responsible for absorbing the ultraviolet radiation from the sun. The ozone is generated by these reactions:

$$O^2 + h\,v \rightarrow O + O$$

$$O_2 + O \rightarrow O_3$$

The troposphere is where all weather takes place; it is the region of rising and falling packets of air. The air pressure at the top of the troposphere is only 10% of that at sea level (0.1 atmospheres). There is a thin buffer zone between the troposphere and the next layer called the tropopause.

The major components in the region close to the surface of the Earth are N_2 (78%), O_2 (21%), Ar (1%) with variable amounts of H_2O, CO_2, CH_4, NO_2, NO_2, CO, N_2O, and O_3. The ozone concentration in this layer is low, about 8% of the total ozone in the atmosphere is in the troposphere.

All the weather phenomena happen in the troposphere. If you want to know the current weather con dition, check the AccuWeather network for a good view of the troposphere.

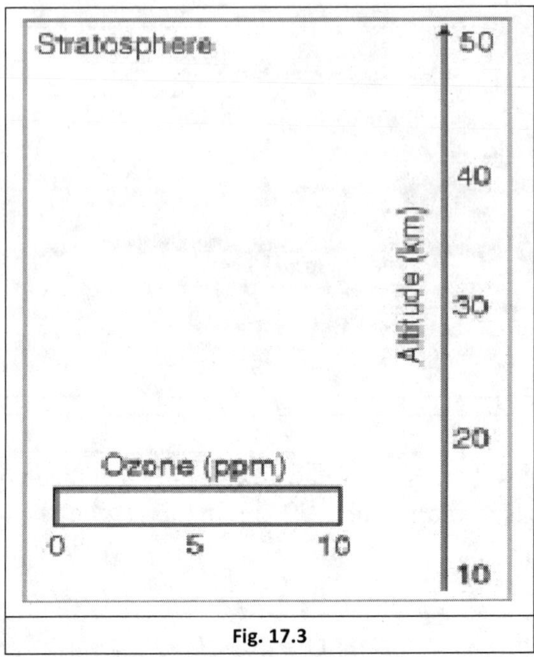

Fig. 17.3

GASES IN ATMOSPHERE

From the atmospheric science viewpoint, interactions of all gasses amoung themselves and their interaction with the environmental elements are of interest. However, for identification purposes, we need to identify the gases produced by man-made process (industry).

Some of the gases due to human activities are:

- Carbon dioxide result from the excess burning of carbon-containing fuel.
- Carbon monoxide produced by automobiles. This orderless and Colourless gas is very toxic.
- Ozone produced in the exhaust of internal combustion engine, and the variation of ozone concentration in the stratosphere.
- Nitrogen oxides such as NO, NO_2, N2O4; due to the production of NO in the internal combustion engin.
- Methane gas produced due to treatments of large amount of waste.
- Sulfur oxides produced in mining operation and in the combustion of sulfur containing fuel. Sulfur oxide causes the so called acid rain problem.
- Chlorofluorocarbons (CFC) are gases used as refrigurant. When disposed into the atmosphere, they cause the ozone concentration to decrease.

Water vapour is also considered a greenhouse gas, but it is also generated by nature contineously due to radiation from the Sun. Of course, when water vapour condense into a liquid, much energy is released in the exthermal process. Condensation of watervapour causes storms and many of the weather phenomena.

STRATOSPHERE AND TROPOSPHERE

One of the most important developments in the past several years has been the realization that the next phase in atmospheric chemistry, dynamics, radiation, and climate research will demand a far more integrated application of modeling, satellite retrievals, and high-resolution, in situ observations. This realization stems from two different causes:

The first is the increasing awareness that scientific research must develop forecasts that are tested and trusted internationally in such areas as climate change, UV dosage levels resulting from ozone changes in the stratosphere, toxicity levels stemming from the urban-to-regional emission of nitrates, sulfates, organics, soot and heavy metals, etc. This demands an integration of transport and photochemical mechanisms that go well beyond the normal alliance that stems from curiosity-driven studies that operate without regard for timetables or societal demands.

The second is the increasing level of sophistication in modeling, data assimilation, and technology related to observational systems, whether they be remote or in situ. This enables the scientific community to investigate in finer granularity the cause and effect relationships in the study of atmospheric chemistry and dynamics. In response to this new realization, the Anderson Group has pursued a course of action based upon a set of objectives of critical importance in today's world.

TROPOSPHERE AND STRATOSPHERE

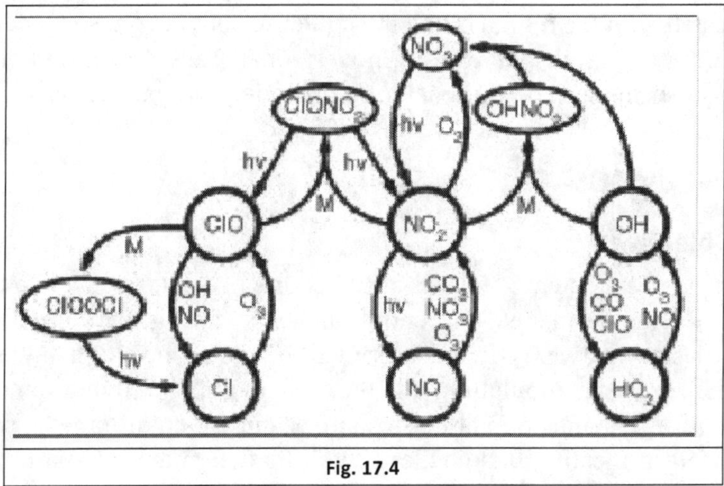

Fig. 17.4

Hypothesis

The boundary condition for water vapour entering the stratosphere in the tropics (irreversible passage through the 390 K è surface) is set by the vertical/horizontal temperature structure of the cold trap region over the Western Tropical Pacific through which the quasi-horizontal passage of air provides adequate residence time (coupled potentially with multiple traversals of these cold pools) to desiccate the air to a vapour pressure corresponding to the local, seasonally dependent, temperature minimum.

Understanding the processes governing stratospheric water vapour is broadly recognized as a dominating scientific need. Changes in stratospheric water vapour under future climate regimes would strongly affect both radiative balance and ozone chemistry.

In particular, increases in stratospheric water content could dramatically increase ozone losses by:

- Increasing the cold aerosol surface area that draws down the NOx/NOy ratio thereby increasing the fraction of reactive halogens and hydrogen species portioned into free radical form (ClO, BrO, IO, OH and HO_2)
- Promoting formation of the polar stratospheric clouds on which catalytic ozone loss chemistry occurs
- Increasing the source of HOx radicals
- Radiative cooling of the lower stratosphere

Because of a lack of measurements, however, debate continues about a wide array of potential scenarios by which air entering the stratosphere is dehydrated. As an illustrative example, we note that while the average saturation mixing ratio of the tropical tropopause in January is 3 ppmv, the average water vapour content of air entering the stratosphere is 4 ppmv. Yet, nearly one quarter of the inner tropics in January has a cold point tropopause temperature corresponding to 2 ppmv (or less) co-located with the region of the most intense convection; so what fraction of the air entering the stratosphere (as it moves zonally and quasi-horizontally with convection superimposed on this background motion) has passed through this cold region?

OZONE AND UV DOSAGE

Societal Objectives

The mission statement identifies as a first order societal issue: "To understand and protect our home planet." A central challenge facing the United States and other countries in the twenty-first century will be to enhance human well-being in a world where growing populations and the drive to improve living standards place potentially huge demands on natural resources and the environment. Whether we succeed or fail in meeting this challenge will be determined, in part, by how we respond to immediate demands to address human health and economic growth in the context of the wide range of crucial, environmentally related decisions made every day by insurance companies, water resource managers, agribusiness, households, city planners, public health officials, and countless others. Rising to this challenge will entail using natural resources as efficiently as possible, devising practical solutions that meet our immediate needs and also provide for long-run economic growth, while maintaining the environmental systems on which life depends.

As we have seen, the Earth's system can be fragile. Stratospheric ozone loss has had profound implications for human health. It has lead to disfiguring and deadly skin cancers that occur as a result of increased ultra-violet radiation reaching the earth's surface. Losses over northern mid-latitudes will affect the population Centres of Europe and North America.

Isolating the cause is complicated by the fact that the photochemical lifetime of ozone in this region is comparable to the dynamical residence time and the dynamical residence time is highly variable depending on season, altitude, longitude, and latitude. Careful measurements and analyses are required to identify and quantify the role of both chemistry and dynamics in determining mid-latitude ozone conce-ntrations.

To guide wise public policy decisions that continue to improve human and economic conditions and to clarify public debate, it is necessary to restructure the science and engineering framework addressing the biological, chemical, and physical integrity of our surroundings. Private-sector and governmental decisions will be made regarding air, water, and living systems that will fundamentally affect our nation's health and its economic and environmental vitality.

AIRBORNE MISSIONS

Airborne Platform

The Anderson Group instruments constitute an important component of the observational array required to address both the AURA calibration/validation requirements and the hypotheses that constitute the scientific objectives of the Programmes. Consideration of AURA colLabourative science and the scientific objectives of TC4 serves to identify required characteristics of the platform used in the airborne field operation that constitutes the core of this effort:

- Ability to carry the proper complement of instruments
- Ability to contiguously cover the altitudes required from the boundary layer to the lower stratosphere
- Ability to operate in convectively active regions and to engage in sophisticated storm avoidance
- Ability to minimize operational complexity for foreign deployments
- Ability to accept new instruments and to accept multiple instruments for intercomparison
- Ability to combine in situ and remote observations to increase the colLabourative overlap between aircraft and satellites.

LASER DEVELOPMENT

A problem of fundamental importance to the NASA vision and mission to improve life here and to understand and protect our home planet, involves a new strategic plan for establishing the quantitative link between rapidly increasing urban/regional emission of aerosol precursors (nitrate, sulfate, heavy metal, organic and elemental carbon) and changes in the optical properties of the atmosphere that constitute a powerful climate forcing.

Dissecting the mechanisms underlying these changes requires measurements that span temporal scales from seconds to years and spatial scales from tens of meters to thousands of Kilometres, to provide an effective union with A-Train satellites.

This research will advance an array of innovative new developments in cavity enhanced spectroscopic techniques employing new miniaturized laser systems coupled with high finesse cavities that will provide a new generation of observations from NASA UAVs.

Central to these efforts will be the advancement of:

- Difference frequency generation (DFG), both singly resonant and intracavity,
- Pulsed fibre lasers using emerging master oscillator power amplifier (MOPA) configurations, micro-etalons, and double-clad fibres, and
- Tailored quantum cascade lasers using newly developed beam epitaxy approaches.

In conjunction with these laser developments, we will advance developments in flight electronics with emphasis on scalable distributed processing and data acquisition necessary for the inclusion of a scientifically potent instrument array within the weight and volume restrictions of UAVs. A key component of this research endeavor is the objective of unifying thc sub-orbital and satellite components of the Programmes.

COMBUSTION RELATED AIR POLLUTION

AIR POLLUTION

Air pollution, both indoor and outdoor, is a significant cause of health problems worldwide. Urban and rural outdoor environments contain infections, allergens, irritants and chemical toxins that can reduce the quality of life and cause disease. Inhaled air pollution is directed at the the nose, throat and lungs. The exposed airway allows hazardous pollutants to enter the body and all tissues are ultimately exposed.

The tolerance for environmental destruction is ancient and human history is littered with civilizations that failed because humans indiscriminately exploited natural resources and spoiled their own nest. Humans adapt easily to deteriorating conditions and will persist in following daily routines even when air pollution is severe, traffic is congested, water and food supplies are at risk, and social order is unstable.

Fine-particulate air pollution is known to contribute to cardiovascular and lung disease, increasing the risk of heart attack and a heart-related death. Researchers at Brigham Young University and Harvard School of Public Health compared changes in air pollution from 1980 to 2000 with residents' life expectancies, They concluded that a reduction in air pollutants in 51 U.S. cities between 1980 and 2000 added an average of five months to life expectancy.

Residents in cities that made the most significant improvements in air quality, such as Pittsburgh, PA, lived almost 10 months longer. For every microgram per cubic meter decrease in fine-particulate air pollution, life expectancies rose by more than seven months. Airborne chemicals contaminate food and water. They may be ingested and are also collected in the nose and throat and swallowed, often in mucus that attempts to protect exposed surfaces.

Airborne chemicals entering the digestive system include well-known toxins such as pesticides, organophosphate, PCBs, dioxin, arsenic, cadmium, lead, and mercury. In addition occupational exposures to airborne pathogens can be intense and can cause cancer.

While ambient air pollution is a major concern, indoor air can be more polluted than outdoor air. Building materials and furnishings are a source of volatile chemicals. A decrease in indoor air quality is the result of reduced ventilation and efficient construction practices, sealing homes, stores and office buildings from the outdoor environment.

Smoking tobacco remains a personal method of producing air pollution that remains popular worldwide, despite overwhelming evidence that tobacco smoke produces a long list of disabling and fatal diseases. It is estimated that 30% of all fatal cancers could be prevented if tobacco smoking were eliminated from the list of air pollutants.

The really sad part of our current predicament is that all the right ideas for creating a healthy environment have been around for decades and have been clearly articulated in many forms by a host of intelligent people. The right ideas involve unselfish and compassionate Behaviour. The right ideas involve long-term planning, conservation and deep commitment to preserving the natural world. Without a healthy natural environment, there will be few or no healthy humans.

Our big environmental problems are built from many small, personal decisions - little mistakes that add up over time. If there is a solution, it will emerge from the collective value of millions of better decisions made by individuals all over the globe. The environmental action plan is to think globally and act locally - it does make sense.

AIR POLLUTANTS

In cities across the globe, the personal automobile is the single greatest polluter, as emissions from a billion vehicles on the road add up to a planet- wide problem. Driving a private car is a typical citizen's most air polluting activity. The negative effects of automotive emissions are maximum when you sit in traffic surrounded by cars, their engines idling. Everyone in a traffic jam is getting poisoned.

The Combustion Process Gasoline and diesel fuels are mixtures of hydrocarbons (made of hydrogen, oxygen and carbon atoms.) Hydrocarbons are burned by combining with oxygen. Nitrogen and sulphur atoms are also present and combine with oxygen when burned to produce gases. Automotive engines emit several types of pollutants.

Typical Engine Combustion

Fuel + Air ® Hydrocarbons + Nitrogen Oxides + Carbon Dioxide + Carbon Monoxide + water Hydrocarbon emissions are fragments of fuel molecules, only partially burned. See Toxicity of Benzene and other Hydrocarbons in exhaust.

Hydrocarbons react in the presence of nitrogen oxides and sunlight to form ground-level ozone, a major component of smog. Ozone irritates the eyes, nose, throat and damages the lungs. A number of exhaust hydrocarbons are also toxic, some with the potential to cause cancer.

Nitrogen Oxides

Under high pressure and temperature conditions in an engine, nitrogen and oxygen atoms react to form nitrogen oxides. Nitrogen oxides, like hydrocarbons, are precursors to the formation of ozone and contribute to acid rain. Catalytic converters in car exhaust systems break down heavier nitrogen gases, forming nitrous oxide (NO_2) - 300 times more potent than carbon dioxide as a greenhouse gas. Nitrous oxidc makes up about 7.2 per cent of the gases that cause global warming. Vehicles with catalytic converters produced nearly half of that nitrous oxide. Nitrous oxide also comes from nitrogen-based fertilizers and manure from farm animals.

Carbon Monoxide

Carbon monoxide (CO) is a Colourless, odourless, poisonous gas, a product of incomplete burning of hydrocarbon-based fuels. Carbon monoxide consists of a single carbon atom and a single oxygen atom linked together (CO), the product Carbon monoxide of incomplete combustion of fuel. Most CO is produced when air-to-fuel ratios are too low in the engine during vehicle starting, when cars are not tuned properly, and at higher altitudes, where thin air reduces the amount of oxygen available for combustion. Two-thirds of the carbon monoxide emissions come from transportation sources, with the largest contribution coming from cars. In urban areas, the passenger vehicle contribution to carbon monoxide pollution can exceed 90%.

Carbon Dioxide

Environmental Protection Agency (EPA) originally viewed carbon dioxide as a product of "perfect" combustion, but now views CO_2 as a pollution concern. Carbon dioxide is a greenhouse gas that traps the earth's heat and contributes to Global Warming.

Evaporative Emissions

Hydrocarbon pollutants also escape into the air through fuel evaporation - evaporation causes significant hydrocarbon pollution from cars on hot days when ozone levels are highest. Evaporative emissions occur several ways:

Diurnal

Gasoline evaporation increases as the temperature rises during the day, heating the fuel tank and venting gasoline vapors.

Running Loses

The hot engine and exhaust system can vaporize gasoline when the car is running.

Sitting Evaporation

The engine remains hot for a period of time after the car is turned off, and gasoline evaporation continues when the car is parked.

Adding Fuel

Gasoline vapors are always present in fuel tanks. These vapors are forced out when the tank is filled with liquid fuel.

Benzene is the main toxin in the hydrocarbon fraction of exhaust. Benzene and other less known hydrocarbons are produced in petroleum refining, and are widely used as solvents and as materials in the production of various industrial products and pesticides. Benzene also is found in gasoline and in cigarette smoke. Other environmental sources of benzene include gasoline (filling) stations, underground storage tanks that leak, wastewater from industries that use benzene, chemical spills, and groundwater next to landfills containing benzene. Exposure to benzene is related to the development of leukemia and lymphoma. Benzene has a suppressive effect on bone marrow and it impairs blood cell maturation and amplification.

Polycyclic Aromatic Hydrocarbon (PAH)

PAHs are a group of chemicals that are formed during the incomplete burning of coal, oil and gas, garbage, or other organic substances. PAHs can be man-made or occur naturally. A few of the PAHs are used in medicines and to make dyes, plastics, and pesticides. They are found throughout the environment in the air, water and soil. There are more than 100 different PAH compounds. Although the health effects of the individual PAHs vary, the following 15 PAHs are considered as a group with similar toxicity: acenaphthene, acenaphthylene, anthracene, benzanthracene, benzopyrene, benzofluoranthene, benzoperylene, benzofluoranthene, chrysene dibenzanthracene, fluoranthene, fluorene, indenopyrene, phenanthrene, pyrene.

Long term solutions require reduced combustion of all kinds. While vehicles with new energy sources such ethanol, biofuels, propane and natural gas can contribute to reduced air pollution, their benefit is limited if vehicle use continues at current intensities. If you pay more money to buy a hybrid car, but drive it more, you have contributed little to solving air pollution problems. If you buy a gas guzzling clunker and use only one gallon of gas to 20 miles each week, you have contributed more to the solution.

The problem with all alternative fuels is that the manufacture of fuels requires energy, distribution with a manufacturing infrastructure that consume energy, often derived from burning fossil fuels. No alternative fuel is ideal. While ethanol has been championed as an alternative to petroleum fuels, it mainly helps to reduce dependency on oil producing countries. When ethanol is made from corn, more than 75% of its energy value must be spent on its production. Burning ethanol still produces carbon dioxide. Climate change with extreme weather events threatens corn production in the US. The new competition between hastily constructed ethanol plants and food production became an international issue in 2008. Other approaches to new fuels made from non-food plant sources are part of the solution.

Hydrogen

Ultimately cars might burn hydrogen in fuel cells, but despite working prototypes, a hydrogen economy is a distant fantasy. There are many problems to be solved before hydrogen can replace fossil fuels as a portable energy source. The biggest problem is that producing hydrogen requires a large amount of energy. In Canada, there are opportunities to dam rivers and produce electricity with falling water, a non polluting, renewable energy resource. A more problematic energy source would be be nuclear reactors that "burn"uranium or plutonium.

Even if new non-polluting energy sources are developed, hydrogen storage and distribution requires a new, very expensive infrastructure that could replace gasoline and diesel fuels. With once rich countries such as the USA on the verge of bankruptcy and facing the extensive repairs of already aging, derelict infrastructures, adding a new, unprecedented development costs seems unlikely. Unless, of course the priorities in the US shift dramatically. The US could adopt a sane, smart strategy, reduce its military budget by 60% and invest the money and skills in rebuilding the country's infrastructure with new sustainable energy sources.

GLOBAL ENVIRONMENTAL PROBLEMS

INTRODUCTION

Ethics is understood to be the domain of enquiry that explores what is right or wrong, obligatory or non-obligatory, or when responsibility attaches to human Behaviour. Are there features of global environmental problems that call for classifying them as essentially ethical problems with even greater force than some local or regional environmental problems? If so, what are these features?

Why is this important? If some global environmental problems are essentially ethical problems, then some of the excuses that nations often use to justify not reducing their contributions to the global environmental problems are ethically problematic. This is so because ethical obligations entail duties and responsibilities to others with the result that national policies may not be justified on national interests alone. That is, if there are ethical obligations to others to cease causing harm, then policy options must be responsive to these obligations.

The features of global environmental problems that strongly call for classification as ethical problems. Some global environmental problems are caused by people in one part of the world but are most harshly experienced by others who are separated from those causing the problem by great space and time.

In addition, those most vulnerable to global environ-mental problems are often least responsible for causing them. Climate change is a strong example of this feature of global environmental problems. Those most vulnerable to climate include many of the poorest people in the world who are among those least responsible for emitting the greenhouse gases that are causing the problem. The same separation between victim and cause is also true of other global environmental problems, including upper atmospheric ozone depletion, some ocean degradation problems including destruction of global fisheries, some biodiversity and forest problems, and exposure to toxic substances that are deposited around the world by long-range air transport.

In the case of global fisheries depletion, for instance, fishing trawlers from some countries often go great distances to harvest fish for domestic consumption, yet the harshest consequences of depleted global fisheries are often experienced by subsistence fisherman in poor countries far away from the home port of the fishing boats.

This separation of space and time and between cause and impact of course can also happen in more local environmental problems such as water pollution, but the distances between those causing the problem and those who suffer the consequences are most problematic at the global scale because of the magnitude of the distances at this scale.. With the rise in truly global environmental problems in the last few decades, it is now abundantly clear that Behaviour of people in one part of the world can gravely affect the human health, economic viability and overall welfare of others far away, but their suffering is almost invisible to those who cause the problem.

Several implications for ethical reasoning follow directly from this separation of distance and time between cause and effect. First, our intuitive sense of ethical responsibility for harms that we cause others is harder to trigger when those who will be harmed by our action are people we will not ever know and are from a different culture. Yet ethics and morality require us to refrain from harming all others who may suffer because of our Behaviour even if they are separated from us in time and space.

Secondly, because the actual harm that we are causing to others may be in a particular part of the world that is uniquely vulnerable to global environmental change but is unknown to us, we not only don't know the people we are harming but have no idea where the locus of the worst harm is located. For instance, particularly vulnerable villages to climate change impacts may turn on the geographical features where the village is located such as dependence on glacier water flows in the summer coupled with vulnerability to monsoon irregularities. Yet few of us know where these places are. And so in the case of global environmental problems we may be causing great harm to others who we not only do not know personally, but about whom we neither have an idea of their location nor why they are particularly vulnerable to actions we are taking. For this reason, global scale environmental problems create a huge challenge to trigger our sense of ethical responsibility, yet ethics requires us to not harm others through our actions none the less.

A related reason why the global environmental problems demand classification as ethical problems stems from the fact that some actions that cause some environmental damage are often justified on the basis of cost-benefit analysis. Yet this justification is ethically problematic when those who benefit from an action that causes harm have no relation with those who are harmed. For instance, government may tolerate inevitable pollution from a land fill by identifying the benefits that citizens receive by having a place to take their garbage.

But in the case of global environmental problems, those who are harmed by actions do not enjoy the benefits that come from taking the action. A harms-benefits justification for environmental degradation, if ethically justifiable at all, only works where those who will suffer the harms and those who benefit from the action that causes the harm are in the same community of interests.

In the case of global environmental problems, those who are harmed are rarely made better by the benefits of the action that cause the harm.

The second reason why global environmental problems call for classification as ethical problems stems from the fact that their consequences are often catastrophic to those who are most harmed by them. Climate change, for instance, directly threatens human life and health and resources to sustain life as well as species of plants and animals and ecosystems around the world.

The harms include deaths from disease, droughts, floods, heat and intense storms and damage to homes and villages from rising oceans and intense storms, adverse impacts on agriculture, social disputes caused by diminishing natural resources, sickness from a variety of diseases, the inability to rely upon traditional sources of food, the destruction of water supplies, and the inability to live where one has lived to sustain life. In addition, the very existence of some small island nations is threatened by climate change caused seal level rise. Clearly these impacts are catastrophic for some.

Loss of global fisheries can be devastating to substance fishermen, global desertification and drought are likely to cause mass starvation, loss of atmospheric ozone can cause skin cancer in millions, and long-range transport of persistent organic chemicals can cause deadly diseases and birth defects. Because ethics requires people to be particularly careful to not to harm others when the harm is great, global environmental problems need to be seen as ethical issues.

Governments are expected to protect their citizens from life-threatening dangers. At the local, regional or national scale, citizens can petition their governments to protect them from environmental harms. But at the global level, no government exists whose jurisdiction matches the scale of global environmental problems.

And so, although existing national, regional and local governments have jurisdiction over activities within their boundaries coupled to responsibilities to their citizens, they have no responsibility to those outside their boarders in the absence of binding international law. For this reason, ethical appeals to get governments to require that their citizens not harm non-citizens in other nations are necessary to prompt action.

In fact, according to a leading foreign policy theory called "realism," national governments should limit the scope of their international engagements to the pursuit of national interest alone. According to some "realist" proponents, governments should refrain from acting at the international level for ethical reasons that do not coincide with national interests. Yet global environmental problems require people in one part of the world to consider the interests of others outside the jurisdiction of their national governments. Because there is no duty of national governments to protect foreigners from global environmental problems unless they bind themselves voluntarily in treaties, an appeal to ethical and moral responsibility is particularly important to motivate national actions in regard to global scale environmental problems. Moreover, if global environmental problems actually create ethical obligations, they have ethical duties to mitigate global environmental problems

CFCS IN ATMOSPHERE

There has been a movement since the late 1970s to ban CFCs because of their destructive effect on the ozone layer. This damage was discovered by Sherry Rowland and Mario Molina, who first published a paper suggesting the connection in 1974. It turns out that one of CFCs' most attractive features-their unreactivity-has been instrumental in making them one of the most significant pollutants. CFCs' lack of reactivity gives them a lifespan which can exceed 100 years in some cases. This gives them time to diffuse into the upper stratosphere. Here, the sun's ultraviolet radiation is strong enough to break off the chlorine atom, which on its own is a highly reactive free radical. This catalyses the break up of ozone into oxygen:

$$Cl + O_3 \rightarrow ClO + O_2$$

$$ClO + O \rightarrow Cl + O_2$$

CFCs are a problem because the chlorine is regenerated at the end of these reactions, making it able to repeat this reaction millions of times. This disrupts the ionic structure of oxygen atoms, causing their lowest ionisation energy to drop. This reaction is believed to be the cause of the ozone hole observed over the poles and upper latitudes of the Earth. Decreases in stratospheric ozone have been linked to increases in skin cancer. In 1975, Oregon enacted the world's first ban of CFCs. By 1985, scientists observed a dramatic seasonal depletion of the ozone layer over Antarctica, which led the United Nations to take action in 1987. International attention to CFCs resulted in a meeting of world diplomats in Montreal in 1987. They forged a treaty, the Montreal Protocol, which called for drastic reductions in the production of CFCs. On March 2, 1989, 12 European Community nations agreed to ban the production of all CFCs by the end of the century. In 1990, diplomats met in London and voted to significantly strengthen the Montreal Protocol by calling for a complete elimination of CFCs by the year 2000. By the year 2010 CFCs should be completely eliminated from developing countries as well

FREON

Freon is a trade name for a group of chlorofluorocarbons used primarily as a refrigerant. The word Freon is a registered trademark belonging to Dupont. Prior to their implication as a major contributor to ozone depletion (and a minor contributor to the greenhouse effect), freons were also used as aerosol propellants and to foam up polymer materials.

Freon was developed as a safer alternative to toxic gases such as, ammonia (NH_3), methyl chloride (CH_3Cl), and sulfur dioxide (SO_2), that were used as refrigerants prior to the discovery of freon. Despite the controversy, R12 works well and is very safe. Its inventor, Thomas Midgley, actually breathed in a lungful and exhaled over a candle to demonstrate its nontoxicity and nonflammability. Before CFCs were banned by developed nations, a 12 ounce (340 g) can of freon (R12) cost about $1; following the ban the cost of freon rose dramatically. People now go through considerable trouble to recycle freon from existing sources. As of 2002, freon cost about $85 per lb ($187 per kg).

Freon is most frequently sold in 12 ounce (340 g) cans and 30 pound (13.6 kg) canisters.

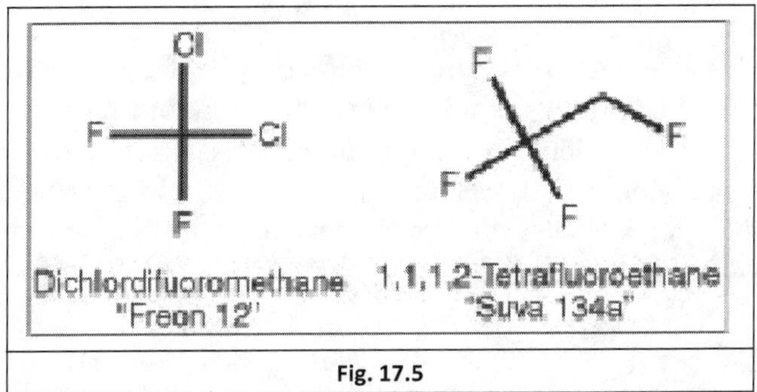

Fig. 17.5

Freon 12 →(also known as R–12) was a popular re-frigerant before it was banned. The chemical formula is CCl2F2and its proper chemical name is dichloro-difluoromethane. When people refer to freon without specifying a number, it is likely that they mean freon-12.

Suva 134a →(also known as R–134 or R–134a) is the most common replacement for R–12. Unlike R–12, which is a chlorofluorocarbon, R–134 is a hydro fluorocarbon. The chemical formula is CH2FCF3 and the proper chemical name 1,1,1,2-tetrafluoroethane.

TABLE

	Freon 41	**Fluoromethane**
R-11	CFC-11	Trichlorofluoromethane
R-12	CFC-12	Dichlorodiflucoromethane
R-13	CFC-13	Chlorotrifluoromethane
R-22	HCFC-22	Chlorodifluoromethane
R-23	HFC-23	Trifluoromethane
R-113	CFC-113	Trichlorotrifluoroethane
R-114	CFC-114	1,2-dichloro-1,1,2,2-Tetraflucoroethane
R-115	CFC-115	1-chloro-1,1,2,2,2-Penfatafluoroethane
R-116	CFC-116	Hexachloroethane
R-134a	HFC-134a	1,1,1,2-tetrafluor oethane
R-227 ea	HFC-227ea	1, 1,1,2,3,3,3-heptafluoropropane

Freon in copper tubing open to the environment can turn into phosgene gas after coming in contact with extreme heat, such as, while brazing or in a fire situation. Phosgene is a substance that was used as a chemical weapon in World War I. Low exposure can cause irritation, but high levels cause fluid to collect in the lungs, possibly resulting in death.

HALON

Halons are a group of compounds obtained by replacing the hydrogen atoms of a hydrocarbon with halogen atoms, such as bromine or fluorine.

Halon 1211 is bromochlo-rodifluoromethane (CF_2BrCl) and Halon 1301 is bromotrifl-uoromethane (CF_3Br). Halons are very stable and are widely used in fire extinguishers where water and other alternatives would be ineffective and dangerous (e.g. when dealing with oil fires) or cause unacceptable collateral damage (e.g. with electronic equipment.)

Introduction

At high temperatures, halons decompose to release halogen atoms that combine readily with active hydrogen atoms, quenching the flame propagation reaction even when adequate fuel, oxygen and heat remains. The chemical reaction in a flame proceeds as a free radical chain reaction; by sequestering the radicals which propagate the reaction, halons are able to "poison" the fire at much lower concentrations than are required by fire suppressants using the more traditional methods of cooling, oxygen deprivation, or fuel dilution. For example, Halon 1301 flood systems are typically used at concentrations no higher than 6% v/v in air, and can suppress many fires at 2.9% v/v. By contrast, carbon dioxide fire suppression flood systems are operated from 34% (surface-only combustion of liquid fuels) up to 75% (dust traps). Carbon dioxide can cause severe distress at concentrations of 3 to 6%, and has caused death by respiratory paralysis in a few minutes at 10% concentration.

Halon 1301 causes only slight giddiness at its effective concentration of 5%, and even at 15% persons remain conscious but impaired, and suffer no long term effects. (Experimental animals have also been exposed to 2% concentrations of Halon 1301 for 30 hours per week for 4 months, with no discernible health effects at all.) Halon 1211 also has low toxicity, although it is more toxic than Halon 1301, and thus considered unsuitable for flooding systems.

Fig.17.6: Halon Canisters Used in a Fire-suppression System

However, halon fire suppression is not completely non-toxic; very high temperature flame, or contact with red-hot metal, can cause decomposition to toxic byproducts. The presence of such byproducts is readily detected because they include hydrobromic acid and hydrofluoric acid, which are intensely irritating. Halons are very effective on Class A (organic solids), B (flammable liquids and gases) and C (electrical) fires, but they are totally unsuitable for Class D (metal) fires, as they will not only produce toxic gas and fail to halt the fire, but in some cases pose a risk of explosion. Halons can be used on Class K (kitchen oils and greases) fires, but offer no advantages over specialised foams.

Nomenclature

Halons are referred to by a system of code numbers similar to (but simpler than) the system used for freons. The first digit specifies the number of carbon atoms in the molecule, the second is the number of fluorine atoms, the third is the chlorine atoms, and the fourth is the number of bromine atoms. If the number includes a fifth digit, the fifth number indicates the number of iodine atoms (though iodine in Halon is rare). (Note that the various atomic species are listed in order of increasing atomic number.) Any bonds not taken up by halogen atoms are then allocated to hydrogen atoms.

For example, consider Halon 1211: C F Cl Br 1 2 1 1

Halon 1211 has one carbon atom, two fluorine atoms, one chlorine atom and one bromine atom. A single carbon only has four bonds, all of which are taken by the halogen atoms, so there is no hydrogen. Thus its formula is CF2BrCl, and its IUPAC name is therefore bromochlorodi fluoromethane.

Application

Halon 1211 is typically used in hand-held extinguishers, in which a stream of liquid halon is directed at a smaller fire by a user. The stream evaporates under reduced pressure, producing strong local cooling, as well as a high concentration of halon in the immediate vicinity of the fire. In this mode, extinguishment is achieved by cooling and oxygen deprivation at the core of the fire, as well as radical quenching over a larger area. After fire suppression, the halon moves away with the surrounding air, leaving no residue.

Halon 1301 is more usually employed in flooding systems. In these systems, banks of halon cylinders are kept pressurised to about 4 MPa with compressed nitrogen, and ducts lead to the protected enclosure. On triggering, the entire measured contents of one or more cylinders are dumped into the enclosure in a few seconds, through nozzles designed to ensure uniform mixing throughout the room. The quantity dumped is pre-calculated to achieve the desired concentration, typically 3-5% v/v. This level is maintained for some time, typically twenty minutes, to ensure all items have cooled so reignition is unlikely to occur, then the air in the enclosure is purged. During this time the enclosure may be entered by persons wearing SCBA. (There exists a common myth that this is because halon is highly toxic; in fact it is because it can cause giddiness and mildly impaired perception, and also due to the risk of combustion byproducts.) Flooding systems may be manually operated or automatically triggered by a VESDA. In the latter case, a warning siren and strobe lamp will first be activated for a few seconds to warn personnel to evacuate the area. The rapid dumping of halon and consequent rapid cooling fills the air with fog, and is accompanied by a loud, disorienting noise.

Environmental Issues

There is concern that halons are being broken down in the atmosphere to bromine, which reacts with ozone, leading to depletion of the ozone layer (this is similar to the case of chlorofluorocarbons such as freon). These issues are complicated: the kinds of fires that require halon extinguishers to be put out will typically cause more damage to the ozone layer than the halon itself, not to mention human and property damage. However, fire extinguisher systems must be tested regularly, and these tests may lead to damage. As a result, some regulatory measures have been taken, and halons are being phased out in most of the world.

History

Carbon tetrachloride was used in fire extinguishers and glass (anti)-"fire grenades"from the latenineteenth century until around the end of World War II. Experimentation with halogenated hydrocarbons for fire suppression on military aircraft began at least as early as the 1920s. During World War II, various early halons were in standard use in aircraft by some combatants, but these early halons suffered from excessive toxicity. Nevertheless after the war they slowly became more common in civil aviation as well.

In the 1960s, fluorinated halons became available, and were quickly recognised as one of the most effective fire fighting materials discovered. Much early research with Halon 1301 was conducted under the auspices of the US Armed Forces, while Halon 1211 was initially mainly developed in the UK. By the late 1960s, they were standard in many applications where water and dry powder extinguishers posed a threat of damage to the protected property, including computer rooms, telecommunications switches, Labouratories, museums and art collections. Beginning with warships in the 1970s, halons also progressively came to be associated with rapid knockdown of severe fires in confined spaces with minimal risk to personnel.

By the early 1980s, halons were in common use on aircraft, ships and large vehicles, as well as in computer facilities and galleries. However, concern was beginning to be felt about the possible impact of chlorinated and brominated hydrocarbons on the ozone layer. The Vienna Convention on Ozone Layer Protection did not cover halons, as it was felt that emergency discharge of systems was too small in volume to produce a significant impact, and too important to human safety for restriction. However, by the time of the Montreal Protocol it was realised that discharges during system tests and maintenance accounted for substantially larger volumes than emergency discharges, and so halons were brought into the treaty, but with many exceptions. Halons have been generally phased out and the possession of halon equipment is prohibited in some countries like the Netherlands and Belgium from January 1, 2004, based on the Montreal Protocol and guidelines of the European Union. Production of new stocks has ceased in most (probably all) countries as of 1994. However many countries still require aircraft to be fitted with halon fire suppression systems, as no safe and completely satisfactory alternative has been discovered for this application. There are also a few other highly specialised users.

OZONE DEPLETION

The ozone layer protects the Earth from the ultraviolet rays sent down by the sun. If the ozone layer is depleted by human action, the effects on the planet could be catastrophic.

Ozone is present in the stratosphere. The stratosphere reaches 30 miles above the Earth, and at the very top it contains ozone. The suns rays are absorbed by the ozone in the stratosphere and thus do not reach the Earth.

Ozone is a bluish gas that is formed by three atoms of oxygen. The form of oxygen that humans breathe in consists of two oxygen atoms, O_2. When found on the surface of the planet, ozone is considered a dangerous pollutant and is one substance responsible for producing the greenhouse effect.

The highest regions of the stratosphere contain about 90% of all ozone. In recent years, the ozone layer has been the subject of much discussion. And rightly so, because the ozone layer protects both plant and animal life on the planet. The fact that the ozone layer was being depleted was discovered in the mid-1980s. The main cause of this is the release of CFCs, chlorofluorocarbons.

Ozone is a bluish gas that is formed by three atoms of oxygen. The form of oxygen that humans breathe in consists of two oxygen atoms, O_2. When found on the surface of the planet, ozone is considered a dangerous pollutant and is one substance responsible for producing the greenhouse effect.

The highest regions of the stratosphere contain about 90% of all ozone. In recent years, the ozone layer has been the subject of much discussion. And rightly so, because the ozone layer protects both plant and animal life on the planet. The fact that the ozone layer was being depleted was discovered in the mid-1980s. The main cause of this is the release of CFCs, chlorofluorocarbons.

Antarctica was an early victim of ozone destruction. A massive hole in the ozone layer right above Antarctica now threatens not only that continent, but many others that could be the victims of Antarctica's melting icecaps. In the future, the ozone problem will have to be solved so that the protective layer can be conserved.

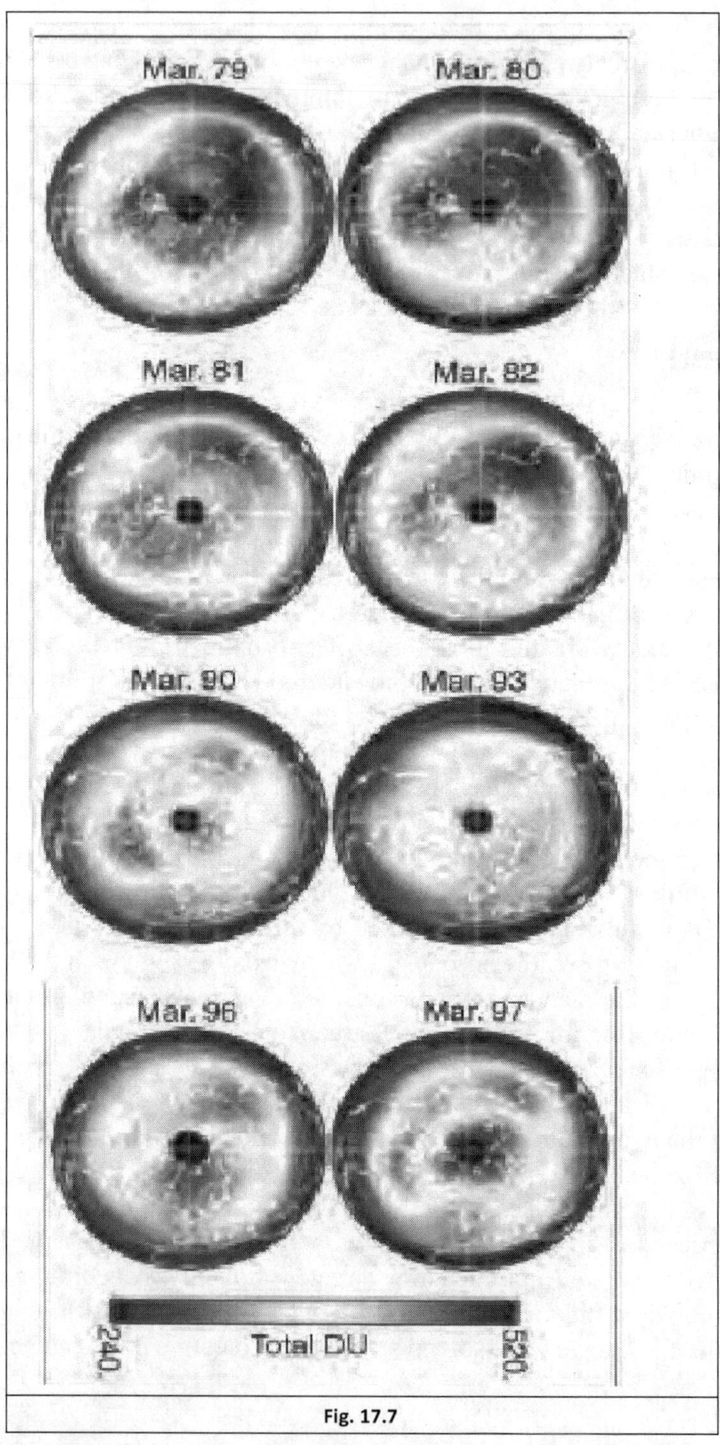
Mar. 79
Mar. 80
Mar. 81
Mar. 82
Mar. 90
Mar. 93
Mar. 96
Mar. 97
240.
Total DU
520.

Fig. 17.7

Only a few factors combine to create the problem of ozone layer depletion. The production and emission of CFCs, chlorofluorocarbons, is by far the leading cause.

Many countries have called for the end of CFC production because only a few produce the chemical. However, those industries that do use CFCs do not want to discontinue usage of this highly valuable industrial chemical.

CFCs are used in industry in a variety of ways and have been amazingly useful in many products. Discovered in the 1930s by American chemist Thomas Midgley, CFCs came to be used in refrigerators, home insulation, plastic foam, and throwaway food containers. Only later did people realise the disaster CFCs caused in the stratosphere. There, the chlorine atom is removed from the CFC and attracts one of the three oxygen atoms in the ozone molecule.

The process continues, and a single chlorine atom can destroy over 100,000 molecules of ozone.

In 1974, Sherwood Rowland and Mario Molina followed the path of CFCs. Their research proved that CFCs were entering the atmosphere, and they concluded that 99% of all CFC molecules would end up in the stratosphere.

When the ozone layer hole was discovered over Antarctica, was the proof truly conclusive. At that point, it was hard to question the destructive capabilities of CFCs.

Even if CFCs were banned, problems would remain. There would still be no way to remove the CFCs that are now present in the environment. Clearly though, something must be done to limit this international problem in the future

EFFECTS

Even minor problems of ozone depletion can have major effects. Every time even a small amount of the ozone layer is lost, more ultraviolet light from the sun can reach the Earth.

Every time 1% of the ozone layer is depleted, 2% more UV-B is able to reach the surface of the planet. UV-B increase is one of the most harmful consequences of ozone depletion because it can cause skin cancer. The increased cancer levels caused by exposure to this ultraviolet light could be enormous. The EPA estimates that 60 million Americans born by the year 2075 will get skin cancer because of ozone depletion. About one million of these people will die.In addition to cancer, some research shows that a decreased ozone layer will increase rates of malaria and other infectious diseases. According to the EPA, 17 million more cases of cataracts can also be expected.

The environment will also be negatively affected by ozone depletion. The life cycles of plants will change, disrupting the food chain. Effects on animals will also be severe, and are very difficult to foresee. Oceans will be hit hard as well. The most basic microscopic organisms such as plankton may not be able to survive. If that happened, it would mean that all of the other animals that are above plankton in the food chain would also die out. Other ecosystems such as forests and deserts will also be harmed.

The planet's climate could also be affected by depletion of the ozone layer. Wind patterns could change, resulting in climatic changes throughout the world.

SOLUTIONS

He discovery of the ozone depletion problem came as a great surprise. Now, action must be taken to ensure that the ozone layer is not destroyed. Because CFCs are so widespread and used in such a great variety of products, limiting their use is hard. Also, since many products already contain components that use CFCs, it would be difficult if not impossible to eliminate those CFCs already in existence.

The CFC problem may be hard to solve because there are already great quantities of CFCs in the environment. CFCs would remain in the stratosphere for another 100 years even if none were ever produced again.

Despite the difficulties, international action has been taken to limit CFCs. In the Montreal Protocol, 30 nations worldwide agreed to reduce usage of CFCs and encouraged other countries to do so as well. However, many environmentalists felt the treaty did "too little, too late", as the Natural Resources Defence Council put it. The treaty asked for CFC makers to only eliminate half of their CFC production, making some people feel it was inadequate.

By the year 2000, the US and twelve nations in Europe have agreed to ban all use and production of CFCs. This will be highly significant, because these countries produce three quarters of the CFCs in the world. Many other ountries have signed treaties and written laws restricting the use of CFCs.

Companies are finding substitutes for CFCs, and people in general are becoming more aware of the dangers of ozone depletion.

CHEMICAL PROCESSES

About 90 per cent of the Earth's protective ozone layer resides in the stratosphere between 15km and 50km altitude. Molecular oxygen is broken down in the stratosphere by solar radiation to yield atomic oxygen, which then combines with molecular oxygen to produce ozone. Ozone is destroyed naturally through a series of catalytic cycles involving oxygen, nitrogen, hydrogen and to a lesser extent chlorine and bromine species.

The abundance of stratospheric ozone is therefore chemically controlled by the stratospheric abundances of compounds containing hydrogen, nitrogen, chlorine and bromine. Increases in the abundances of methane and nitrous oxide (sources of hydrogen and nitrogen oxides respectively) thus affect the abundance and distribution of stratospheric ozone.

Stratospheric ozone is also affected by the abundance of carbon dioxide (CO_2), because the rates of the chemical reactions that control the abundance of ozone are temperature-dependent, and the abundance of CO_2 plays a key role in determining the temperature structure of the stratosphere.

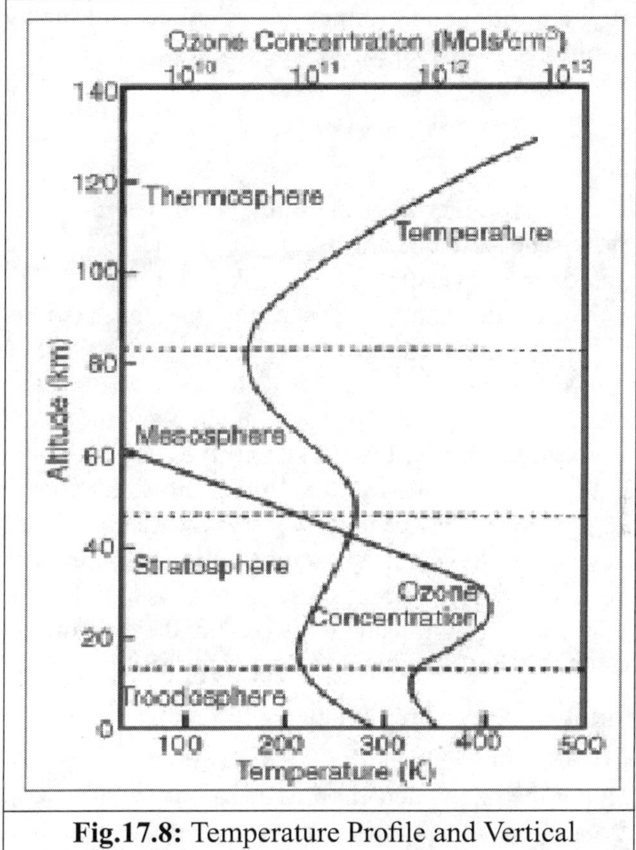

Fig.17.8: Temperature Profile and Vertical Distribution of Ozone in the Atmosphere

During the 1970s and early 1980s, theoretical model calculations focused on predicting the response of stratospheric ozone to changes in chlorine, assuming that the atmospheric abundances of other trace gases remained constant.

However, since the early 1980s, with advances in understanding of trace gas trends and model formulation, model calculations have been used to predict the response of stratospheric ozone to simultaneous increases in chlorine (from chlorofluorocarbons, hydrochlorofluorocarbons, carbon tetrachloride, methylchloroform and methyl chloride) and bromine (from halons and methyl bromide), as well as methane, nitrous oxide and carbon dioxide.

During the 1970s and early 1980s, theoretical model calculations focused on predicting the response of stratospheric ozone to changes in chlorine, assuming that the atmospheric abundances of other trace gases remained constant.

However, since the early 1980s, with advances in understanding of trace gas trends and model formulation, model calculations have been used to predict the response of stratospheric ozone to simultaneous increases in chlorine (from chlorofluorocarbons, hydrochlorofluorocarbons, carbon tetrachloride, methylchloroform and methyl chloride) and bromine (from halons and methyl bromide), as well as methane, nitrous oxide and carbon dioxide.

RESPONSES

Policy Responses

Summarizes the main international policy responses in the 1970s and 1980s to the ozone layer issue. Established by UNEP in the 1970s, the Coordinating Committee on the ozone Layer (CCOL) has played a major role in establishing scientific research priorities, and in describing the implications of the latest scientific findings in terms understandable to policy-makers. In 1980, for example, the CCOL reported that continued releases of CFCs would eventually deplete the ozone layer, and that this could have serious impacts on the health of people and the biosphere. Subsequently, the CCOL synthesized the knowledge base to provide the scientific basis for the Montreal Protocol. The policy implications of the scientific findings of the last decade have been profound, leading, for example, to a significant revision of the control measures of the Montreal Protocol.

The most recent scientific findings also have substantial policy implications and could lead to further significant revisions in the control measures of the Montreal Protocol at the Fourth meeting of the Contracting Parties.

Simulation Models in Exploring Policy

Nowadays, simulation models are used more and more to explore policy options for resolving the ozone depletion issue. A particular example, performed, was a sensitivity study of atmospheric chlorine-loading scenarios for a range of global emissions of compounds currently regulated under the Montreal Protocol, and for possible new HCFC substitutes. The results are summarized in Figure below, which shows that:

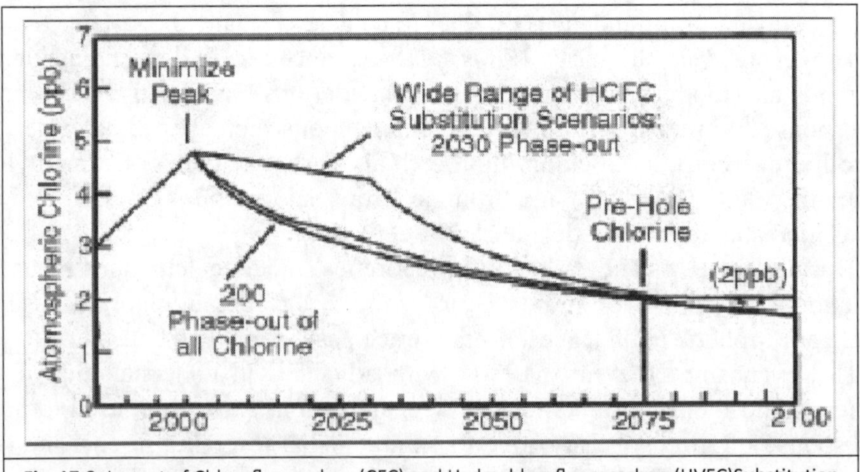

Fig. 17.9: Impact of Chloroflurocarbon (CFC) and Hydrochlorofluorocarbon (HVFC)Substitution Policies on Atmospheric Abundance of Chlorine (Cl).

- Peak chlorine loading is controlled by the date of the global phase-out (assumed to be the year 2000) for long-lived CFCs. and by emission rates between 1990 and 2000;
- Even with a complete phase-out of all chlorine emissions in 2000, the atmospheric level of chlorine will not decrease to 2 ppbv (the level of chlorine prior to the formation of the Antarctic ozone hole) until about 2075;
- A wide range of HCFC substitution policies can be implemented without delaying recovery of the Antarctic ozone hole, assuming that there is a phase-out of HCFC substitutes before the middle of the next century (the required phase-out date for the HCFCs is dependent upon their lifetime); and The additional atmospheric loading of chlorine due to HCFC substitution depends upon the HCFC emission rates and the lifetimes of the HCFCs, but is typically limited to less than 1 ppbv even with significant initial substitution rates and subsequent growth in these emission rates.
- The study concluded that in order to minimize future ozone layer depletion and to facilitate the most rapid elimination of the Antarctic ozone hole (i.e. by about 2075):
- Production and emission of long-lived halocarbons (CFCs and CCl4) should be phased out as soon as possible, with the greatest possible global compliance:
- Halocarbon substitutes should be phased out sometime in the middle of the next century:
- Atmospheric concentrations of bromine should be stabilized at current levels or below.
- Peak chlorine loadings should be reduced, most practicably by significant cutbacks in the emissions of abundant, short-lived halocarbons, particularly methyl chloroform.

CONCLUDING REMARKS

Depletion of the stratospheric ozone layer by halogenated chemicals is a global problem. While the issue was primarily a matter of scientific curiosity during the 1970s and early 1980s, it has now become an urgent policy question for governments in both developed and developing countries. The weight of scientific evidence strongly indicates that chlorinated and brominated chemicals (largely man-made) are primarily responsible for the substantial decreases in stratospheric ozone (greater than 50 per cent in the total column) that occur over Antarctica every Southern Hemisphere springtime.

In addition, satellite data supported by ground-based observations have demonstrated that ozone has been decreasing since the late 1960s at middle and high latitudes in both the northern and southern hemispheres, and that this decrease cannot be explained by known natural processes. Indeed, the conclusion of the

1991 International Scientific Assessment was that 'the weight of evidence suggests that the observed middle- and high-latitude ozone losses are largely due to chlorine and bromine'. The current and historic use of CFCs and other such chemicals in developed countries is the primary cause of the problem. However, adequate protection of the ozone layer will require a full partnership between the developed countries, whose industries and consumers have caused the problem, and the developing countries, whose people wish to use these chemicals in refrigeration and other ways.

As in the case of atmospheric concentrations of lead, which are now declining, even in Greenland, as a result of tighter emission controls, the stratosphere would gradually recover if emissions of CFCs and similar chemicals were to cease.

Thus, the most urgent task is to strengthen the Montreal Protocol and to develop an international method of monitoring national Programmemes to control CFCs and related gases. However scientific research and monitoring should continue into the next century to provide a basis for better policies.

GREENHOUSE EFFECT

The greenhouse effect is a naturally occurring process that aids in heating the Earth's surface and atmosphere. It results from the fact that certain atmospheric gases, such as carbon dioxide, water vapour, and methane, are able to change the energy balance of the planet by absorbing longwave radiation emitted from the Earth's surface. Without the greenhouse effect life on this planet would probably not exist as the average temperature of the Earth would be a chilly -18° Celsius, rather than the present 15° Celsius.

As energy from the Sun passes through the atmosphere a number of things take place. A portion of the energy (26% globally) is reflected or scattered back to space by clouds and other atmospheric particles. About 19% of the energy available is absorbed by clouds, gases (like ozone), and particles in the atmosphere. Of the remaining 55% of the solar energy passing through the Earth's atmosphere, 4% is reflected from the surface back to space. On average, about 51% of the Sun's radiation reaches the surface. This energy is then used in a number of processes, including the heating of the ground surface; the melting of ice and snow and the evaporation of water; and plant photosynthesis.

The heating of the ground by sunlight causes the Earth's surface to become a radiator of energy in the longwave band (sometimes called infrared radiation).

Absorption of longwave radiation by the atmosphere causes additional heat energy to be added to the Earth's atmospheric system. The now warmer atmospheric greenhouse gas molecules begin radiating longwave energy in all directions. Over 90% of this emission of longwave energy is directed back to the Earth's surface where it once again is absorbed by the surface. The heating of the ground by the longwave radiation causes the ground surface to once again radiate, repeating the cycle described above, again and again, until no more longwave is available for absorption.

The amount of heat energy added to the atmosphere by the greenhouse effect is controlled by the concentration of greenhouse gases in the Earth's atmosphere. All of the major greenhouse gases have increased in concentration since the beginning of the Industrial Revolution. As a result of these higher concentrations, scientists predict that the greenhouse effect will be enhanced and the Earth's climate will become warmer.

Predicting the amount of warming is accomplished by computer modeling. Computer models suggest that a doubling of the concentration of the main greenhouse gas, carbon dioxide, may raise the average global temperature between 1 and 3° Celsius. However, the numeric equations of computer models do not accurately simulate the effects of a number of possible negative feedbacks. For example, many of the models cannot properly simulate the negative effects that increased cloud cover would have on the radiation balance of a warmer Earth.

Increasing the Earth's temperature would cause the oceans to evaporate greater amounts of water, causing the atmosphere to become cloudier. These extra clouds would then reflect a greater proportion of the Sun's energy back to space reducing the amount of solar radiation absorbed by the atmosphere and the Earth's surface. With less solar energy being absorbed at the surface, the effects of an enhanced greenhouse effect may be counteracted. Artificially created chlorofluorocarbons are the strongest greenhouse gas per molecule.

However, low concentrations in the atmosphere reduce their overall importance in the enhancement of the greenhouse effect. Current measurements in the atmosphere indicate that the concentration of these chemicals may soon begin declining because of reduced emissions.

Reports of the development of ozone holes over the North and South Poles and a general decline in global stratospheric ozone levels over the last two decades has caused many nations to cutback on their production and use of these chemicals. In 1987, the signing of the Montreal Protocol agreement by forty-six nations established an immediate timetable for the global reduction of chlorofluorocarbons production and use.

Since 1750, methane concentrations in the atmosphere have increased by more than 150%. The primary sources for the additional methane added to the atmosphere (in order of importance) are rice cultivation, domestic grazing animals, termites, landfills, coal mining, and oil and gas extraction. Anaerobic conditions associated with rice paddy flooding results in the formation of methane gas. However, an accurate estimate of how much methane is being produced from rice paddies has been difficult to obtain. More than 60% of all rice paddies are found in India and China where scientific data concerning emission rates are unavailable.

Nevertheless, scientists believe that the contribution of rice paddies is large because this form of crop production has more than doubled since 1950. Grazing animals release methane to the environment as a result of herbaceous digestion. Some researchers believe the addition of methane from this source as more than quadrupled over the last century. Termites also release methane through similar processes. Land-use change in the tropics, due to deforestation, ranching, and farming, may be causing termite numbers to expand. If this assumption is correct, the contribution from these insects may be important. Methane is also released from landfills, coal mines, and gas and oil drilling. Landfills produce methane as organic wastes decompose over time. Coal, oil, and natural gas deposits release methane to the atmosphere when these deposits are excavated or drilled.

The average concentration of nitrous oxide in the atmosphere is now increasing at a rate of 0.2 to 0.3% per year. Sources for this increase include land-use conversion; fossil fuel combustion; biomass burning; and soil fertilization. Most of the nitrous oxide added to the atmosphere each year comes from deforestation and the conversion of forest, savanna and grassland ecosystems into agricultural fields and rangeland.

Both of these processes reduce the amount of nitrogen stored in living vegetation and soil through the decomposition of organic matter. Nitrous oxide is also released into the atmosphere when fossil fuels and biomass are burned. However, the combined contribution of these sources to the increase of this gas in the atmosphere is thought to be minor.

The use of nitrate and ammonium fertilizers to enhance plant growth is another source of nitrous oxide. Accurate measurements of how much nitrous oxide is being released from fertilization have been difficult to obtain. Estimates suggest that the contribution from this source may represent from 50% to 0.2% of nitrous oxide added to the atmosphere annually. Ozone's role in the enhancement of the greenhouse effect has been difficult to determine scientifically.

Accurate measurements of past long-term (more than 25 years in the past) levels of this gas in the atmosphere are currently unavailable. Concentrations of ozone gas are found in two different regions of the Earth's atmosphere. The majority of the ozone (about 97%) found in the atmosphere is localized in the stratosphere at an altitude of 15.to 55 Kilometres above the Earth's surface.

In recent years, the concentration of the stratospheric ozone has been decreasing because of the buildup of chlorofluorocarbons in the atmosphere. Since the late 1970s, scientists have discovered that total column ozone amounts over Antarctica in the springtime have decreased by as much as 70%. Satellite measurements have indicated that the zone from 65° North to 65° South atitude has had a 3% decrease in stratospheric ozone since 1978. Ozone is also highly concentrated at the Earth's surface.

Most of this ozone is created as an artificial by product of photochemical smog. In summary, the greenhouse effect causes the atmosphere to trap more heat energy at the Earth's surface and within the atmosphere by absorbing and re-emitting longwave energy. Of the longwave energy emitted back to space, 90% is intercepted and absorbed by greenhouse gases. Without the greenhouse effect the Earth's average global temperature would be -18° Celsius, rather than the present 15° Celsius.

In the last few centuries, the activities of humans have directly or indirectly caused the concentration of the major greenhouse gases to increase. Scientists predict that this increase mayenhance the greenhouse effect making the planet warmer. Some experts estimate that the Earth's average global temperature has already increased by 0.3 to 0.6° Celsius, since the beginning of this century, because of this enhancement. Predictions of future climates indicate that by the middle of the next century the Earth's global temperature may be 1 to 3° Celsius higher than today.

BASIC MECHANISM

The Earth receives energy from the Sun mostly in the form of visible light. The bulk of this energy is not absorbed by the atmosphere since the atmosphere is transparent to visible light. 50% of the sun's energy reaches the Earth which is absorbed by the surface as heat.

Because of its temperature, the Earth's surface radiates energy in infrared range.The Greenhouse gases are not transparent to infrared radiation so they absorb infrared radiation. Infrared radiation is absorbed from all directions and is passed as heat to all gases in the atmosphere.

The atmosphere also radiates in the infrared range (because of its temperature, in the same way the Earth's surface does) and does so in all directions. The surface and lower atmosphere are warmed because of the greenhouse gases and makes our life on earth possible.

DETAILED EXPLANATION

The Earth receives energy from the Sun in the form of radiation. Most of the energy is in visible wavelengths and in infrared wavelengths that are near the visible range (often called "near infrared"). The Earth reflects about 30% of the incoming solar radiation. The remaining 70% is absorbed, warming the land, atmosphere and ocean.

For the Earth's temperature to be in steady state so that the Earth does not rapidly heat or cool, this absorbed solar radiation must be very closely balanced by energy radiated back to space in the infrared wavelengths. Since the intensity of infrared radiation increases with increasing temperature, one can think of the Earth's temperature as being determined by the infrared flux needed to balance the absorbed solar flux.

The visible solar radiation mostly heats the surface, not the atmosphere, whereas most of the infrared radiation escaping to space is emitted from the upper atmosphere, not the surface. The infrared photons emitted by the surface are mostly absorbed in the atmosphere by greenhouse gases and clouds and do not escape directly to space.

The reason this warms the surface is most easily understood by starting with a simplified model of a purely radiative greenhouse effect that ignores energy transfer in the atmosphere by convection (sensible heat transport, Sensible heat flux) and by the evaporation and condensation of water vapour (latent heat transport, Latent heat flux). In this purely radiative case, one can think of the atmosphere as emitting infrared radiation both upwards and downwards. The upward infrared flux emitted by the surface must balance not only the absorbed solar flux but also this downward infrared flux emitted by the atmosphere. The surface temperature will rise until it generates thermal radiation equivalent to the sum of the incoming solar and infrared radiation.

A more realistic picture taking into account the convective and latent heat fluxes is somewhat more complex. But the following simple model captures the essence. The starting point is to note that the opacity of the atmosphere to infrared radiation determines the height in the atmosphere from which most of the photons are emitted into space. If the atmosphere is more opaque, the typical photon escaping to space will be emitted from higher in the atmosphere, because one then has to go to higher altitudes to see out to space in the infrared. Since the emission of infrared radiation is a function of temperature, it is the temperature of the atmosphere at this emission level that is effectively determined by the requirement that the emitted flux balance the absorbed solar flux.

But the temperature of the atmosphere generally decreases with height above the surface, at a rate of roughly 6.5 °C per Kilometre on average, until one reaches the stratosphere 10 15 km above the surface. (Most infrared photons escaping to space are emitted by the troposphere, the region bounded by the surface and the stratosphere, so we can ignore the stratosphere in this simple picture.) A very simple model, but one that proves to be remarkably useful, involves the assumption that this temperature profile is simply fixed, by the non-radiative energy fluxes.

Given the temperature at the emission level of the infrared flux escaping to space, one then computes the surface temperature by increasing temperature at the rate of 6.5 °C per Kilometre, the environmental lapse rate, until one reaches the surface.

The more opaque the atmosphere, and the higher the emission level of the escaping infrared radiation, the warmer the surface, since one then needs to follow this lapse rate over a larger distance in the vertical. While less intuitive than the purely radiative greenhouse effect, this less familiar radiative-convective picture is the starting point for most discussions of the greenhouse effect in the climate modeling literature.

LA NINO

The term El Niño (Spanish for "the little boy" or "the Christ Child") was originally used by fishermen along the coasts of Ecuador and Peru to refer to above-normal sea-surface temperatures that typically appear around Christmas time in the eastern Pacific Ocean and last for several months. But El Niño's effects are not limited to Peru and Ecuador. They can impact weather patterns around the world, and the disruption of the normal climate can have profound and even tragic consequences.

WORKING OF EL NINO

In normal years, winds tend to blow from east to west across the waters of the tropical Pacific. The easterly winds (remember that wind directions refer to the direction FROM which the wind comes: an easterly trade wind comes from the EAST and blows WEST) push the surface waters westward across the ocean. In turn, this causes deeper, colder waters to rise to the surface. This "upwelling" of deep ocean waters brings with it nutrients that otherwise would remain near the bottom. Fish populations living in the upper waters are dependent on these nutrients for survival.

During El Nino years, the winds weaken, reducing or even choking off the upwelling of deep water. The consequent warming of the ocean surface further weakens the winds and strengthens El Niño .

As the Pacific continues to heat up, the warmer waters shift eastward, and so do the clouds and thunderstorms that produce heavy rainfall along the equator. This results in changes in jet streams (the winds high aloft), which lead to dry conditions in Indonesia and Australia, and floods in Peru and Ecuador. El Niño events occur on average every 3 to 5 years, though there can be periods of up to a decade without an El Niño.

EL NIÑO'S EFFECTS

The 1982-83 El Niño was unusually strong. In Ecuador and northern Peru, up to 100 inches of rain fell during a six-month period, transforming the coastal desert into grassland dotted with lakes.

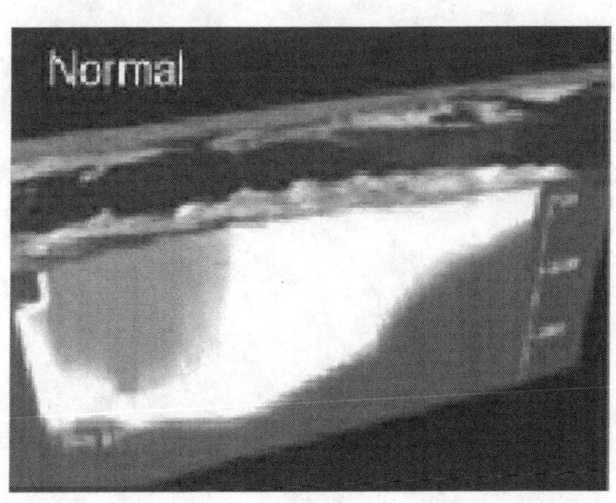

Abnormal wind patterns also caused the monsoon rains to fall over the central Pacific instead of on the western shores, which led to droughts and disastrous forest fires in Indonesia and Australia.

Overall, the loss to the global economy as a result of the El Niño amounted to more than $8 billion.

Likewise, the winter of 1997-1998 was marked by a record-breaking El Niño event. The result was unusual weather in parts of the world, including the U.S. Severe weather events included flooding in the southeastern United States, major storms in the Northeast, and flooding in California.

❒

20

Microbiology

OVERVIEW

Microbiology is the study of microorganisms, that is the organisms which are of microscopic dimensions. These organisms are too small to be clearly perceived by the unaided human eye. If an object has a diameter of less than 0.1 mm, the eye can not perceive (or more correctly resolve) it at all, and very little detail can be perceived in an object with a diameter of I mm.

Roughly speaking organisms with a diameter of 1 mm or less are microorganisms and fall into the broad domain of microbiology. Since most microorganisms are only a few thousandths of a millimeter in size. they can only be seen with the aid of microscope.

As a direct consequence of the invisibility of micros to the naked eye and the need for special techniques to study them, microbiology was the last of the three major divisions in biology (the other two are botany and zoology to develop.

Microbiology is the study of organisms that are too small to be clearly seen by the unaided eye. Roughly speaking, organisms with a diameter of 1 mm or less which can not be seen by the human eye unaided, are calIed microorganisms.

These include protozoa, algae, fungi and bacteria Viruses are ultramicroscopic and have an obligate parasitic relationship, but for practical purposes these still come under the domain of microbiology. At present, there is general agreement to include five major groups as microorganisms: the subdivisions of virology, bacteriology, mycology, phycology and protozoology (the studies of viruses, bacteria, fungi, algae and protozoa respectively). Traditionally, all organisms had been included under the disciplines of either botany or zoology.

Bacteria, algae and fungi have in the past been considered to be the part of the Plant Kingdom while protozoa have been included in the Animal Kingdom. This view cannot be supported from a taxonomic standpoint. Since microbiology encompasses the study of groups of organisms in all three of these divisions of biology it may be argued that it covers a greater biological diver itv than the other two divisions.

Although microorganisms have existed for a long time, their existence was unknown until the invention of the microscope in the 17th century. The year 1674, marks the birth of microbiology when Antony van Leeu wenhoek a Dutch cloth merchant, looked at a drop of lake water through a glass lens which he had ground. What he observed through this simple magnifying lens was an amazing sight since that was perhaps the first time that man ever had a glimpse of the world of the microbes.

Subsequently, in a series of letters to the Royal Society of London, Leeuwenhoek describ ed a variety of microorganisms such as protozoa, algae, yeast and bacteria and the description was so precise that it is now possible to assign them into specific genera without any additicnal descrip-tion. Leeuwenhoek had little formal education but his keen interest in nature made him to examine a variety of materials.

Glass grinding and preparation of lenses was his hobby and this led him to the assembly of about 400 simple microscopes. The earliest microscope that he constructed consisted of a spherical lens mounted on two plates This process of unilinear inheritance of the transferred gene continues leaving a trail of nonmotile celled colonies until the gene is lost for reasons not well known. Using single cell studies, this mechanism was confirmed as abortive transduction.

Later in 1956, Ozeki also observed the same type of phenomenon in' transduction of nutritional characters in Salmonella diagrammatically summarises linear inheritance as it occurs in abortive transduction.

SCOPE

The aim of this component on microbiology is to enable you to understand the concepts underlying the biology of microbes. We will achieve this aim by examining aspects of their biology that differentiate microbes from larger organisms. We revise and build on previous material on the structure and evolution of the cell. This component finishes with some extension material on competition. By the end of the unit you will be expected to apply these concepts to any practical situation and develop a response to a problem or issue that involves microbes. This means you need to be able to recall, integrate and apply information and concepts from this part of the unit. Please be aware that the structure of this electronic learning resource does not strictly follow the lectures, though the underlying concepts are exactly the same.

CONCEPTS

The concepts covered are:

- The enormous functional diversity within microbes, despite morphological similarity.
- The different methods microbes use to acquire energy and oxygen.
- The potential for microbes to increase the size of their populations.
- The mechanisms underlying microbial competition and the consequences of competition.

Other concepts in microbiology useful for understanding aspects of environmental microbiology are:

- Growth Rates of bacteria.
- Elemental composition and structure of bacterial cells
- General physiological functions of these elements in cells.
- General characteristics of bacteria (haploid cells, high growth rates, binary fission, biochemical diversity, high variation rate, grow in extreme environments, can survive via sporulation, etc).
- Generalized growth physiology (response to environmental factors, growth in extreme environments, oligotrophy, etc.) and kinetics of growth (lag phase, log phase, stationary phase and decline or death phase of growth).
- Main groups of microorganisms in terms of physiology - aerobic, anaerobic, aerotolerant, facultative anaerobe, obligate anaerobe,
- Groupings by trophic type: Photoautotrophs, photohete-rotrophs, chemoautotrophs, chemoheterotrophs, and mixotrophs
- General patterns of metabolism - energy reactions to "building block reactions", to polymerization to cellular structures.

CLASSIFICATIONS

In general, a classification is a method for organizing information. Human beings classify things spontaneously. For example, after seeing enough examples of chairs, we form a classification in which any given object is either a chair or a non-chair. Given a new object we do remarkably well at deciding whether it is a chair (in essence, whether or not it is practical and socially acceptable sit on it). Although there is little scientific value in this particular classification, it illustrates their ubiquity.

CLASSIFICATIONS GROUP

In general, classifications group similar things together. This definition is necessarily vague; there are many reasonable ways of defining similarity, and hence many alternative classifications for the same things. In the example of chairs, we considered the relationship (similarity) of objects to an abstract definition of chair. This definition was based on a potential function of the object. In choosing to classify by "chairishness", we assign less importance to other properties—Colour, hardness, age, etc.—that can vary without altering the utility of the object as a chair. There is no "right way" to classify things. For example, words can be classified:

- By their spelling (as in a dictionary).
- By the spelling of a foreign language synonym (as in a bilingual dictionary).
- By their meanings (as in a thesaurus).

The intended use of a classification is an integral part of deciding what properties it should be based upon. Here are some other classifications that you have encountered: the Library of Congress book classifications (organized by subject); the fiction section in a library (usually organized by the author's last name); and a food store (organized primarily by storage temperature).

An early biological classification is implicit in the ancient Greeks' definition of the three kingdoms: animal, vegetable and mineral. This classification split animate matter into (what we would now call) plants and animals. More detailed classifications also existed. For example, Aristotle grouped humans and birds together, since they walk on two feet.

Classifications are Often Hierarchical

A key property of classifications is that they can be nested within one another, creating an hierarchy. Thus, any group within a classification can be split in still greater detail. For example, objects that have been classified as chairs might be subdivided into stools, rocking chairs, recliners, etc. Similarly, listings in the "yellow pages" of a telephone directory are first classified by the product or service, and then within each of these categories they are further classified alphabetically.

There is no limit to the depth of a hierarchical classification. Most only have a few levels, but there are some that are quite deep. When alphabetical order is considered carefully, it is seen that the first level of classification deals only with the first letter of each word.

Then all words starting with the same first letter (for example "c") are categorized by their second letter, adding a second level to the alphabetical classification. This process continues one letter at a time until all words have been distinguished.

Biological Classification

Linnaeus defined the biological classification system that we still use for plants and animals, and, with relatively minor modifications, for fungi and microorganisms. It is a hierarchical system that starts with a few categories at the highest level, and further subdivides them at each lower level.

The levels in the hierarchy were given names that are already familiar to you:

- Domain
- Kingdom
- Phylum

- Class
- Order
- Family
- Genus
- Species

Using this system, a fruit fly might be classified as:

- Eucarya
- Animalia
- ArthropodaInsecta
- Diptera
- Drosophilidae
- Drosophila
- Melanogaster

This means that this fly is the species melanogaster in the genus Drosophila in the Family Drosophilidae in the Order Diptera in the Class Insecta in the Phylum Arthropoda in the Kingdom Animalia in the Domain Eucarya. In the Linnaean system, to uniquely name a species it is necessary to supply both the genus and species, Drosophila melanogaster.

In the case of animals, Linnaean classifications often reflect our intuitive sense of similarity. Thus in the following classifications species 2 and 3 have more levels of classification in common than either has with species 1. This is a reflection of the greater similarity of the fly (Drosophila melanogaster) and the mosquito (Aedes aegypti), than of fly and lobster (Homarus americanus) or mosquito and lobster.

TABLE: Classifications of Three Species

	Species		
Level	**1**	**2**	**3**
Domain	Eucarya	Eucarya	Eucarya
Kingdom	Animalia	Animalia	Animalia
Phylum	Arthropoda	Arthropoda	Arthropoda
Class	Crustacea	Insecta	Insecta
Order	Decapoda	Diptera	Diptera
Family	Caridea	Drosophilidae	Nematocera
Genus	Homarus	Drosophila	Aedes
Species	Americanus	mulanogaster	Aegypti

The shared levels of classification can be accentuated by slightly reformatting the data in the table.

TABLE: Classifications of Three Species

	Species		
Level	**1**	**2**	**3**
Domain			Eucarya
Kingdom			Animalia
Phylum			Arthropoda
Class	Crustacea		Insecta
Order	Decapoda		Diptera
Family	Caridea	Drosophilidae	Nematocera Genus Homarus
Drosophilia	Aedes	Species	americanus melanogaster Aegypti

NATURAL SYSTEM

Although there are no right and wrong ways to classify things, when the natures of objects are defined by a common history then there is a "natural" way to classify them. For most objects, their natures are largely independent of their histories; but organisms are products of their genetic history.

Similarity in Organisms

For millennia it has been appreciated that organisms give rise to new individuals of the same kind (i.e., same species). Thus, humans give rise to more humans, dandelions give rise to more dandelions, Escherichia coli gives rise to more E. coli, and so forth. In essence, like begets like begets like begets like etc. ad nauseam. The continuity of a species is a consequence of the genes passed from each generation to the next. This is a static view of the biological world; without other processes, life could not change.

Species are Dynamic

In the 19th century, biologists began to recognize that species are dynamic, not static; that is, species change over time. Several discoveries necessitated this view and overcame seemingly conflicting observations. For example, the "like begets like" view does not accommodate the variation evident among individuals of a species. Also, the discovery of fossils indicated that ancient life forms were not the same as those alive today. Some of the fossils could even be organized into progressions, with later forms being similar, but not identical, to earlier forms.

The simplest modification to the static view would be to suggest that like begets similar. At the same time, work in geology was revealing that the Earth was old—very much older than had been previously believed. This meant that even if changes between successive generations were small, there had been time for changes to accumulate into the large divergences that distinguish species.

The impact of this new perspective is profound. In the earlier static view, if we followed members of a species (call it A0) over time, their descendants and their descendants' descendants, and so on, would belong to the same species, forever, regardless of the duration of the intervening interval:

$$A0 \rightarrow A0 \rightarrow A0 \rightarrow A0 \ldots \rightarrow A0 \rightarrow A0$$

In contrast, the dynamic view of species implies that, given sufficient time, naturally occurring changes would accumulate, ultimately changing the lineage to such an extent that a descendent would be a member of a new species, A1, that is distinct from that of its distant ancestors:

$$A0 \rightarrow A0' \rightarrow A0'' \rightarrow A0''' \rightarrow \ldots \rightarrow A1' \rightarrow A1$$

This realization was the motivation for Darwin's title: On the Origin of Specie. This could be represented diagramatically as:

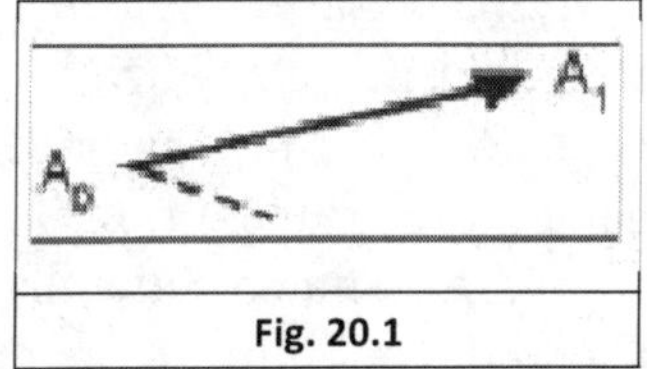

Fig. 20.1

Consequence of Ancestry

Consider the consequences of species A_0 giving rise to two lineages that are prevented from further interbreeding. With sufficient time, the first lineage would change enough to be called a new species, A_1 (as above), while in the same interval of time, the second group would also have changed enough to be a new species. Because the two lineages have accumulated changes independently (there was no interbreeding), most of the changes in the groups will be different. Thus the second group will have become a species that is not the same as either A_0 or A_1, so it needs a name of its own, A_2:

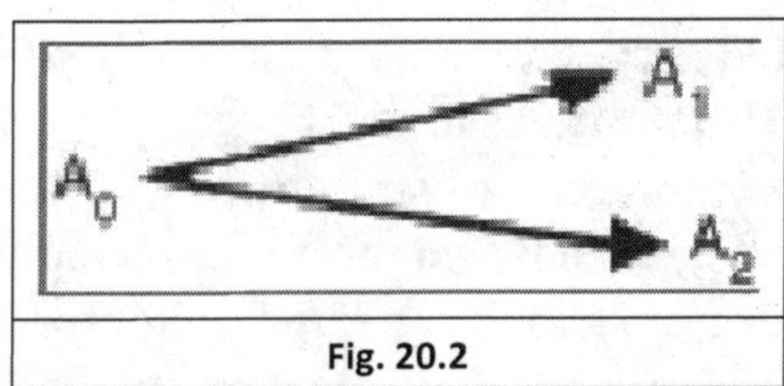

Fig. 20.2

Species A_1 and A_2 must each bear residual similarity to A_0 (their most recent common ancestor), so they must also be similar to each other. Most similarities between present-day species are due to their common ancestry. To first approximation, the more recent their common ancestry, the more similar two species will be (the more properties they will share). Turning this around, the most similar species today tend to be those that share the most recent common ancestry. More distinct species tend to have more remote common ancestry. Darwin (and others) saw that once we accept this perspective, then we inevitably continue until we have accepted the common ancestry of all life.

A Natural System of Classification for Organisms

As noted above, there is no absolute rule for designing a classification; we can classify things by any criterion we choose. This is true in classifying organisms, as well. We can classify by size, shape, Colour, alphabetical order (in any language), or even chronology of published descriptions.

There is however a practical question: what do we learn from our classification? How useful is it? Is there a preferred classification of organisms from which we can learn more than we can from alternative classifications? If we classify by an arbitrary criterion, then knowing an organism's placement within the classification can be remarkably uninformative.

For example, if we classify bacteria on the basis of shape (i.e., rod, coccus, spiral, etc.), then knowing an organism's classification tells us little more than its shape. We cannot begin to guess whether it is pathogenic, whether it makes food taste good, or whether it enhances the growth of our lawn.

In the case of organisms, we can do much better than an arbitrary classification. Ideally, we wish to base our classification on the relationships of the organisms, that is, on their genealogy or phylogeny. This is uniquely useful because knowing that two organisms are closely related tells us that they will have many properties in common. Knowing historical relationships has more predictive potential (more applicability to properties that have not yet been measured) than does an arbitrary classification.

Darwin referred to a classification based on evolutionary history as a natural system. In recognizing the parallel between phylogenetic closeness and species similarity, Darwin explained why the Linnaean system had been so successful for macroscopic plants and animals: they had been placed into the hierarchical groups of the Linnaean classification on the basis of similarities, and because these similarities reflected the evolutionary closeness of the respective species, the classification tended to reflect their historical relationships. In general, when similarities due to common ancestry are easy to recognize and measure, a classification of the organisms tends to approximate a natural system. The original successes of the Linnaean system were largely limited to complex plants and animals. In these cases there are numerous features that can be seen and compared between organisms. In addition, many of the features are sufficiently complex that it is safe to assume that they were only invented once, so that similarities really are the result of inheritance from a common ancestor that also possessed the shared feature.

CLASSIFYING MICROORGANISMS

Determinative Microbiology

The prokaryotes and eukaryotic protists were more difficult to classify. They are much simpler, and there are few obvious features that are sufficiently widespread to permit meaningful measurements of similarity between distantly related organisms. Some of the features that have been used to classify prokaryotes include: Gram stain (cell wall structure) Mole per cent G+C in the genome Growth temperature Ability to form heat stable spores Electron acceptors for respiration (if any) Photosynthetic ability Motility Cell shape Ability to use various carbon and nitrogen sources Special nutritional requirements (e.g., vitamins) Leading microbiologists in the 20th century (e.g., Stanier and van Niel) realised and acknowledged that classifications based upon features such as these were not natural systems, but they also realised that no satisfactory alternatives were available. Consequently, the resulting classification of bacteria (excluding cyanobacteria, which were classified with the eukaryotic algae) was codified in a volume called Bergey's Manual of Determinative Bacteriology. In doing so, it was recognized that the classification could be used to accurately identify organisms that had already been characterized. However, given a new isolate with the following properties: Gram-variable 32% G+C genome thermophilic (85 C optimum) nonspore-forming anaerobic nonphotosynthetic nonmotile rod-shaped prokaryote grows on H_2, CO_2 and N_2 no special growth factors required microbiologists were unable to predict its phylogeny or additional properties. That is, it was not possible to say much about a new organism other than what was directly measured.

Molecular Phylogeny

Although the morphologies and physiologies of prokaryotes are much simpler than those of eukaryotes, there is a large amount of information in the molecular sequences of their DNA, RNAs and proteins. Thus, it is possible to use molecular similarities to infer the relationships of genes, and, by extension, to learn the relationships of the organisms themselves.

DNA-DNA Hybridization

One important technique for comparing prokaryotes at the molecular level is DNA-DNA hybridization. In this test, the genomic DNA from one species is mixed with the DNA from a second species and the similarity of the DNAs is reflected in the extent to which strands of DNA from one organism anneal with strands of DNA from the other organism. The sensitivity of DNA-DNA hybridization declines rapidly as the organisms become more diverged, limiting the method to characterization of closely related strains, species and genera. In addition, testing the relationships of a new organism can require many hybridizations. If no close relative is found by this test, we have only learned what the new organism is not—whereas we wish to know what it is.

Phylogenies Based on Specific Genes

DNA-DNA hybridization gives a measure of relatedness across the whole genome. As mentioned above, the average similarity falls off rapidly when looking at more diverged species. In principle, the range accessible to molecular analysis could be increased by looking at specific genes with above average conservation.

There are several important requirements if we wish to use a gene phylogeny to infer organismal relationships:

- The gene must be present in all organisms of interest. Thus, to infer relationships that span the diversity of prokaryotes (or life), we must look at the central (universal) cellular functions. Examples include genes whose products function in replication, transcription, or translation—the processes constituting the "Central Dogma" of molecular biology.
- The gene cannot be subject to transfer between species (lateral transfer). Since we wish to infer organismal relationships, if a gene is transferred, then the gene history is not the same as the organismal history. If a gene performs a central function, an organism is unlikely to acquire a copy by lateral transfer, since the organism must already have a functional copy to be alive.
- The gene must display an appropriate level of sequence conservation for the divergences of interest. If there is too much change, then the sequences become randomized, and there is a limit to the depth of the divergences that can be accurately inferred. If there is too little change (if the gene is too conserved), then there may be little or no change between the evolutionary branchings of interest, and it will not be possible to infer close (genus or species level) relationships.
- The gene must be sufficiently large to contain a record of the historical information. Thus, although transfer RNA (tRNA) genes are present in all species, they are too small (about 75 nucleotides) to provide an accurate sample of evolutionary history.

Ribosomal RNA Genes and their Sequences

To infer relationships that span the diversity of known life, it is necessary to look at genes conserved through the billions of years of evolutionary divergence. An example of genes in this category are those that define the ribosomal RNAs (rRNAs). Most prokaryotes have three rRNAs, called the 5S, 16S and 23S rRNA.

TABLE: Ribosomal RNAs in Prokaryotes

Namea	Size (nucleotides)	Location
5S	120	Large subunit of ribosome
16S	1500	Small subunit of ribosome
23S	2900	Large subunit of ribosome

Noet: a The name is based on the rate that the molecule sediments (sinks)in water. Bigger molecules sediment faster than small ones.

The 5S has been extensively studied, but it is usually too small for reliable phylogenetic inference. The 16S and 23S rRNAs are sufficiently large to be quite useful.

The extraordinary conservation of rRNA genes can be seen in these fragments of the small subunit rRNA gene sequences from organisms spanning the known diversity of life: As a graduate student at the University of Illinois, Bernadette Pace used the annealing of rRNA with genomic DNA to measure the similarity of rRNAs in various species. These experiments demonstrated that rRNA-based methods arc applicable to directly comparing a broader range of organisms (i.e., spanning greater phylogenetic distances) than is whole genome DNA-DNA hybridization. However, as with DNA-DNA measurements, it was necessary to have DNA and/ or RNA from each species of interest.

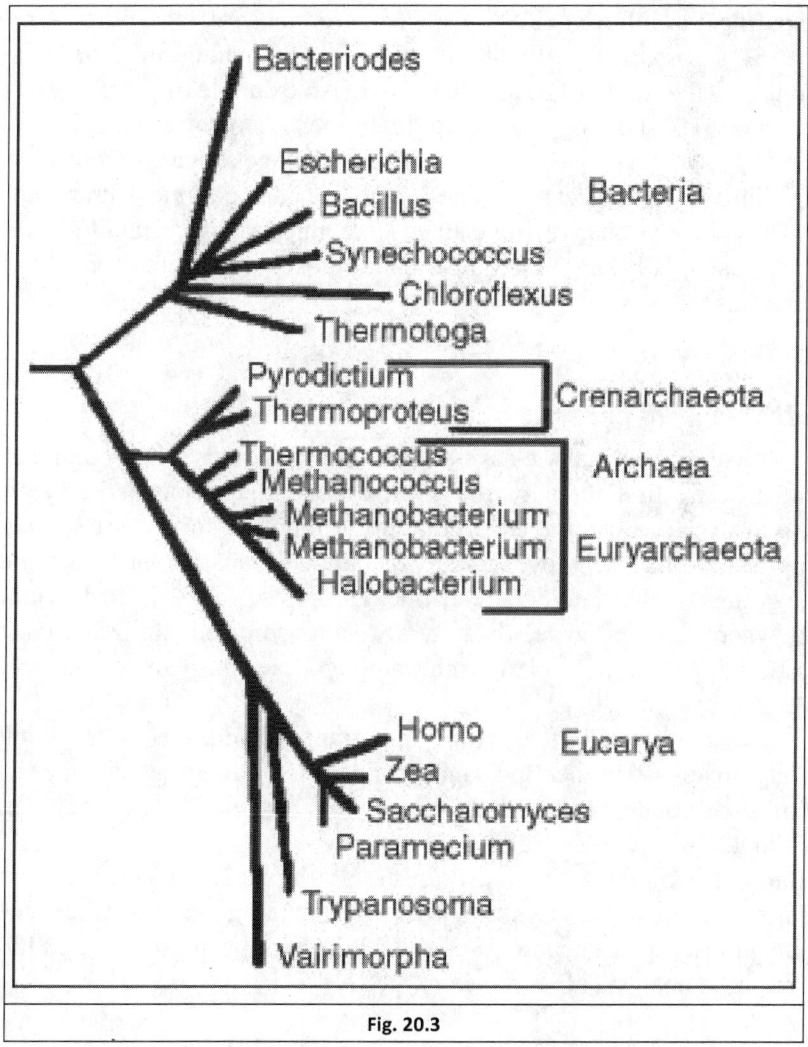

Fig. 20.3

If relationships were Analysed by comparing sequence data, rather than hybridizing the molecules, one could infer relationships without having all of the molecules in hand (only the sequence data from previous studies are necessary).

This was already being done with protein sequences. Carl Woese recognized the full potential of rRNA sequences as a measure of phylogenetic relatedness. He initially used an RNA sequencing method that determined about 1/4 of the nucleotides in the 16S rRNA (the best technology available at the time). This amount of data greatly exceeded anything else then available. Using newer methods, it is now routine to determine the sequence of the entire 16S rRNA molecule. Today, the accumulated 16S rRNA sequences (about 10,000) constitute the largest body of data available for inferring relationships among organisms.

Molecular Phylogenies can Reflect Genealogy and Amount of Change By comparing the inferred rRNA sequences (or those of any other appropriate molecule) it is possible to estimate the historical branching order of the species, and also the total amount of sequence change. An example of a 16S rRNA- based phylogenetic tree showing the three (identified) Domains of life—Bacteria, Archaea and Eucarya—is below. In this tree, lineages diverge from a common ancestral lineage on the far left. The lengths of the individual lines reflect the amount of sequence change (note that some lineages have modified the gene sequence substantially more than others, and thus have accumulated longer total branch lengths).

Eucarya (Eukaryotes)

Major Divisions of Eucarya

The Eucarya are usually classified into four kingdoms: plants, animals, fungi and protists. The first three of these correspond to phylogenetically coherent groups as well. However, the eukaryotic protists do not form a group, but rather are comprised of many phylogenetically disparate groups (including slime molds, multiple groups of algae, and many distinct groups of protozoa). Just as molecular analyses where required to see the natural relationships among prokaryotes, they are also allowing us to infer relationships among these non-plant, non-animal, non-fungus eukaryotes.

Many of the groups of fungi are important. In addition to providing yeast for making bread and fermenting sugar to alcohol, the fungi include the causative agents of such conditions as blight, ergot, rot, rust, smut, wilt, ringworm and athlete's foot.

Although it might seem that it is straightforward to identify an organism as a fungus, this is not always the case. The unambiguous identification of the opportunistic pathogen Pneumocystis (which has not been grown in culture) as a fungus required analysis of its ribosomal RNA.

Archaea

Major Groups within the Archaea

The major physiological groups within the Archaea (informally, archaes) are extreme thermophiles (including thermoacidophiles), sulfate reducers, methanogens, and extreme halophiles. These physiological groups do not correspond to the phylogenetic groups. This is largely because the methanogens have given rise to multiple other groups, including the halophiles, the sulfate reducers, and at least two other lineages of nonmethanogenic thermophiles.

The thermophilic members of the Archaea include the most thermophilic organisms cultivated in the Labouratory, some having optimal growth at >100 *f*C. The aerobic thermophiles are also acidophilic; they oxidize sulfur in their environment to sulfuric acid, creating acid hot springs.

The hot acidic waters can dissolve limestone and some other minerals out of the surrounding rock matrix, creating boiling mud pots of the insoluble residue from the rock. The extreme halophiles include the most salt tolerant organisms known. They are aerobic or microaerophilic. They have also invented a form of photosynthesis that does not use chlorophyll.

The sulfate-reducing Archaea are now known to be a wide spread group. One of the reasons that this group has received attention is that sulfate-reducing organisms are responsible for "souring" oil wells (increasing their sulfur content). In addition to increasing the sulfur emissions when the oil is burned (or increasing the cost of refining it), sulfide attacks the metal in well casings and pipelines.

Methanogens are strict anaerobes, yet they gave rise to at least two separate aerobic groups: the halophiles and a thermoacidophilic lineage. Methanogens produce most of the methane in the Earth's atmosphere. Methane is a significant green-house gas (i.e., it is a cause of global warming). Much of this methane production is associated with activities for which humans are at least partially responsible. This includes the flooding of rice paddies and the raising of domestic cattle.

Archaea are Prokaryotes

The Archaea are not eukaryotes, which makes them prokaryotes. However, the fact that they are related to Eucarya, not to the other prokaryotic organisms (the Bacteria), means that prokaryotes are not a natural group. This is difficult to fully appreciate, since we tend to accord a special status to eukaryotic cells. It is perhaps most easily understood by considering the ability to learn about one group from another:

If we wonder about a biosynthetic reaction in a eukaryote, it makes more sense to use a member of the Archaea as a model than to use a member of the Bacteria.

Even though much more money is being spent on analysis of eukaryotic genomes than prokaryotic genomes, due to their smaller size, some prokaryotic genomes will be completed first.

Of these, it appears that we will learn more about human genes from the archaeal genomes than from the bacterial genomes. If we wonder about transcription initiation in Thermococcus celer (an archaeon), should we look at what is known about transcription in E. coli(another prokaryote) or in yeast (a eukaryote)? It is now known that the transcription apparatus of T. celer seems to be much more similar to that of yeast than to that of E. coli.

Chlorophyll based Photosynthesis

Five of the major phylogenetic groups of Bacteria listed above include photosynthetic genera. Given the complexity of the process, it is unlikely to have been invented multiple times. Instead, it seems much more likely that it originated early in eubacterial evolution, and has been lost (or gained by lateral transfer) in many lineages. Some of the branches that diverge earliest in the Bacterial domain (notably the hydrogen oxidizers and the Thermotogales) have no known photosynthetic members. This suggests, but does not prove, that chlorophyll-based photosynthesis was invented after these groups diverged from the rest of the Bacterial lineages.

Oxygenic Photosynthesis

Although chlorophyll-based photosynthesis is widely distributed among the Bacteria, those that perform oxygenic photosynthesis form a phylogenetically monolithic group: all members of the cyanobacterial group share this physiology, and no organisms outside of the cyanobacteria have it. Photosynthetic eukaryotes all contain chloroplasts (more generally called plastids). One of the great triumphs of molecular phylogenetic analyses has been the demonstration that the chloroplast genome is of bacterial origin. That is, the chloroplast was once a bacterium (more specifically, a cyanobacterium), and, after incorporation into the eukaryotic host cell, it has degenerated into a much simpler cellular organelle (it is an "endosymbiont'). Because of its loss of autonomy, the organelle is maintained by the host cell. In return, it allows the host cell to harvest light energy and to convert carbon dioxide and water into useful organic compounds. The cyanobacterial lineage is responsible for essentially all biologically produced oxygen, whether by eukaryotic plants and algae, or by free-living cyanobacteria.

Molecular Microbial Ecology

One of the most dramatic contributions of molecular phylogeny to our understanding of microorganisms is its application to the area of microbial ecology.

Difficulty of Growth for Characterization

Traditional methods of microbial ecology require that organisms from an environment be cultivated in the Labouratory so that they can be characterized and identified. However, if the number of cells visible in an environmental sample are directly counted and compared with the number of cells successfully cultivated from the sample, it is found that anywhere from 90% to 99.9% of the cells defy cultivation, and therefore evade identification. Thus, most characterizations by cultivation methods have contained a large bias in an unknown direction.

In addition, if a new sequence is related to those from organisms with particular physiological properties, then those properties can be extrapolated onto the organism in the environment that was the source of the particular new sequence. Thus, it is even possible to learn something about the biochemical processes that might be going on in an environment, even without the ability to grow the organisms and make the measurements in the Labouratory. The same methodologies are being applied to human and animal pathogens. These techniques have permitted the phylogenetic identification organisms that have defied cultivation and have provided the basis for rapid detection of specific organisms or groups of organisms in clinical samples.

Branching Order in Phylogenetic Tree

The branching pattern (often called branching order) shows the genealogy of the organisms. That is, it shows which species share more common ancestry than which others.

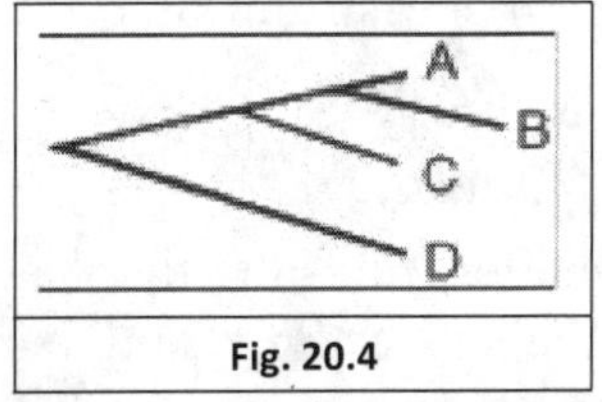

Fig. 20.4

In the tree above, A and B share the most recent common ancestry. Thus, of the species in the tree, A and B are the most closely related. The next most recent common ancestry is C with the group composed of A and B. Notice that the relationship of C is with the group containing A and B. In particular, C is not more closely related to B than to A. This can be emphasized by the following two trees, which are equivalent to each other:

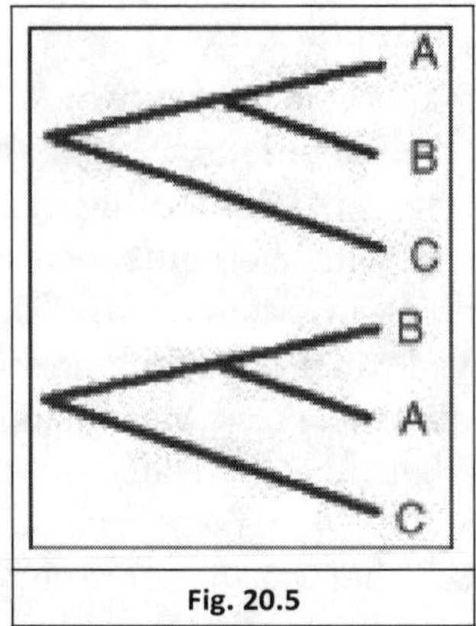

Fig. 20.5

It is only a matter of taste which way it is drawn. In essence, in both versions A and B share a more recent common ancestor than do either of them with C. Thus it does not make sense for either A or B to be closer to C than is the other, since the alternative drawing would seem to imply the opposite. Going back to the first drawing:

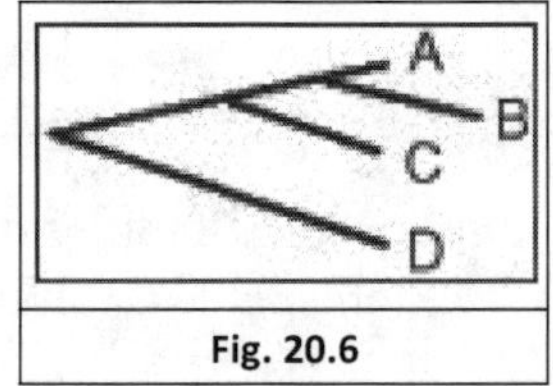

Fig. 20.6

The most ancient common ancestry in this tree is that of D with the group composed of A, B and C.

Branch Lengths in a Phylogenetic Tree

In the above drawing, the length of the line leading from the A+B common ancestry to A is shorter than the line leading from the A+B common ancestry to B. This is intended to represent the accumulation of a different amount of change. Note that the both A and B represent present-day species, and their most recent common ancestor lived some specific period of time ago. That is, the time represented by the branches is the same, even though a different amount of change has accumulated. To get unequal amounts of change in equal time periods, the rate of change must differ. There is no reason why rates of molecular change have to be either constant or equal in all lineages. When we examine molecular sequence data, we often find that some lineages have changed different average rates.

KINDS OF MICROORGANISMS

The types of micro organisms found in milk vary considerably, and are dependent upon the specific conditions associated with a batch of milk. Bacteria, yeasts, moulds, and bacteriophages are commonly encountered. Other viruses and Protozoa are seldom observed in milk products, except as occasional contaminants.

BACTERIA

Bacteria are the most common, and probably the most numerous of microorganisms with which the dairy processing industry is concerned.
They belong to four main groups:

1. Cocci, usually gram positive.
2. Gram positive non-sporeforming rods
3. Gram positive sporeforming rods.
4. Gram negative non-spore forming rods.

YEASTS

The yeasts most frequently encountered in milk and milk products act upon the lactose to producc acid and carbon dioxide. Yeasts are more commonly found in raw, cream during hot weather, but are potential contaminants throughout the year.

MOULDS

Moulds often grow in large concentrations and are visible as a fuzzy or fluffy growth. They are sometimes observed on the surface, of butter, old cream, khoa, or cheese. They are black, grey, green, blue or white. They discolour milk products and often produce undesir able, at times repulsive, odours and flavours. Moulds are essential in production of the certain kinds of cheese.

BACTERIOPHAGES

Bacteriophages are particularly obnoxious in starter cultures used for making cultured milk, butter and cheese. Phages will, kill the bacterial culture and the whole process of fermentation will be lost.

Microorganisms found in milk can also be described on the basis of the following characteristics;

- Biochemical activities.
- Temperature response.
- Ability to cause infection and disease.

BIOCHEMICAL ACTIVITIES

If allowed to stand under condition that permit bacterial growth, raw milk of a good sanitary quality will rapidly undergo a series of chemical changes. The principal change is lactose fermentation to lactic acid. This change is brought about by acid uric lactic organisms, especially Strepotococcus lactis and certain lactobacilli. These include two distinct biochemical types, homo-and heterofermentative. In homofermentation lactic acid is the major product of lactose fermentation.

Heterofermentative organisms, however, produce lactic, acetic, propionic, and some other acids, and some alcohols and gases such as CO_2 and H_2 Organisms continue to form lactic acid until the concentration of acid is itself too great for the organisms to remain live.

Microbacteria, micrococci, coliforn18, etc. also ferment lactose to lactic acid and other products. Many Clostriifiul1J species and, some yeasts such as Torula lactic, and Torula cremoris ferment lactose with acid and gas production. As the acidity continues to increase and reaches a pH of 4.7, it eventually causes a precipitation of casein. Organisms capable of metabolizing lactic and other acids develop especially acid uric, yeasts and moulds.

The acidity of milk is diminished and the alkaline products of protein decomposition such as amines, ammonia and the like are produced. This is accomplished by many species of the genera Bacillus, Clostridium, Pseudomonas, Proteus and numerous other forms.

The action of microorganisms does not involve fat as readily as it does lactose and protein. Lipolysis results from the action of lipase produced by bacteria such as Pseudomonas, Achromobacter and by some yeasts and moulds. Fat is hydrolysed to glycerol and fatty acids. Some of the fatty acids, for example, butyric and caproic acid give milk products, distinctive and usually rancid, odours and flavours.

Several microorganisms also bring about certain objection able changes in the milk which may not be deleterious to health. Rapines in milk is sometimes encountered. The milk become ropy or slimy and may be pulled out into long threads. It is produced by several organisms but the most important species is Alcaligenes viscolactis. A rapid fermentation of lactose in milk is sometimes observed and is known as stormy fermentation. This is brought about by Clostridium perfringens. The curd become torn to shreds by the Vigourous fermentation and gas production. Several organisms have been isolated from milk which impart brilliant colours. Pseudomonas syncyanea imparts blue colour, pseudomonas synxantha yellow colour and Serratia marcescens red colour to the milk.

TEMPERATURE RESPONSE

Microorganisms found in milk can also be described according their optimum temperature for growth and heat resistance. This is a very practical consideration since milk is preserved by employing low temperatures to prevent changes due to microbial activi1y'and by high temperatures to reduce microbial population and destroy pathogens. All the four types of microorganisms i.e. psychrophilic, mesophilic, thermophilic and thermoduric are found in milk.

Psychrophiles grow at temperatures just above freezing and at refrigeration tempratures. They produce a wide variety of spoilage defects. The defects may result in the production bf many "off" flavours and odours. The most commonly encountered psychrophilic bacteria are members of the genera pseudomonas, Achromobacter, Vibrio, Flavobacterium and Alcaligenes, They arc killed in the pasteurization process, but are sometimes found in pasteurized milk. The contamin ation takes place after pasteurization from equipment, cans, bottles, and water.

The most important mesophilic bacteria are streptococci, lactobacilli and coliforms, which produce acid and gas and off flavours. They are killed in the pasteurization process.

Thermophilic bacteria grow well at the temperature used in pasteurization, specially when the low temperature holding method is followed. Most thermophilic forms are found in two genera, Bacillus and Clostridium. Thermoduric organisms are regarded as those which survive pasteurization but do not grow at pasteurization temperatures. The most common thermoduric bacteria are found in the genera Microbacterium, Corynebacterium, Micrococcus, Streptococcus and Bacillus. Excessive numbers of thermoduric bacteria in milk make it difficult to meet the grading standard.

ABILITY TO CAUSE INFECTION AND DISEASE

Pathogenic organisms of both bovine and human origin have been isolated from milk. Milk, therefore, can serve as a carrier of diseases. Many serious epidemics were caused by the consumption of such products before this fact was clearly recognized.

However, this became less common as milk sanitation has improved and pasteurization is being more widely practised.

The disease organisms present in milk may be derived from:

- Diseased animals
- Persons collecting and handling milk:

Thus the danger is due to the inoculum and not to the growth of organisms in the milk. The health of animal is an important factor. Several diseases of cattle including staphylococcal and streptococcal infections, tuberculosis, brucellosis, salmonellosis, Q fever and Foot and mouth disease may be transmitted to man. The organisms causing these diseases may get into the milk either directly from the udder, or indirectly from infected body discharges, which may drop, splash, or be blown into the milk.

Some of the important diseases of human origin that have been transmitted by milk are:

- Typhoid fever
- Diphtheria,
- Scarlet fever,
- Dysentery
- Septic sore throat
- Poliomyelitis.

It is also possible for humans to infect animals. For example, mastitis may be caused by a variety of organisms, including Staphylococcus aureus. The infecting organism, in s9me cases, has been traced to humans.

CHARACTERISTICS OF MICROORGANISMS

MICROORGANISMS AND THEIR ROLE

The activated sludge process is a wastewater treatment method in which the carbonaceous organic matter of wastewater provides an energy source for the production of new cells for a mixed population of microorganisms in an aquatic aerobic environment.

The microbes convert carbon into cell tissue and oxidized end products that include carbon dioxide and water. In addition, a limited number of microorganisms may exist in activated sludge that obtain energy by oxidizing ammonia nitrogen to nitrate nitrogen in the process known as nitrification.

Bacteria constitute the majority of microorganisms present in activated sludge. Bacteria that require organic compounds for their supply of carbon and energy (heterotrophic bacteria) predominate, whereas bacteria that use inorganic compounds for cell growth (autotrophic bacteria) occur in proportion to concentrations of carbon and nitrogen. Both aerobic and anaerobic bacteria may exist in the activated sludge, but the preponderance of species are facultative, able to live in either the presence of or lack of dissolved oxygen.

Fungi, rotifers, and protozoans are also residents of activated sludge. The latter microorganisms are represented largely by ciliated species, but flagellated protozoans and amoebae may also be present. Protozoans serve as indicators of the activated sluge condition, and ciliated species are instrumental in removing Escherichia coli from sewage. Additionally, viruses of human origin may be found in raw sewage influent, but a large percentage appear to be removed by the activated-sludge process.

The success of the activated-sludge process is dependent upon establishing a mixed community of microorganisms that will remove and consume organic waste material, that will aggregate and adhere in a process known as bioflocculation, and that will settle in such a manner as to produce a concentrated sludge (return activated sludge, or RAS) for recycling.

Any of several types of activated sludge solids separations problems indicate an imbalance in the biological component of this process. In the ideal "healthy"system, filamentous organisms grow within a floc (a large aggregate of adherent, or floc-forming, microorganisms, such as bacteria) and give it strength, with few filaments protruding out into the surrounding bulk solution. In such a system, there is no interference with the compaction and settling rates of the activated sludge prior to its recycling.

INTRODUCTION

The activated-sludge process is a biological method of wastewater treatment that is performed by a variable and mixed community of microorganisms in an aerobic aquatic environment. These microorganisms derive energy from carbonaceous organic matter in aerated wastewater for the production of new cells in a process known as synthesis, while simultaneously releasing energy through the conversion of this organic matter into compounds that contain lower energy, such as carbon dioxide and water, in a process called respiration. As well, a variable number of microorganisms in the system obtain energy by converting ammonia nitrogen to nitrate nitrogen in a process termed nitrification. This consortium of microorganisms, the biological component of the process, is known collectively as activated sludge.

The overall goal of the activated-sludge process is to remove substances that have a demand for oxygen from the system. This is accomplished by the metabolic reactions (synthesis-respiration and nitrificaction) of the microor-ganisms, the separation and settling of activated-sludge solids to create an acceptable quality of secondary wastewater effluent, and the collection and recycling of microorganisms back into the system or removal of excess microorganisms from the system.

Physical Components of Sludge Process

Before beginning a discussion of the biological component of the system, an overview of the physical components that comprise the activated-sludge process would seem to be in order. This will help the reader gain a better understanding of the environment in which a mixed community of microorganisms metabolizes organic (or in some cases inorganic) matter, settles to form a thickened sludge, and is recycled back into or removed from the system.

According to Activated Sludge, Manual of Practice #9 (Water Environment Association, 1987), the activated-sludge process contains five essential interrelated equipment components. The first is an aeration tank or tanks in which air or oxygen is introduced into the system to create an aerobic environment that meets the needs of the biological community and that keeps the activated sludge properly mixed. At least seven modifications in the shape and number of tanks exist to produce variations in the pattern of flow.

Fig. 20.7

Second, an aeration source is required to ensure that adequate oxygen is fed into the tank(s) and that the appropriate mixing takes place. This source may be provided by pure oxygen, compressed air or mechanical aeration. Just as there are modifications in the shape and number of aeration tanks that can be used in the activated-sludge process, different equipment systems exist to deliver air or oxygen into aeration tanks.

SOLIDS SEPARATION

Microorganisms play a second important role in the activated-sludge process beyond removing carbonaceous organic material from and nitrifying ammonia in secondary influent wastewater. This is the process of solids separation in which activated-sludge solids separate by flocculation and gravity sedimentation from treated wastewater in secondary clarifers. The goal of this process is to create a secondary effluent low in suspended solids in the upper portion of a clarifier and a thickened activated sludge composed of flocs in the bottom portion of a clarifier that will be recycled back into the system as return activated sludge (RAS).

Activated sludge flocs, agglomerations of particles that may reach sizes of more than 1mm, are composed of the biological component discussed in the previous section and a nonbiological component. According to Jenkins et al. genera of heterotrophic bacteria, including Achromobacter, Alcaligenes,Arthrobacter, Citromonas, Flavobacterium, Pseudomonas, and Zoogloea, appear to be the primary floc-forming microorganisms. The nonbiological component of activated sludge flocs are organic and inorganic particles and extracellular microbial polymers that are generally made up largely of carbohydrates (polysaccharide [complex carbohydrate] and glycoprotein [a conjugated protein in which the nonprotein group is a carbohydrate] fibres).

Two levels of structure exist in activated-sludge flocs: microstructure and macrostructure. Microbial aggregation, adhesion, and bioflocculation are the basis of the microstructure. Although the mechanism of bioflocculation is not well understood, it is felt to be the result of bridging between extracellular microbial polymers functioning as polyelectrolytes (a substance of high molecular weight, such as a protein, that is an ionic conductor).

MICROORGANISMS IN NATURE

DEFINITION

The Theoretical Biology Group of the University of Bonn, Germany, visited the Wessex Institute and gave an overview of microbial survival strategies involving competition, cooperation, and communication, explaining the importance of spatial dynamics for these processes in his talk on "Survival strategies of microorganisms in nature".

Kinetic theory of optimal pathway length assumes that the total concentration of enzymes as well as metabolites in a cell is minimized because of the costs of enzyme synthesis and metabolite toxicity. Extending a given pathway by an extra step while constraining total concentration of enzymes and metabolites would lead to a reduction in the concentration of the already existing enzymes and intermediates. Assuming linear kinetics, where reaction rates are proportional to enzyme and metabolite concentrations, the rate of substrate consumption of a pathway decreases with the square of the pathway length, while the ATP yield of the pathway increases with its length, typically resulting in an optimal length in terms of the rate of ATP production or growth rate.

The predictions of kinetic theory are in full agreement with observed lengths of pathways in nitrification, methane oxidation, aerobic and anaerobic sugar degradation, etc. However, a longer pathway would have a higher ATP or growth yield. This is the reason for the observed trade-off between growth rate and yield.

This trade-off has consequences for fitness, as can be seen by competing two growth strategies based on this trade-off. The strategy of fast growth at low yield, or low efficiency, (HighRate) is due to short pathways, whereas the strategy of slow growth at high yield (HighYield) is due to long pathways. In well-mixed systems, global competition for resources allows the HighRate strategy to win because it has the higher growth rate at any resource concentration. However, in spatially structured systems such as biofilms, the HighYield strategy has a higher fitness when the diffusional influx of substrate into the biofilm limits growth, because under such conditions, the higher efficiency of resource exploitation pays off, provided the neighbours cooperate in using the resource economically. This cooperation among neighbours can evolve because microbes attached to surfaces typically grow and multiply forming clusters of relatives; the offspring from a single ancestor stays together.

Microbes not only cooperate, they often do so only if they are not alone, and they use cell-to-cell communication to find out whether they are alone or in a group. They do this by producing a diffusible extracellular signal that switches on gene expression if a threshold concentration of the signal has been reached.

However, the concentration of the signal not only depends on the density of signal producing cells, but also on diffusional constraints in the physical environment and on the degree of clustering of the cells. This clustering is essential for communication in complex natural environments, since it prevents interference from the multitude of other signalling species (practical problems) and exploitation by cheaters, cells that save costs because they do not produce signals or because they do not respond to the signal but nevertheless benefit from the signal producing and responding other cells (evolutionary problems).

A new, non-obligate system of bacterial predation on bacteria in soil was discovered. The predators comprising this system were a genetically unstable strain of Streptomyces venezuelae; a new kind of budding bacterium (now designated as Ensifer adhaerens, gen. n., sp. n.); and a myxobacterium. The natural population of these organisms in soil is attracted to and rapidly destroys (lyses) added Micrococcus luteus cells. There is also attack of predator on predator in this system. The transitory myceloid stage that occurs in the pleomorphic growth cycle of various Arthrobacter species was expanded (by manganese removal) so that it became the main form of growth. This stage was found to be more sensitive to bacteriophage attack than were the other growth stages. A new procedure for isolating Arthrobacter sp. bacteriophage from soil was developed for these studies. It was found that the encystment of Azotobacter vinelandii does not require prior intracellular production of poly-Beta-hydroxybutyrate, and that it is triggered by hypoxanthine.

❐

Index

❐❐❐